SUGARCANE TRANSFORMATION

Biofuels, Biotechnology, and Beyond

SUGARCANE TRANSFORMATION

Biofuels, Biotechnology, and Beyond

Krishan K. Verma, PhD
Xiu-Peng Song, PhD
Hai-Rong Huang, PhD
Aradhna Kumari, PhD
Yang-Rui Li, PhD
Editors

AAP | APPLE ACADEMIC PRESS

First edition published 2027

Apple Academic Press Inc.
1265 Goldenrod Circle, NE,
Palm Bay, FL 32905 USA

760 Laurentian Drive, Unit 19,
Burlington, ON L7N 0A4, Canada

CRC Press
2385 NW Executive Center Drive,
Suite 320, Boca Raton FL 33431

4 Park Square, Milton Park,
Abingdon, Oxon, OX14 4RN, UK

Apple Academic Press exclusively co-publishes with CRC Press, an imprint of Taylor & Francis Group, LLC

Library and Archives Canada Cataloguing in Publication

CIP data on file with Canada Library and Archives

Library of Congress Cataloging-in-Publication Data

CIP data on file with US Library of Congress

ISBN: 978-1-77964-156-4 (hbk)
ISBN: 978-1-77964-157-1 (ebk)
DOI: 10.1201/9781779641571

About the Editors

Krishan K. Verma, PhD
Sugarcane Research Institute, Guangxi Academy of Agricultural Sciences, Nanning, Guangxi, China

Krishan K. Verma, is working as a Principal Scientist at the Sugarcane Research Institute, Guangxi Academy of Agricultural Sciences/ Sugarcane Research Center, Guangxi Key Laboratory of Sugarcane Genetic Improvement/ Key Laboratory of Sugarcane Biotechnology and Genetic Improvement (Guangxi) of the Ministry of Agriculture and Rural Affairs in Nanning, Guangxi, China. He has 14 years of research experience in the field of biotic and abiotic stresses, such as environmental toxicology, plant physiology, bioengineering, and nanobioscience. His research focuses on environmental toxicology, bioengineering modeling, plant physiology, and molecular biology, and their impacts on growth and physiological and molecular adaptation strategies of plants to stressful environments and how they affect plant structure. He is working with Prof. Yang-Rui Li, a world-renowned plant scientist. He has published more than 120 scientific research articles/reviews, 15 books, and 15 book chapters. He is an internationally recognized reviewer. He received an Award of Excellence-2021 for his contribution to agricultural science from the Sugarcane Society of Guangxi Academy of Agricultural Science, Nanning, Guangxi, China. He has also been elected as a Fellow of the Society for Sugarcane Research & Promotion (FSSRP-2024) for his outstanding contribution to sugarcane crop production. In the year 2023, he was included in the list of 2% of the world's top scientists by Elsevier BV. He is also Regional Editor of the journal Fluoride, published by the International Society for Fluoride Research Inc., New Zealand. He has been serving as an editorial board member of various peer-reviewed journals.

Xiu-Peng Song, PhD
Sugarcane Research Institute, Guangxi Academy of Agricultural Sciences, Nanning, Guangxi, China

Xiu-Peng Song, is working as an Associate Professor at the Sugarcane Research Institute, Guangxi Academy of Agricultural Sciences/ Sugarcane Research Center, Chinese Academy of Agricultural Sciences/Guangxi Key Laboratory of Sugarcane Genetic Improvement/Key Laboratory of Sugarcane Biotechnology and Genetic Improvement (Guangxi) of Ministry of Agriculture and Rural Affairs, Nanning, Guangxi, China. His ongoing research focuses on disease-resistance breeding and molecular basic and comprehensive control of diseases and insect pests in sugarcane. He received his master's and PhD (biochemistry and molecular biology) from Guangxi University, Guangxi, China. He presided over six projects, such as the National Natural Science Foundation China, and participated in 16 projects that involved national and international cooperation. He has won eight awards, such as Guangxi Science and Technology Progress Award. He has published 66 scientific publications, three book chapters, and three books (Springer and CRC Press/Taylor & Francis).

Hai-Rong Huang, PhD
Sugarcane Research Institute, Guangxi Academy of Agricultural Sciences, Nanning, Guangxi, China

Hai-Rong Huang, is working as Scientist at the Sugarcane Research Institute, Guangxi Academy of Agricultural Sciences, Nanning, Guangxi, China. His ongoing research focuses on breeding and molecular biology of sugarcane crop. He has two years of overseas study and visit experience, including serving as a technical consultant in Cambodia in 2011 and participating as a visiting scholar in USA in 2015. He has bred 30 varieties, such as the most popular of GT42/GT44 in China, presided over four projects, such as those of the Natural Science Foundation of Guangxi, and participated in 12 national and international cooperation projects, winning eight awards, such as the Guangxi Science and Technology Progress Award. He has published over 40 scientific publications and has been granted five invention patents.

Aradhna Kumari, PhD
Plant Physiology, College of Agriculture, JNKVV, Ganj Basoda, Madhya Pradesh, India

Dr. (Mrs.) Aradhna Kumari, is presently working as an Assistant Professor (senior scale), Plant Physiology, College of Agriculture, Jawaharlal Nehru Krishi Vishwa Vidyalaya, Ganj Basoda, Madhya Pradesh, India. She obtained her BSc (Agriculture) degree from the Institute of Agricultural Sciences, Banaras Hindu University (2007) with a national talent scholarship from ICAR, MSc (Agriculture) from the College of Basic Sciences and Humanities, Govind Ballabh Pant University of Agriculture & Technology, Pantnagar (2009) with an ICAR-JRF, and PhD from the *Indian Agricultural Research Institute*, New Delhi, with CSIR and DST fellowship in Plant Physiology. She has also qualified for various NET examinations conducted by ASRB, CSIR, and UGC. She has presented research papers at various national and international conferences. She has published three books, two technical bulletins, various research papers in national and international journals, review articles, manuals, quality instructional materials for students, book chapters, and various popular articles in reputed agriculture magazines. She has 12 years of experience in teaching and is a life member of many professional societies. She has been awarded the Raj Bhasha Gaurav Award 2017 for her Hindi book by Rajbhsaha Vibhag, Government of India, and a Young Scientist Award 2015 by SURE, Varanasi, as well as Dr. Abdul Kalam Ajad Best Teacher Award.

Yang-Rui Li, PhD
Sugarcane Research Center, Chinese Academy of Agricultural Sciences; Chief-Expert, National Joint Research Program for Elite Sugarcane Variety Development, China

Yang-Rui Li, is a life-time Professor and has been serving as Director of the Sugarcane Research Center, Chinese Academy of Agricultural Sciences since September 2007. He is also Chief Expert of the National Joint Research Program for Elite Sugarcane Variety Development in China. He worked in various capacities including as Professor in the Department of Agronomy, Guangxi Agricultural University; Vice-President of Guangxi University; President of the Guangxi Academy of Agricultural Sciences, etc. He is the President of the International Association for Professionals of Sugar and Integrated Technologies (IAPSIT); Vice-President of the Society for Sugar Research and Promotion (SSRP); and President of the Chinese Sugarcane Industry Association for Technological Innovation (CSIATI). His research interests include sugarcane physiology, biochemistry and molecular biology, sugarcane genetics, breeding and cultivation, and chemical regulation of growth and development of sugarcane. He has published 16 books and more than 1000 papers and received 23 scientific research achievement awards from the Chinese government to date as well as 14 awards from international organizations, including a Lifetime Achievement Award from the Society for Sugar Research and Promotion (2011), Lifetime Achievement Award from the International Association of Professionals in Sugar and Integrated Technologies (2014), and Leadership Excellence Award from the Thailand Society of Sugarcane and Sugar Technologists (2018). He earned his BA degree at the Department of Agronomy, Guangxi Agricultural University, China, and MSc and PhD degrees from the Department of Agronomy, Fujian Agricultural University, China, respectively.

Contents

Contributors

Shahid Afghan
Sugarcane Research and Development Board, Ayub Agricultural Research Institute, Faisalabad, Punjab, Pakistan

Noé Aguilar-Rivera
Facultad de Ciencias Biológicas y Agropecuarias Región Orizaba-Córdoba de la Universidad Veracruzana, Amatlán de los Reyes, Veracruz, México

S. Alarmelu
Division of Crop Improvement, ICAR-Sugarcane Breeding Institute, Coimbatore, Tamil Nadu, India

Jericó Jabín Bello-Bello
SECIHTI-Colegio de Postgraduados Campus Córdoba, Carretera Federal Córdoba-Veracruz, Veracruz, México

B. M. Bhalerao
Department of Biochemistry, Post Graduate Institute, Mahatma Phule Krishi Vidyapeeth, Rahuri, Maharashtra, India

Nurmalasari Darsono
Research Center for Genetic Engineering, Research Organization for Life Sciences and Environment, National Research and Innovation Agency, Cibinong, Bogor, Indonesia

Mayur Darvhankar
Department of Agronomy, School of Agriculture, Lovely Professional University, Phagwara, Punjab, India

Munmun Dash
Odisha University of Agriculture and Technology, Bhubaneswar, Odisha, India

Mattaparthi Lakshmi Durga
Agricultural College, Naira, Acharya N.G. Ranga Agricultural University, Andhra Pradesh, India

Arti Goel
Amity Institute of Microbial Technology, Amity University, Noida, Uttar Pradesh, India

Rikno Harmoko
Research Center for Genetic Engineering, Research Organization for Life Sciences and Environment, National Research and Innovation Agency, Cibinong, Bogor, Indonesia

Hai-Rong Huang
Sugarcane Research Institute, Guangxi Academy of Agricultural Sciences/Key Laboratory of Sugarcane Biotechnology and Genetic Improvement (Guangxi), Ministry of Agriculture and Rural Affairs/Guangxi Key Laboratory of Sugarcane Genetic Improvement, Nanning, Guangxi, China

Aparna Jaiswal
College of Agriculture, Jawaharlal Nehru Krishi Vishwa Vidyalaya, Ganj Basoda, Vidisha, Madhya Pradesh, India

Sweta Kataria
Amity Institute of Microbial Technology, Amity University, Noida, Uttar Pradesh, India

Simran Kaur
Amity Institute of Microbial Technology, Amity University, Noida, Uttar Pradesh, India

Muhammad Ehsan Khan
Sugarcane Research and Development Board, Ayub Agricultural Research Institute, Faisalabad, Punjab, Pakistan

Prasann Kumar
Department of Agronomy, School of Agriculture, Lovely Professional University, Phagwara, Punjab, India

Dharmendra Kumar
Department of Botany, University of Delhi, New Delhi, India

Aradhna Kumari
College of Agriculture, Jawaharlal Nehru Krishi Vishwa Vidyalaya, Ganj Basoda, Vidisha, Madhya Pradesh, India

Yang-Rui Li
Sugarcane Research Institute, Guangxi Academy of Agricultural Sciences/Key Laboratory of Sugarcane Biotechnology and Genetic Improvement (Guangxi), Ministry of Agriculture and Rural Affairs/Guangxi Key Laboratory of Sugarcane Genetic Improvement, Nanning, Guangxi, China

Qiang Liang
Sugarcane Research Institute, Guangxi Academy of Agricultural Sciences/Key Laboratory of Sugarcane Biotechnology and Genetic Improvement (Guangxi), Ministry of Agriculture and Rural Affairs/Guangxi Key Laboratory of Sugarcane Genetic Improvement, Nanning, Guangxi, China

Bhumika Singh Lodhi
Department of Plant Breeding and Genetics, Jawaharlal Nehru Krishi Vishwa Vidyalaya, Jabalpur, Madhya Pradesh, India

H. K. Mahadevaswamy
Division of Crop Improvement, ICAR-Sugarcane Breeding Institute, Coimbatore, Tamil Nadu, India

Jyotirmay Mahapatra
Indian Agricultural Research Institute, Pusa, New Delhi, India; College of Agricultural Engineering and Technology, Odisha University of Agriculture and Technology, Bhubaneswar, Odisha, India

Bhupendra Mathpal
Department of Agronomy, School of Agriculture, Lovely Professional University, Phagwara, Punjab, India

Eid. M. Mehareb
Sugar Crops Research Institute, Agricultural Research Center, Giza, Egypt

Chandra Mohan Mehta
Department of Agronomy, School of Agriculture, Lovely Professional University, Phagwara, Punjab, India

Mahendra Kumar Mohanty
College of Agricultural Engineering and Technology, Odisha University of Agriculture and Technology, Bhubaneswar, Odisha, India

Madhuri More
Arvind Gavali College of Engineering, Satara, Maharashtra, India

Amin Nikpay
Department of Plant Protection, Sugarcane & By-Products Development Company, Salman Farsi Agro-Industry, Ahwaz, Iran

Chandra Shekhar Seth
Department of Botany, University of Delhi, New Delhi, India

R. M. Shanthi
Division of Crop Improvement, ICAR-Sugarcane Breeding Institute, Coimbatore, Tamil Nadu, India

S. Sheelamary
Division of Crop Improvement, ICAR-Sugarcane Breeding Institute, Coimbatore, Tamil Nadu, India

Santosh Kumar Singh
Department of Soil Science, Dr. Rajendra Prasad Central Agricultural University, Pusa, Samastipur, Bihar, India

Xiu-Peng Song
Sugarcane Research Institute, Guangxi Academy of Agricultural Sciences/Key Laboratory of Sugarcane Biotechnology and Genetic Improvement (Guangxi),
Ministry of Agriculture and Rural Affairs/Guangxi Key Laboratory of Sugarcane Genetic Improvement, Nanning, Guangxi, China

Jose Luis Spinoso-Castillo
Colegio de Postgraduados Campus Córdoba, Veracruz, México

Bambang Sugiharto
Laboratory of Molecular Biology and Biotechnology, Center for Development of Advanced Science and Technology (CDAST), Jember University, Jember, Indonesia, and Department of Biology, Faculty of Mathematics and Natural Sciences, Jember University, Jember, Indonesia

Kirti Suryawanshi
Arvind Gavali College of Engineering, Satara, Maharashtra, India

Krishan K. Verma
Sugarcane Research Institute, Guangxi Academy of Agricultural Sciences/Key Laboratory of Sugarcane Biotechnology and Genetic Improvement (Guangxi),
Ministry of Agriculture and Rural Affairs/Guangxi Key Laboratory of Sugarcane Genetic Improvement, Nanning, Guangxi, China

Lin Xu
Sugarcane Research Institute, Guangxi Academy of Agricultural Sciences/Key Laboratory of Sugarcane Biotechnology and Genetic Improvement (Guangxi),
Ministry of Agriculture and Rural Affairs/Guangxi Key Laboratory of Sugarcane Genetic Improvement, Nanning, Guangxi, China

About the Book

Energy plays a significant role in the socioeconomic development of a nation. The current energy sources around the globe are unsustainable from atmospheric and socioeconomic perspectives. Governments have started the application of alternative renewable energy sources to ensure energy security, create employment, and reduce carbon emissions worldwide. Biofuels have proven to be the most effective approach to fulfilling these necessities. For over a decade, the biofuel industries, particularly bioethanol and biodiesel, have been increasing in conjunction with agricultural crop production. The development of first-generation biofuels mostly depends on agricultural crops like canola, soybeans, sugarcane, corn, and sugar beets. As a result, the natural competitiveness between food and fuel has been debated in society for the past several years. The production of next-generation biofuels from various feedstocks, such as agricultural waste products, crop residues, and cellulosic biomass from high-yielding grass species, is now possible because of current technological advancements in research and development.

Acknowledgment

The authors would like to thank the Sugarcane Research Institute, Guangxi Academy of Agricultural Sciences, Nanning, Guangxi, China, for providing the necessary facilities. This research was financially supported by the Guangxi Key R&D Program (Guike AB24010140), Guangxi Natural Science Foundation (2021GXNSFAA220022; 2023GXNSFAA026460), Guangxi Innovation Teams of Modern Agriculture Technology (nycytxgxcxtd-2021-03), Guangxi Characteristic Crop Experimental Station (GTS2022022), National Key Research and Development Project (2022YFD2301102-07), The National Natural Science Foundation of China (31760415), Fund of Guangxi Academy of Agricultural Sciences (2021YT011), and Guangxi Key Laboratory of Sugarcane Genetic Improvement Project (21-238-16-K-03-09). Science and Technology Major Project of Guangxi (Guike AA22117002-1). The book content expresses the authors' opinions and future research directions for the sugarcane transformation.

Abbreviations

4CL	4-coumarate:CoA ligase
AFLP	amplified fragment length polymorphism
AGO	argonaute
AMF	arbuscular mycorrhizal fungi
AMT	agrobacterium-mediated transformation
ARS	Agricultural Research Service
BC	biomass conversion
C3H	4-hydroxycinnamate 3-hydroxylase
CAD	cinnamyl alcohol dehydrogenase
CBP	consolidated bioprocessing
CCS	commercial cane sugar
cDNA	complementary DNA
CI	compression ignition
CNNSGR	China National Nursery of Sugarcane Germplasm Resources
CNTs	carbon nanotubes
COMT	caffeic acid 3-*O*-methyltransferase
CRISPR	clustered regularly interspaced short palindromic repeats
CWI	cell wall integrity
CY	cane yield
DGAT	diacylglycerol:acyl-CoA acyltransferase
DSB	double-strand breaks
DSS	decision support systems
EBP	ethanol-blended petrol
EST	expressed sequence tag
F5H	ferulate 5-hydroxylase
F6P	fructose 6-phosphate
FGVs	flexible gasoline vehicles
GE	genetically engineered
GHGs	greenhouse gas emissions
GIS	geographic information systems
GISH	genomic *in situ* hybridization
GMO	genetically modified organism
GPS	global positioning systems
gRNA	guide RNA
GS	genomic selection
GT	green technology

GWAS	genome-wide association study
HDR	homology-directed repair
HGs	homology groups
HSBS	Hainan Sugarcane Breeding Station
IEA	International Energy Agency
IFY	Indian fiscal year
IPM	integrated pest management
ISSR	inter simple sequence repeats
KTs	key technologies
LCA	life cycle assessment
LGs	linkage groups
LHV	lower heating value
MAS	marker-assisted selection
MC	minimally truncated cassettes
MDH	malate dehydrogenase
MEG	monoethyleneglycol
MSIRI	Mauritius Sugar Industry Research Institute
MWCNTs	ultiwalled carbon nanotubes
NGS	next-generation sequencing
NHEJ	nonhomologous end joining
NI	neutral invertase
OAA	oxaloacetate
PAL	phenylalanine ammonia-lyase
PAM	proto-adjacent motif
PCR	polymerase chain reaction
PEPC	phosphoenolpyruvate carboxylase
PFP	1-phosphotransferase
POJ	Proefstation Oost Java
PPP	purchasing power parity
PTC	plant tissue culture
RAPD	random-amplified polymorphic DNA
RFLP	restriction fragment length polymorphism markers
ROS	reactive oxygen species
RS	regional station
Rubisco	ribulose 1,5-bisphosphate carboxylase/oxygenase
SAI	soluble acid invertase
SASRI	South African Sugarcane Research Institute
SCoT	start codon targeted polymorphism
SDP1	sugar-dependent 1
SES	Spontaneum Expedition Scheme
SHF	separate hydrolysis and fermentation
SI	spark ignition

SNP	single-nucleotide polymorphism
SPS	sucrose phosphate synthase
SRAP	sequence-related amplified polymorphism
SSF	saccharification and fermentation
SSR	simple sequence repeats
SuSy	sucrose synthase
SWCNTs	single-walled carbon nanotubes
TAG	triacylglycerol
TALENs	transcription activator-like effector nucleases
TRAP	targeted region amplified polymorphism
USDA	United States Department of Agriculture
VRT	Variable rate technology
WCSRG	World Collection of Sugarcane and Related Grasses

Preface

In the ever-evolving landscape of global energy production, the search for sustainable and renewable sources has never been more critical. As the world grapples with the challenges of climate change and dwindling fossil fuel reserves, sugarcane has emerged as a promising crop with vast potential to contribute to the green energy revolution. Renewable energy resources are crucial to minimizing the dependency on fossil fuels and reducing greenhouse gas emissions. Various types of biofuels can restore traditional fossil fuels with favorable ecological implications and produce net-zero carbon emissions. Therefore, biofuel is essential to combating the present global warming disaster resulting from our dependency on fossil fuels.

This book, *Sugarcane Transformation: Biofuels, Biotechnology, and Beyond*, explores the multifaceted roles of sugarcane in addressing these global challenges. The future direction of agriculture and energy industries has been analyzed to feed the increasing global population and to fuel the global optimum energy-demanding sector, transportation.

The chapters in this book provide a comprehensive examination of the current state and future prospects of sugarcane as a key player in the bioenergy sector. Drawing on the expertise of leading researchers and practitioners, we delve into the scientific, technological, and socioeconomic aspects of sugarcane transformation. This book aims to serve as a comprehensive resource for researchers, policymakers, and industry professionals interested in the transformative potential of sugarcane. By presenting a thorough analysis of the latest scientific advancements and practical applications, we hope to inspire further innovations and collaborations that will help unlock the full potential of this remarkable crop.

We extend our deepest gratitude to the contributing authors for their invaluable insights and dedication to advancing the field of sugarcane biotechnology and biofuels. We also thank the reviewers and the publishing team for their efforts in bringing this work to fruition.

CHAPTER 1

Advancing the Transformation of Sugarcane Crops in the Era of Green Technology

JOSE LUIS SPINOSO-CASTILLO,[1] JERICÓ JABÍN BELLO-BELLO,[2*] and NOÉ AGUILAR-RIVERA[3*]

[1]*Colegio de Postgraduados Campus Córdoba, Veracruz, Mexico*

[2]*SECIHTI-Colegio de Postgraduados Campus Córdoba, Carretera Federal Córdoba-Veracruz, Veracruz, México*

[3]*Facultad de Ciencias Biológicas y Agropecuarias Región Orizaba-Córdoba de la Universidad Veracruzana, Amatlán de los Reyes, Veracruz, Mexico*

*Corresponding author

ABSTRACT

Green technology (GT) involves developing and applying products, equipment, and systems used to preserve the environment and natural resources to mitigate the negative impact of human activities. It aims to produce safe agricultural products that are free of pollutants, high-quality, nutritious, and environmentally friendly. Applying GTs for sustainable production is an important strategy to solve food insecurity. Therefore, transforming sugarcane production using plant biotechnology represents an alternative to reduce the labor force, increase plant development, and improve plant quality. The types of GT described in this chapter are plant biotechnology, plant tissue culture (PTC), biofertilizers, and nano-technology. PTC is fundamental to plant biotechnology and is the best approach to be exploited by numerous practical applications, including GT.

Sugarcane Transformation: Biofuels, Biotechnology, and Beyond.
Krishan K. Verma, Xiu-Peng Song, Hai-Rong Huang, Aradhna Kumari, and Yang-Rui Li (Eds.)

DOI: 10.1201/9781779641571-1

PTC offers technologies for the genetic improvement of crops, *in vitro* propagation or micropropagation, plant health, *in vitro* conservation of germplasm, and production of secondary metabolites. In conclusion, there is great potential for using GTs to produce sugarcane and its industrial derivatives. However, it is important to explore the applications of these innovations and the technological development that rapidly advances toward the consolidation of sustainable agri-food systems.

1.1 INTRODUCTION

Sugarcane (*Saccharum* spp. hybrids), a member of the Poaceae family, is a highly productive crop thanks to its C_4 photosynthetic metabolism that efficiently converts solar energy into biomass (Lima and Bearcon 2022; Huang et al., 2024). Commercially, sugarcane is grown in tropical or subtropical countries and some South Asian countries for the production of sugar and other by-products and coproducts, such as bagasse, molasses, syrup, and brown sugar (Jamir et al., 2021; Kane et al., 2022; Verma et al., 2023a, 2023b). Thus, *Saccharum* spp. has an important role in the socioeconomic development of rural areas by mobilizing resources and generating greater income and job opportunities. However, the current technological model in most sugarcane-producing countries is based on monoculture farming, with no crop rotation or intercropping (Solanki et al., 2023; Bhatt et al., 2023a, 2023b). Additionally, there is limited use of compost derived from sugarcane by-products or sludge and filter waste. Furthermore, most sugarcane-producing areas have edaphic problems such as erosion, nutrient depletion, low soil pH (Aguilar-Rivera et al., 2022; Verma et al., 2023c, 2023d), and climate-change effects. Hence, there is a high dependence on agrochemicals, which elevates production costs. Therefore, it is crucial to develop innovative and sustainable solutions to address these pressing issues in sugarcane production.

Green technologies (GTs) stand out as a long-term sustainable strategy to boost social and economic growth, conserve natural resources, and reduce pollutants that cause harmful environmental and social well-being effects (Simon and Joshi 2021; Verma et al., 2024a, 2024b; Jagadesh et al., 2024). Adopting GT in sugarcane production would improve producers' yield, income, and sustainability. These technological advances have increased agricultural systems' scale, speed, and productivity, resulting in a more efficient crop (Bhatt et al., 2023a; Verma et al., 2024c). Additionally, GTs such as plant biotechnology, plant tissue culture (PTC), biofertilizers, tolerance to abiotic factors, and nanotechnology have improved considerably, helping producers to increase yields (Bradu et al., 2023; Chen et al., 2023). This chapter compiles the most recent knowledge and progress made in various GTs in the context of agriculture that have been implemented in sugarcane production systems. Additionally, this chapter highlights different GTs that allow the reduction of the carbon footprint and the efficient use of water and biopesticides in sugarcane production to improve the yield and productivity of the crop.

1.2 KEY TECHNOLOGIES FOR ADVANCING THE TRANSFORMATION OF SUGARCANE

By 2050, it will be necessary to feed a projected population of 10 billion people (Kumari et al., 2023a, 2023b; Sabbaghi, 2024). Notably, the demographic growth will be higher in developing countries. Sugarcane is one of the most cost-effective energy sources for humans. Thus, enhancing its production through sustainable approaches becomes paramount (Verma et al., 2019a, 2020a, 2020b). An avenue for this enhancement would be through the acquisition of cutting-edge technologies that, until now, have not been applied commercially at all in the agricultural sector (Verma et al., 2021a, 2022a, 2022b; Bradu et al., 2023). Growing populations and climate change represent a threat to food supply and demand. Food security is also threatened by the increased use of crops to produce biofuels, bioenergy, and other industrial uses (Tyczewska et al., 2023). The demand for biofuels and bioenergy has exacerbated the strain on agricultural resources (Joshi et al., 2022, 2023).

Recent practices in active agriculture provide environmental and social benefits to the sugarcane industry by effectively reducing environmental pollution caused by the industry and improving the ecological environment. Key technologies (KTs) in the agricultural sector promote agricultural modernization and support the development of GTs for sugarcane cultivation. These KTs secure sugarcane production and effectively reduce agricultural energy consumption to save energy and reduce emissions based on the efficient use of new energy sources (Iwuozor et al., 2022; Li et al., 2023; Singh et al., 2023). Therefore, innovative, precise, and advanced technologies are required to boost the agricultural sector. Additionally, for sustainable agriculture, the effects of environmental and climatic conditions should be minimal. This section discusses different GTs and their application in the sugarcane industry (Figure 1.1). These technologies include plant biotechnology, PTC, abiotic stress management, and nanotechnology. Currently, these emerging technologies address various problems in traditional and industrial crops, including disease management, efficient use of water, agroecological management, and different responses to various types of abiotic stress.

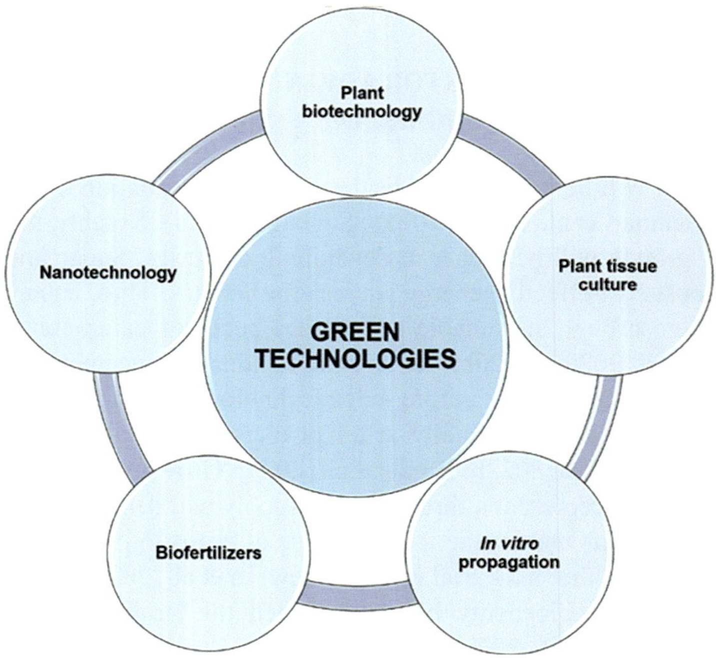

FIGURE 1.1 Green technologies and their application in the sugarcane industry.

1.3 TYPES OF GREEN TECHNOLOGIES

1.3.1 PLANT BIOTECHNOLOGY AS A TOOL FOR CROPS TRANSFORMATION

A wide range of "biotechnologies" with different techniques and applications existS. The Convention on Biological Diversity defines biotechnology as "any technological application that uses biological systems, live organisms, or their derivatives, to produce or modify products or processes for a specific use" (SCBD, 2024). Advances in plant biotechnology, especially those associated with *in vitro* culture and molecular biology, have also provided powerful tools to support and improve the management of sugarcane diversity and productivity (Vickers et al., 2005; Deng et al., 2023; Guo et al., 2021, 2023). Since then, the productivity and quality of sugarcane can only be improved by using innovative technologies like GT. PTC and genetic engineering have great potential to solve the problems faced by this crop. Transgenic technology can significantly address all performance limitations. Additionally, omics approaches have contributed to the understanding and exploration of the sugarcane genome in order to develop superior genotypes (Barnabas et al., 2015; Tian et al., 2023; Zheng et al., 2023). Similarly, different molecular markers have been developed to accelerate genetic improvement through the early identification of elite genotypes. Thus, biotechnological interventions can greatly promote sugarcane as a future energy crop and a biofactory for producing various industrial compounds and by-products.

1.3.2 PLANT TISSUE CULTURE TECHNIQUES

Plant biotechnology is based on PTC, which consists of a series of techniques that allow the establishment, conservation, manipulation, and development of plant cells, tissues, or organs under controlled and artificial conditions. PTC offers technologies for the genetic improvement of crops, *in vitro* propagation or micropropagation, plant health, *in vitro* conservation of germplasm, and production of secondary metabolites (Figure 1.2). PTC has produced plants that increase the performance of different crops, including sugarcane, by providing farmers with healthier and homogeneous planting material (Sengar et al., 2009; Anas et al., 2021a, 2021b; Song et al., 2021). Within PTC, micropropagation offers alternatives to low crop productivity. Micropropagation refers to the asexual propagation of plants using PTC techniques. This plant propagation system has different advantages, such as plant health (disease-free plants), genetic homogeneity (clonal lines), and reinvigoration (*in vitro* rejuvenation). *In vitro* culture techniques have had a huge impact on basic and commercial research, which includes the maintenance of elite germplasm, the production of agronomically superior somaclones, the micropropagation of elite clones, healthy planting material, and plants with tolerance/resistance to abiotic/biotic stress (Chen et al., 2020; Guo et al., 2022; Chaudhary et al., 2023).

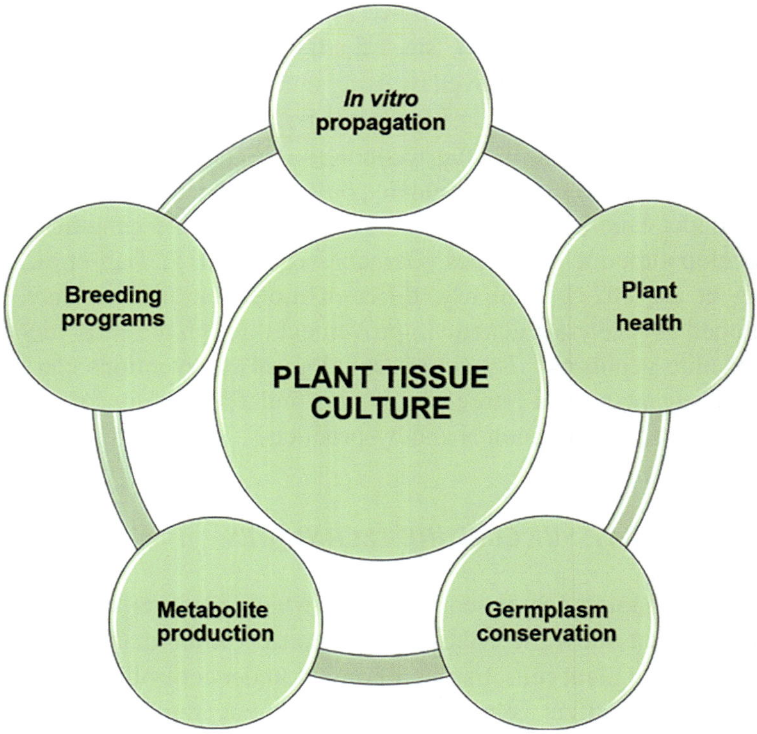

FIGURE 1.2 PTC applications can be divided into five categories: (1) obtaining pathogen-free plants by meristem culture, thermotherapy, and cryotherapy; (2) *in vitro* conservation of germplasm using minimum growth and cryopreservation techniques; (3) genetic improvement via genetic transformation, somatic hybridization, mutagenesis, and *in vitro* selection; (4) the production of secondary metabolites with high industrial value; and (5) *in vitro* production of plants.

Sugarcane has a lengthy breeding cycle and requires 10–15 years to complete the selection cycle. Furthermore, in recently released sugarcane varieties, the fuzz multiplication rate (true seed) is always slow, and diseases accumulate in groups of seeds during multiplication (Jalaja et al., 2008; Song et al., 2019, 2021; Sharma et al., 2021). The lack of disease-free seeding material is a major limitation in improving sugarcane crops. Normally, sugarcane reproduces vegetatively. However, seed propagation also occurs under specific climatic conditions. Nodal cuttings are used for the commercial propagation of sugarcane (Verma et al., 2019b; Wu et al., 2022; Huang et al., 2022). This multiplication method uses meristematic tissue as explants. Sugarcane explants have regenerated directly from apical and axillary meristems, as well as from immature foliar tissues (Ali et al., 2010; Singh et al., 2020). However, the *in vitro* propagation of sugarcane using meristematic tissues has yielded better results than other types of plant tissues. Therefore, important efforts have been made to explore meristematic tissues for their massive multiplication.

These advantages make *in vitro* propagation a profitable commercial practice with high added value. Commercial micropropagation has five well-defined stages: Stage 0: mother plant selection, Stage I: *in vitro* culture establishment, Stage II: multiplication, Stage III: elongation and rooting, and Stage IV: acclimatization. Plant micropropagation has great productive potential and is being used in commercial and research laboratories, mainly for the propagation of plants with ornamental, agri-food, and forestry importance. However, during the micropropagation stages, problems associated with contamination and ethylene accumulation in culture flasks may occur (Aragón et al., 2014; Malviya et al., 2021; Bello-Bello et al., 2021; Xu et al., 2022). Sugarcane micropropagation has been carried out conventionally in a semisolid medium and semiautomatedly on a commercial scale using a liquid medium (Lorenzo et al., 1998; Mordocco et al., 2008; Sahoo et al., 2011; Kaur et al., 2015; Bello-Bello et al., 2018; Salokhe, 2021; Gupta et al., 2022).

1.3.3 BIOFERTILIZERS AS AN INNOVATIVE TOOL

Sugarcane is considered a nutrient-demanding crop because it requires significant amounts for higher yields. Sugarcane needs about 205 kg N, 55 kg P, 275 kg K/ha, and a large amount of micronutrients to produce sugarcane yields of 100 t/ha (Yaduvanshi and Yadav, 1990; Krishna et al., 2022; Oliveira et al., 2022). However, for decades, the suggested complete N, P_2O_5, and K_2O doses were 150, 60, and 60 kg/ha. These doses have not been updated despite the introduction of high-yielding varieties. Applying old and obsolete fertilizer recommendations could not sustain sugarcane productivity and reduced nutrient use efficiency, ranging from 16% to 45%, as more of the applied N is carried through soil layers (Yadav and

Prasad, 1992; Verma et al., 2020c, 2021b, 2022c). The excessive and continuous use of inorganic biofertilizers deteriorates soil quality and crop productivity (Juntahum et al., 2022; Kumari et al., 2022). Due to this, beneficial microorganisms have been isolated and characterized, and various formulations have been developed. Therefore, the ecological interactions between different species can be applied to crops of agricultural and industrial interest, such as sugarcane (Juntahum et al., 2021; Malviya et al., 2022).

Mycorrhizae play an important role in the development, survival, and success of the species that comprise it (plants and fungi) (Wang et al., 2022; Sharma et al., 2022). Arbuscular mycorrhizal fungi (AMF) are obligate biotrophs that form endosymbiotic associations with the roots of 85% of vascular plants, including various crops (Smith and Read, 2008; Di Martino et al., 2019). AMF improve the absorption of mineral nutrients by the host plant (Vessey 2003; Ortas 2019; Song et al., 2022). They also have positive effects on photosynthetic activity (Al-Karaki and Williams, 2021), indirectly contribute to reducing the attack of herbivorous insects, decrease disease invasion, and confer tolerance to different types of abiotic stress, such as drought, salinity, abrupt temperature changes, and heavy metals (Kumalawati et al., 2014; Pontes et al., 2017; Chandrasekaran et al., 2021; Gupta et al., 2021; Wu et al., 2021; Verma et al., 2020d, 2021c, 2022d, 2022e, 2022f). For example, *Trichoderma* spp. can maximize soil organic carbon and improve crop growth and yield (Shukla et al., 2008). The main role of AMF, like *Trichoderma* spp., is to improve nitrogen absorption and soil immobile nutrients, such as phosphorus and zinc, protect the roots of the host plant from soil pathogens, improve tolerance to abiotic stresses, maintain the stability of soil aggregation, and increase disease resistance (Evelin et al., 2009). Similarly, *Rhizophagus intraradices* is an obligate endomycorrhiza that acts as a natural root symbiont and could be used to improve plant growth and nutrient acquisition. Furthermore, *R. intraradices* probably induces tolerance to abiotic stresses such as salinity (Baltazar-Bernal et al., 2022) and drought (Spinoso-Castillo et al., 2023; Verma et al., 2020e, 2021d, 2021f, 2022g, 2022h). Other plant growth-promoting microorganisms are bacteria belonging to the *Azospirillum* genus. Their use has been reported in various grasses, such as sugarcane. In particular, *Azospirillum brasilense* is widely distributed in tropical and subtropical soils. These bacteria associate with plants' roots, stimulate the production of growth-regulating hormones, such as auxins, and improve the absorption of nutrients like phosphorus. Using this microorganism as GT is promising and could increase the competitiveness of the sugarcane agroindustrial sector, even by reducing the environmental impacts derived from the uncontrolled use of fertilizers.

Acclimatization, the final stage of micropropagation, involves transferring *in vitro* seedlings to external conditions and beginning their *ex-vitro* growth (Gómez-Kosky et al., 2020). Additionally, acclimatization is a process that allows the early inoculation of mycorrhizae prior to transplanting into the field (Vergara et al., 2019). In sugarcane (*Saccharum* spp. cv Mex 69-290), del Rosario et al. (2021) demonstrated the effect of *R. intraradices* on the survival and colonization percentages of *in vitro* obtained seedlings. Overall, the survival percentage tended to decrease as the mycorrhizal colonization percentage increased. Regarding the effect of mycorrhizae on development, del Rosario et al. (2021) reported that the recommended dose per plant was between 50 and 100 spores. The authors also mentioned that the positive effect of these doses on development could be because, with 30% and 58% colonization, an adequate symbiotic interaction is achieved that could favor some physiological processes, such as photosynthetic efficiency and nutrient and water absorption, which allows the proper development of the seedling for its subsequent transfer to the field. del Rosario et al. (2021) also mentioned that the increase in dry matter in the plants treated with *R. intraradices* could be because this AMF is an endomycorrhiza. This type of fungus lives between plant cells and tissues and could cause a relative increase in root dry matter compared to the control without colonies in the roots. Moreover, da Silva Viana et al. (2020) reported that sugarcane (*Saccharum* sp. cv RB86-7515) plants treated with *A. brasilense* (100 mL/ha) in the planting furrow, 7 and 14 days after sowing, were 30.5%, 34.5%, and 27.1% taller than the control.

1.3.4 MANAGEMENT OF ABIOTIC STRESSES THROUGH PTC AND BIOFERTILIZER APPLICATIONS

In the last century, the mean temperature of the planet has increased by 0.89°C due to the increasing anthropogenic emissions of greenhouse gases derived from burning fossil fuels and agricultural activities (IPCC 2018, 2021; Bhatt 2020). In addition to biotic agents, abiotic stress can result in the partial or total loss of crops. Water scarcity, saline soils, and high or low temperatures reduce sugarcane yields (Preet et al., 2024). In sugarcane, the drought caused by climate change directly affects crop productivity (Gomathi et al., 2020; Verma et al., 2020f, 2022i, 2022j, 2022k). Plant biotechnology techniques, such as PTC, represent an alternative to address this problem (Bednarek and Orłowska, 2020). *In vitro* selection using PTC techniques is an alternative to identifying sugarcane cultivars tolerant to hydric stress. The *in vitro* selection of hydric stress-tolerant plants using polyethylene glycol (PEG) is a common strategy in plant biotechnology (Islam et al., 2019). PEG is an osmotic agent that reduces water availability in the culture medium. High-molecular-weight PEG (6000–8000) is commonly used as a selective agent because it does not penetrate plant tissues, making it ideal for hydric stress generation in culture media under *in vitro* conditions (Steuter and Mozafar, 1981). The physiologic effects caused by hydric stress are decreased photosynthesis, loss of turgor, reduced nutrient absorption, and production of reactive oxygen species (ROS). These effects result in oxidative damage, leading to the cell death of plant tissues (Diagne et al., 2020). However, plants can tolerate hydric stress through physiologic, biochemical, and genetic mechanisms, such as changes in stomatal and photosynthesis regulation, expression of genes related to the synthesis of chaperon, channel, and transporter proteins (Yang et al., 2021), osmoprotection through the accumulation of proline, betaine glycine, and sugar polyols (Kumaria et al., 2022), and increased antioxidant activity, which decreases ROS (Malhi et al., 2020).

Moreover, chemical fertilizers are commonly applied to increase sugarcane yield and productivity due to their ease of use and efficiency. Large amounts of nitrogen and potassium are required, and their absorption corresponds to the biomass accumulation pattern (Gavito et al., 2019; Bautista et al., 2019). Due to the limited rate of phosphorous mobility around the root zone, the circulation of available phosphorus has become a serious concern in agriculture. Therefore, one of the alternative methods for soil amendment with phosphorus is using microorganisms capable of using this element, such as AMF and plant growth-promoting bacteria from the *Azospirillum* genus. The most common microorganisms are *R. intraradices* and *A. brasilense*, respectively. The use of these microorganisms has been increasing as an alternative tool to improve sugarcane production and as a GT simultaneously. Sugarcane is a C_4 plant that can tolerate changing climatic conditions to a certain degree, compared to C_3 plants. However, the changing climatic conditions worldwide can affect sugarcane yield and productivity (Huang et al., 2022; Verma et al., 2023a, 2023b). Various studies have reported the positive effects of inoculating *A. brasilense* and *R. intraradices* on antioxidant and polyamine production, osmotic adjustment and water transport, plant development, and drought tolerance (Moutia et al., 2010; Hashem et al., 2019; Zhang et al., 2020; He et al., 2020).

Additionally, the positive effects of *A. brasilense* and *R. intraradices* have been demonstrated in sugarcane under field and greenhouse conditions during acclimatization (Moutia et al., 2010; Fors et al., 2020; del Rosario et al., 2021; Spinoso-Castillo et al., 2023). In sugarcane (*Saccharum* spp. cv M1176/77 and R 570), Moutia et al. (2010) observed that the M1176/77 cultivar responded positively to inoculation, with 15% more growth and 75% more root dry mass when subjected to drought stress, while the R 570 responded negatively, particularly in the absence of drought stress. Spinoso-Castillo et al. (2023) reported that the total content of chlorophyll, protein, proline, betaine glycine, and antioxidant capacity of sugarcane (*Saccharum* spp. cv Mex 69-290) increased after inoculation with *R. intraradices*. Furthermore, the soluble phenolic content was higher in the control plants than in the treated plants during drought.

1.3.5 NANOTECHNOLOGY

The potential areas in which nanotechnology could play an important role are plant genetic engineering, performance enhancement, and protection against pests and diseases (Salem, 2023; Verma et al., 2024a, 2024b). Nanotechnology has allowed the application of different materials on a nanometric scale (0–10 nm). Nanomaterials offer unique properties that depend on their size. These materials also have a higher surface-to-volume ratio, which changes their physical properties and chemical reactivity (Khan et al., 2022). These exceptional properties can be used to develop systems, devices, or other novel materials that are impossible to obtain using conventional materials (Hornyak et al., 2008).

Various nanomaterials and nanoparticles exert antibacterial and antifungal properties without compromising the normal physiology of the host, such as silver nanoparticles (AgNPs). The use of AgNPs in PTC is part of developing bionanotechnology as a GT (Castro-González et al., 2019; Salama et al., 2021). AgNPs can eliminate microbial contaminants, such as fungi and bacteria (Crisan et al., 2021; Villarreal-Gómez et al., 2021). These particles can also decrease the effects of ethylene accumulation in *in vitro* cultures (Manh-Cuong et al., 2021; Tung et al., 2021). Additionally, AgNPs have the advantages of not generating antimicrobial resistance, being easy to acquire, being nontoxic at adequate concentrations, and being economical compared to other products. AgNPs have been used in plants to induce germination, increase crop yields, and promote the development of *in vitro* cultures (Spinoso-Castillo et al., 2017; Castro-González et al., 2019; Manh-Cuong et al., 2021; Salama et al., 2021; Tung et al., 2021). However, AgNPs have some disadvantages, including the need to evaluate the adequate concentration for each study model, the cytotoxicity in different organisms, and the possible effect on beneficial environmental microorganisms. Some studies have focused on evaluating the effects of AgNPs on microbicidal activity (Sarmast and Salehi, 2016; Spinoso-Castillo et al., 2017; Kale et al., 2021), ethylene inhibition, and hormesis under *in vitro* conditions (Bello-Bello et al., 2017; Ramírez-Mosqueda et al., 2020; Hu and Xianyu, 2021; Manh-Cuong et al., 2021; Tung et al., 2021). Hormesis is characterized by the stimulation of development at low concentrations and inhibition at high concentrations (Calabrese, 2008; Calabrese and Mattson, 2011; Calabrese et al., 2021). Due to the importance of micropropagation, it is important to study the effects of AgNPs on this *in vitro* plant production system.

On the other hand, carbon nanotubes (CNTs) are rolled sheets of graphene with a hollow cylindrical shape. CNTs are classified as single-walled carbon nanotubes (SWCNTs) and multiwalled carbon nanotubes (MWCNTs) (Ahmad and Silva, 2020; Paramo et al., 2020). MWCNTs, unlike SWCNTs, have higher density, tensile strength, and electrical conductivity (Shoukat and Khan, 2021) and lower toxicity in plant cells (Samadi et al., 2021). CNTs significantly affect plant development, nutritional state, and photosynthesis. However, these effects depend on the plant species, the type and concentration of CNT, the plant development stage, and culture conditions (Martínez-Ballesta et al., 2020). The potential use of MWCNTs during the *in vitro* propagation of sugarcane demonstrates their important role in plant development (Sorcia-Morales et al., 2021). Lahiani et al. (2013) mention that CNTs can enter the cells and act as a specific stress factor, affecting aquaporin expression. Aquaporins are transmembrane channels important for transporting water and small solutes, such as hydrogen peroxide, urea, ammonia, metalloids, gases, and ions.

Aquaporins are regulated in response to environmental and hormonal stimuli, contributing to various plant growth and development processes (Li et al., 2014; Verdoucq and Maurel, 2018).

According to Zhao et al. (2017), the internalization of MWCNTs into cells could be through cell wall pores, causing the expansion of the nanotubes with different effects on plant development depending on their concentration. Mathew et al. (2020) mention that the effect of CNTs could be related to plant species, developmental stage, growth conditions, nanotube type, size, and concentration. Selvaraj and Gobikrishnan (2020) reported that the diameter of pores in different genotypes of wild sugarcane (*Saccharum spontaneum*) ranges between 3.65 and 9.71 nm. The MWCNTs used by Sorcia-Morales et al. (2021) had a 6–13 nm diameter. This indicates that a percentage of MWCNTs could enter the cell through the pores and expand inside the cells of *Saccharum* spp., resulting in the effects previously described. Additionally, the results obtained by Sorcia-Morales et al. (2021) suggest a hormetic effect. However, evaluating concentrations higher than 200 mg/L of MWCNTs is crucial. Other authors have reported that the MWCNTs could act as growth regulators because, at low concentrations, they affect plant development (Khodakovskaya et al., 2009). In the studies conducted by Sorcia-Morales et al. (2021), MWCNTs may affect *in vitro* multiplication, photosynthetic pigment concentration, and macro- and microelement contents.

Moreover, fertilizers at a nanometric scale are known as "nanofertilizers." Using nanofertilizers (NFs) in agriculture aims to reduce mineral losses during fertilization and increase yields through efficient management. NFs reduce the need for agricultural fertilizers, improve nutrient absorption efficiency, and reduce the loss of fertilizers due to leaching. CNTs are nanomaterials frequently used in the production of NFs. NFs can directly deliver macro (nitrogen, phosphorus, potassium, sulfur, magnesium, and calcium) and microelements (zinc, iron, boron, and manganese) to the roots. In wheat (*Triticum aestivum*), Zhang et al. (2020) reported that using biocarbon and carbon nanoparticles as nanofertilizers significantly improved development and productivity. In corn (*Zea mays* L.), Zhao et al. (2021) observed that carbon nanoparticles (up to 200 mg/kg) improved the activities of urease, dehydrogenase, and phosphatase. They also improved the soil's capacity to store nitrogen and phosphorus. Lastly, it is important to mention that using nanotechnology as GT opens the possibility of increasing the productivity and phytosanitary quality of sugarcane and other plants of agricultural interest.

1.4 CONCLUSION

Increasing the productivity of sugarcane as an energy source is crucial in agriculture due to the rapid growth of the world population and decreasing resources. There is great potential for using GTs to produce sugarcane and its industrial derivatives. It is important to explore the role of these innovations and the technological development that rapidly advances toward the consolidation of sustainable agri-food systems. Furthermore, compared to conventional technologies, most of these emerging technologies require less energy and could improve yields, ensuring a more sustainable use of resources. However, some of these technologies have been barely used and should be optimized for commercial implementation in the sugar industry.

KEYWORDS

- **green technologies**
- **plant biotechnology**
- **plant tissue culture**
- **micropropagation**
- **productivity**

REFERENCES

Aguilar-Rivera N, Khan MT, Khan IA, et al. (2022). Sustainable conversion of wastes into green bioproducts to introduce diversification and green economy in the sugar industry. A review. *Sugar Tech.* 24, 1198–1211. https://doi.org/10.1007/s12355-022-01167-7.

Ahmad M, Silva SRP. (2020). Low temperature growth of carbon nanotubes—A review. *Carbon.* 158, 24–44.

Ali S, Iqbal J, Khan MS. (2010). Genotype independent in vitro regeneration system in elite varieties of sugarcane. *Pak. J. Botany.* 42, 3783–3790.

Al-Karaki GN, Williams M. (2021). Mycorrhizal mixtures affect the growth, nutrition, and physiological responses of soybean to water deficit. *Acta Physiol. Plant.* 43, 75.

Anas M, Verma KK, Haq IU, Naeem M, Li Q, Liao F, Xiang YX, Li YR. (2021a). Characterization of exotic, native and wild-type genotypes of sugarcane (*Saccharum* spp. hybrids) for internal nitrogen use efficiency under different nitrogen levels. *Sugar Tech.* 23, 1258–1267. https://doi.org/10.1007/s12355-021-01014-1.

Anas M, Verma KK, Riaz M, Qiang L, Liao F, Liu Y, Li YR. (2021b). Physio-morphological and biochemical mechanism of nitrogen use efficiency in sugarcane (*Saccharum* spp.) genotypes under different growth stages and nitrogen levels. *J. Plant Interact.* 16, 332–343.

Aragón CE, Sánchez C, González-Olmedo J, Escalona M, Carvalho L, Amâncio S. (2014). Comparison of plantain plantlets propagated in temporary immersion bioreactors and gelled medium during in vitro growth and acclimatization. *Biol. Plant.* 58, 29–38. https://doi.org/10.1007/s10535-013-0381-6.

Baltazar-Bernal O, Spinoso-Castillo JL, Mancilla-Álvarez E, Bello-Bello JJ. (2022). Arbuscular mycorrhizal fungi induce tolerance to salinity stress in taro plantlets (Colocasia esculenta L. Schott) during acclimatization. *Plants.* 11, 1780. https://doi.org/10.3390/plants11131780.

Barnabas L, Ramadass A, Amalraj RS, Palaniyandi M, Rasappa V. (2015). Sugarcane proteomics: An update on current status, challenges, and future prospects. *Proteomics.* 15, 1658–1670.

Bautista EAP, López IG, Salgado-Garcia S. et al. (2019). Fertilization alternatives for sugarcane crop in Pujiltic sugarcane mill, Chiapas, Mexico. *Sugar Tech.* 21, 756–764. https://doi.org/10.1007/s12355-018-0692-1.

Bednarek PT, Orłowska R. (2020). Plant tissue culture environment as a switch-key of (epi) genetic changes. *Plant Cell Tissue Organ Cult.* 140, 245–257.

Bello-Bello JJ, Chávez-Santoscoy RA, Lecona-Guzmán CA. et al. (2017). Hormetic response by silver nanoparticles on in vitro multiplication of sugarcane (*Saccharum* spp. Cv. Mex 69-290) using a temporary immersion system. *Dose Resp.* 15(4), 1559325817744945. https://doi.org/10.1177/1559325817744945.

Bello-Bello JJ, Mendoza-Mexicano M, Pérez-Sato JA. (2018). In vitro propagation of sugarcane for certified seed production. *Sugarcane Technology and Research.* IntechOpen. https://doi.org/10.5772/intechopen.74037.

Bello-Bello JJ, Schettino-Salomón S, Ortega-Espinoza J. et al. (2021). A temporary immersion system for mass micropropagation of pitahaya (*Hylocereus undatus*). *3 Biotech.* 11, 437. https://doi.org/10.1007/s13205-021-02984-5.

Bhatt R, Verma KK, Jamwal NS. (2023a). Can paddy-compost eliminate N fertilization for climate-smart sugarcane? *Sugar Tech.* 25, 1036–1046. https://doi.org/10.1007/s12355-023-01270-3.

Bhatt R, Verma KK, Kumar R, Sanghera GS. (2023b). Foliar application of potassium salt of active phosphorus (PSAP) mitigates insect pests and improves yield along with sugarcane quality in response to agroclimatic conditions of Punjab. *Sugar Tech.* 25(3), 660–669.

Bhatt R. (2020). Resources management for sustainable sugarcane production. In: Kumar S, Meena RS, Jhariya MK (Eds.), *Resources use Efficiency in Agriculture.* Springer: Singapore; pp. 647–693.

Bhatt R, Majumder D, Tiwari AK. et al. (2023). Climate-smart technologies for improving sugarcane sustainability in India—A review. *Sugar Tech*. 25, 1–14. https://doi.org/10.1007/s12355-022-01198-0.

Bradu P, Biswas A, Nair C. et al. (2023). Recent advances in green technology and industrial revolution 4.0 for a sustainable future. *Environ. Sci. Pollut. Res*. 30, 124488–124519. https://doi.org/10.1007/s11356-022-20024-4.

Calabrese EJ, Agathokleous E, Kapoor R, Dhawan G, Calabrese V. (2021). Luteolin and hormesis. *Mech. Ageing Dev*. 199, 111559.

Calabrese EJ, Mattson MP. (2011). Hormesis provides a generalized quantitative estimate of biological plasticity. *J. Cell Commun. Signal*. 5, 25–38. https://doi.org/10.1007/s12079-011-0119-1.

Calabrese EJ. (2008). Converging concepts: Adaptive response, preconditioning, and the Yerkes–Dodson law are manifestations of hormesis. *Ageing Res. Rev*. 7, 8–20.

Castro-González CG, Sánchez-Segura L, Gómez-Merino FC, Bello-Bello JJ. (2019). Exposure of stevia (Stevia rebaudiana B) to silver nanoparticles in vitro: Transport and accumulation. *Sci. Rep*. 9, 10372. https://doi.org/10.1038/s41598-019-46828-y.

Chandrasekaran M, Boopathi T, Manivannan P. (2021). Comprehensive assessment of ameliorative effects of AMF in alleviating abiotic stress in tomato plants. *J Fungi*. 7, 303. https://doi.org/10.3390/jof7040303.

Chaudhary P, Xu M, Ahamad L, Chaudhary A, Kumar G, Adeleke BS, Verma KK, Hu D-M, Siric I, Kumar P, Popescu SM, Fayssal SA. (2023). Application of synthetic consortia for improvement of soil fertility, pollution remediation, and agricultural productivity: A review. *Agronomy*. 13, 643. https://doi.org/10.3390/agronomy13030643.

Chen GL, Zheng FJ, Lin B, Lao SB, He J, Huang Z, Zeng Y, Sun J, Verma KK. (2020). Phenolic and volatile compounds in the production of sugarcane vinegar. *ACS Omega*. 5, 30587−30595.

Chen GL, Zheng F-J, Lin B, Yang Y-X, Fang X-C, Verma KK, Yang L-F. (2023). Vinegar: A potential source of healthy and functional food with special reference to sugarcane vinegar. *Front. Nutr*. 10, 1145862. https://doi.org/10.3389/fnut.2023.1145862.

Crisan CM, Mocan T, Manolea M, Lasca LI, Tabaran FA, Mocan L. (2021). Review on silver nanoparticles as a novel class of antibacterial solutions. *Appl. Sci*. 11, 1120. https://doi.org/10.3390/app11031120.

da Silva Viana R, de Almeida Moreira BR, Lisboa LAM, et al. (2020). Morphological changes in sugarcane crop induced by the plant growth-promoting bacterium *Azospirillum brasilense*. *Sugar Tech*. 22, 241–249. https://doi.org/10.1007/s12355-019-00773-2.

del Rosario MHM, Luis SCJ, Lino SS, et al. (2022). Arbuscular mycorrhizal fungi: Inoculum dose affects plant development and performance of sugarcane (*Saccharum* spp.) plantlets during acclimatization stage. *J Soil Sci. Plant Nutr*. 22, 4847–4856. https://doi.org/10.1007/s42729-022-00964-z.

Deng ZN, Huang HR, Huang CM, Verma KK, Luo HB, Jiang SL, Li YJ, Cao HQ, Wu KC, Xu L. (2023). Short-term limited water irrigation influences the photosynthetic pigments, enzymatic and non-enzymatic activities in *Saccharum spontaneum* L. at vegetative stage. *Sugar Tech*. 25, 110–118. https://doi.org/10.1007/s12355-022-01175-7.

Di Martino C, Fioretto A, Palmieri D, Torino V, Palumbo G. (2019). Influence of tomato plant mycorrhization on nitrogen metabolism, growth and fructification on P-limited soil. *J. Plant Growth Regul*. 38, 1183–1195.

Diagne N, Ngom M, Djighaly PI, Fall D, Hocher V, Svistoonoff S. (2020). Roles of arbuscular mycorrhizal fungi on plant growth and performance: Importance in biotic and abiotic stressed regulation. *Diversity*. 12, 370.

Evelin H, Kapoor R, Giri B. (2009). Arbuscular mycorrhizal fungi in alleviation of salt stress: A review. *Ann. Bot.* 104(7), 1263–1280. https://doi.org/10.1093/aob/mcp251.

Fors RO, Saggin OJ, Carneiro MAC, Berbara RLL. (2020). Selection of arbuscular mycorrhizal fungi for sugarcane in four soils with the presence of dark septate endophytes. *Acta Sci. Agron.* 42, e42477.

Gavito MEI, Jakobsen TN, Mikkelsen Mora F. (2019). Direct evidence for modulation of photosynthesis by an arbuscular mycorrhiza-induced carbon sink strength. *New Phytol.* 223, 896–907.

Gomathi R, Krishnapriya V, Arunkumar R, Govindaraj P, Ram B. (2020). Physiological traits imparting drought stress tolerance to promising sugarcane (*Saccharum* spp.) clones. *Plant Physiol. Rep.* 25, 509–515.

Gómez-Kosky R, Jaramillo DN, Esquiro CR, Villegas AB, Calimano MB, Armas PM, Daniels DD. (2020). Effect of VIUSID Agro® and FitoMas-E® on the ex vitro acclimatization of sugarcane plants (*Saccharum* spp.) Cultivar C90–469. *Sugar Tech.* 22, 42–51. https://doi.org/10.1007/s12355-019-00752-7.

Guo D-J, Li D-P, Chen Z-D, Yang B, Verma KK, Singh RK, Singh P, Khan Q, Sharma A, Qin Y, Zhang B-Q, Song X-P, Li Y-R. (2023). Effect of endophytic diazotroph *Enterobacter roggenkampii* ED5 on nitrogen-metabolism-related microecology in the sugarcane rhizosphere at different nitrogen levels. *Front. Microbiol.* 14, 1132016. https://doi.org/10.3389/fmicb.2023.1132016.

Guo D-J, Li D-P, Singh RK, Singh P, Sharma A, Verma KK, Qin Y, Khan Q, Lu Z, Malviya MK, Song X-P, Xing Y-X, Li Y-R. (2021). Differential protein expression analysis of two sugarcane varieties in response to diazotrophic plant growth-promoting endophyte *Enterobacter roggenkampii* ED5. *Front. Plant Sci.* 12, 727741. https://doi.org/10.3389/fpls.2021.727741.

Guo DJ, Li DP, Singh RK, Singh P, Verma KK, Sharma A, Qin Y, Khan Q, Song XP, Malviya MK, Xing YX, Li YR. (2022). Comparative transcriptome analysis of two sugarcane varieties in response to diazitrophic plant growth promoting endophyte *Enterobacter roggenkampii* ED5. *J. Plant Interact.* 17(1), 75–84.

Gupta S, Singh A, Yadav K. et al. (2022). Micropropagation for multiplication of disease-free and genetically uniform sugarcane plantlets. *Advances in Plant Tissue Culture: Current Developments and Future Trends*. Academic Press, pp. 31–49.

Gupta S, Thokchom SD, Kapoor R. (2021). Arbuscular mycorrhiza improves photosynthesis and restores alteration in sugar metabolism in *Triticum aestivum* L. grown in arsenic contaminated soil. *Front. Plant Sci.* 12, 334. https://doi.org/10.3389/fpls.2021.640379.

Hashem A, Kumar A, Al-Dbass AM, Alqarawi AA, Al-Arjani AF, Singh G, Farooq M, Abd_Allah EF. (2019). Arbuscular mycorrhizal fungi and biochar improves drought tolerance in chickpea. *Saudi J. Biol. Sci.* 26, 614–624.

He JD, Zou YN, Wu QS, Ku˘ca K. (2020). Mycorrhizas enhance drought tolerance of trifoliate orange by enhancing activities and gene expression of antioxidant enzymes. *Sci. Hortic.* 262, 108745.

Hornyak GL, Tibbals HF, Dutta J, Moore JJ. (2008). Perspectives. In: Hornyak GL, Tibbals HF, Dutta J, Moore JJ. (Eds.), *Introduction to Nanoscience and Nanotechnology*. CRC Press: Boca Raton, FL, USA, pp. 1–104.

Huang CM, Wu KC, Verma KK, Luo HB, Deng ZN, Shi GY, Cao HQ, Wu XJ, Ye LP, Wei YW, Xu L. (2024). Isolation and expression analysis of cellulose synthase 3 (Ces3) genes from sugarcane (*Saccharum officinarum* L.). *Chem. Biol. Technol. Agric*. 11, 71. https://doi.org/10.1186/s40538-024-00597-7.

Huang HR, Gao YJ, Malviya MK, Verma KK, Solanki MK, Huang YX, Li X, Deng Y, Yan J, Tang SY, Wang LW, Xu L. (2022). Genetic diversity analysis of sugarcane (*Saccharum* spp. Hybrids) among high-sucrose clones of GT series and commonly used parents by using microsatellite markers in Guangxi, *China. Sugar Tech*, 24, 397–407. https://doi.org/10.1007/s12355-021-01033-y.

IPCC-(Intergovernmental Pannel on Climate Change). 2018. Summary for policymakers of IPCC special report on global warming of 1.5°C approved by governments https://www.ipcc.ch/2018/10/08/summary-for-policymakers-of-ipcc-special-report-on-globalwarming-of-1-5c-approved-by-governments/.

Islam NU, Ali G, Dar ZA, Maqbool S, Khulbe RK, Bhat A. (2019). Effect of peg induced drought stress on maize (*Zea mays* L.) INBREDS. *Plant Arch*. 19, 1677–1681.

Iwuozor KO, Emenike EC, Ighalo JO, et al. (2022). Valorization of sugar industry's by-products: A perspective. *Sugar Tech*. 24, 1052–1078. https://doi.org/10.1007/s12355-022-01143-1.

Jagadesh M, Dash M, Kumari A, Singh SK, Verma KK, Kumar P, Bhatt R, Sharma SK. (2024). Revealing the hidden world of soil microbes: Metagenomic insights into plant, bacteria, and fungi interactions for sustainable agriculture and ecosystem restoration. *Microbiol. Res*. 285, 127764. https://doi.org/10.1016/j.micres.2024.127764.

Jalaja NC, Neelamathi D, Sreenivasan TV, (2008). *Micropropagation for Quality Seed Production in Sugarcane in Asia and the Pacific*. Food and Agriculture Organization of the United Nations: India.

Jamir, L, Kumar V, Kaur J, Kumar S, Singh H. (2021). Composition, valorization, and therapeutical potential of molasses: A critical review. *Environ. Technol. Rev*. 10 (1), 131–142. https://doi.org/10.1080/21622515.2021.1892203.

Joshi A, Verma KK, Rajput VD, Arora J. (2023). Metabolic engineering of microorganisms in advancing biofuel production. *Adv. Lignocellulosic Biofuel Prod. Syst*. 181–202. https://doi.org/10.1016/B978-0-323-91192-4.00016-X.

Joshi A, Verma KK, Rajput VD, Minkina T, Arora J. (2022). Recent advances in metabolic engineering of microorganisms for advancing lignocellulose-derived biofuels. *Bioengineered*. 13(4), 8135–8163. https://doi.org/10.1080/21655979.2022.2051856.

Juntahum S, Ekprasert J, Boonlue S. (2022). Efficiency of arbuscular mycorrhizal fungi for the growth promotion of sugarcane under pot conditions. *Sugar Tech*. 24, 1738–1747. https://doi.org/10.1007/s12355-022-01129-z.

Juntahum S, Jongrungklang N, Kaewpradit W, et al. (2021). Improved physiological performances of sugarcane during maturation and ripening phase by inoculation of arbuscular mycorrhizal fungi. *Sugar Tech*. 23, 336–342. https://doi.org/10.1007/s12355-020-00887-y.

Kale S, Parishwad GV, Husainy ASN, Patil AS. (2021). Emerging agriculture applications of silver nanoparticles. *Eng. Sci. Food Agrofor.* 3, 17–22. https://doi.org/10.30919/esfaf438.

Kane AO, Pellergini VOA, Espirito Santo MC. et al. (2022). Evaluating the potential of culms from sugarcane and energy cane varieties grown in Argentina for second-generation ethanol production. *Waste Biomass Valor*. 13, 329–343. https://doi.org/10.1007/s12649-021-01528-5.

Kaur A, Sandhu JS. (2015). High throughput in vitro micropropagation of sugarcane (*Saccharum officinarum* L.) from spindle leaf roll segments: Cost analysis for agri-business industry. *Plant Cell Tiss. Organ Cult*. 120(1), 339–350.

Khan F, Shariq M, Asif M, Siddiqui MA, Malan P, Ahmad F. (2022). Green nanotechnology: Plant-mediated nanoparticle synthesis and application. *Nanomaterials (Basel)*. 12, 673. https://doi.org/10.3390/nano12040673

Khodakovskaya M, Dervishi E, Mahmood M, Xu Y, Li Z, Watanabe F, Biris AS. (2009). Carbon nanotubes are able to penetrate plant seed coat and dramatically affect seed germination and plant growth. *ACS Nano*. 3, 3221–3227.

Krishna D, Sachan HK, Verma KK. (2022). Abiotic stresses in plants: An overview. In: *Understanding Abiotic Stresses*. Nova Science Publishers, Inc.: New York.

Kumalawati Z, Musa Y, Amin A, Asrul L, Ridwan I. (2014). Exploration of arbuscular mycorrhizal fungi from sugarcane rhizosphere in south Sulawesi. *Int. J. Sci. Technol. Res.* 3(1), 201–203.

Kumari A, Dash M, Singh SK, Jagadesh M, Mathpal B, Mishra PK, Pandey SK, Verma KK. (2023a). Soil microbes: A natural solution for mitigating the impact of climate change. *Environ. Monit. Assess.* 195, 1436. https://doi.org/10.1007/s10661-023-11988-y.

Kumari A, Lakshmi GA, Krishna GK, Patni B, Prakash S, Bhattacharyya M, Singh SK, Verma KK. (2022). Climate change and its impact on crops: A comprehensive investigation for sustainable agriculture. *Agronomy*. 12, 3008. https://doi.org/10.3390/agronomy12123008.

Kumari A, Singh SK, Mathpal B, Verma KK, Garg VK, Bhattacharyya M, Bhatt R. (2023b). The biosynthesis, mechanism of action, and physiological functions of melatonin in horticultural plants: A review. *Horticulturae.* 9, 913. https://doi.org/10.3390/horticulturae9080913.

Kumaria J, Mahatmana KK, Sharmaa S, Singhb AK, Adhikaric S, Bansala R, Yadavb MC. (2022). Recent advances in differentomics mechanism for drought stress tolerance in rice. *Russ. J. Plant Physiol.* 69, 18.

Lahiani MH, Dervishi E, Chen J., Nima Z, Gaume A., Biris AS, Khodakovskaya MV. (2013). Impact of carbon nanotube exposure to seeds of valuable crops. *ACS Appl. Mater. Interfaces*. 5, 7965–7973.

Li A-M, Liao F, Wang M, Chen Z-L, Qin C-X, Huang R-Q, Verma KK, Li Y-R, Que Y-X, Pan Y-Q, et al. (2023). Transcriptomic and proteomic landscape of sugarcane response to biotic and abiotic stressors. *Int. J. Mol. Sci.* 24, 8913. https://doi.org/10.3390/ijms24108913.

Li G, Santoni V, Maurel C. (2014). Plant aquaporins: Roles in plant physiology. *Biochim. Biophys. Acta (BBA)*. 1840, 1574–1582.

Lima IM, Beacorn JA. (2022). Targeting a sustainable sugar crops processing industry: A review (part I)—By-product applications. *Sugar Tech.* 24, 970–991. https://doi.org/10.1007/s12355-022-01169-5.

Lorenzo JC, Gonzalez BL, Escalona M, Teisson C, Borroto C. (1998). Sugarcane shoot formation in an improved temporary immersion system. *Plant Cell Tiss. Organ Cult.* 54(3), 197–200.

Malhi GS, Kaur M, Kaushik P, Alyemeni MN, Alsahli AA, Ahmad P. (2021). Arbuscular mycorrhiza in combating abiotic stresses in vegetables: An eco-friendly approach. *Saudi J. Biol. Sci.* 28, 1465–1476.

Malviya MK, Li C-N, Lakshmanan P, Solanki MK, Wang Z, Solanki AC, Nong Q, Verma KK, Singh RK, Singh P, Sharma A, Guo D-J, Dessoky ES, Song X.-P, Li Y.-R. (2022). High-throughput sequencing-based analysis of rhizosphere and diazotrophic bacterial diversity among wild progenitor and closely related species of sugarcane (*Saccharum* spp. inter-specific hybrids). *Front. Plant Sci.* 13, 829337. https://doi.org/10.3389/fpls.2022.829337.

Malviya MK, Solanki MK, Li CN, Wang Z, Zeng Y, Verma KK, Singh RK, Singh P, Huang HR, Yang LT, Song XP, Li YR. (2021). Sugarcane-legume intercropping can enrich the soil microbiome and plant growth. *Front. Sustain. Food Syst.* 5, 606595. https://doi.org/10.3389/fsufs.2021.606595.

Manh-Cuong D, Cong Du P, Tung HT. et al. (2021). Silver nanoparticles as aneffective stimulant in micropropagation of *Panax vietnamensis*—A valuable medicinal plant. *Plant Cell Tiss. Organ Cult.* 146, 577–588. https://doi.org/10.1007/s11240-021-02095-2.

Martinez-Ballesta MC, Chelbi N, Lopez-Zaplana A, Carvajal M. (2020). Discerning the mechanism of the multiwalled carbon nanotubes effect on root cell water and nutrient transport. *Plant Physiol. Biochem.* 146, 23–30.

Mathew S, Tiwari DK, Tripathi D. (2020). Interaction of carbon nanotubes with plant system: A review. *Carbon Lett.* 31, 1–10.

Mordocco AM, Brumbley JA, Lakshmanan P. (2008). Development of a temporary immersion system (RITA®) for mass production of sugarcane (*Saccharum* spp. interspecific hybrids). *In Vitro Cell. Dev. Biol. Plant.* 45, 450–457. https://doi.org/10.1007/S11627-008-9173-7.

Moutia JFY, Saumtally S, Spaepen S. et al. (2010). Plant growth promotion by Azospirillum sp. in sugarcane is influenced by genotype and drought stress. *Plant Soil.* 337, 233–242. https://doi.org/10.1007/s11104-010-0519-7.

Oliveira MA, Verma KK, Bhatt R, Oliveira TBA. (2022). Impact of green and organic fertilizers on soil fertility and sugarcane productivity. In: Verma, KK, et al. (Eds.), *Agro-Industrial Perspectives on Sugarcane Production under Environmental Stress.* pp. 193–213.

Ortas I. (2019). Under filed conditions, mycorrhizal inoculum effectiveness depends on plant species and phosphorus nutrition. *J. Plant Nutr.* 42, 2349–2362.

Paramo LA, Feregrino-Pérez AA, Guevara R, Mendoza S, Esquivel K. (2020). Nanoparticles in agroindustry: Applications, toxicity, challenges, and trends. *Nanomaterials.* 10, 1654.

Pontes JS, Oehl F, Marinho F, Coyne D, da Silva DKA, Yano-Melo AM, Maia LC. (2017). Diversity of arbuscular mycorrhizal fungi in Brazil's Caatinga and experimental agroecosystems. *Biotropica.* 49(3), 413–427.

Preet K, Dhansu P, Sehrawat N, et al. (2024). Morpho-physiological analysis of salinity tolerance in sugarcane genotypes. *Plant Physiol. Rep*. 29, 356–366. https://doi.org/10.1007/s40502-024-00782-8.

Ramírez-Mosqueda MA, Sánchez-Segura L, Hernández-Valladolid SL. et al. (2020). Influence of silver nanoparticles on a common contaminant isolated during the establishment of Stevia rebaudiana Bertoni culture. *Plant Cell Tiss. Organ Cult.* 143, 609–618. https://doi.org/10.1007/s11240-020-01945-9.

Sabbaghi O. (2024). The United Nations sustainable development goals and global finance: Recent evidence. *Corp. Soc. Responsib. Environ. Manag.* 31, 4020–4033. https://doi.org/10.1002/csr.2785.

Sahoo DP, Samantrai D, Rout G.R. (2011). Rapid clonal propagation of *Saccharum officinarum* L. Vars. CO-6907 and CO-86249 and to assess the genetic uniformity through molecular markers. *Plant Biosyst*. 145(2), 445–451.

Salama DM, Abd El-Aziz ME, Rizk FA, Abd Elwahed MSA. (2021). Applications of nanotechnology on vegetable crops. *Chemosphere*. 266, 129026.

Salem SS. (2023). A mini review on green nanotechnology and its development in biological effects. *Arch. Microbiol*. 205, 128. https://doi.org/10.1007/s00203-023-03467-2.

Salokhe S. (2021). Development of an efficient protocol for production of healthy sugarcane seed cane through Meristem culture. *J. Agric. Food Res*. 4, 100126. https://doi.org/10.1016/J.JAFR.2021.100126.

Samadi S, Lajayer BA, Moghiseh E, Rodríguez-Couto S. (2021). Effect of carbon nanomaterials on cell toxicity, biomass production, nutritional and active compound accumulation in plants. *Environ. Technol. Innov*. 21, 101323.

Sarmast MK, Salehi H. (2021). Sub-lethal concentrations of silver nanoparticles mediate a phytostimulatory response in tobacco via the suppression of ethylene biosynthetic genes and the ethylene signaling pathway. *In Vitro Cell. Dev. Biol. Plant*. 58, 1–14https://doi.org/10.1007/s11627-021-10193-1.

Secretariat of the Convention on Biological Diversity. Convention on Biological Diversity: Text and Annexes. Secretariat for the Convention on Biological Diversity: Montreal, QC, Canada, 2011. Retrieved January 14, 2024, from https://www.cbd.int/doc/legal/cbd-en.pdf

Selvaraj A, Gobikrishnan S. (2020). Physicochemical characterization of native and steam explosion pretreated wild sugarcane (*Saccharum spontaneum*). *Int. J. Renew. Energy Dev.* 9, 3.

Sengar AS, Thind KS, Kumar B, Pallavi M, Gosal SS. (2009). In vitro selection at cellular level for red rot resistance in sugarcane (*Saccharum* sp.). *Plant Growth Regul.* 58, 201–209.

Sharma A, Singh RK, Singh P, Vaishnav A, Guo DJ, Verma KK, Li DP, Song XP, Malviya MK, Khan N, Lakshmanan P, Li YR. (2021). Insights into the bacterial and nitric oxide-induced salt tolerance in sugarcane and their growth-promoting abilities. *Microorganisms*. 9, 2203. https://doi.org/10.3390/microorganisms9112203.

Sharma A, Song X-P, Singh RK, Vaishnav A, Gupta S, Singh P, Guo D-J, Verma KK, Li Y-R. (2022). Impact of carbendazim on cellular growth, defence system and plant growth promoting traits of Priestia megaterium ANCB-12 isolated from sugarcane rhizosphere. *Front. Microbiol*. 13, 1005942. https://doi.org/10.3389/fmicb.2022.1005942.

Shoukat R, Khan MI. (2021). Carbon nanotubes: A review on properties, synthesis methods and applications in micro and nanotechnology. *Microsyst. Technol.* 27, 4183–4192.

Shukla SK, Yadav RL, Suman A, Singh PN. (2008). Improving rhizospheric environment and sugarcane ratoon yield through bioagents amended farm yard manure in *udic ustochrept* soil. *Soil Tillage Res.* 99(2), 158–168. https://doi.org/10.1016/j.still.2008.02.007.

Simon M, Joshi H. (2021). A review on green technologies for the rejuvenation of polluted surface water bodies: Field-scale feasibility, challenges, and future perspectives. *J. Environ Chem. Eng.* 9(4), 105763.

Singh P, Singh RK, Li HB, Guo DJ, Sharma A, Verma KK, Solanki MK, Upadhyay SK, Lakshmanan P, Yang LT, Li YR. (2023). Nitrogen fixation and phytohormone stimulation of sugarcane plant through plant growth promoting diazotrophic Pseudomonas. *Biotechnol. Genet. Eng. Rev.* 40, 15–35. https://doi.org/10.1080/02648725.2023.2177814.

Singh P, Singh RK, Song QQ, Li HB, Guo DJ, Malviya MK, Verma KK, Song XP, Lakshmanan P, Yang LT, Li YR. (2021). Comparative analysis of protein and differential responses of defense-related gene and enzyme activity reveals the long-term molecular responses of sugarcane inoculated with *Sporisorium scitamineum*. *J. Plant Interact.* 16(1), 12–29.

Singh RK, Singh P, Li HB, Song QQ, Guo DJ, Solanki MK, Verma KK, Malviya MK, Song XP, Lakshmanan P, Yang LT, Li YR. (2020). Diversity of nitrogen-fixing rhizobacteria associated with sugarcane: A comprehensive study of plant-microbe interactions for growth enhancement in *Saccharum* spp. *BMC Plant Biol.* 20, 220. https://doi.org/10.1186/s12870-020-02400-9.

Smith SE, Read D. (2008). The symbionts forming arbuscular mycorrhizas. In: *Mycorrhizal Symbiosis*, Elsevier, pp. 13–41. https://doi.org/10.1016/B978-012370526-6.50003-9.

Solanki MK, Verma KK, Dastogeer KM.G, Mora-Poblete F, Mundra S. (2023). Microbial resilience in plant nutrient management towards sustainable farming. *Front. Microbiol.* 14, 1280811. https://doi.org/10.3389/fmicb.2023.1280811.

Song X-P, Mo F, Verma KK, Wei J, Zhang X, Yang L, Li YR. (2019). Effect of sugarcane smut (*Ustilago scitaminea* Syd.) on ultrastructure and biochemical indices of sugarcane. *Biomed. J. Sci. Tech. Res.* 17, 12546–12551. https://doi.org/10.26717/BJSTR.2019.17.002950.

Song XP, Mo F, Yan M, Zhang X, Zhang B, Huang X, Huang D, Pan Y, Verma KK, Li YR. (2022). Effect of smut infection on the photosynthetic physiological characteristics and related defense enzymes of sugarcane. *Life.* 12, 1201. https://doi.org/10.3390/life12081201.

Song XP, Verma KK, Tian DD, Zhang XQ, Liang YJ, Huang X, Li CN, Li YR. (2021). Exploration of silicon functions to integrate with biotic stress tolerance and crop improvement. *Biol. Res.* 54, 19. https://doi.org/10.1186/s40659-021-00344-4.

Sorcia-Morales M, Gómez-Merino FC, Sánchez-Segura L, Spinoso-Castillo JL, Bello-Bello JJ. (2021). Multi-walled carbon nanotubes improved development during in vitro multiplication of sugarcane (*Saccharum* spp.) in a semi automated bioreactor. *Plants.* 10, 2015. https://doi.org/10.3390/plants10102015.

Spinoso-Castillo JL, Chávez-Santoscoy RA, Bogdanchikova N, Pérez-Sato JA, Morales-Ramos V, Bello-Bello JJ. (2017). Antimicrobial and hormetic effects of silver nanoparticles on in vitro regeneration of vanilla (Vanilla planifolia Jacks. ex Andrews) using a temporary immersion system. *Plant Cell Tiss. Organ Cult.* 129, 195–207. https://doi.org/10.1007/s11240-017-1169-8.

Spinoso-Castillo JL, Moreno-Hernández MdR, Mancilla-Álvarez E, Sánchez-Segura L, Sánchez-Páez R, Bello-Bello JJ. (2023). Arbuscular mycorrhizal symbiosis improves ex vitro acclimatization of sugarcane plantlets (*Saccharum* spp.) under drought stress conditions. Plants. 12, 687. https://doi.org/10.3390/plants12030687.

Steuter AA, Mozafar A, Goodin JR. (1981). Water potential of aqueous polyethylene glycol. *Plant Physiol.* 67, 64–67.

Tian D, Qin L, Verma KK, Wei L, Li J, Li B, Zhou W, He Z, Wei D, Huang S, Long S, Huang Q, Li C, Wei S. (2023). Transcriptomic and metabolomic differences between banana varieties which are resistant or susceptible to Fusarium wilt. *PeerJ.* 11, e16549. https://doi.org/10.7717/peerj.16549.

Tung HT, Thuong TT, Cuong DM. et al. (2021). Silver nanoparticles improved explant disinfection, in vitro growth, runner formation and limited ethylene accumulation during micropropagation of strawberry (Fragaria × ananassa). *Plant Cell Tiss. Organ Cult.* 145, 393–403. https://doi.org/10.1007/s11240-021-02015-4.

Tyczewska A, Twardowski T, Woźniak-Gientka E. (2023). Agricultural biotechnology for sustainable food security. *Trends Biotechnol.* 41, 331–341. https://doi.org/10.1016/j.tibtech.2022.12.013.

Verdoucq L, Maurel C. (2018). Plant aquaporins. *Adv. Bot. Res.* 87, 25–56.

Vergara C, Araujo K, de Souza SR, Schultz N, Saggin-Júnior OJ, Sperandio M, Zilli JE. (2019). Plant-mycorrhizal fungi interaction and response to inoculation with different growth-promoting fungi. *Pesquisa Agropecuaria Brasileria.* 54. e25140. https://doi.org/10.1590/S1678-3921.pab2019.v54.25140.

Verma KK, Anas M, Chen Z, Rajput VD, Malviya MK, Verma CL, Singh RK, Singh P, Song XP, Li YR. (2020f). Silicon supply improves leaf gas exchange, antioxidant defense system and growth in sugarcane responsive to water limitation. *Plants*. 9, 1032, https://doi.org/10.3390/plants9081032.

Verma KK, Joshi A, Song X.-P, Liang Q, Xu L, Huang H.-R., Wu K.-C, Seth CS, Arora J, Li Y.-R. (2024a). Regulatory mechanisms of plant rhizobacteria on plants to the adaptation of adverse agroclimatic variables. *Front. Plant Sci*. 15, 1377793. https://doi.org/10.3389/fpls.2024.1377793.

Verma KK, Joshi A, Song X-P, Singh S, Kumari A, Arora J, Singh SK, Solanki MK, Seth CS, Li Y-R. (2024b). Synergistic interactions of nanoparticles and plant growth promoting rhizobacteria enhancing soil-plant systems: A multigenerational perspective. *Front. Plant Sci*. 15, 1376214. https://doi.org/10.3389/fpls.2024.1376214.

Verma KK, Li DM, Singh M, Rajput VD, Malviya MK, Minkina T, Singh RK, Singh P, Song XP, Li YR. (2020a). Interactive role of silicon and plant-rhizobacteria mitigating abiotic stresses: A new approach for sustainable agriculture and climate change. *Plants*. 9, 1055. https://doi.org/10.3390/plants9091055.

Verma KK, Rajput VD, Singh M, Verma CL, Minkina T. (2022a). Influence of waterlogging on plant physiology and molecular biology. In: *Understanding Abiotic Stresses*. Nova Science Publishers, Inc.: New York.

Verma KK, Siddiqui N, Singh M. (2022b). Understanding the plant physiological and molecular adaptation strategies to soil salinity. In: *Understanding Abiotic Stresses*: Nova Science Publishers, Inc.: New York. https://doi.org/10.52305/UBUJ4024.

Verma KK, Singh P, Song X.-P, Malviya MK, Singh RK, Chen G.-L, Solomon S, Li YR. (2020b). Mitigating climate change for sugarcane improvement: Role of silicon in alleviating abiotic stresses. *Sugar Tech.* 22(5), 741–749. https://doi.org/10.1007/s12355-020-00831-0.

Verma KK, Singh RK, Song QQ, Singh P, Zhang B.-Q, Song X.-P, Chen G.-L, Li YR. (2019a). Silicon alleviates drought stress of sugarcane plants by improving antioxidant responses. *Biomedical J. Sci. Tech. Res*. 002957. https://doi.org/10.26717/BJSTR.2019.17.002957

Verma, KK, Song X, Degu HD, Guo D, Joshi A, Huang H, Xu L, Singh M, Huang D, Rajput VD, Li YR. (2023a). Recent advances in nitrogen and nano-nitrogen fertilizers for sustainable crop production: A mini-review. *Chem. Biol. Technol. Agric*. 10, 1–14. https://doi.org/10.1186/s40538-023-00488-3.

Verma KK, Song X, Verma CL, Huang H, Singh M, Xu L, Wu J, Khan ZH, Zhang B, Li Y. (2023b). Mathematical modeling of climate and fluoride effects on sugarcane photosynthesis with silicon nanoparticles. *Plant Physiol. Biochem.* 204, 108089. https://doi.org/10.1016/j.plaphy.2023.108089.

Verma KK, Song XP, Budeguer F, Nikpay A, Enrique R, Singh M, Zhang B, Wu JM, Li YR. (2022c). Genetic engineering: An efficient approach to mitigating biotic and abiotic stresses in sugarcane cultivation. *Plant Signal. Behav*. 17(1), e2108253.

Verma KK, Song XP, Joshi A, Rajput VD, Singh M, Sharma A, Singh RK, Li DM, Arora J, Minkina T, Li YR. (2022d). Nanofertilizer possibilities for healthy soil, water and food in future: An overview. *Front. Plant Sci.* 13, 865048. https://doi.org/10.3389/fpls.2022.865048.

Verma KK, Song XP, Joshi A, Tian DD, Rajput VD, Singh M, Arora J, Minkina TM, Li YR. (2022e). Recent trends in nano-fertilizer for advancing sustainable agriculture under the era of climate change to ensure food security for future generations. *Nanomaterials*. 12, 173. https://doi.org/10.3390/nano12010173.

Verma KK, Song XP, Kumari A, Jagadesh M, Singh SK, Bhatt R, Singh M, Seth CS, Li YR. (2024c). Climate change adaptation: Challenges for agricultural sustainability. *Plant Cell Environ.* 48, 2522–2533. https://doi.org/10.1111/pce.15078.

Verma KK, Song XP, Li DM, Singh M, Wu JM, Singh RK, Sharma A, Zhang BQ, Li YR. (2022f). Silicon and soil microorganisms improves rhizospheric soil health with bacterial community, plant growth, performance and yield. *Plant Signal. Behav.* 17(1), e2104004.

Verma KK, Song XP, Lin B, Guo DJ, Singh M, Rajput VD, Singh RK, Singh P, Sharma A, Malviya MK, Chen GL, Li YR. (2022g). Silicon induced drought tolerance in crop plants: Physiological adaptation strategies. *Silicon.* 14(6), 2473–2487. https://doi.org/10.1007/s12633-021-01071-x.

Verma KK, Song XP, Rajput VD, Boldyreva V, Zhang BQ, Minkina T, Li YR. (2022h). Morpho-physiological, biochemical and ultrastructural modifications on sugarcane to prolonged water deficit. In: Verma KK, et al. (Eds.), *Agro-Industrial Perspectives on Sugarcane Production Under Environmental Stress*. https://doi.org/10.1007/978-981-19-3955-6_11.

Verma KK, Song XP, Rajput VD, Solomon S, Li YR, Rao GP. (2022i). *Agro-Industrial Perspectives on Sugarcane Production Under Environmental Stress*. Springer Nature Singapore, https://doi.org/10.1007/978-981-19-3955-6.

Verma KK, Song XP, Singh M, Tian DD, Rajput VD, Minkina T, Li YR. (2023c). Association of silicon and soil microorganisms induces stress mitigation, increasing plant productivity. In: de Mello Prado R (Ed.), *Benefits of Silicon in the Nutrition of Plants*. pp. 299–328. https://doi.org/10.1007/978-3-031-26673-7_17.

Verma KK, Song XP, Tian DD, Guo DJ, Chen ZL, Zhong CS, Nikpay A, Singh M, Rajput VD, Singh RK, Minkina T, Li YR. (2021a). Influence of silicon on biocontrol strategies to manage biotic stress for crop protection, performance and improvement. *Plants*. 10, 2163. https://doi.org/10.3390/plants10102163.

Verma KK, Song XP, Tian DD, Singh M, Verma CL, Rajput VD, Singh RK, Sharma A, Singh P, Malviya MK, Li YR. (2021b). Investigation of defensive role of silicon during drought stress induced by irrigation capacity in sugarcane: Physiological and biochemical characteristics. *ACS Omega*. 6, 19811–19821.

Verma KK, Song XP, Verma CL, Chen ZL, Rajput VD, Wu KC, Liao F, Chen GL, Li YR. (2021c). Functional relationship between photosynthetic leaf gas exchange in response to silicon application and water stress mitigation in sugarcane. *Biol. Res.* 54, 15. https://doi.org/10.1186/s40659-021-00338-2.

Verma KK, Song XP, Verma CL, Malviya MK, Guo DJ, Rajput VD, Sharma A, Wei KJ, Chen GL, Solomon S, Li YR. (2021d). Predication of photosynthetic leaf gas exchange of sugarcane (*Saccharum* spp.) leaves in response to leaf positions to foliar spray of potassium salt of active phosphorus under limited water irrigation. *ACS Omega*. 6, 2396–2409.

Verma KK, Song X-P, Xu L, Huang H-R, Liang Q, Seth CS, Li Y-R. (2023d). Nano-microplastic and agro-ecosystems: A mini-review. *Front. Plant Sci.* 14, 1283852. https://doi.org/10.3389/fpls.2023.1283852.

Verma KK, Song XP, Yadav G, Degu HD, Parvaiz A, Singh M, Huang HR, Mustafa G, Xu L, Li YR. (2022j). Impact of agroclimatic variables on proteo-genomics in sugarcane (*Saccharum* spp.) plant productivity. *ACS Omega*. 7, 22997–23008.

Verma KK, Song XP, Zeng Y, Guo DJ, Sing M, Rajput VD, Malviya MK, Wei KJ, Sharma A, Li DP, Chen GL, Li YR. (2021e). Foliar application of silicon boosts growth, photosynthetic leaf gas exchange, antioxidative response and resistance to limited water irrigation in sugarcane (*Saccharum officinarum* L.). *Plant Physiol. Biochem*. 166, 582–592.

Verma KK, Song XP, Zengm Y, Li DM, Guo DJ, Rajput VD, Chen GL, Barakhov A, Minkina TM, Li YR. (2020c). Characteristics and correlation of leaf stomata and its relationship with photosynthesis on *Saccharum* spp. under different irrigation and silicon application. *ACS Omega*. 5, 24145–24153.

Verma KK, Vatsal S, Singh M, Rajput VD, Minkina T. (2022k). Coping with water stress: Effects and adaptive mechanisms to plants. In: *Understanding Abiotic Stresses*. Nova Science Publishers, Inc.: New York.

Verma KK, Wu KC, Verma CL, Li DM, Malviya MK, Singh RK, Singh P, Chen GL, Song XP, Li YR. (2020d). Developing mathematical model for diurnal dynamics on photosynthetic responses in sugarcane responsive to different irrigation and silicon application. *PeerJ*. 8, e10154. https://doi.org/10.7717/peerj.10154.

Verma KK, Liu X-H, Wu K-C, Singh RK, Song QQ, Malviya MK, Song X-P, Singh P, Verma CL, Li YR. (2020e). The impact of silicon on photosynthetic and biochemical responses of sugarcane under different soil moisture levels. *Silicon*. 12, 1355–1367.

Verma KK, Wu K-C, Singh P, Malviya MK, Singh RK, Song X-P, Li YR. (2019b). The protective role of silicon in sugarcane under water stress: Photosynthesis and antioxidant enzymes. *Biomed. J. Sci. Tech. Res.* 15, 1–7. https://doi.org/0.26717/BJSTR.2019.15.002685.

Vessey KJ. (2003). Effect of Rhizobacterial biofertilizer on plant growth. *Plant Soil.* 255(2), 571–586. https://doi.org/10.1023/A:1026037216893.

Vickers JE, Grof CPL, Bonnett GD, Jackson PA, Morgan TE. (2005). Effects of tissue culture, biolistic transformation, and introduction of PPO and SPS gene constructs a performance of sugarcane clones in the field. *Aust. J. Agric. Res.* 56, 57–68.

Villarreal-Gómez LJ, Pérez-González GL, Bogdanchikova N, Pestryakov A, Nimaev V, Soloveva A, Cornejo-Bravo JM, Toledaño-Magaña Y. (2021). Antimicrobial effect of electrospun nanofibers loaded with silver nanoparticles: Influence of Ag incorporation method. *J. Nanomater.* https://doi.org/10.1155/2021/9920755.

Wang L, Yang D, Ma F, Wang G, You Y. (2022). Recent advances in responses of arbuscular mycorrhizal fungi-Plant symbiosis to engineered nanoparticles. *Chemosphere*. 286, 131644.

Wu K-C, Huang C-M, Verma KK, Deng Z-N, Huang H-R, Pang T, Cao H-Q, Luo H-B, Jiang S-L, Xu L. (2022). Transcriptomic responses of *Saccharum spontaneum* roots in response to polyethylene glycol–6000 stimulated drought stress. *Front. Plant Sci.* 13, 992755. https://doi.org/10.3389/fpls.2022.992755.

Wu M, Yan Y, Wang Y, Mao Q, Fu Y, Peng X, Yang Z, Ren J, Liu A, Chen S, Ahammed GJ. (2021). Arbuscular mycorrhizal fungi for vegetable (VT) enhance resistance to Rhizoctonia solani in watermelon by alleviating oxidative stress. *Biol. Control*. 152, 104433. https://doi.org/10.1016/j.biocontrol.2020.104433.

Xu L, Deng Z-N, Wu K-C, Malviya MK, Solanki MK, Verma KK, Pang T, Li Y-J, Liu X-Y, Kashyap BK, Dessoky ES, Wang W-Z, Huang H-R. (2022). Transcriptome analysis reveals a gene expression pattern that contributes to sugarcane bud propagation induced by indole-3-butyric acid. *Front. Plant Sci*. 13, 852886. https://doi.org/10.3389/fpls.2022.852886.

Yadav R, Prasad SR. (1992). Conserving the organic matter contents of the soil to sustain sugar cane yield. *Exp. Agric.* 28, 57–62. https://doi.org/10.1017/S0014479700023012.

Yaduvanshi NPS, Yadav DV. (1990). Effects of sulphitation press mud and nitrogen fertilizer on biomass, nitrogen economy and plant composition in sugarcane and on soil chemical properties. *J. Agric. Sci.* 114, 259–263. https://doi.org/10.1017/S0021859600072646.

Yang X, Lu M, Wang Y, Wang Y, Liu Z, Chen S. (2021). Response mechanism of plants to drought stress. *Horticulturae*. 7, 50.

Zhang F, Zou YN, Wu QS, Kuˇca K. (2020). Arbuscular mycorrhizas modulate root polyamine metabolism to enhance drought tolerance of trifoliate orange. *Environ. Exp. Bot*. 171, 103962.

Zhang P, Huang P, Xu X, Sun H, Jiang B, Liao Y. (2020). Spectroscopic and molecular characterization of biochar-derived dissolved organic matter and the associations with soil microbial responses. *Sci. Total Environ*. 708, 134619.

Zhao F, Xin X, Cao Y, Su D, Ji P, Zhu Z, He Z. (2021). Use of carbon nanoparticles to improve soil fertility, crop growth and nutrient uptake by corn (Zea mays L.). *Nanomaterials*. 11, 2717.

Zhao Q, Ma C, White JC, Dhankher OP, Zhang X, Zhang S, Xing B. (2017). Quantitative evaluation of multi-wall carbon nanotube uptake by terrestrial plants. *Carbon.*, 114, 661–670.

Zheng F-J, Lin B, Yang Y-X, Fang X-C, Verma KK, Chenf G-L. (2023). Efficacy and functionality of sugarcane original vinegar on mice. *Front. Microbiol.* 14, 1224666. https://doi.org/10.3389/fmicb.2023.1224666.

CHAPTER 2

Recent Progress and Challenges for the Production of Sugarcane Ethanol

JYOTIRMAY MAHAPATRA,[1,2*] MADHURI MORE,[3]
MATTAPARTHI LAKSHMI DURGA,[4] MAHENDRA KUMAR MOHANTY,[2]
and KIRTI SURYAWANSHI[3]

[1]*Indian Agricultural Research Institute, Pusa, New Delhi, India*

[2]*College of Agricultural Engineering and Technology, Odisha University of Agriculture and Technology, Bhubaneswar, Odisha, India*

[3]*Arvind Gavali College of Engineering, Satara, Maharashtra, India*

[4]*Agricultural College, Naira, Acharya N.G. Ranga Agricultural University, Andhra Pradesh, India*

*Corresponding author

ABSTRACT

The rising cost of fossil fuels and the favorable policies of several nations to lower greenhouse gas emissions (GHGs) are driving up demand for bioethanol. Bioethanol is an organic compound that is a clear liquid, flammable, and has a characteristic odor. It can be produced through the fermentation of feedstock sugars by yeast. At present, 60% of bioethanol is derived from corn, 25% from sugarcane, and 3% from wheat and other grains. Brazil primarily produced 110 billion liters of ethanol from sugarcane in 2019. This chapter mainly elucidates the extraction methods (diffusion and crushing) for sugarcane juice and their advancements; however, fermentation approaches like continuous and batch/fed-batch are discussed in detail. Most importantly, it clearly states about existing methods and advanced technology to produce ethanol from sugarcane. It

Sugarcane Transformation: Biofuels, Biotechnology, and Beyond.
Krishan K. Verma, Xiu-Peng Song, Hai-Rong Huang, Aradhna Kumari, and Yang-Rui Li (Eds.)

DOI: 10.1201/9781779641571-2

discusses the various challenges, including variations in sugarcane production, requirements for water and power for sugarcane-based bioethanol production, nonavailability of postproduction distribution channel, and inefficient suitable engine.

2.1 INTRODUCTION

The sustained expansion of the global economy has amplified both the utilization of energy and apprehension regarding the accumulation of greenhouse gases in the atmosphere, along with their ramifications for climate change (Wu et al., 2022; Verma et al., 2024a, 2024b, 2024c; Jagadesh et al., 2024). In this connection, developing countries are advancing renewable energy initiatives, which encompass the production of biofuels (Li et al., 2023; Huang et al., 2024). Here, biofuels refer to fuels derived from biomass, which includes organic waste materials. These fuels often exhibit a substantially diminished ecological impact when compared to conventional fossil fuels. Bioethanol is one of the most important biofuels because of its environmental friendliness; therefore, it has the potential to replace fossil fuels in the future (Chen et al., 2020; Zheng et al., 2023). Bioethanol, with the chemical formula C_2H_6O, is generated through the fermentation of sugars through the utilization of starch-rich cereals as a substrate. Critically, corn is estimated to provide 60% of the ethanol extracted, followed by sugarcane (≈25%), wheat (≈3%), and other grains for the remaining. A portion of the juice is used to make sugar in a nearby sugar factory, while the remainder is used to make ethanol. Diffusion (diffuser) and crushing (roll mills) are two methods for extracting juice. The solid residue remains after the juice extraction, known as "bagasse," is usually burned to provide energy for the production process. Additionally, it can be utilized as a lignocellulosic feedstock to produce more ethanol. The details of the bioethanol production from sugarcane, current status, and limitations are elucidated in the following sections.

2.2 CURRENT STATUS OF SUGARCANE-BASED ETHANOL PRODUCTION

At present, ≈40% of the world's bioethanol is derived from sugarcane and sugar beet, whereas the remaining 60% is sourced from feedstocks containing starch. In 2023, the United States led with 15.6 billion gallons, primarily sourced from corn, while Brazil followed with 8.3 billion gallons, predominantly derived from sugarcane (Guo et al., 2021; Bhatt et al., 2023a). Despite the uptick in production, it still falls short of prepandemic levels. The global bioethanol output witnessed a decline in 2020 due to the COVID-19 pandemic. In 2023, the United States led with 15.6 billion gallons, primarily sourced from corn, while Brazil followed with 8.3 billion gallons, predominantly derived from sugarcane. Despite the uptick in production, it still falls short of prepandemic levels. India holds sixth place in ethanol production, globally. With Prime Minister Modi's focus on bolstering self-sufficiency, India sets its sights on achieving an ambitious E-20 target by 2025, while concurrently striving to meet its short-term objective of E-10 by 2022. In India, ethanol majorly produced from molasses. Maharashtra and Uttar Pradesh account for around 60% of sugar production, with Karnataka and Tamil Nadu also playing significant roles. Previously, ethanol production in India was limited to molasses or sugar juice. However, the country has expanded its feedstock base to include cereals such as maize, rice, barley, wheat, and sorghum. Yet, in 2020, around 2.98 billion liters of ethanol stemmed from molasses production. Projections for 2021 anticipated a surge to 3.17 billion liters, reflecting a 7% rise over the previous year, buoyed by an excess in sugarcane yields. Moreover, the average blending rate of bioethanol in gasoline (India) was approximated at 7.5% in 2021, showcasing heightened government endeavors to channel additional feedstock toward ethanol (Hoang and Nghiem, 2021).

2.3 SUGARCANE PRODUCT AND BY-PRODUCTS, THEIR COMPOSITION, AND SUITABILITY FOR ETHANOL PRODUCTION

Sugarcane presents notable advantages as a base material for the production of bioethanol. Sugarcane proves economically advantageous over other materials (Verma et al., 2019a, 2019b; Xu et al., 2022; Chen et al., 2023). The sugar processing industry does not just cater to food consumption; its by-products also play crucial roles in energy generation. The industry primarily yields three main by-products: bagasse (25–30% of cane) following sugarcane crushing, pressmud (3–5% of cane) postclarification (not suitable for bioethanol production), and molasses (3.5–5% of cane) after centrifugation. On the other hand, sugarcane juice serves as a crucial resource for ethanol production (Anas et al., 2021a; Malviya et al., 2021; Joshi et al., 2023), a practice observed notably in countries like Brazil. Sugarcane juice contains an ample supply of organic nutrients, minerals, and free sugars, making it an ideal raw material for bioethanol production. In Brazil, it serves as the primary precursor for bioethanol production, while molasses takes precedence as the principal feedstock in India. In this process, a portion of the juice is allocated for sugar production in nearby manufacturing plants, while the remainder is dedicated to ethanol production. Juice extraction techniques commonly involve either crushing methods, employing roll mills, or diffusion processes utilizing a diffuser.

Bagasse, the solid residue available after juice extraction, has traditionally been burned to produce energy for internal purposes. However, it also holds promise as a lignocellulosic composition, suitable for bioethanol production (Table 2.1) (Sahu, 2018).

TABLE 2.1 Chemical Compositions of the Sugarcane Bagasse

Component	Chemical Composition
Cellulose	42–48
Hemicellulose	19–28
Lignin	17–25

Additionally, molasses, another by-product of sugarcane processing, can serve as a source for ethanol production. Sugarcane molasses is regarded as a superb substrate for bioethanol production because of its abundance in fermentable sugars. Molasses stands out as an economical substrate that does not compete with food stocks and is readily accessible. Around 2.5–4% of molasses is generated from one ton of sugarcane. Normally, sugarcane comprises 12–17% total sugars (wet weight), with almost 68–72% moisture content. Approximately 90% of these sugars consist of sucrose, while glucose and fructose constitute the remaining portion. This composition renders the substrate suitable for large-scale bioethanol production.

2.4 OVERVIEW OF METHODS USED IN SUGARCANE ETHANOL PLANTS

Bioethanol derived from edible raw materials such as sugarcane is categorized as first-generation (1G) bioethanol, while if the raw material comes from lignocellulose, it is termed as second-generation (2G) bioethanol. In comparison to 1G bioethanol, 2G bioethanol holds greater promise in reducing GHGs.

2.4.1 1 G ETHANOL

The 1G ethanol distillers are designed based on the average composition of sugarcane on a mass basis, which includes: 13.3% sucrose, 4.77% cellulose, 4.53% hemicellulose, 2.62% lignin, 0.62% reduced sugars, 0.20% minerals, 1.79% impurities, 0.60% soil, and 71.57% water. The process design encompasses the following steps.

2.4.2 EXTRACTION OF SUGARCANE JUICE AND TREATMENT

Sugarcane undergoes dry cleaning, in which air blowers are employed to eliminate soil and other sizable particles that might adversely affect the industrial operation. Following that, a crushing unit is employed to separate fibers and sugarcane juice through the use of knives. Later, water is introduced to enhance the extraction of the juice. The bagasse, which retains a moisture content of 50% by weight, is isolated from the liquid stream. It primarily comprises cellulose, hemicellulose, lignin, and ash. Before concentration, the juice undergoes treatment to eliminate impurities and enhance the fermentation process.

2.4.3 CONCENTRATION AND STERILIZATION

To elevate the juice concentration for fermentation, a portion of the processed juice undergoes concentration in multieffect evaporators until reaching 65° Brix. It is subsequently blended with the remaining diverted juice to achieve a glucose level of 20% for ethanol conversion. Sterilization is carried out using a high-temperature short-time treatment. This treatment involves heating the must to 130°C and keeping it there for around 30 min. In order to prevent contamination by microorganisms that could negatively affect fermentation, this treatment is followed by rapid cooling to 32°C.

2.4.4 FERMENTATION

Sucrose, the most common sugar, is hydrolyzed by a reaction catalyzed by *Saccharomyces cerevisiae* to glucose and fructose, which are then converted to ethanol and carbon dioxide. Considering the stoichiometric reactor to simulate the fermentation process, in the first reaction, all the sucrose is converted to glucose, and the glucose is converted to ethanol and CO_2 with 89% efficiency. By-products such as succinic acid, acetic acid, and glycerol are obtained in this process (Viana Ensinas et al., 2013).

2.4.5 DISTILLATION

Monoethyleneglycol (MEG) extractive distillation is considered a distillation process. However, in this process, two columns are used: the subtraction column and the return column. The initial column receives the vapor phase of hydrated ethanol, which leads to distillation near the bottom and MEG solvent near the top. In the inverted tower, water is at the top and good MEG is at the bottom. The column comprises roughly 10 stages and operates under a pressure of 20 kPa, ensuring an average temperature of 150°C at the column's base to prevent solvent separation.

2.4.6 2G ETHANOL

Secondary ethanol could be produced in many ways. Pretreatment with SO_2 catalytic and then enzymatic hydrolysis is the most adopted technique. The pretreatment purpose is to eliminate lignin and hemicellulose and cellulose crystallinity, increase the porosity and contact area of the product, and allow cellulose to enter the cellulose molecules.

2.5 DIFFERENT PROPOSED PROCESSES FOR SUGARCANE ETHANOL PLANTS

2.5.1 STEPS INVOLVED IN BIOMASS TO ETHANOL CONVERSION PROCESS

The ethanol production from sugarcane stands as a pivotal process in South America, notably in countries like Brazil, Colombia, and Ecuador.

In this process, sugarcane undergoes juice extraction to yield syrup rich in sugars. Subsequently, the syrup is sterilized to facilitate the inoculation of yeast, particularly *S. cerevisiae*, which is extensively researched and cultivated for industrial applications. The ethanol conversion process from sugarcane by-products typically involves four primary stages: pretreatment, enzymatic hydrolysis, fermentation, and distillation (Figure 2.1). Therefore, selecting appropriate pretreatment methods is crucial, as they influence the subsequent hydrolysis, fermentation, and distillation steps in the process design.

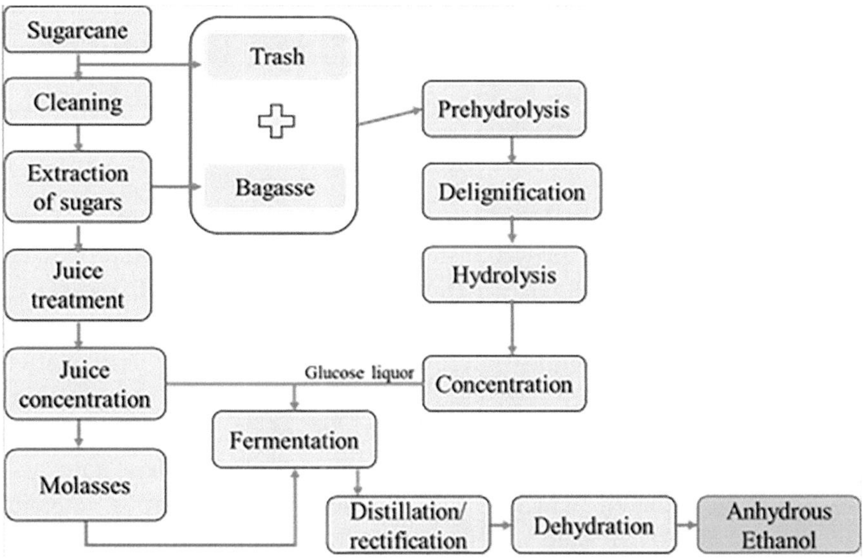

FIGURE 2.1 Systematic presentation of bioethanol production from by-products of sugarcane.

Initially, biomass is transported to the ethanol plant, stored in warehouses, and conditioned to avoid premature fermentation and Infection (Figure 2.2). By pretreatment, carbohydrates are extracted or facilitated for further extraction. In this step, monosaccharides can be separated into biomass and pretreatment. Most of the fiber will remain sacrificial and then fermented by hydrolysis or other techniques to yield the simple sugars (Radhakumari et al., 2015; Guo et al., 2021; Oliveira et al., 2022).

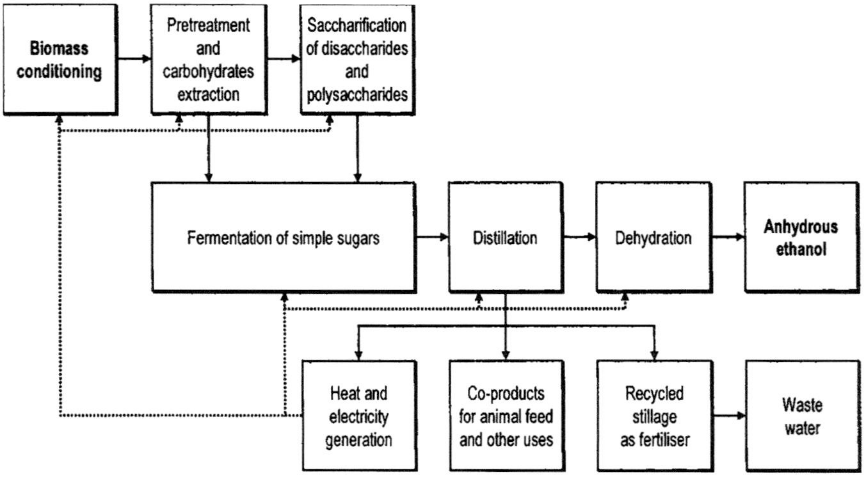

FIGURE 2.2 Process flowchart of biomass to ethanol production.

In batch fermentation, hydrolysates, yeast, nutrients, and other ingredients were mixed from the initial step. However, in batch feed processes, more inputs are added as fermentation increases. In the effective process, cell density is increased by recovering or immobilizing the yeast to increase its activity and improve fermentation efficiency. The fermentation reaction occurs at 25–30°C and takes 6–72 h, depending on compositions of the hydrolysate, density, type, and activity of the yeast. Distillation produces a zeotrope (95.5% alcohol, 4.5% water), often called "aqueous" or "hydrated" ethanol, which is then divided into "anhydrous" ethanol (99.6% alcohol, 0.4% water). The residual flow from the distillation line, called vinous wine, is processed for animal feed and other commercial products. Concentrated vinous wine can also be used as fertilizer.

2.5.2 DISACCHARIDE FOR ETHANOL PROCESS

The primary disaccharide utilized in bioethanol production is sucrose, derived from glucose and fructose. The fermentation of sucrose is facilitated by commercial yeast strains like *S. cerevisiae*. The process involves enzymatic hydrolysis of sucrose and subsequent fermentation of the resulting simple sugars. Initially, invertase, an enzyme present in yeast, catalyzes the hydrolysis of sucrose, breaking it down into glucose and fructose.

$$\underset{\text{Sucrose}}{C_{12}H_{22}O_{11} + H_2O} \xrightarrow{\textbf{Invertase}} \underset{\text{Glucose or fructose}}{C_6H_{12}O_6 + C_6H_{12}O_6}$$

Another enzyme present in yeast catalyzes the conversion of glucose and fructose into ethanol and CO_2.

$$\underset{\text{Glucose}}{C_6H_{12}O_6} \xrightarrow{\textbf{Zymase}} \underset{\text{Ethanol or fructose}}{2CH_3CH_2OH + 2CO_2}$$

The 1 t of hexose (glucose or fructose) produces 511 kg of ethanol. The actual fermentation efficiency is approximately 92% of the yield. Sucrose comes mainly from sugarcane and sugar beets. The detailed composition of different sugarcane stems, extracted juice, and bagasse is shown in Figure 2.3. The extracted juice typically comprises 84% water and 14% sugar. Subsequent steps vary depending on whether ethanol is solely produced or alongside sugar (Figure 2.4). Following purification and evaporation, the juice proceeds to the crystallization stage. This process yields molasses alongside crystallized sugar, both of which can be subjected to fermentation, distillation, and dehydration to produce anhydrous ethanol.

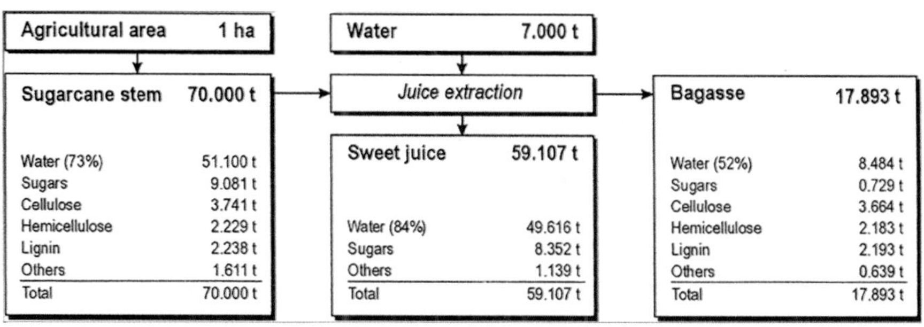

FIGURE 2.3 Mass balance representation for sugarcane juice extraction.

2.5.3 *PRETREATMENT*

Pretreatment consists of delignification, which uses physical, physico-chemical, chemical, and biological treatment to purify the raw material so that the cellulose is easier in the hydrolysis step. The effectiveness

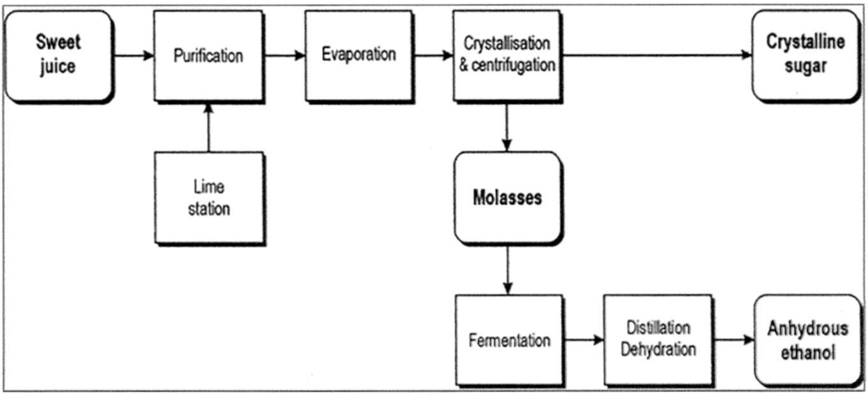

FIGURE 2.4 A straightforward representation of the combined production of sugar and ethanol.

of some methods was evaluated based on fermentable sugar, inhibitors, recovery chemicals, waste production, and investment results. Comparative findings indicate that carbonic acid extraction and alkali extraction yield optimal results. Nonetheless, the prevailing methods involve fractionated and dilute acid prehydrolysis followed by enzymatic hydrolysis. Steam explosion, on the other hand, utilizes high-energy saturated steam (0.69–4.83 MPa) to treat lignocellulosic materials at elevated temperatures (160–260°C) for several minutes. The subsequent rapid reduction in pressure to atmospheric levels induces an explosive effect. In this procedure, efficiency hinges on temperature and residence time, with the majority of hemicelluloses dissolving. Optimal effectiveness is achieved with lower temperatures and extended residence times. The adition of sulfuric acid or carbon dioxide is common to minimize inhibitor production and enhance hemicellulose dissolution. However, vapor decomposition presents limitations: (1) incomplete decomposition of the lignin–carbohydrate matrix, (2) formation of degradation by-products hindering hydrolysis and fermentation efficiency, and (3) damage to some xylan components.

2.6 WATER AND ENERGY REQUIREMENT FOR SUGARCANE ETHANOL PRODUCTION

Water plays a vital role in energy production, necessary for both raw material extraction and fuel processing (Verma et al., 2020a, 2021a, 2021b; Sharma et al., 2021). Consequently, an upsurge in energy generation from conventional, nonconventional, and renewable sources would amplify the demand for water (Singh et al., 2020; Anas et al., 2021b; Bhatt et al., 2023b). In recent years, heightened public awareness regarding the United States' reliance on foreign oil, which compromises energy security, impedes economic growth, and exacerbates climate change, has spurred increased attention and backing for alternative and renewable fuels. As depicted in Figure 2.5, water consumption in fuel production encompasses various aspects, including (1) water losses attributable to evaporation, drift (wherein downward-flowing cooling water in cooling towers contacts ascending ambient air), and blow down from the cooling tower, (2) incorporation into products and by-products, (3) discharge of process water, and (4) water losses stemming from evaporation and blow down from the boiler (Pimentel et al., 2007; Verma et al., 2020b, 2020c, 2022a, 2022b, 2022c).

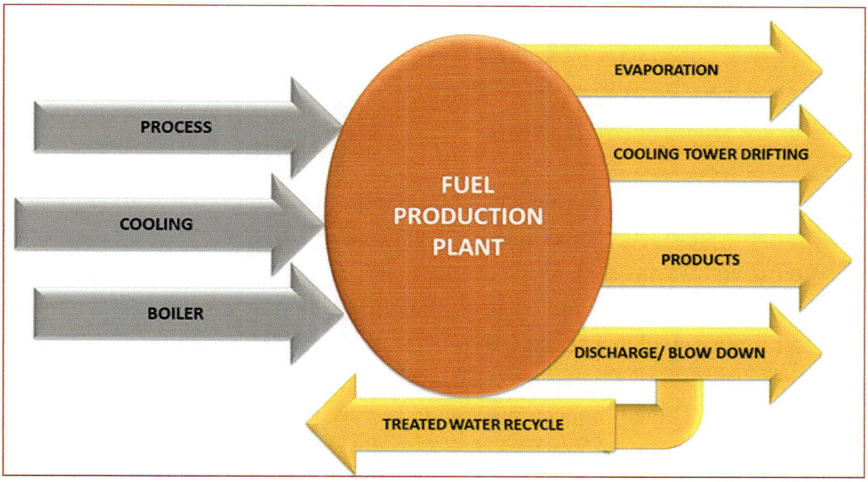

FIGURE 2.5 Water usage at the fuel processing production plant.

The volume of water utilized in ethanol production varies depending on the specific production process and the extent to which water is recycled and reused. Due to disparities in by-products, energy consumption, and capital and operational expenses, comparing processes can be challenging. However, gasification and pyrolysis typically entail relatively low water consumption. The biomass conversion (BC) process necessitates additional water for pretreatment to break down the cellulosic feedstock. Previously estimated water consumption for producing 1 gallon of cellulosic ethanol via a BC process (such as dilute acid pretreatment followed by enzymatic hydrolysis) ranged from 9.8 to 5.9 gallons; water consumption diminishes as yield increases (Wu and Xu, 2018; Verma et al., 2021c, 2022d, 2022e, 2022f; Singh et al., 2023).

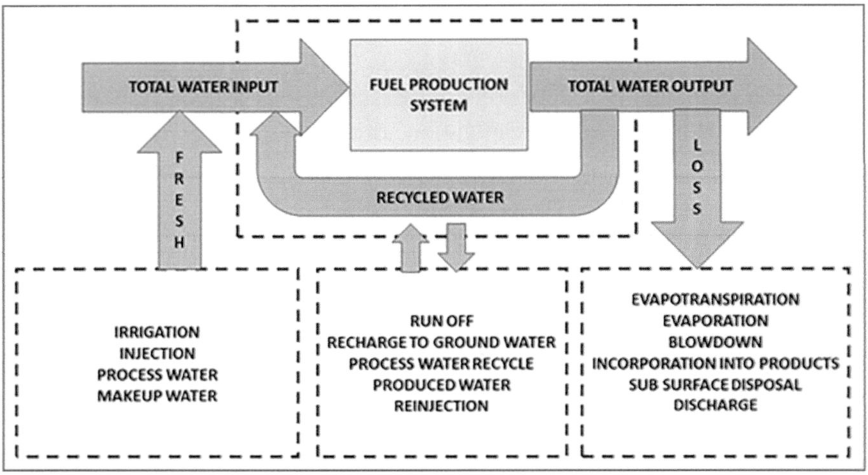

FIGURE 2.6 Fuel production systems.

In the ethanol refineries, water can be an input or an output stream in the fuel generation system. Recycled and fresh water are included in the total amount of water used (Verma et al., 2022g, 2022h, 2022i; Guo et al., 2023; Deng et al., 2023). The production of fuel or raw materials is supported by the total amount of water input, which is used as make-up water for process heating and cooling, irrigation water, or injection water for crude oil extraction. Water losses happen as a result of ET, evaporation, discharge, disposal, and product inclusion. The system's primary throughput is water recycling (Figure 2.6) (Wu et al. 2009; Song et al., 2019, 2021; Singh et al., 2021). Water consumption must be taken into consideration while modeling a robust ethanol plant. The sugar ethanol industry requires large amounts of water (Verma et al., 2022j, 2022k, 2023a, 2023b, Solanki et al., 2023). For example, a mill producing 50% ethanol and 50% sugar consumes on average 22 m^3 of water per ton of sugarcane, and about 10% of this amount is used to wash the sugarcane. In the second-generation process, the consumption of washing water is even higher, as after most pretreatments, a postwashing step of the pretreated material is required to recover dissolved sugars or remove potential inhibitors for enzymatic saccharification and/or alcoholic fermentation (Figure 2.7) (Verma et al., 2020d, 2020e, 2020f; Nakasu et al., 2021; Huang et al., 2022).

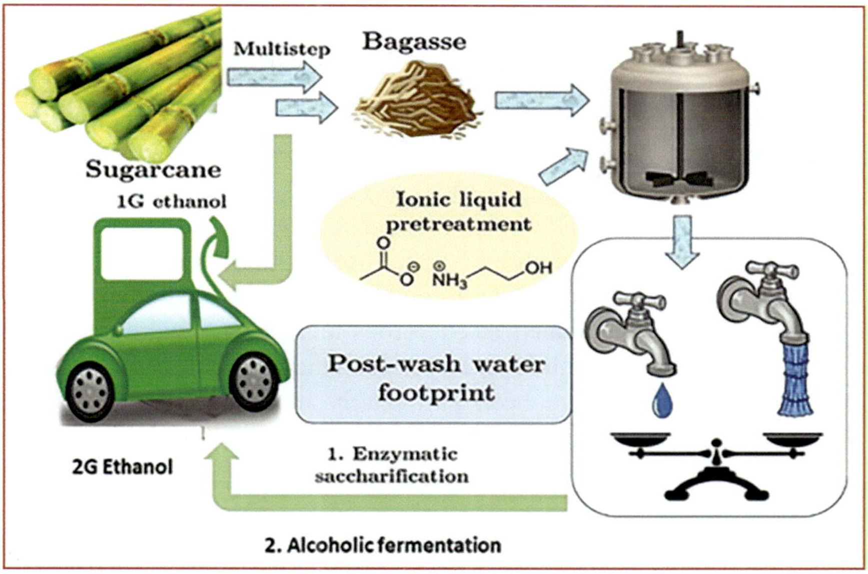

FIGURE 2.7 Sugarcane ethanol (1G and 2G) production process.

With a comparatively low need for nitrogen (N) fertilizer, sugarcane ethanol has the most energy-to-volume ratio of any energy crop and is a highly efficient converter of solar energy. Furthermore, sugarcane is a low-energy feedstock, and it is predicted that the avoided GHGs from its cultivation and use equal 2100 kg CO_2 equivalent per m^3 of ethanol. Even in cases where life cycle analyses are used to assess the energy balance of sugarcane ethanol, and the energy balance is stated in terms of energy input and output (Martinelli et al. 2013; Joshi et al., 2022; Tian et al., 2023).

The kind of mill used in the sugarcane industry affects how much water is used. A mill of the first kind only yields sugar, of the second kind only ethanol, and of the third kind, both. A poll indicates that 75% of the mills generate both sugar and ethanol, 21% just ethanol, and the remaining mills only sugar. Ninety percent of the harvested sugarcane (measured in tons) is utilized in mills that also create ethanol, 7% in mills that just produce sugar, and 3% in mills that produce sugar alone. More particularly, compared to mills generating only sugar, those producing only ethanol or both consume 50% and 70% less water, respectively. The following ratios of water are utilized in each case: 36% is used to clean by washing the cane stalks before fermentation, 27% in the fermentation process, and almost 27% in the distillation process. About 10% is used in additional processes that occur during both ethanol and sugar production (Martinelli et al. 2013; Verma et al., 2021d, 2021e; Sharma et al., 2022; Song et al., 2022).

Figure 2.8 illustrates the flow of water consumption in the conventional industrial process of ethanol production. The main points are as follows: the sugarcane washing phase; the ethanol and sugar production processes' evaporation and condensation phases; the distillation (water heater, recycling, bottom and tank washing, and disposal); the gas scrubbers; the cooling of the machinery; and the water used in the boilers (de Martino Jannuzzi, 2014). It displays two different forms of ethanol: anhydrous and hydrated. Both are generated in enormous numbers; anhydrous ethanol can be used to make gasohol, and hydrated ethanol is utilized as fuel for ethanol engines that have been particularly modified (Moreira, 2007; Malviya et al., 2022; Verma et al., 2023c, 2023d).

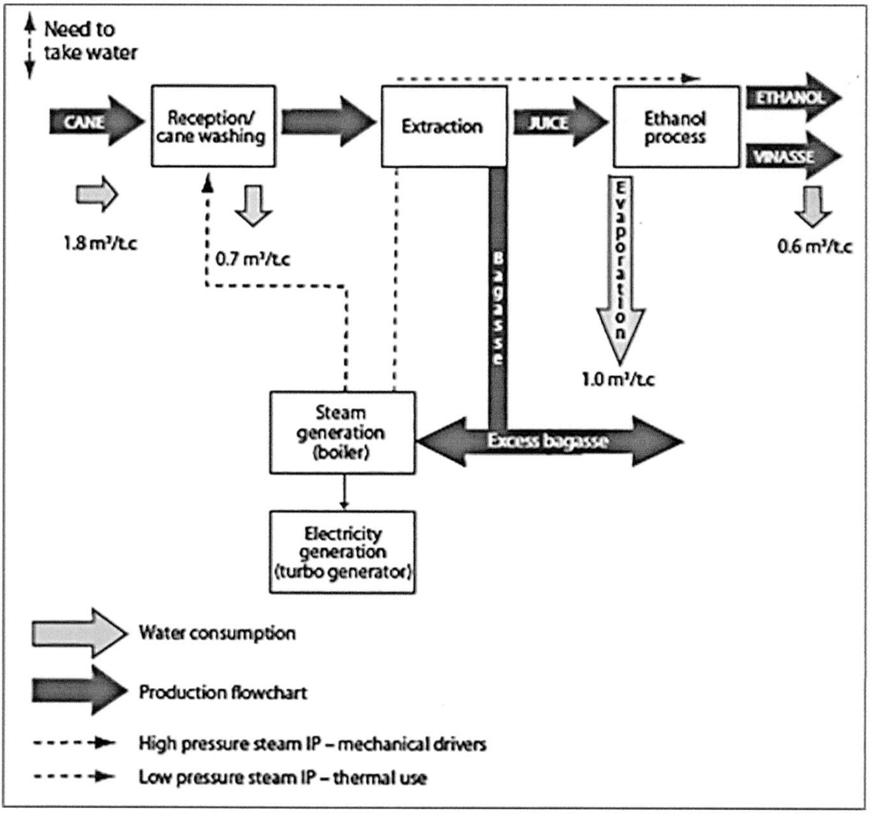

FIGURE 2.8 Flowchart of the conventional industrial process of water consumption in the sugarcane ethanol production.

2.6.1 *WATER WITHDRAWAL FOR ETHANOL PRODUCTION*

Table 2.2 summarizes the water consumption rate for the industrial sugarcane production process. It is assumed that the sugarcane is used to produce sugar (50%) and sugarcane ethanol (50%). As per the report, final consumption of 21 m^3 tons of sugarcane consists of a lower level of extraction of water, consumption of water, and water reuse or recycle. The four processes of sugarcane water consumption are washing, condenser in evaporation and vacuum, fermentation cooling, and alcohol condenser cooling. The result was 1.8 m^3 water per tons sugarcane, and if we exclude the sugar mills production with the highest specific consumption, the average value for the sugar factories, which account for 92% of sugarcane processing, is 1.23 m^3 water per tons sugarcane. These overall results indicate the exceptional progress in water management.

TABLE 2.2 Water Consumption for Industrial Processing of Sugarcane (Moreira, 2007).

Sections	Processes	Mean Use (Total m^3 Per Tons of Sugarcane)	Distribution of Water Withdrawal
Feeding sector	Sugarcane washing	5.4	25.40
Extraction	Inhibition	0.2	1.20
	Cooling	0.5	0.70
Juice treatment	Lime mixture preparation	0.010	0.10
	Cooling sulfating (in sugar production only)	0.050	0.20
	Filter inhibition process	0.040	1.40
	Filter condenser process	0.3	~10
Juice concentration section	Condensers evaporation (in sugar production only)	2	19
	Condenser heaters (in sugar production only)	4	0.10
	Molasses dilution process	0.030	0.20
	Crystallizer cooling (in sugar production only)	0.050	0
	Sugar washing (in sugar production only)	0.010	2.40
Electrical power generation sector	Steam production process	0.5	1
	Turbo generator cooling process	0.2	~5
Fermentation section	Juice cooling process (in ethanol production only)	1	14.30

TABLE 2.2 *(Continued)*

Sections	Processes	Mean Use (Total m^3 Per Tons of Sugarcane)	Distribution of Water Withdrawal
	Fermentation cooling (in ethanol production only)	3	19
Distillery process	Condenser cooling (in ethanol production only)	4	0.020
Other	Floor/equipment cleaning process	0.050	0.20
	Drinking purpose	0.030	0.10
Total		21 m^3/t of sugarcane	100%

2.6.2 ENERGY REQUIREMENT IN SUGARCANE ETHANOL PRODUCTION

The farm's intake of fossil fuels has a significant impact on the net pricing. Differences in energy yield may occur due to differences in sugar yield, fertilizer, water use, ethanol conversion technology, and fertilizer use. Figure 2.9 shows the impact of fossil energy inputs on sugar production, surpassing ethanol's renewable energy. Approximately 16 MJ/L of fossils were used in the ethanol production, which has an energy content of 22 MJ/L. Analysis shows that sugar-based ethanol has a positive energy value. Approximately 56% of fossil energy inputs are used in agricultural irrigation. Approximately 40% of fossil energy use comes from chemical fertilizers. About 3% of fossil energy comes from agricultural machinery (tractors), and about 2% comes from transporting sugar from the sugarcane farm to the sugar factory. On the other hand, it was found that 1 L of ethanol produces 40% residual energy (Shelar et al., 2023). The comparison of the analysis of net fossil fuels per liter of ethanol is shown in Figure 2.9.

Table 2.3 lists the utilization energy of different materials in agriculture. Equivalent taxes include labor, machinery, diesel, fertilizer, fungicides, and electricity. These materials are calculated based on the functionality of the products as well as their energy balance.

The average energy used for 1 ha of sugarcane cultivation in the study area was 63,208 MJ/ha, and the energy received was 33,974 MJ/ha. The energy storage value was 19.86 MJ/ha. Labor and nitrogenous fertilizers account for the greatest portion of energy storage in production, which is 42% and 32%, respectively. Other agrochemicals and phosphate fertilizers

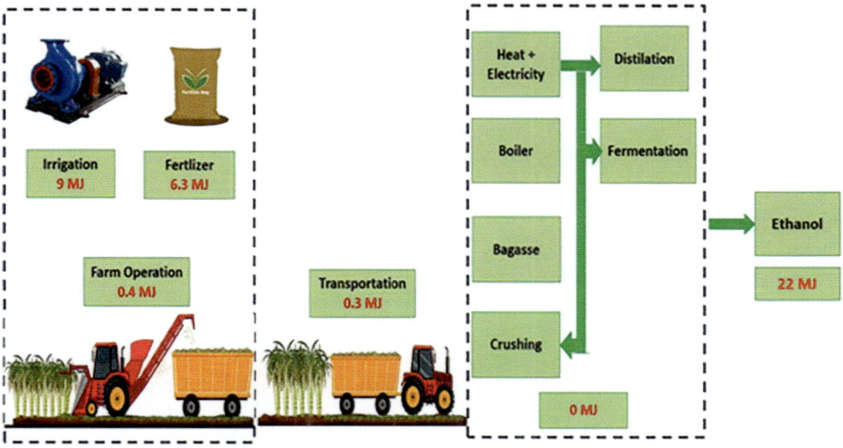

FIGURE 2.9 Analysis of net fossil fuels per liter of ethanol.

TABLE 2.3 Energy Coefficients, Input–Output Relationships, and Sugarcane Production (Shapouri and Salassi, 2006).

Parameter	Energy value (MJ)
Energy inputs	2.00
Human labor (h)	63
Machineries	56
Diesel fuel (L)	–
Chemical fertilizers in kg: nitrogen and phosphate (P_2O_5)	66.10 and 12.44
Biocides (kg)	120
Electricity (kJ)	12
Sugarcane cutting plants (kg)	5.30
Energy outputs	
Sugarcane (kg)	11

account for the least amount of energy storage in production (Shapouri and Salassi, 2006; Krishna et al., 2022). Table 2.4 illustrates how optimal techniques in sugar production have reduced energy use. If the energy output of 14,659.56 MJ/ha is constant, 51,916.93 MJ can be saved per hectare, which is equivalent to a savings of 28.23%. By predicting sugar yield, this strategy maintains yield levels while increasing energy efficiency. This implies that sugarcane produces more energy if the input of these resources is optimized. In order to further improve energy efficiency, the study examined average energy use and found promising outcomes

(Kumari et al., 2023a; Chaudhary et al., 2023). Energy saving is defined as employing the most effective method and adhering to the control, resulting in 14,660 MJ energy savings.

TABLE 2.4 Optimum Energy Requirement for Sugarcane Production (Shapouri and Salassi, 2006).

Parameters	Optimum Energy Requirement (MJ/L)	Saving Energy (MJ/L)	Saving Energy (%)
Human/Labor	41	2	32
Useful machineries	715	265	27
Diesel	4833	1083	18
Nitrogen use	6923	3262	42
Phosphate use	2235	364	14
Useful biocides	1870	233	11
Electricity use	8417	1560	16
Energy from sugarcane cutting plant	35,300	7891	18
Total energy input	51,916	14,660	28

2.6.3 ENERGY ANALYSIS

The ratio of the energy of ethanol to the total amount of nonsustainable energy consumed over the entire process, from biomass production to manufacturing, is known as the net energy balance of biomass to ethanol. Table 2.5 shows that the typical ratio of nonrenewable energy use (output/input) to ethanol production is 1.2 to 3.0 or higher. The ratio of the energy generated in the form of ethanol to the final energy (in relation to the starting energy) used in the production process of ethanol (8.3–10.2 MJ output per MJ input) is measured by the energy balance of sugar ethanol (Gnansounou and Dauriat, 2005).

TABLE 2.5 Mean Net Energy Balance for Several Methods of Ethanol Production

Feedstock	Regions	Net Energy Balance
Sugar beet	EU	1.86
Lignocellulosic biomass	Global	2.17
Wheat	EU	2.17
Corn	USA	1.36

2.7 ECONOMIC COST OF SUGARCANE ETHANOL PLANT

An effective method for assessing the economics of the many steps involved in the manufacture of ethanol is technoeconomic analysis. In order to select the best approach, there is a need to assess the technical and financial features of many ethanol manufacturing techniques. Project and capital cost estimation, lower cash flow, and minimum ethanol selling price calculation are some essential components of it. The sugar machine converts sugar into raw sugar that must be sent to the factory to convert to free glucose. The by-products of sugarcane factories that convert sugar into raw sugar are cane molasses and bagasse. Unlike corn, sugar in the form of molasses, processed sugar, or raw sugar does not require prehydrolysis before fermentation. Therefore, it is simpler to produce ethanol from sugar compared to corn. The Bureau of Labour Statistics Manufacturing Price Index is used to update prices for sugar manufacturing. Wages and benefits are estimated by updating labor expenses using the labor cost index. Cost estimates for energy, capital equipment, and other intermediate products and materials are used to update other operating expenses. An attachment to this report contains information on annual costs. According to updated cost estimates from 2003 to 2005, the average overall cost of turning cane into raw sugar was 24.3 cents/pound. Seed, fertilizer, chemicals, and fuel are examples of variable production costs. According to current and future ethanol prices, it seems profitable to convert sugar and sugar beets into ethanol; otherwise, there will be no return on investment. Futures prices suggest that ethanol prices would likely decline over the coming months due to the intense competition in the market for producing ethanol from sugar and sugarcane (Shapouri and Salassi, 2006; Kumari et al., 2023b).

The projected annual capital investment per gallon of yearly capacity for producing ethanol utilizing alternative meals is displayed in Table 2.6. Over a 20-year period, these anticipated yearly charges indicate the cost of the initial capital investment at an annual interest rate of 7.0%. The annual capital cost of a plant using cane/beet juice or molasses is similar to the cost of a plant using maize for food. Ethanol plants using sugarcane or sugar beet as ethanol feedstocks require more capital as equipment is needed to process and prepare the food, thus requiring a large reinvestment every year. Capital expenditures for ethanol plants using sugarcane may vary depending on many factors such as technology, plant size, and location (Kumari et al., 2022). Additional capital expenditures for ethanol plants near existing sugar mills or sugarcane fields are less than for standalone plants.

TABLE 2.6 Annual Capital Investment Expense for Alternative Feedstocks (Shapouri and Salassi 2006)

Feedstock	Plant Size (Million Gallons per Unit) 20 MGY (Dollar Per Gallon Capacity)	40 MGY (Dollar Per Gallon Capacity)
Corn	0.14	0.1
Sugarcane	0.2	0.16
Sugarbeets	0.2	0.16
Cane juice	0.13	0.1
Cane molasses	0.1340 MGY	0.1

Note: Assume a 20-year investment with an interest rate of 7%.

In 2023, India's annual average national ethanol blend was expected to be 11.5%. This was a record high, up 13% from the previous year. India's monthly average pricing is predicted to stay around 12% for the remainder of the year after crossing 11% for the first time in April 2023. Ethanol blended with gasoline is expected to rise up to 10.2% by 2022 due to the high conversion rate of sugar and flour, heavy molasses, and damaged rice by the Food Corporation of India.

In the last five years, India has emerged as a major sugar producer by working to stabilize the price and make good raw material under the ethanol-blended petrol (EBP) scheme. Government policy aims to increase domestic production while continuing to restrict ethanol imports for fuel blending. India is also expected to limit sugar exports as it is more important for ethanol to reach E-20 by 2025 due to lower-than-expected levels last year. Oil marketing companies have invested $91 million (INR 750 crore) to increase their ethanol capacity to 344 million liters, providing ethanol storage in their warehouse with a turnaround time of 20 days. Government subsidies for sugarcane ethanol production increased by 54% in the Indian fiscal year (IFY) 2023–2024 compared to the IFY 2022–2023, which is mentioned in Table 2.7.

TABLE 2.7 Budget Allocation by the Indian Government for the Sugar Industry (in USD Million or INR Billion) (Shapouri and Salassi, 2006).

Parameter	IFY 2021/2022	IFY 2022/2023		IFY 2023/2024	Change (%)
	Realized Outlays	Initial Budget	Revised Budget	Budget	
Proposal for providing sugar mills with financial support to increase and expand their capacity to produce ethanol	$19.35 million (INR 1.6 billion)	$36.29 million (INR 3 billion)	$31.43 million (INR 2.6 billion)	$48.38 million (INR 4 billion)	54

2.7.1 ECONOMIC FEASIBILITY OF BIOTECHNOLOGY—ETHANOL PRODUCTION IN INDIA

The conversion rate, which takes purchasing power parity (PPP) into account, determines the minimum and variable operating costs, which comprise raw material costs and product credits. Standard deviation is used to calculate investment costs, which are presumed to be independent of PPP. Prices in US dollars (2000) can be expressed in Indian rupees (2000). Data presented in Table 2.8 show different costs for ethanol plants producing 125 million liters (or 100 t) annually (Gnansounou and Dauriat, 2005).

TABLE 2.8 Details of Ethanol Production Cost in the Context of India (Gnansounou and Dauriat, 2005)

Capacity of Ethanol Plant	Molasses
Annual feed capacity in t/y	405,000
Annual production capacity in ML/y	125
Annual cost elements	Cost
Feedstock costs, US$/y	33.40
Variable operating costs, US$/y	6.30
Fixed operating costs, US$/y	1.40
Investment costs, US$/y	2.10
Sales of by-products and production cost per liter, US$/y	−10.30
Gross production cost, US$/L	0.350
Credit due to by-products, US$/L	−0.080
Net production cost, US$/L	0.260

2.7.1.1 IMPORTS AND EXPORTS

India's ethanol imports may drop as domestic production rises to sustain trade. With the exception of fuel blending, India continues to be the world's top exporter of ethanol. The amount of ethanol imported was predicted to drop to 370 million liters in 2022. The United States has been India's top exporter of denatured medical and commercial ethanol for the past ten years. Brazil is typically among the top exporters, followed by China and Canada. India's exports of ethanol in 2022 increased by 25% from 2021 to 109 million liters (mostly unlabeled). A gradual increase in exports is expected, looking at the demand from many countries. African countries have strong domestic demand for goods related to the drinking water industry and EBP projects. Angola, Congo, and Cameroon are among the countries with ethanol export goals in 2022 (Figure 2.10). The Indian government said in March 2023 that if the biofuels are produced for use in food crops, it will permit export facilities and special economic zones to export petroleum and nonoil biofuels without any limitations.

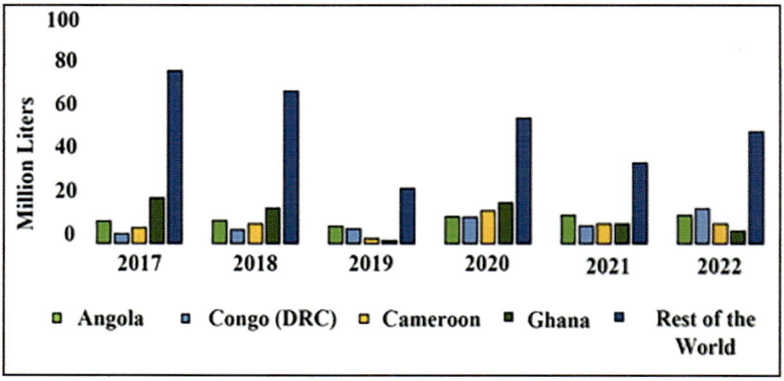

FIGURE 2.10 India, ethanol exports (million liters), 2017–2022.

2.7.2 ETHANOL PRODUCTION, SUPPLY, AND DISTRIBUTION

Based on the data given by Gnansounou and Dauriat (2005), Figures 2.11–2.15 graphically present the details of ethanol production and their distribution from the year 2014 to 2023.

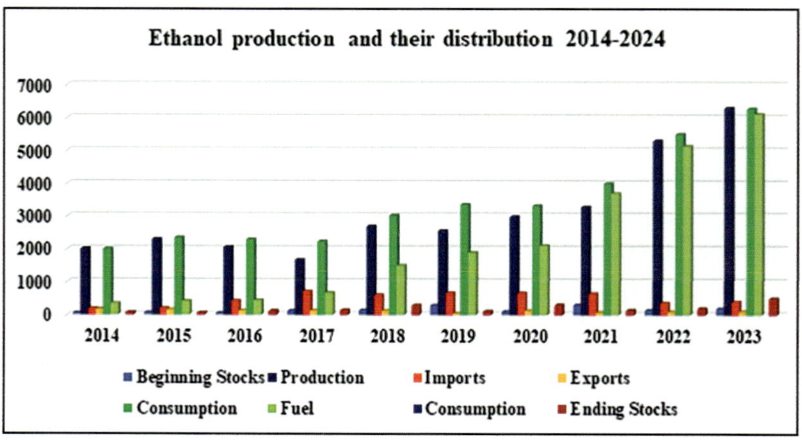

FIGURE 2.11 Ethanol production and its distribution from the year 2014 to 2023.

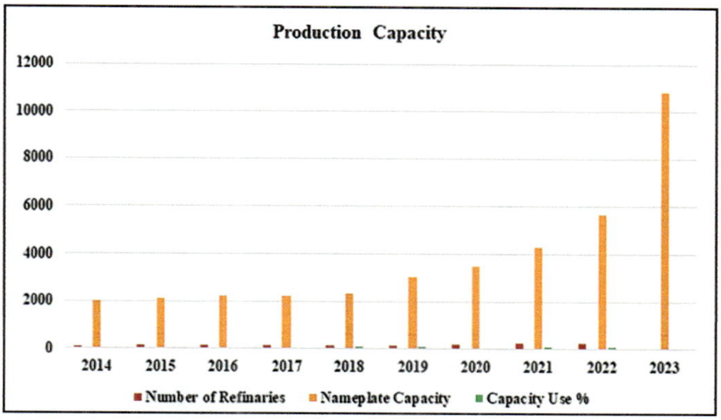

FIGURE 2.12 Production capacity.

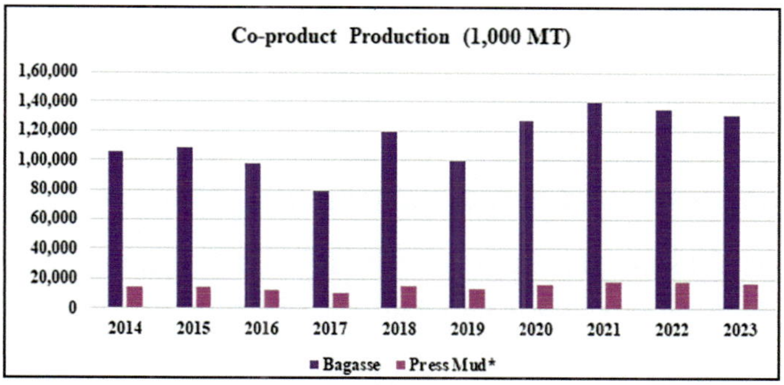

FIGURE 2.13 Coproduct production (1000 MT).

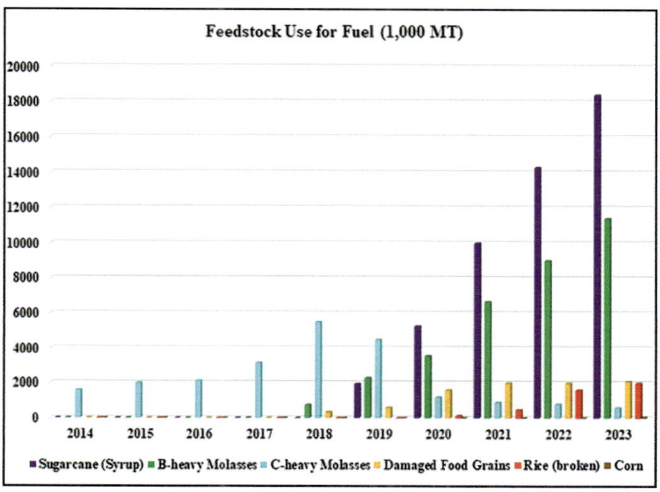

FIGURE 2.14 Feedstock use for fuel (1000 MT).

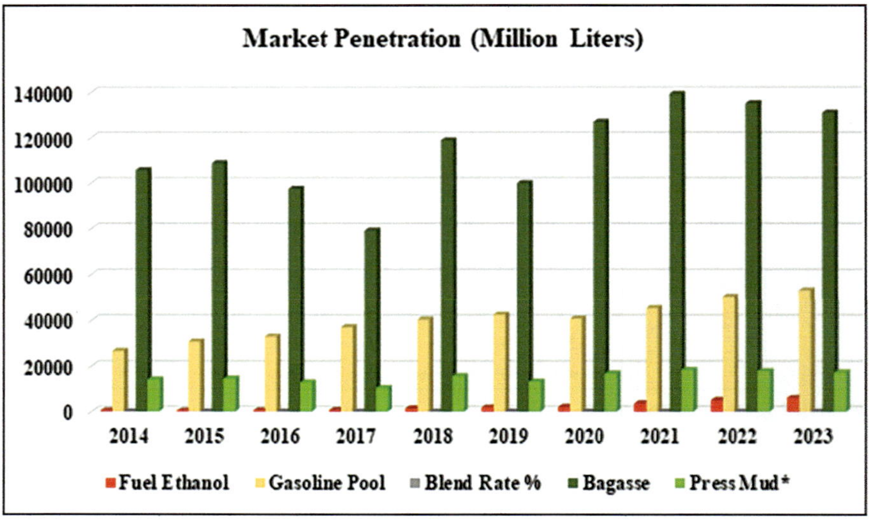

FIGURE 2.15 Market penetration (million liters).

2.7.3 USE OF BIOETHANOL IN SPARK IGNITION (SI) ENGINES

Ethanol exhibits favorable characteristics for internal combustion engines (IC engines). Its research octane number is 109, while its motor octane number is 90. This translates to an average octane rating of 99, as opposed to 88 for regular gasoline. The lower heating value (LHV) of ethanol (21.2 MJ/L) is two-thirds that of gasoline (30.1 MJ/L). The bioethanol fuel used in the combustion engine is either pure or a mixture of 5–26% anhydrous ethanol with gasoline (more than 5% in Europe and India, 10% in the United States, and 22–26% in Brazil) or hydrated ethanol (100%). When anhydrous bioethanol is mixed into gasoline in small amounts (up to 15%), the impact of low LHV is negligible, and at higher mixture levels, fuel economy declines in comparison to natural gas.

In order to ensure ideal combustion, certain cars known as flexible gasoline vehicles (FGVs) are fitted with in-line sensors that detect the amount of ethanol in the gasoline and modify the air–fuel ratio accordingly. In Brazil and Sweden, where numerous businesses are producing them, these vehicles, which have the ability to burn fuel containing ethanol (0–85%) in gasoline, are becoming more and more popular. The bioethanol component of the fuel mixture contributes to the decline in fuel economy in the case of FFV, as the compression ratio cannot be changed.

2.7.4 ETHANOL USE IN COMPRESSION IGNITION (CI) ENGINES

Because of its low cetane number, ethanol cannot be burnt efficiently using compression ignition. Furthermore, ethanol and diesel fuel do not mix well. When developing the ethanol utilized in CI vehicles, some criteria must be satisfied. To make ethanol and diesel more miscible, an emulsifier should be added. To raise the cetane number, additional chemicals like di-*tert*-butyl peroxide or ethylhexyl nitrate are added. The majority of ethanol–diesel blends, or "e-diesel," have 15% or more ethanol and 5% or more emulsifier. The second kind is a dual-fuel market, in which cylinder 15 is filled with both diesel and ethanol. Ultimately, the engine is changed to conform to its self-ignition features to make it capable of operating on fuel mixtures.

The following drawbacks apply to the usage of bioethanol in internal combustion engines in addition to the issues discussed above.

- Lower energy content: Bioethanol has a lower energy content compared to gasoline, meaning vehicles running on bioethanol may experience reduced mileage per gallon. This can result in decreased driving range and increased fuel consumption.
- Corrosion and material compatibility: Bioethanol can be corrosive to certain materials commonly used in fuel systems, such as rubber and some types of plastics. This can lead to damage to engine components and fuel system parts over time, necessitating the use of corrosion-resistant materials or modifications to existing systems.
- Cold starting issues: Bioethanol has a higher vapor pressure than gasoline, which can lead to difficulties in cold starting, particularly in colder climates. This may require engine modifications or the addition of special fuel blends to improve cold-starting performance.
- Emission: Low ethanol combined with gasoline raises gas pressure and produces evaporative emissions, leading to smog.

2.7 CONCLUSION

As the global energy demand escalates, bioethanol derived from sugarcane emerges as a valuable and environmentally friendly alternative to nonrenewable fuels. Typically, bioethanol, represented by the chemical formula C_2H_6O, is produced via the fermentation of sugars. Moreover, India holds sixth position globally in ethanol production, majorly produced from sugar industry-based molasses. This particular chapter drew attention to the critical steps involved in bioethanol production from sugarcane and its by-products. 1G and 2G ethanol production processes and requirements are described. It also highlighted the water consumption, energy requirements, and economic aspects of the bioethanol production process. According to this chapter, the total water consumption is estimated to be 21 m^3/t sugarcane. When the energy requirement was analyzed from sugarcane farming to ethanol production, a net gain of 6 MJ/L of ethanol was found. There exists a number of issues regarding the application of ethanol as engine fuel. More studies on the development of a suitable engine as well as modification techniques for available CI/SI engines need to focus. This chapter provides the fundamental pathway for future studies on sugarcane-based bioethanol.

KEYWORDS

- **sugarcane by-products**
- **sugarcane ethanol**
- **ethanol production**
- **process**
- **limitations**

REFERENCES

Anas M, Verma KK, Haq IU, Naeem M, Li Q, Liao F, Xiang YX, Li YR. (2021a). Characterization of exotic, native and wild-type genotypes of sugarcane (*Saccharum* spp. hybrids) for internal nitrogen use efficiency under different nitrogen levels. *Sugar Tech.* 23, 1258–1267. https://doi.org/10.1007/s12355-021-01014-1.

Anas M, Verma KK, Riaz M, Qiang L, Liao F, Liu Y, Li YR. (2021b). Physio-morphological and biochemical mechanism of nitrogen use efficiency in sugarcane (*Saccharum* spp.) genotypes under different growth stages and nitrogen levels. *J. Plant Interact.* 16, 332–343.

Bhatt R, Verma KK, Jamwal NS. (2023a). Can paddy-compost eliminate N fertilization for climate-smart sugarcane? *Sugar Tech.* 25, 1036–1046. https://doi.org/10.1007/s12355-023-01270-3.

Bhatt R, Verma KK, Kumar R, Sanghera GS. (2023b). Foliar application of potassium salt of active phosphorus (PSAP) mitigates insect pests and improves yield along with sugarcane quality in response to agroclimatic conditions of Punjab. *Sugar Tech.* 25(3), 660–669.

Biofuel Annual-2023 retrieved from https://fas.usda.gov/. Retrieved March 22, 2024, from https://apps.fas.usda.gov/newgainapi/api/Report/DownloadReportByFileName?fileName=Biofuels%20Annual_New%20Delhi_India_IN2023-0039

Chaudhary P, Xu M, Ahamad L, Chaudhary A, Kumar G, Adeleke BS, Verma KK, Hu D-M, Siric I, Kumar P, Popescu SM, Fayssal SA. (2023). Application of synthetic consortia for improvement of soil fertility, pollution remediation, and agricultural productivity: A review. *Agronomy*. 13, 643. https://doi.org/10.3390/agronomy13030643.

Chen GL, Zheng FJ, Lin B, Lao SB, He J, Huang Z, Zeng Y, Sun J, Verma KK. (2020). Phenolic and volatile compounds in the production of sugarcane vinegar. *ACS Omega*. 5, 30587−30595.

Chen G-L, Zheng F-J, Lin B, Yang Y-X, Fang X-C, Verma KK, Yang L-F. (2023). Vinegar: A potential source of healthy and functional food with special reference to sugarcane vinegar. *Front. Nutr.* 10, 1145862. https://doi.org/10.3389/fnut.2023.1145862.

de Martino Jannuzzi, G. (2014). USe of the water in the production of ethanol from sugarcane. In: Luis Augusto Barbosa Cortez (Coord.). *Sugarcane bioethanol—R&D for Productivity and Sustainability*. Editora Edgard Blücher: São Paulo, pp. 271–278. http://dx.doi.org/10.5151/BlucherOA-Sugarcane-SUGARCANEBIOETHANOL_27.

Deng ZN, Huang HR, Huang CM, Verma KK, Luo HB, Jiang SL, Li YJ, Cao HQ, Wu KC, Xu L. (2023). Short-term limited water irrigation influences the photosynthetic pigments, enzymatic and non-enzymatic activities in *Saccharum spontaneum* L. at vegetative stage. *Sugar Tech.* 25, 110–118. https://doi.org/10.1007/s12355-022-01175-7

Gnansounou, E, Dauriat, A. (2005). Ethanol fuel from biomass: A review. J. Sci. Ind. Res. 64, 809–821.

Guo D-J, Li D-P, Chen Z-D, Yang B, Verma KK, Singh RK, Singh P, Khan Q, Sharma A, Qin Y, Zhang B-Q, Song X-P, Li Y-R. (2023). Effect of endophytic diazotroph *Enterobacter roggenkampii* ED5 on nitrogen-metabolism-related microecology in the sugarcane rhizosphere at different nitrogen levels. *Front. Microbiol.* 14, 1132016. https://doi.org/10.3389/fmicb.2023.1132016.

Guo D-J, Li D-P, Singh RK, Singh P, Sharma A, Verma KK, Qin Y, Khan Q, Lu Z, Malviya MK, Song X-P, Xing Y-X and Li Y-R (2021). Differential protein expression analysis of two sugarcane varieties in response to diazotrophic plant growth-promoting endophyte *Enterobacter roggenkampii* ED5. *Front. Plant Sci.* 12, 727741. https://doi.org/10.3389/fpls.2021.727741.

Guo DJ, Li DP, Singh RK, Singh P, Verma KK, Sharma A, Qin Y, Khan Q, Song XP, Malviya MK, Xing YX, Li YR. (2022). Comparative transcriptome analysis of two sugarcane varieties in response to diazitrophic plant growth promoting endophyte *Enterobacter roggenkampii* ED5. *J. Plant Interact.* 17(1), 75–84.

Hoang TD, Nghiem, N. (2021). Recent developments and current status of commercial production of fuel ethanol. *Fermentation*. 7(4), 314. https://doi.org/10.3390/fermentation7040314.

Huang CM, Wu KC, Verma KK, Luo HB, Deng ZN, Shi GY, Cao HQ, Wu XJ, Ye LP, Wei YW, Xu L. (2024). Isolation and expression analysis of cellulose synthase 3 (Ces3) genes from sugarcane (*Saccharum officinarum* L.). *Chem. Biol. Technol. Agric.* 11, 71. https://doi.org/10.1186/s40538-024-00597-7.

Huang HR, Gao YJ, Malviya MK, Verma KK, Solanki MK, Huang YX, Li X., Deng Y, Yan J, Tang SY, Wang LW, Xu L. (2022). Genetic diversity analysis of sugarcane (*Saccharum* spp. Hybrids) among high-sucrose clones of GT series and commonly used parents by using microsatellite markers in Guangxi, China. *Sugar Tech.* 24, 397–407. https://doi.org/10.1007/s12355-021-01033-y.

Jagadesh M, Dash M, Kumari A, Singh SK, Verma KK, Kumar P, Bhatt R, Sharma SK. (2024). Revealing the hidden world of soil microbes: Metagenomic insights into plant, bacteria, and fungi interactions for sustainable agriculture and ecosystem restoration. *Microbiol. Res.* 285, 127764. https://doi.org/10.1016/j.micres.2024.127764

Joshi A, Verma KK, Rajput VD, Arora J. (2023). Metabolic engineering of microorganisms in advancing biofuel production. In: *Advances in Lignocellulosic Biofuel Production Systems*. pp. 181–202. https://doi.org/10.1016/B978-0-323-91192-4.00016-X.

Joshi A, Verma KK, Rajput VD, Minkina T, Arora J. (2022). Recent advances in metabolic engineering of microorganisms for advancing lignocellulose-derived biofuels, *Bioengineered.* 13(4), 8135–8163. https://doi.org/10.1080/21655979.2022.2051856.

Krishna D, Sachan HK, Verma KK. (2022). Abiotic stresses in plants: An overview. In: *Understanding Abiotic Stresses*. Nova Science Publishers, Inc.: New York.

Kumari A, Dash M, Singh SK, Jagadesh M, Mathpal B, Mishra PK, Pandey SK, Verma KK. (2023a). Soil microbes: A natural solution for mitigating the impact of climate change. *Environ. Monit. Assess.* 195, 1436. https://doi.org/10.1007/s10661-023-11988-y.

Kumari A, Lakshmi GA, Krishna GK, Patni B, Prakash S, Bhattacharyya M, Singh SK, Verma KK. (2022). Climate change and its impact on crops: A comprehensive investigation for sustainable agriculture. *Agronomy*. 12, 3008. https://doi.org/10.3390/agronomy12123008.

Kumari A, Singh SK, Mathpal B, Verma KK, Garg VK, Bhattacharyya M, Bhatt R. (2023b). The biosynthesis, mechanism of action, and physiological functions of melatonin in horticultural plants: A review. *Horticulturae*. 9, 913. https://doi.org/10.3390/horticulturae9080913.

Li A-M, Liao F, Wang M, Chen Z-L, Qin C-X, Huang R-Q, Verma KK, Li Y-R, Que Y-X, Pan Y-Q, et al. (2023). Transcriptomic and proteomic landscape of sugarcane response to biotic and abiotic stressors. *Int. J. Mol. Sci.* 24, 8913. https://doi.org/10.3390/ijms24108913.

Malviya MK, Li C-N, Lakshmanan P, Solanki MK, Wang Z, Solanki AC, Nong Q, Verma KK, Singh RK, Singh P, Sharma A, Guo D-J, Dessoky ES, Song X-P, Li Y-R. (2022). High-throughput sequencing-based analysis of rhizosphere and diazotrophic bacterial diversity among wild progenitor and closely related species of sugarcane (*Saccharum* spp. inter-specific hybrids). *Front. Plant Sci*. 13, 829337. https://doi.org/10.3389/fpls.2022.829337.

Malviya MK, Solanki MK, Li CN, Wang Z, Zeng Y, Verma KK, Singh RK, Singh P, Huang HR, Yang LT, Song XP, Li YR. (2021). Sugarcane-legume intercropping can enrich the soil microbiome and plant growth. *Front. Sustain. Food Syst.* 5, 606595. https://doi.org/10.3389/fsufs.2021.606595.

Martinelli LA, Filoso S, de Barros Aranha C, Ferraz SF, Andrade TM, Ravagnani EDC, Della Coletta L, de Camargo PB. (2013). Water use in sugar and ethanol industry in the state of São Paulo (Southeast Brazil). *J. Sustain. Bioenergy Syst.* 3(2), 135–142. https://doi.org/10.4236/jsbs.2013.32019.

Moreira JR. (2007). Water use and impacts due ethanol production in Brazil. *World.* 41(6960), 3–414.

Nakasu PYS, Ienczak JL, Rabelo SC; Costa AC. (2021). The water consumption of sugarcane bagasse post-washing after protic ionic liquid pretreatment and its impact on 2G ethanol production. *Ind. Crops Prod.* 169, 113642. https://doi.org/10.1016/j.indcrop.2021.113642

Oliveira MA, Verma KK, Bhatt R, Oliveira TBA. (2022). Impact of green and organic fertilizers on soil fertility and sugarcane productivity. In: Verma KK, et al. (Eds.), *Agro-Industrial Perspectives On Sugarcane Production Under Environmental Stress*. https://doi.org/10.1007/978-981-19-3955-6_11.

Pimentel D, Patzek T, Cecil G. (2007). Ethanol production: energy, economic, and environmental losses. *Rev. Environ. Contam. Toxicol.* 189, 25–41. https://doi.org/10.1007/978-0-387-35368-5_2.

Radhakumari M, Prakash DJ, Satyavathi B. (2015). Pyrolysis characteristics and kinetics of algal biomass using tga analysis based on ICTAC recommendations. *Biomass Convers. Biorefin.* 6, 189–195. https://doi.org/10.1007/s13399-015-0173-7.

Sahu O. (2018). Assessment of sugarcane industry: Suitability for production, consumption, and utilization. *Ann. Agrar. Sci.* 16, 389–395. https://doi.org/10.1016/j.aasci.2018.08.001.

Shapouri H, Salassi M. (2006). *The Economic Feasibility of Ethanol Production from Sugar in the United States*. USDA.

Sharma A, Singh RK, Singh P, Vaishnav A, Guo DJ, Verma KK, Li DP, Song XP, Malviya MK, Khan N, Lakshmanan P, Li YR. (2021). Insights into the bacterial and nitric oxide-induced salt tolerance in sugarcane and their growth-promoting abilities. *Microorganisms*. 9, 2203. https://doi.org/10.3390/microorganisms9112203.

Sharma A, Song X-P, Singh RK, Vaishnav A, Gupta S, Singh P, Guo D-J, Verma KK, Li Y-R. (2022). Impact of carbendazim on cellular growth, defence system and plant growth promoting traits of Priestia megaterium ANCB-12 isolated from sugarcane rhizosphere. *Front. Microbiol.* 13, 1005942. https://doi.org/10.3389/fmicb.2022.1005942.

Shelar MN, Matsagar VK, Patil VS, Barahate SD. (2023). Net energy analysis of sugarcane based ethanol production. *Clean. Energy Syst.* 4, 100059. https://doi.org/10.1016/j.cles.2023.100059.

Singh P, Singh RK, Li HB, Guo DJ, Sharma A, Verma KK, Solanki MK, Upadhyay SK, Lakshmanan P, Yang LT, Li YR. (2023). Nitrogen fixation and phytohormone stimulation of sugarcane plant through plant growth promoting diazotrophic Pseudomonas. *Biotechnol. Genet. Eng. Rev.* 40(1), 15–35. https://doi.org/10.1080/02648725.2023.2177814.

Singh P, Singh RK, Song QQ, Li HB, Guo DJ, Malviya MK, Verma KK, Song XP, Lakshmanan P, Yang LT, Li YR. (2021). Comparative analysis of protein and differential responses of defense-related gene and enzyme activity reveals the long-term molecular responses of sugarcane inoculated with *Sporisorium scitamineum*. *J. Plant Interact.* 16(1), 12–29.

Singh RK, Singh P, Li HB, Song QQ, Guo DJ, Solanki MK, Verma KK, Malviya MK, Song XP, Lakshmanan P, Yang LT, Li YR. (2020). Diversity of nitrogen-fixing rhizobacteria associated with sugarcane: A comprehensive study of plant-microbe interactions for growth enhancement in *Saccharum* spp. *BMC Plant Biol.* 20, 220. https://doi.org/10.1186/s12870-020-02400-9.

Solanki MK, Verma KK, Dastogeer KMG, Mora-Poblete F, Mundra S. (2023). Microbial resilience in plant nutrient management towards sustainable farming. *Front. Microbiol.* 14, 1280811. https://doi.org/10.3389/fmicb.2023.1280811.

Song X-P, Mo F, Verma KK, Wei J, Zhang X, Yang L, Li YR. (2019). Effect of sugarcane smut (*Ustilago scitaminea* Syd.) on ultrastructure and biochemical indices of sugarcane. *Biomed J. Sci. Tech. Res.* 17(1), 12546–12551. https://doi.org/10.26717/BJSTR.2019.17.002950.

Song XP, Mo F, Yan M, Zhang X, Zhang B, Huang X, Huang D, Pan Y, Verma KK, Li YR. (2022). Effect of smut infection on the photosynthetic physiological characteristics and related defense enzymes of sugarcane. *Life*. 12, 1201. https://doi.org/10.3390/life12081201.

Song XP, Verma KK, Tian DD, Zhang XQ, Liang YJ, Huang X, Li CN, Li YR. (2021). Exploration of silicon functions to integrate with biotic stress tolerance and crop improvement. *Biol. Res.* 54, 19. https://doi.org/10.1186/s40659-021-00344-4

Tian D, Qin L, Verma KK, Wei L, Li J, Li B, Zhou W, He Z, Wei D, Huang S, Long S, Huang Q, Li C, Wei S. (2023). Transcriptomic and metabolomic differences between banana varieties which are resistant or susceptible to Fusarium wilt. *PeerJ*. 11, e16549. https://doi.org/10.7717/peerj.16549

Verma KK, Anas M, Chen Z, Rajput VD, Malviya MK, Verma CL, Singh RK, Singh P, Song XP, Li YR. (2020f). Silicon supply improves leaf gas exchange, antioxidant defense system and growth in sugarcane responsive to water limitation. *Plants*. 9, 1032, https://doi.org/10.3390/plants9081032.

Verma KK, Joshi A, Song X-P, Liang Q, Xu L, Huang H-r, Wu K-C, Seth CS, Arora J, Li Y-R. (2024a). Regulatory mechanisms of plant rhizobacteria on plants to the adaptation of adverse agroclimatic variables. *Front. Plant Sci.* 15, 1377793. https://doi.org/10.3389/fpls.2024.1377793.

Verma KK, Joshi A, Song X-P, Singh S, Kumari A, Arora J, Singh SK, Solanki MK, Seth CS, Li Y-R. (2024b). Synergistic interactions of nanoparticles and plant growth promoting rhizobacteria enhancing soil-plant systems: A multigenerational perspective. *Front. Plant Sci.* 15, 1376214. https://doi.org/10.3389/fpls.2024.1376214.

Verma KK, Li DM, Singh M, Rajput VD, Malviya MK, Minkina T, Singh RK, Singh P, Song XP, Li YR. (2020a). Interactive role of silicon and plant-rhizobacteria mitigating abiotic stresses: A new approach for sustainable agriculture and climate change. *Plants.* 9, 1055. https://doi.org/10.3390/plants9091055.

Verma KK, Rajput VD, Singh M, Verma CL, Minkina T. (2022a). Influence of waterlogging on plant physiology and molecular biology. In: *Understanding Abiotic Stresses*. Nova Science Publishers, Inc.: New York.

Verma KK, Siddiqui N, Singh M. (2022b). Understanding the plant physiological and molecular adaptation strategies to soil salinity. In: *Understanding Abiotic Stresses*. Nova Science Publishers, Inc.: New York. https://doi.org/10.52305/UBUJ4024.

Verma KK, Singh P, Song X-P, Malviya MK, Singh RK, Chen G-L, Solomon S, Li YR. (2020b). Mitigating climate change for sugarcane improvement: role of silicon in alleviating abiotic stresses. *Sugar Tech.* 22(5), 741–749. https://doi.org/10.1007/s12355-020-00831-0.

Verma KK, Singh RK, Song QQ, Singh P, Zhang B-Q, Song X-P, Chen G-L, Li YR. (2019a). Silicon alleviates drought stress of sugarcane plants by improving antioxidant responses. *Biomed J. Sci. Tech. Res.* 17(1), 12580–12586. https://doi.org/10.26717/BJSTR.2019.17.002957.

Verma KK, Song X, Degu HD, Guo D, Joshi A, Huang H, Xu L, Singh M, Huang D, Rajput VD, Li YR. (2023a). Recent advances in nitrogen and nano-nitrogen fertilizers for sustainable crop production: A mini-review. *Chem. Biol. Technolog. Agricult.* 10, 1–14. https://doi.org/10.1186/s40538-023-00488-3.

Verma KK, Song X, Verma CL, Huang H, Singh M, Xu L, Wu J, Khan ZH, Zhang B, Li Y. (2023b). Mathematical modeling of climate and fluoride effects on sugarcane photosynthesis with silicon nanoparticles. *Plant Physiol. Biochem.* 204, 108089. https://doi.org/10.1016/j.plaphy.2023.108089.

Verma KK, Song XP, Budeguer F, Nikpay A, Enrique R, Singh M, Zhang B, Wu, JM, Li YR. (2022c). Genetic engineering: An efficient approach to mitigating biotic and abiotic stresses in sugarcane cultivation. *Plant Signal. Behav.* 17(1), e2108253.

Verma KK, Song XP, Joshi A, Rajput VD, Singh M, Sharma A, Singh RK, Li DM, Arora J, Minkina T, Li YR. (2022d). Nanofertilizer possibilities for healthy soil, water and food in future: An overview. *Front. Plant Sci.* 13, 865048. https://doi.org/10.3389/fpls.2022.865048.

Verma KK, Song XP, Joshi A, Tian DD, Rajput VD, Singh M, Arora J, Minkina TM, Li YR. (2022e). Recent trends in nano-fertilizer for advancing sustainable agriculture under the era of climate change to ensure food security for future generations. *Nanomaterials.* 12, 173. https://doi.org/10.3390/nano12010173.

Verma KK, Song XP, Kumari A, Jagadesh M, Singh SK, Bhatt R, Singh M, Seth CS, Li YR. (2024c). Climate change adaptation: Challenges for agricultural sustainability. *Plant Cell Environ.* 48(4), 2522–2533. https://doi.org/10.1111/pce.15078.

Verma KK, Song XP, Li DM, Singh M, Wu JM, Singh RK, Sharma A, Zhang BQ, LI YR. (2022f). Silicon and soil microorganisms improves rhizospheric soil health with bacterial community, plant growth, performance and yield. *Plant Signal. Behav.* 17(1), e2104004.

Verma KK, Song XP, Lin B, Guo DJ, Singh M, Rajput VD, Singh RK, Singh P, Sharma A, Malviya MK, Chen GL, Li YR. (2022g). Silicon induced drought tolerance in crop plants: physiological adaptation strategies. *Silicon.* 14(6), 2473–2487. https://doi.org/10.1007/s12633-021-01071-x.

Verma KK, Song XP, Rajput VD, Boldyreva V, Zhang BQ, Minkina T, Li YR. (2022h). Morpho-physiological, biochemical and ultrastructural modifications on sugarcane to prolonged water deficit. In: Verma KK, et al. (Eds.), *Agro-Industrial Perspectives on Sugarcane Production Under Environmental Stress.* https://doi.org/10.1007/978-981-19-3955-6_11.

Verma KK, Song XP, Rajput VD, Solomon S, Li YR, Rao GP. (2022i). *Agro-Industrial Perspectives on Sugarcane Production under Environmental Stress.* Springer Nature: Singapore, https://doi.org/10.1007/978-981-19-3955-6.

Verma KK, Song XP, Singh M, Tian DD, Rajput VD, Minkina T, Li YR. (2023c). Association of silicon and soil microorganisms induces stress mitigation, increasing plant productivity. R. de Mello Prado (Eds.), *Benefits of Silicon in the Nutrition of Plants.* pp. 299–328. https://doi.org/10.1007/978-3-031-26673-7_17.

Verma KK, Song XP, Tian DD, Guo DJ, Chen ZL, Zhong CS, Nikpay A, Singh M, Rajput VD, Singh RK, Minkina T, Li YR. (2021a). Influence of silicon on biocontrol strategies to manage biotic stress for crop protection, performance and improvement. *Plants*. 10, 2163. https://doi.org/10.3390/plants10102163.

Verma KK, Song XP, Tian DD, Singh M, Verma CL, Rajput VD, Singh RK, Sharma A, Singh P, Malviya MK, Li YR. (2021b). Investigation of defensive role of silicon during drought stress induced by irrigation capacity in sugarcane: Physiological and biochemical characteristics. *ACS Omega*. 6, 19811–19821.

Verma KK, Song XP, Verma CL, Chen ZL, Rajput VD, Wu KC, Liao F, Chen GL, Li YR. (2021c). Functional relationship between photosynthetic leaf gas exchange in response to silicon application and water stress mitigation in sugarcane. *Biol. Res*. 54, 15. https://doi.org/10.1186/s40659-021-00338-2.

Verma KK, Song XP, Verma CL, Malviya MK, Guo DJ, Rajput VD, Sharma A, Wei KJ, Chen GL, Solomon S, Li YR. (2021d). Predication of photosynthetic leaf gas exchange of sugarcane (*Saccharum* spp.) leaves in response to leaf positions to foliar spray of potassium salt of active phosphorus under limited water irrigation. *ACS Omega*. 6, 2396–2409.

Verma KK, Song X-P, Xu L, Huang H-R, Liang Q, Seth CS, Li Y-R. (2023d). Nano-microplastic and agro-ecosystems: A mini-review. *Front. Plant Sci*. 14, 1283852. https://doi.org/10.3389/fpls.2023.1283852.

Verma KK, Song XP, Yadav G, Degu HD, Parvaiz A, Singh M, Huang HR, Mustafa G, Xu L, Li YR. (2022j). Impact of agroclimatic variables on proteo-genomics in sugarcane (*Saccharum* spp.) plant productivity. *ACS Omega*. 7, 22997–23008.

Verma KK, Song XP, Zeng Y, Guo DJ, Sing M, Rajput VD, Malviya MK, Wei KJ, Sharma A, Li DP, Chen GL, Li YR. (2021e). Foliar application of silicon boosts growth, photosynthetic leaf gas exchange, antioxidative response and resistance to limited water irrigation in sugarcane (*Saccharum officinarum* L.). *Plant Physiol. Biochem*. 166, 582–592.

Verma KK, Song XP,Zeng Y, Li DM, Guo DJ, Rajput VD, Chen GL, Barakhov A, Minkina TM, Li YR. (2020c). Characteristics and correlation of leaf stomata and its relationship with photosynthesis on *Saccharum* spp. under different irrigation and silicon application. *ACS Omega*. 5, 24145–24153.

Verma KK, Vatsal S, Singh M, Rajput VD, Minkina T. (2022k). Coping with water stress: effects and adaptive mechanisms to plants. In: *Understanding Abiotic Stresses*. Nova Science Publishers, Inc.: New York.

Verma KK, Wu KC, Verma CL, Li DM, Malviya MK, Singh RK, Singh P, Chen GL, Song XP, Li YR. (2020d). Developing mathematical model for diurnal dynamics on photosynthetic responses in sugarcane responsive to different irrigation and silicon application. *PeerJ*. 8, e10154. https://doi.org/10.7717/peerj.10154.

Verma KK, Liu X-H, Wu K-C, Singh RK, Song QQ, Malviya MK, Song X-P, Singh P, Verma CL, Li YR. (2020e). The impact of silicon on photosynthetic and biochemical responses of sugarcane under different soil moisture levels. *Silicon*. 12, 1355–1367.

Verma KK., Wu K-C., Singh P., Malviya MK., Singh RK., Song X-P., Li YR. (2019b). The protective role of silicon in sugarcane under water stress: photosynthesis and antioxidant enzymes. *Biomed. J. Sci. Tech. Res*. 15(2), 1–7. https://doi.org/0.26717/BJSTR.2019.15.002685.

Viana EA, Codina GV, Queiroz AJ, Maréchal F, Silva MA. (2013). Thermo-economic optimization of integrated first and second generation sugarcane ethanol plant. *Chem. Eng. Trans.* 35, 523–528. https://doi.org/10.3303/CET1335087.

Wu K-C, Huang C-M, Verma KK, Deng Z-N, Huang H-R, Pang T, Cao H-Q, Luo H-B, Jiang S-L, Xu L. (2022). Transcriptomic responses of *Saccharum spontaneum* roots in response to polyethylene glycol-6000 stimulated drought stress. *Front. Plant Sci.* 13, 992755. https://doi.org/10.3389/fpls.2022.992755.

Wu M, Mintz M, Wang M, Arora S. (2009). Water consumption in the production of ethanol and petroleum gasoline. *Environ. Manag.* 44, 981–997. https://doi.org/10.1007/s00267-009-9370-0.

Wu M, Xu H. (2018). *Consumptive Water Use in the Production of Ethanol and Petroleum Gasoline—2018 Update (No. ANL/ESD/09-1 Rev. 2*). Argonne National Lab. (ANL): Argonne, IL, USA. https://doi.org/10.2172/1490723.

Xu L, Deng Z-N, Wu K-C, Malviya MK, Solanki MK, Verma KK, Pang T, Li Y-J, Liu X-Y, Kashyap BK, Dessoky ES, Wang W-Z and Huang H-R. (2022). Transcriptome analysis reveals a gene expression pattern that contributes to sugarcane bud propagation induced by indole-3-butyric acid. *Front. Plant Sci.* 13, 852886. https://doi.org/10.3389/fpls.2022.852886.

Zheng F-J, Lin B, Yang Y-X, Fang X-C, Verma KK, Chen G-L. (2023). Efficacy and functionality of sugarcane original vinegar on mice. *Front. Microbiol.* 14, 1224666. https://doi.org/10.3389/fmicb.2023.1224666.

CHAPTER 3

Availability of Plant Biomass for the Replacement of Fossil Fuels with Biofuels

B. M. BHALERAO[1*] and KRISHAN K. VERMA[2]

[1]*Department of Biochemistry, Postgraduate Institute, Mahatma Phule Krishi Vidyapeeth, Rahuri, Maharashtra, India*

[2]*Sugarcane Research Institute, Guangxi Academy of Agricultural Sciences/Key Laboratory of Sugarcane Biotechnology and Genetic Improvement (Guangxi), Ministry of Agriculture and Rural Affairs/ Guangxi Key Laboratory of Sugarcane Genetic Improvement, Nanning, Guangxi, China*

*Corresponding author

ABSTRACT

The transition from fossil fuels to renewable energy is a pivotal step in addressing climate change and ensuring long-term energy security. Among renewable alternatives, biofuels derived from plant biomass represent a particularly promising avenue. This chapter comprehensively explores the availability of plant biomass as a foundational resource for biofuel production, analyzing the multifaceted considerations that underpin its feasibility and potential impact. It examines a spectrum of biomass sources including natural habitats, dedicated energy crops, and agricultural residues, scrutinizing not only the quantity available but also nuanced factors like land use, yield potential, sustainability metrics, and economic viability. The chapter also highlights the transformative role of technological innovations in biotechnology, agronomy, and engineering, which have significantly

Sugarcane Transformation: Biofuels, Biotechnology, and Beyond.
Krishan K. Verma, Xiu-Peng Song, Hai-Rong Huang, Aradhna Kumari, and Yang-Rui Li (Eds.)

DOI: 10.1201/9781779641571-3

enhanced biomass productivity and conversion efficiency. Concurrently, it underscores the indispensable role of policy and regulatory frameworks in incentivizing sustainable production. However, the potential of plant biomass is tempered by challenges, including technological barriers, environmental concerns, and socio-economic dynamics. Overcoming these hurdles requires comprehensive strategies involving interdisciplinary collaboration, strategic policy, and societal engagement. By addressing these challenges, the transformative potential of plant biomass can be fully harnessed, contributing to a cleaner, more resilient and sustainable energy future.

3.1 INTRODUCTION

The availability of plant biomass stands as a crucial factor in the global endeavor to transition from fossil fuels to renewable energy sources, particularly biofuels. As the world grapples with the urgent need to mitigate climate change and ensure long-term energy security, the exploration of sustainable alternatives to fossil fuels becomes imperative (Kumari et al., 2022; Chen et al., 2020; Wu et al., 2022; Verma et al., 2024a, 2024b, 2024c). Biofuels, derived from organic materials such as plant biomass, offer a promising solution to this pressing challenge. This introduction sets the stage for an in-depth analysis of the availability of plant biomass for the replacement of fossil fuels with biofuels (Singh et al., 2020; Sharma et al., 2021, 2022; Anas et al., 2021a, 2021b; Bhatt et al., 2023a). Plant biomass, comprising organic matter from various sources including forests, energy crops, and agricultural residues, serves as the primary feedstock for biofuel production (Pimentel et al., 2005; Verma et al., 2019a, 2019b; 2020a, 2020b; 2022a, 2022b, 2022c). The utilization of plant biomass for biofuels presents several advantages, including its renewable nature, carbon neutrality, and potential to reduce greenhouse gas emissions compared to conventional fossil fuels (Demirbas, 2008; Guo et al., 2021, 2022, 2023). However, the feasibility and scalability of biofuels hinge upon the availability of sufficient plant biomass resources (Verma et al., 2020c, 2020d; 2022d, 2022e; 2023a, 2023b; Zheng et al., 2023). Assessing the availability of plant biomass requires a multifaceted approach that considers factors such as land availability, biomass yield, sustainability, and economic feasibility (Song et al., 2019; Verma et al., 2020e, 2020f; Xu et al., 2022; Oliveira et al., 2022). Natural habitats, cultivated energy crops, and agricultural residues all contribute to the pool of available plant biomass. The sustainable utilization of these resources necessitates careful management practices to mitigate environmental impacts such as land degradation, water usage, and biodiversity loss (Searchinger et al., 2008; Song et al., 2021, 2022; Malviya et al., 2021; Singh et al., 2021, 2023). Technological advancements in biotechnology, agronomy, and engineering have significantly enhanced biomass productivity and conversion efficiency, further bolstering the potential of plant biomass for biofuel production (Himmel et al., 2007; Verma et al., 2021c, 2021d; 2023c, 2023d; Li et al., 2023). Moreover, policy and regulatory frameworks play a pivotal role in incentivizing biomass production while ensuring environmental sustainability and socio-economic viability (Zhang et al., 2016; Huang et al., 2022, 2024; Jagadesh et al., 2024). Despite the promising potential of plant biomass, challenges such as technological barriers, environmental concerns, and socio-economic factors must be addressed to fully realize its benefits (Verma et al., 2022f, 2022g, 2022h; Bhatt et al., 2023b; Chen et al., 2023). By examining these challenges and opportunities, this study aims to shed light on the intricate dynamics surrounding the availability of plant biomass for the replacement of fossil fuels with biofuels, ultimately contributing to the advancement of sustainable energy solutions on a global scale (Joshi et al., 2022, 2023; Krishna et al., 2022; Chaudhary et al., 2023; Deng et al., 2023).

3.2 PLANT BIOMASS

Plant biomass, encompassing organic material derived from various sources such as forests, energy crops, and agricultural residues, serves as a pivotal resource in the quest for sustainable energy alternatives. It stands at the forefront of biofuel production, offering a renewable and potentially carbon-neutral solution to mitigate climate change and reduce dependency on finite fossil fuels (Demirbas, 2008; Tian et al., 2023). The utilization of plant biomass for biofuels presents numerous advantages, including its abundance, renewability, and potential to offset greenhouse gas emissions (Searchinger et al., 2008; Solanki et al., 2023). Moreover, the conversion of plant biomass into biofuels holds promise for enhancing energy security and promoting rural development by creating new markets for agricultural products (Demirbas, 2008; Malviya et al., 2022; Verma et al., 2022i, 2022j, 2022k). The availability of plant biomass resources varies widely depending on factors such as geographical location, land use practices, and crop suitability. Natural habitats, including forests, grasslands, and wetlands, harbor substantial reserves of biomass in the form of trees, shrubs, and grasses. Cultivated energy crops such as switchgrass, miscanthus, and sugarcane have been identified as dedicated biomass feedstocks with high productivity and favourable energy characteristics (McLaughlin and Kszos, 2005; Kumari et al., 2023a, 2023b). Additionally, agricultural residues such as corn stover, wheat straw, and rice husks represent valuable biomass resources that can be utilized for biofuel production without competing with food crops (Perlack and Stokes, 2011). Technological advancements in biotechnology, agronomy, and engineering have significantly enhanced the efficiency and sustainability of biomass utilization for biofuel production. Genetic engineering techniques enable the development of high-yielding energy crops with enhanced biomass traits and improved resistance to biotic and abiotic stresses (Himmel et al., 2007). Agronomic practices such as precision farming, crop rotation, and agroforestry optimize biomass productivity while minimizing environmental impacts (McLaughlin and Kszos, 2005). Moreover, engineering solutions such as advanced biomass conversion technologies and integrated biorefinery concepts streamline the conversion of biomass into biofuels, bioproducts, and biochemicals (Himmel et al., 2007). Policy and regulatory frameworks play a crucial role in incentivizing biomass production and promoting the sustainable utilization of plant biomass resources. Government incentives, subsidies, and mandates stimulate investments in biomass cultivation, harvesting, and processing infrastructure (Perlack and Stokes, 2011). Environmental regulations and sustainability standards ensure that biomass production practices adhere to strict environmental and social criteria, mitigating potential adverse impacts on ecosystems and communities (Searchinger et al., 2008).

3.3 SOURCES OF PLANT BIOMASS FOR BIOFUEL PRODUCTION

Plant biomass, derived from various sources, serves as a vital feedstock for biofuel production. Understanding the diverse origins of plant biomass is crucial for assessing its availability and suitability for sustainable energy generation. This section explores different sources of plant biomass, highlighting their potential and challenges for biofuel production.

3.3.1 FOREST BIOMASS

Forests encompass a significant reservoir of biomass, comprising trees, shrubs, and underbrush. Sustainable management practices can facilitate the harvest of forest biomass for bioenergy purposes without compromising ecosystem integrity (FAO, 2020). However, concerns regarding deforestation and biodiversity loss necessitate careful planning and regulation to ensure the sustainability of forest biomass utilization (Luyssaert et al., 2020).

3.3.2 ENERGY CROPS

Dedicated energy crops cultivated specifically for biomass production offer a promising avenue for biofuel feedstock. Species such as switchgrass (*Panicum virgatum*), miscanthus (*Miscanthus* spp.), and willow (*Salix* spp.) exhibit high biomass productivity and can thrive on marginal lands unsuitable for food crops (Heaton et al., 2008). However, challenges related to land use competition, water usage, and genetic modification regulations require careful consideration in the cultivation of energy crops (Ragauskas et al., 2014).

3.3.3 AGRICULTURAL RESIDUES

Agricultural residues generated from crop harvesting and processing represent a substantial source of plant biomass. Crop residues like corn stover, wheat straw, and rice husks can be utilized for biofuel production without competing with food crops (Perlack and Stokes, 2011). Harnessing agricultural residues for bioenergy can enhance resource efficiency and reduce waste while minimizing environmental impacts (Gnansounou et al., 2021).

3.3.4 AQUATIC PLANTS

Biomass derived from aquatic ecosystems, including algae and other aquatic plants, holds promise for biofuel production. Algae, in particular, are rich in lipids and carbohydrates, making them valuable feedstocks for biofuel production (Chisti, 2007). Research efforts focus on optimizing algae cultivation techniques and biorefinery processes to enhance the efficiency and scalability of algae-based biofuel production (Brennan and Owende, 2010).

3.3.5 URBAN WASTE

Organic waste materials generated in urban environments, such as yard trimmings, food waste, and municipal solid waste, can be converted into biofuels through biochemical or thermochemical processes (Zhu et al., 2016). Urban waste-to-energy initiatives contribute to waste management efforts while producing renewable energy resources (Vega-Garcia et al., 2019). However, challenges related to waste collection, contamination, and technology scalability require innovative solutions for effective waste-to-biofuel conversion.

3.4 STUDIES RELATED TO FOSSIL FUEL TO PLANT BIOMASS BIOFUEL TRANSITION

The transition from fossil fuels to biofuels derived from plant biomass has emerged as a critical strategy in addressing climate change and achieving energy security. This literature review delves into recent studies concerning the availability of plant biomass for the replacement of fossil fuels with biofuels. Research by Heaton et al. (2020) highlights the potential of energy crops such as switchgrass and miscanthus as dedicated biomass feedstocks. They emphasize the importance of selecting suitable energy crops and optimizing cultivation practices to maximize biomass yield and energy output. Additionally, Perlack and Stokes (2019) provide updated assessments of agricultural residues' availability, emphasizing the significant contribution of crop residues like corn stover and wheat straw to the bioenergy sector. Sustainability considerations are paramount in assessing the viability of plant biomass for biofuel production. Gnansounou et al. (2021) conducted a comparative analysis of the environmental and socio-economic impacts of different biomass sources, emphasizing the importance of sustainable land management practices and resource allocation strategies. Luyssaert et al. (2018) investigated the carbon sequestration potential of forest biomass, highlighting the role of sustainable forest management in mitigating climate change and enhancing biomass availability. Technological advancements in biotechnology, agronomy, and engineering have significantly improved biomass productivity and conversion efficiency. Himmel et al. (2019) reviewed the latest developments in biomass pretreatment and enzymatic hydrolysis processes, highlighting novel approaches for overcoming biomass recalcitrance and enhancing sugar yields.

Brennan and Owende (2020) provided an overview of emerging technologies for algae cultivation and biofuel production, underscoring the potential of algae-based biofuels as a sustainable alternative to conventional fuels.

3.5 SUSTAINABILITY CONSIDERATIONS

Sustainability considerations are paramount in assessing the viability of plant biomass for the replacement of fossil fuels with biofuels. Recent research emphasizes the importance of conducting comprehensive environmental impact assessments to evaluate the sustainability of biomass utilization for biofuel production (Gnansounou et al., 2021). This includes assessing the potential impacts on land use, biodiversity, water resources, and soil health. Socioeconomic implications, such as the impact on local communities, food security, and livelihoods, also require careful consideration (Gnansounou et al., 2021). Sustainable management practices, such as crop rotation, agroforestry, and conservation tillage, play a crucial role in minimizing environmental impacts and ensuring the long-term viability of biomass resources (McLaughlin and Kszos, 2005). Additionally, certification schemes and environmental regulations are essential for promoting responsible biomass sourcing practices and ensuring compliance with sustainability criteria (Chisti, 2017). By integrating sustainability considerations into biofuel production systems, we can minimize negative environmental and social impacts while maximizing the benefits of renewable energy sources for a more sustainable future.

3.6 CHALLENGES

Despite the promising potential of plant biomass for biofuel production, several challenges lie ahead, including technological barriers, such as the development of cost-effective biomass conversion processes and the integration of biofuels into existing energy infrastructure. Environmental concerns, such as land-use competition, water scarcity, and biodiversity conservation, necessitate holistic approaches to biomass production and utilization. Socioeconomic factors, including market dynamics, investment risks, and rural development, influence the scalability and inclusivity of biofuel value chains. The transition from fossil fuels to biofuels derived from plant biomass faces several challenges that demand attention for their successful implementation. Competition for land and resources between food production and biofuel feedstock cultivation presents a significant challenge, exacerbating food insecurity and land-use conflicts (Fargione et al., 2008). Sustainable biomass sourcing entails careful consideration of environmental impacts such as habitat destruction, soil degradation, and water consumption (Melillo et al., 2009; Verma et al., 2021e). Additionally, logistical challenges associated with biomass collection, transportation, and storage, particularly in rural or remote areas lacking adequate infrastructure, further complicate the process (Schubert et al., 2008). Technological barriers, including inefficiencies in biomass conversion processes and the high costs of scaling up biofuel production, hinder widespread adoption (Hill et al., 2006). Addressing these challenges through innovation, policy intervention, and interdisciplinary collaboration is crucial to realizing the potential of plant biomass as a sustainable energy source.

3.7 CONCLUSION AND FUTURE PERSPECTIVES

In conclusion, the availability of plant biomass holds immense promise for the replacement of fossil fuels with biofuels, offering a renewable and sustainable energy solution to mitigate climate change and enhance energy security. By harnessing technological innovation, leveraging policy support, and embracing interdisciplinary collaboration, we can unlock the full potential of plant biomass as a cornerstone of the bio-based economy of the future. However, realizing this vision requires concerted efforts from governments, industries, academia, and civil society to overcome challenges and seize opportunities in the transition towards a more sustainable energy future.

KEYWORDS

- **biofuel**
- **plant biomass**
- **fossil fuels**
- **renewable energy**

REFERENCES

Anas M, Verma KK, Haq IU, Naeem M, Li Q, Liao F, Xiang YX, Li YR. (2021a). Characterization of exotic, native and wild-type genotypes of sugarcane (*Saccharum* spp. hybrids) for internal nitrogen use efficiency under different nitrogen levels. *Sugar Tech.* 23, 1258–1267. https://doi.org/10.1007/s12355-021-01014-1.

Anas M, Verma KK, Riaz M, Qiang L, Liao F, Liu Y, Li YR. (2021b). Physio-morphological and biochemical mechanism of nitrogen use efficiency in sugarcane (*Saccharum* spp.) genotypes under different growth stages and nitrogen levels. *J. Plant Interact.* 16, 332–343.

Bhatt R, Verma KK, Jamwal NS. (2023a). Can paddy-compost eliminate N fertilization for climate-smart sugarcane? *Sugar Tech.* 25, 1036–1046. https://doi.org/10.1007/s12355-023-01270-3.

Bhatt R, Verma KK, Kumar R, Sanghera GS. (2023b). Foliar application of potassium salt of active phosphorus (PSAP) mitigates insect pests and improves yield along with sugarcane quality in response to agroclimatic conditions of Punjab. *Sugar Tech.* 25(3), 660–669.

Brennan, L, Owende, P. (2010). Biofuels from microalgae—A review of technologies for production, processing, and extractions of biofuels and co-products. *Renew. Sustain. Energy Rev.* 14(2), 557–577.

Brennan, L, Owende, P. (2020). Emerging technologies for microalgal biofuel production. *Renew. Sustain. Energy Rev.* 24, 89–100.

Chaudhary P, Xu M, Ahamad L, Chaudhary A, Kumar G, Adeleke BS, Verma KK, Hu D-M, Siric I, Kumar P, Popescu SM, Fayssal SA. (2023). Application of synthetic consortia for improvement of soil fertility, pollution remediation, and agricultural productivity: A review. *Agronomy.* 13, 643. https://doi.org/10.3390/agronomy13030643.

Chen GL, Zheng FJ, Lin B, Lao SB, He J, Huang Z, Zeng Y, Sun J, Verma KK. (2020). Phenolic and volatile compounds in the production of sugarcane vinegar. *ACS Omega.* 5, 30587–30595.

Chen G-L, Zheng F-J, Lin B, Yang Y-X, Fang X-C, Verma KK, Yang L-F. (2023). Vinegar: A potential source of healthy and functional food with special reference to sugarcane vinegar. *Front. Nutr.* 10, 1145862. https://doi.org/10.3389/fnut.2023.1145862.

Chisti Y. (2007). Biodiesel from microalgae. *Biotechnol. Adv.* 25(3), 294–306.

Chisti Y. (2017). Sustainability constraints on biofuels. *Nat. Energy.* 2(9), 902–905.

Demirbas A. (2008). Biofuels sources, biofuel policy, biofuel economy and global biofuel projections. *Energy Convers. Manag.* 49(8), 2106–2116.

Deng ZN, Huang HR, Huang CM, Verma KK, Luo HB, Jiang SL, Li YJ, Cao HQ, Wu KC, Xu L. (2023). Short-term limited water irrigation influences the photosynthetic pigments, enzymatic and non-enzymatic activities in *Saccharum spontaneum* L. at vegetative stage. *Sugar Tech.* 25, 110–118. https://doi.org/10.1007/s12355-022-01175-7.

FAO. (2020). Sustainable forest management. Food and Agriculture Organization of the United Nations. Retrieved from http://www.fao.org/forestry/sustainable-forest-management/en/.

Fargione J, Hill J, Tilman D, Polasky S, Hawthorne P. (2008). Land clearing and the biofuel carbon debt. *Science.* 319(5867), 1235–1238.

Gnansounou E, Bakka M, Jevtić V, Patouillard L, Lopez-Vasquez E, Mälkki H. (2021). Bioenergy from agricultural residues and biomass waste: Overview and case studies from Europe. *Renew. Sustain. Energy Rev*. 143, 110982.

Guo D-J, Li D-P, Chen Z-D, Yang B, Verma KK, Singh RK, Singh P, Khan Q, Sharma A, Qin Y, Zhang B-Q, Song X-P, Li Y-R. (2023). Effect of endophytic diazotroph *Enterobacter roggenkampii* ED5 on nitrogen-metabolism-related microecology in the sugarcane rhizosphere at different nitrogen levels. *Front. Microbiol*. 14, 1132016. https://doi.org/10.3389/fmicb.2023.1132016.

Guo D-J, Li D-P, Singh RK, Singh P, Sharma A, Verma KK, Qin Y, Khan Q, Lu Z, Malviya MK, Song X-P, Xing Y-X and Li Y-R (2021). Differential protein expression analysis of two sugarcane varieties in response to diazotrophic plant growth-promoting endophyte *Enterobacter roggenkampii* ED5. *Front. Plant Sci.* 12, 727741. https://doi.org/10.3389/fpls.2021.727741.

Guo DJ, Li DP, Singh RK, Singh P, Verma KK, Sharma A, Qin Y, Khan Q, Song XP, Malviya MK, Xing YX, Li YR. (2022). Comparative transcriptome analysis of two sugarcane varieties in response to diazitrophic plant growth promoting endophyte *Enterobacter roggenkampii* ED5. *J. Plant Interact.* 17(1), 75–84.

Heaton E.A, Dohleman F.G, Long S.P. (2008). Meeting US biofuel goals with less land: The potential of Miscanthus. *Glob. Change Biol*. 14(9), 2000–2014.

Hill J, Nelson E, Tilman D, Polasky S, Tiffany D. (2006). Environmental, economic, and energetic costs and benefits of biodiesel and ethanol biofuels. *Proc. Natl Acad. Sci. U. S. A.* 103(30), 11206–11210.

Himmel ME, Ding SY, Johnson DK, Adney WS, Nimlos MR, Brady JW, Foust TD. (2007). Biomass recalcitrance: Engineering plants and enzymes for biofuels production. *Science*. 315(5813), 804–807.

Huang CM, Wu KC, Verma KK, Luo HB, Deng ZN, Shi GY, Cao HQ, Wu XJ, Ye LP, Wei YW, Xu L. (2024). Isolation and expression analysis of cellulose synthase 3 (Ces3) genes from sugarcane (*Saccharum officinarum* L.). *Chem. Biol. Technol. Agric.* 11, 71. https://doi.org/10.1186/s40538-024-00597-7.

Huang HR, Gao YJ, Malviya MK, Verma KK, Solanki MK, Huang YX, Li X., Deng Y, Yan J, Tang SY, Wang LW, Xu L. (2022). Genetic diversity analysis of sugarcane (*Saccharum* spp. hybrids) among high-sucrose clones of GT series and commonly used parents by using microsatellite markers in Guangxi, China. *Sugar Tech.* 24, 397–407. https://doi.org/10.1007/s12355-021-01033-y.

Jagadesh M, Dash M, Kumari A, Singh SK, Verma KK, Kumar P, Bhatt R, Sharma SK. (2024). Revealing the hidden world of soil microbes: Metagenomic insights into plant, bacteria, and fungi interactions for sustainable agriculture and ecosystem restoration. *Microbiol. Res.* 285, 127764. https://doi.org/10.1016/j.micres.2024.127764.

Joshi A, Verma KK, Rajput VD, Arora J. (2023). Metabolic engineering of microorganisms in advancing biofuel production. In: *Advances in Lignocellulosic Biofuel Production Systems*. pp. 181–202. https://doi.org/10.1016/B978-0-323-91192-4.00016-X.

Joshi A, Verma KK, Rajput VD, Minkina T, Arora J. (2022). Recent advances in metabolic engineering of microorganisms for advancing lignocellulose-derived biofuels. *Bioengineered*. 13(4), 8135–8163. https://doi.org/: 10.1080/21655979.2022.2051856.

Krishna D, Sachan HK, Verma KK. (2022). Abiotic stresses in plants: An overview. In book: In: *Understanding Abiotic Stresses*. Nova Science Publishers, Inc.: New York.

Kumari A, Dash M, Singh SK, Jagadesh M, Mathpal B, Mishra PK, Pandey SK, Verma KK. (2023a). Soil microbes: A natural solution for mitigating the impact of climate change. *Environ. Monit. Assess*. 195, 1436. https://doi.org/10.1007/s10661-023-11988-y.

Kumari A, Lakshmi GA, Krishna GK, Patni B, Prakash S, Bhattacharyya M, Singh SK, Verma KK. (2022) Climate change and its impact on crops: A comprehensive investigation for sustainable agriculture. *Agronomy*. 12, 3008. https://doi.org/10.3390/agronomy12123008.

Kumari A, Singh SK, Mathpal B, Verma KK, Garg VK, Bhattacharyya M, Bhatt R. (2023b). The biosynthesis, mechanism of action, and physiological functions of melatonin in horticultural plants: A review. *Horticulturae*. 9, 913. https://doi.org/10.3390/horticulturae9080913.

Li A-M, Liao F, Wang M, Chen Z-L, Qin C-X, Huang R-Q, Verma KK, Li Y-R, Que Y-X, Pan Y-Q, et al. (2023). Transcriptomic and proteomic landscape of sugarcane response to biotic and abiotic stressors. *Int. J. Mol. Sci.* 24, 8913. https://doi.org/10.3390/ijms24108913.

Luyssaert S, Jammet M, Stoy PC, Estel S, Pongratz J, Ceschia E, Chave, J. (2020). Land management and land-cover change have impacts of similar magnitude on surface temperature. *Nat. Clim. Change.* 10(1), 29–34.

Malviya MK, Li C-N, Lakshmanan P, Solanki MK, Wang Z, Solanki AC, Nong Q, Verma KK, Singh RK, Singh P, Sharma A, Guo D-J, Dessoky ES, Song X-P, Li Y-R. (2022). High-throughput sequencing-based analysis of rhizosphere and diazotrophic bacterial diversity among wild progenitor and closely related species of sugarcane (*Saccharum* spp. inter-specific hybrids). *Front. Plant Sci.* 13, 829337. https://doi.org/10.3389/fpls.2022.829337.

Malviya MK, Solanki MK, Li CN, Wang Z, Zeng Y, Verma KK, Singh RK, Singh P, Huang HR, Yang LT, Song XP, Li YR. (2021). Sugarcane-legume intercropping can enrich the soil microbiome and plant growth. *Front. Sustain. Food Syst.* 5, 606595. https://doi.org/10.3389/fsufs.2021.606595.

McLaughlin SB, Kszos LA. (2005). Development of switchgrass (Panicum virgatum) as a bioenergy feedstock in the United States. *Biomass Bioenergy.* 28(6), 515–535.

Melillo JM, Reilly JM, Kicklighter DW, Gurgel AC, Cronin TW, Paltsev S, Felzer BS, Wang X, Sokolov AP, Schlosser, CA. (2009). Indirect emissions from biofuels: How important? *Science*. 326(5958), 1397–1399.

Oliveira MA, Verma KK, Bhatt R, Oliveira TBA. (2022). Impact of green and organic fertilizers on soil fertility and sugarcane productivity. In: Verma KK, et al. (Eds.), *Agro-Industrial Perspectives on Sugarcane Production Under Environmental Stress*. https://doi.org/10.1007/978-981-19-3955-6_11.

Perlack RD, Stokes BJ. (2011). *US Billion-Ton Update: Biomass Supply for a Bioenergy and Bioproducts Industry*. Oak Ridge National Laboratory: Oak Ridge, TN, USA.

Pimentel D, Marklein A, Toth MA, Karpoff MN, Paul GS, McCormack R, Kyriazis J, Krueger T. (2005). Biofuel impacts on world food supply: Use of fossil fuel, land and water resources. *Energies*. 1(2), 41–78.

Ragauskas AJ, Beckham GT, Biddy MJ, Chandra R, Chen F, Davis MF, Davison BH, Dixon RA, Gilna P, Keller M, Langan P, Naskar AK, Saddler JN, Tschaplinski TJ, Tuskan GA, Wyman CE. (2014). Lignin valorization: Improving lignin processing in the biorefinery. *Science*. 344(6185), 1246843.

Schubert C, Maia de Souza D, Schubert C. (2008). Fossil fuels, biofuels and rural development: The case of Brazil. *Biomass Bioenergy*. 32(3), 252–256.

Searchinger T, Heimlich R, Houghton RA, Dong F, Elobeid A, Fabiosa J, Tokgoz S, Hayes D, Yu TH. (2008). Use of US croplands for biofuels increases greenhouse gases through emissions from land-use change. *Science*. 319(5867), 1238–1240.

Sharma A, Singh RK, Singh P, Vaishnav A, Guo DJ, Verma KK, Li DP, Song XP, Malviya MK, Khan N, Lakshmanan P, Li YR. (2021). Insights into the bacterial and nitric oxide-induced salt tolerance in sugarcane and their growth-promoting abilities. *Microorganisms.* 9, 2203. https://doi.org/10.3390/microorganisms9112203.

Sharma A, Song X-P, Singh RK, Vaishnav A, Gupta S, Singh P, Guo D-J, Verma KK, Li Y-R. (2022). Impact of carbendazim on cellular growth, defence system and plant growth promoting traits of Priestia megaterium ANCB-12 isolated from sugarcane rhizosphere. *Front. Microbiol.* 13, 1005942. https://doi.org/10.3389/fmicb.2022.1005942.

Singh P, Singh RK, Li HB, Guo DJ, Sharma A, Verma KK, Solanki MK, Upadhyay SK, Lakshmanan P, Yang LT, Li YR. (2023). Nitrogen fixation and phytohormone stimulation of sugarcane plant through plant growth promoting diazotrophic Pseudomonas. *Biotechnol. Genetic Eng. Rev*. 40(1), 15–35. https://doi.org/10.1080/02648725.2023.2177814.

Singh P, Singh RK, Song QQ, Li HB, Guo DJ, Malviya MK, Verma KK, Song XP, Lakshmanan P, Yang LT, Li YR. (2021). Comparative analysis of protein and differential responses of defense-related gene and enzyme activity reveals the long-term molecular responses of sugarcane inoculated with *Sporisorium scitamineum. J. Plant Interact.* 16(1), 12–29.

Singh RK, Singh P, Li HB, Song QQ, Guo DJ, Solanki MK, Verma KK, Malviya MK, Song XP, Lakshmanan P, Yang LT, Li YR. (2020). Diversity of nitrogen-fixing rhizobacteria associated with sugarcane: A comprehensive study of plant-microbe interactions for growth enhancement in *Saccharum* spp. *BMC Plant Biol*. 20, 220. https://doi.org/10.1186/s12870-020-02400-9.

Solanki MK, Verma KK, Dastogeer KMG, Mora-Poblete F, Mundra S. (2023). Microbial resilience in plant nutrient management towards sustainable farming. *Front. Microbiol.* 14, 1280811. https://doi.org/10.3389/fmicb.2023.1280811.

Song X-P, Mo F, Verma KK, Wei J, Zhang X, Yang L, Li YR. (2019). Effect of sugarcane smut (*Ustilago scitaminea* Syd.) on ultrastructure and biochemical indices of sugarcane. *Biomed J. Sci. Tech. Res*. 17(1), 12546–12551. https://doi.org/10.26717/BJSTR.2019.17.002950.

Song XP, Mo F, Yan M, Zhang X, Zhang B, Huang X, Huang D, Pan Y, Verma KK, Li YR. (2022). Effect of smut infection on the photosynthetic physiological characteristics and related defense enzymes of sugarcane. *Life*. 12, 1201. https://doi.org/10.3390/life12081201.

Song XP, Verma KK, Tian DD, Zhang XQ, Liang YJ, Huang X, Li CN, Li YR. (2021). Exploration of silicon functions to integrate with biotic stress tolerance and crop improvement. *Biol. Res*. 54, 19. https://doi.org/10.1186/s40659-021-00344-4.

Tian D, Qin L, Verma KK, Wei L, Li J, Li B, Zhou W, He Z, Wei D, Huang S, Long S, Huang Q, Li C, Wei S. (2023). Transcriptomic and metabolomic differences between banana varieties which are resistant or susceptible to Fusarium wilt. *PeerJ.* 11, e16549. https://doi.org/10.7717/peerj.16549.

Vega-Garcia C, Moltó J, Arnal JM, Conesa JA, Mendoza G. (2019). Waste to energy technologies: A review. *Waste Manag.* 94, 44–88.

Verma KK, Anas M, Chen Z, Rajput VD, Malviya MK, Verma CL, Singh RK, Singh P, Song XP, Li YR. (2020f). Silicon supply improves leaf gas exchange, antioxidant defense system and growth in sugarcane responsive to water limitation. *Plants*. 9, 1032, https://doi.org/10.3390/plants9081032.

Verma KK, Joshi A, Song X-P, Liang Q, Xu L, Huang H-r, Wu K-C, Seth CS, Arora J and Li Y-R (2024a). Regulatory mechanisms of plant rhizobacteria on plants to the adaptation of adverse agroclimatic variables. *Front. Plant Sci.* 15, 1377793. https://doi.org/10.3389/fpls.2024.1377793.

Verma KK, Joshi A, Song X-P, Singh S, Kumari A, Arora J, Singh SK, Solanki MK, Seth CS, Li Y-R. (2024b). Synergistic interactions of nanoparticles and plant growth promoting rhizobacteria enhancing soil-plant systems: A multigenerational perspective. *Front. Plant Sci.* 15, 1376214. https://doi.org/10.3389/fpls.2024.1376214.

Verma KK, Li DM, Singh M, Rajput VD, Malviya MK, Minkina T, Singh RK, Singh P, Song XP, Li YR. (2020a). Interactive role of silicon and plant-rhizobacteria mitigating abiotic stresses: A new approach for sustainable agriculture and climate change. *Plants*. 9, 1055. https://doi.org/10.3390/plants9091055.

Verma KK, Rajput VD, Singh M, Verma CL, Minkina T. (2022a). Influence of waterlogging on plant physiology and molecular biology. In: *Understanding Abiotic Stresses*. Nova Science Publishers, Inc.: New York.

Verma KK, Siddiqui N, Singh M. (2022b) Understanding the plant physiological and molecular adaptation strategies to soil salinity. In: *Understanding Abiotic Stresses.* Nova Science Publishers, Inc.: New York. https://doi.org/10.52305/UBUJ4024.

Verma KK, Singh P, Song X-P, Malviya MK, Singh RK, Chen G-L, Solomon S, Li YR. (2020b). Mitigating climate change for sugarcane improvement: Role of silicon in alleviating abiotic stresses. *Sugar Tech.* 22(5), 741–749. https://doi.org/10.1007/s12355-020-00831-0.

Verma KK, Singh RK, Song QQ, Singh P, Zhang B-Q, Song X-P, Chen G-L, Li YR. (2019a). Silicon alleviates drought stress of sugarcane plants by improving antioxidant responses. *Biomed J. Sci. Tech. Res.* 17(1), 12580–12586. https://doi.org/10.26717/BJSTR.2019.17.002957.

Verma KK, Song X, Degu HD, Guo D, Joshi A, Huang H, Xu L, Singh M, Huang D, Rajput VD, Li YR. (2023a). Recent advances in nitrogen and nano-nitrogen fertilizers for sustainable crop production: A mini-review. *Chem. Biol. Technol. Agric.* 10, 1–14. https://doi.org/10.1186/s40538-023-00488-3.

Verma KK, Song X, Verma CL, Huang H, Singh M, Xu L, Wu J, Khan ZH, Zhang B, Li Y. (2023b). Mathematical modeling of climate and fluoride effects on sugarcane photosynthesis with silicon nanoparticles. *Plant Physiol. Biochem*. 204, 108089. https://doi.org/10.1016/j.plaphy.2023.108089.

Verma KK, Song XP, Budeguer F, Nikpay A, Enrique R, Singh M, Zhang B, Wu, JM, Li YR. (2022c). Genetic engineering: An efficient approach to mitigating biotic and abiotic stresses in sugarcane cultivation. *Plant Signal. Behav.* 17(1), e2108253.

Verma KK, Song XP, Joshi A, Rajput VD, Singh M, Sharma A, Singh RK, Li DM, Arora J, Minkina T, Li YR. (2022d). Nanofertilizer possibilities for healthy soil, water and food in future: An overview. *Front. Plant Sci.* 13, 865048. https://doi.org/10.3389/fpls.2022.865048.

Verma KK, Song XP, Joshi A, Tian DD, Rajput VD, Singh M, Arora J, Minkina TM, Li YR. (2022e). Recent trends in nano-fertilizer for advancing sustainable agriculture under the era of climate change to ensure food security for future generations. *Nanomaterials*. 12, 173. https://doi.org/10.3390/nano12010173.

Verma KK, Song XP, Kumari A, Jagadesh M, Singh SK, Bhatt R, Singh M, Seth CS, Li YR. (2024c). Climate change adaptation: Challenges for agricultural sustainability. *Plant, Cell Environ.* https://doi.org/10.1111/pce.15078.

Verma KK, Song XP, Li DM, Singh M, Wu JM, Singh RK, Sharma A, Zhang BQ, LI YR. (2022f). Silicon and soil microorganisms improves rhizospheric soil health with bacterial community, plant growth, performance and yield. *Plant Signal. Behav.* 17(1), e2104004.

Verma KK, Song XP, Lin B, Guo DJ, Singh M, Rajput VD, Singh RK, Singh P, Sharma A, Malviya MK, Chen GL, Li YR. (2022g). Silicon induced drought tolerance in crop plants: Physiological adaptation strategies. *Silicon*. 14(6), 2473–2487. https://doi.org/10.1007/s12633-021-01071-x.

Verma KK, Song XP, Rajput VD, Boldyreva V, Zhang BQ, Minkina T, Li YR. (2022h). Morpho-physiological, biochemical and ultrastructural modifications on sugarcane to prolonged water deficit. In: Verma KK, et al. (Eds.), *Agro-Industrial Perspectives on Sugarcane Production Under Environmental Stress*. https://doi.org/10.1007/978-981-19-3955-6_11.

Verma KK, Song XP, Rajput VD, Solomon S, Li YR, Rao GP. (2022i). *Agro-Industrial Perspectives on Sugarcane Production under Environmental Stress*. Springer Nature: Singapore, https://doi.org/10.1007/978-981-19-3955-6.

Verma KK, Song XP, Singh M, Tian DD, Rajput VD, Minkina T, Li YR. (2023c). Association of silicon and soil microorganisms induces stress mitigation, increasing plant productivity. In: de Mello Prado R. (Ed.), *Benefits of Silicon in the Nutrition of Plants*. pp. 299–328. https://doi.org/10.1007/978-3-031-26673-7_17.

Verma KK, Song XP, Tian DD, Guo DJ, Chen ZL, Zhong CS, Nikpay A, Singh M, Rajput VD, Singh RK, Minkina T, Li YR. (2021a). Influence of silicon on biocontrol strategies to manage biotic stress for crop protection, performance and improvement. *Plants*. 10, 2163. https://doi.org/10.3390/plants10102163.

Verma KK, Song XP, Tian DD, Singh M, Verma CL, Rajput VD, Singh RK, Sharma A, Singh P, Malviya MK, Li YR. (2021b). Investigation of defensive role of silicon during drought stress induced by irrigation capacity in sugarcane: Physiological and biochemical characteristics. *ACS Omega*. 6, 19811–19821.

Verma KK, Song XP, Verma CL, Chen ZL, Rajput VD, Wu KC, Liao F, Chen GL, Li YR. (2021c). Functional relationship between photosynthetic leaf gas exchange in response to silicon application and water stress mitigation in sugarcane. *Biol. Res.* 54, 15. https://doi.org/10.1186/s40659-021-00338-2.

Verma KK, Song XP, Verma CL, Malviya MK, Guo DJ, Rajput VD, Sharma A, Wei KJ, Chen GL, Solomon S, Li YR. (2021d). Predication of photosynthetic leaf gas exchange of sugarcane (*Saccharum* spp.) leaves in response to leaf positions to foliar spray of potassium salt of active phosphorus under limited water irrigation. *ACS Omega*. 6, 2396–2409.

Verma KK, Song X-P, Xu L, Huang H-R, Liang Q, Seth CS, Li Y-R. (2023d). Nano-microplastic and agro-ecosystems: A mini-review. *Front. Plant Sci.* 14, 1283852. https://doi.org/10.3389/fpls.2023.1283852.

Verma KK, Song XP, Yadav G, Degu HD, Parvaiz A, Singh M, Huang HR, Mustafa G, Xu L, Li YR. (2022j). Impact of agroclimatic variables on proteo-genomics in sugarcane (*Saccharum* spp.) plant productivity. *ACS Omega*. 7, 22997–23008.

Verma KK, Song XP, Zeng Y, Guo DJ, Sing M, Rajput VD, Malviya MK, Wei KJ, Sharma A, Li DP, Chen GL, Li YR. (2021e). Foliar application of silicon boosts growth, photosynthetic leaf gas exchange, antioxidative response and resistance to limited water irrigation in sugarcane (*Saccharum officinarum* L.). *Plant Physiol. Biochem*. 166, 582–592.

Verma KK, Song XP,Zeng Y, Li DM, Guo DJ, Rajput VD, Chen GL, Barakhov A, Minkina TM, Li YR. (2020c). Characteristics and correlation of leaf stomata and its relationship with photosynthesis on *Saccharum* spp. under different irrigation and silicon application. *ACS Omega*. 5, 24145–24153.

Verma KK, Vatsal S, Singh M, Rajput VD, Minkina T. (2022k). Coping with water stress: Effects and adaptive mechanisms to plants. In: *Understanding Abiotic Stresses*. Nova Science Publishers, Inc.: New York.

Verma KK, Wu KC, Verma CL, Li DM, Malviya MK, Singh RK, Singh P, Chen GL, Song XP, Li YR. (2020d). Developing mathematical model for diurnal dynamics on photosynthetic responses in sugarcane responsive to different irrigation and silicon application. *PeerJ*. 8, e10154. https://doi.org/10.7717/peerj.10154.

Verma KK, Liu X-H, Wu K-C, Singh RK, Song QQ, Malviya MK, Song X-P, Singh P, Verma CL, Li YR. (2020e). The impact of silicon on photosynthetic and biochemical responses of sugarcane under different soil moisture levels. *Silicon*. 12, 1355–1367.

Verma KK, Wu K-C, Singh P, Malviya MK, Singh RK, Song X-P, Li YR. (2019b). The protective role of silicon in sugarcane under water stress: Photosynthesis and antioxidant enzymes. *Biomed J. Sci. Tech. Res*. 15(2), 11301–11307. https://doi.org/0.26717/BJSTR.2019.15.002685.

Wu K-C, Huang C-M, Verma KK, Deng Z-N, Huang H-R, Pang T, Cao H-Q, Luo H-B, Jiang S-L, Xu L. (2022). Transcriptomic responses of *Saccharum spontaneum* roots in response to polyethylene glycol-6000 stimulated drought stress. *Front. Plant Sci*. 13, 992755. https://doi.org/10.3389/fpls.2022.992755.

Xu L, Deng Z-N, Wu K-C, Malviya MK, Solanki MK, Verma KK, Pang T, Li Y-J, Liu X-Y, Kashyap BK, Dessoky ES, Wang W-Z and Huang H-R. (2022). Transcriptome analysis reveals a gene expression pattern that contributes to sugarcane bud propagation induced by Indole-3-Butyric Acid. *Front. Plant Sci*. 13, 852886. https://doi.org/10.3389/fpls.2022.852886.

Zhang Y, Wei X, Zhuang D, He Y, Su Y. (2016). Dynamic analysis of energy input–output and energy efficiency for biodiesel production from soybean oil. *Energy Convers Manag.* 129, 47–56.

Zheng F-J, Lin B, Yang Y-X, Fang X-C, Verma KK, Chen G-L. (2023). Efficacy and functionality of sugarcane original vinegar on mice. *Front. Microbiol*. 14, 1224666. https://doi.org/10.3389/fmicb.2023.1224666.

Zhu X, Thakker C, Santra S, Zhang Y. (2016). Review on landfill gas emissions and its energy utilization: Focus on low carbon generation. *Renew. Sustain. Energy Rev*. 61, 317332.

CHAPTER 4

Improvement of Sugarcane through Advanced Breeding Approaches

RIKNO HARMOKO,[1] NURMALASARI DARSONO,[1] and BAMBANG SUGIHARTO[2,3*]

[1]*Research Center for Genetic Engineering, Research Organization for Life Sciences and Environment, National Research and Innovation Agency, Cibinong, Bogor, Indonesia*

[2]*Laboratory of Molecular Biology and Biotechnology, Center for Development of Advanced Science and Technology (CDAST), Jember University, Jember, Indonesia*

[3]*Department of Biology, Faculty of Mathematics and Natural Sciences, Jember University, Jember, Indonesia*

*Corresponding author

ABSTRACT

Sugarcane (*Saccharum* spp.) is a primary source of sugar and the leading feedstock for biofuel production. As a C_4 plant, sugarcane exhibits high efficiency in photosynthesis, water, and nitrogen use, leading to substantial accumulation of sugar and biomass. The industry generates significant by-products, primarily bagasse and straw, which are potential sources of lignocellulose. Soluble sugars from sugarcane juice and molasses are currently fermented to produce first-generation bioethanol. Rising global fuel demand has spurred the development of 2G biofuels derived from lignocellulosic biomass. However, large-scale production remains economically challenging due to the economically pretreatment required to release cellulose from the complex biomass. Developing sugarcane

Sugarcane Transformation: Biofuels, Biotechnology, and Beyond.
Krishan K. Verma, Xiu-Peng Song, Hai-Rong Huang, Aradhna Kumari, and Yang-Rui Li (Eds.)

DOI: 10.1201/9781779641571-4

cultivars with high sugar yields and easily degradable biomass is crucial for meeting biofuel demand. Conventional cross-breeding is hindered by the plant's complex polyploid genome and poor seed fertility, making it a lengthy and complex process. Advances in molecular biology have enabled advanced breeding approaches, including gene overexpression, RNA interference, and genome editing, which allow for precise modifications of the plant's DNA. This chapter summarizes the application of these techniques to develop sugarcane with traits beneficial for biofuel production, such as increased sugar, biomass, and lipid content, and improved biomass degradability.

4.1 INTRODUCTION

Sugarcane (*Saccharum* spp.) is among the valuable industrial crops, contributing nearly 80% of sugar production worldwide. The sugarcane-based industry also generates valuable by-products such as waxes, fibers, paper, plywood, board, molasses, rum, and bagasse (Chen et al., 2020; Guo et al., 2021, 2022; Verma et al., 2023a, 2023b; Huang et al., 2024). Interestingly, sugarcane is among the top 10 crops that can serve as biofuel sources. Sugarcane juice and molasses are the substrates of fermentation that produce the first generation of bioethanol. However, the bagasse and leaves biomass are sources of cellulose that can be substrates for the second generation of bioethanol (Huang et al., 2022; Zheng et al., 2023; Bhatt et al., 2023a, 2023b).

Physiologically, sugarcane is a C_4 plant that uses light, water, and nitrogen more efficiently than the C_3 plant. The C_4 plant is recognized by the Kranz anatomy, a specialized cell structure in which bundle-sheath cells are surrounded by mesophyll cells in a ring-like fashion in the cross-section of its leaves (Bosabalidis et al., 1994; Singh et al., 2020; Verma et al., 2020a, 2020b; Chen et al., 2023; Deng et al., 2023). In the C_4 plant, CO_2 fixation and CO_2 decarboxylation are spatially separated in different compartments. CO_2 is prefixed in mesophyll cells in the form of HCO_3^- by the activity of carbonic anhydrases and then converted to C_4 acid oxaloacetate (OAA) by the activity of phosphoenolpyruvate carboxylase (PEPC). Subsequently, aspartate aminotransferase or malate dehydrogenase (MDH) reduces OAA to aspartate or malate (Li et al., 2023; Guo et al., 2023; Singh et al., 2023). The C_4 carbon, aspartate or malate, is translocated to the bundle sheath cells via diffusion to plasmodesmata and then undergo decarboxylation by MDH or nicotinamide adenine dinucleotide phosphate-malic enzyme to provide high CO_2 concentration around ribulose 1,5-bisphosphate carboxylase/oxygenase (Rubisco). The concentrated CO_2 in bundle sheath cells provides an abundant substrate for Rubisco to perform a carboxylase rather than an oxygenase reaction, allowing the C_4 plant to continuously assimilate CO_2 into carbohydrates even under full sunlight in the daytime. The spatial separation of CO_2 fixation and decarboxylation in C_4 plants suppress photorespiration, making C_4 plants have higher photosynthesis efficiency than C_3 plants (Schlüter and Weber, 2020; Verma et al., 2019a, 2019b; 2021a; 2023c, 2023d).

Moreover, PEPC displays a high affinity for HCO_3^- and lacks oxygenase activity, enabling C_4 plants to fix CO_2 even at low atmospheric concentrations when stomata are nearly closed in the daytime. This unique capability allows C_4 plants to assimilate CO_2 and simultaneously minimize transpiration, increasing water use efficiency (Bräutigam and Gowik, 2016; Taylor et al., 2010; Malviya et al., 2021, 2022; Oliveira et al., 2022; Li et al., 2023). Because Rubisco works efficiently under the CO_2 concentrating mechanism, C_4 plants require less Rubisco in their leaves, leading to nitrogen saving (Bräutigam and Gowik, 2016; Sharma et al., 2021, 2022; Song et al., 2019, 2021, 2022). Rubisco is well known as the most abundant protein in leaves, where it may contribute up to 50% and 30% of soluble protein in the C_3 and C_4 plants, respectively (Spreitzer and Salvucci, 2002; Sugiyama et al., 1984; Wu et al., 2022; Xu et al., 2022). The leaves of the C_4 plant contain about 40% less Rubisco than C_3 plants. This lower number of photosynthetic proteins, especially Rubisco, contributes to high nitrogen use efficiency in the C_4 plant (Bräutigam and Gowik, 2016; Verma et al., 2020c, 2020d, 2022a). The CO_2 concentrating mechanism that represses photorespiration gives an advantage to C_4 plants by increasing radiation use efficiency (Verma et al., 2021b, 2021c, 2022b, 2022c, 2022d). The efficiency of solar energy conversion to biomass is 6% for C_4 plants and only 4.6% for C_3 plants, indicating that C_4 plants are superior biomass accumulators (Zhu et al., 2010; Singh et al., 2021; Anas et al., 2021a, 2021b).

The unique characteristics of C_4 plants, such as their conversion efficiency of light into biomass energy, high photosynthesis efficiency, high water and nitrogen-use efficiency, and relative tolerance to marginal land, make them ideal for energy crops (Verma et al., 2020e, 2020f; 2021d, 2021e). Energy crops are plant species cultivated for biofuel production, including biogas, biodiesel, and bioethanol (Dida, 2024). Among the C_4 plants that stand out for bioenergy production are sugarcane, maize (*Zea mays*), elephant grass (*Pennisetum purpureum*), sweet sorghum (*Sorghum bicolor*), miscanthus (*Miscanthus giganteus*), foxtail millet (*Setaria italica*), aleman grass (*Echinochloa polystachya*), and switchgrass (*Panicum virgatum*) (Byrt et al., 2011; Tian et al., 2023; Solanki et al., 2023).

Sugarcane is probably the most suitable energy crop due to its high conversion efficiency of light into biomass and stored energy. It has a potential yield of up to 98 t of biomass/ha/year (Irvine, 1978; Waclawovsky et al., 2010; Verma et al., 2022e, 2022f, 2022g). The sugarcane biomass is a combination of sucrose and lignocellulose. Sucrose, which accounts for 14–42% of dry weight, is stored in the internodal tissues (Whittaker and Botha, 1997). The remaining biomass is lignocellulose, which forms the sugarcane cell wall (Pereira et al., 2015). The estimated sugar yield from sugarcane stems can reach 17.4 t/ha (Renouf et al., 2008; Verma et al., 2022h, 2022i). The first generation of bioethanol is produced from the soluble sugar in sugarcane, while the second generation can be derived from the cellulosic components of bagasse and leaves trash. The bioethanol yield from sugarcane soluble sugar ranges from 6000 to 9381 L/ha (Lee and Bressan, 2006; Sánchez and Cardona, 2008), while bioethanol from bagasse and leaves trash is about 3000 L/ha (Somerville et al., 2010). Globally, ethanol production exceeds 66 billion L/year, 60% of which are derived from sugarcane (Balat and Balat, 2009; Goldemberg and Guardabassi, 2010).

Population growth, industrialization, and transportation have driven up global energy demand (Kumari et al., 2022, 2023a, 2023b; Chaudhary et al., 2023). Fossil fuels such as coal and oil have become primary sources of energy production to meet this demand (Yadav et al., 2019; Moodley, 2021). However, the extensive use of coal and oil has led to the depletion of fossil fuel reserves, significantly increased greenhouse gas emissions, and elevated global temperatures, leading to environmental degradation (Aditiya et al., 2016; Day and Day, 2017). In light of this concern, biofuels have gained significant interest in replacing fossil fuels due to their sustainability and environmental safety (Moodley, 2021). Commercial bioethanol production from sugarcane is the world's most successful biofuel production system, and 60% of global bioethanol production is derived from sugarcane (Balat and Balat, 2009; Goldemberg and Guardabassi, 2010).

Soluble sugars from sugarcane stems are the dominant source of bioethanol production. Recently, lignocellulosic biomass from bagasse and leaf straw is also used for bioethanol production. However, the production cost of bioethanol from lignocellulosic biomass is relatively higher than soluble sugar because pretreatment to remove recalcitrant components, like lignin, increases the cost of bioethanol production, which hinders large-scale lignocellulosic bioethanol production and commercialization (Hailing and Simms-Borre, 2008; Yuan et al., 2008; Joshi et al., 2022, 2023). The challenges of increasing bioethanol production to meet global demand from both sugarcane soluble sugar and lignocellulosic biomass are urgent and require immediate attention. Generating sugarcane cultivars with high biomass yield and easy biomass degradability is urgently required to handle this challenge (Verma et al., 2022j, 2022k).

The first sweet sugarcane cultivar is *Saccharum officinarum*, domesticated from *Saccharum robustum* at 8000–10,000 years ago in New Guinea (D'Hont et al., 2002; Healey et al., 2024). Together with *Saccharum sinense* and *Saccharum barberi*, *S. officinarum* is cultivated commercially as a sugar-producing plant until the early 20th century (D'Hont et al., 2002). The genus *Saccharum* is well known to have highly polyploid and aneuploid with a complex genome structure. For example, the genome of *Saccharum spontaneum* is octoploid ($x = 8$), containing a total chromosome number varying from 40 to 128. In comparison, *S. officinarum* is a decaploid ($x = 10$) with a total chromosome number of 80 (Piperidis and D'Hont, 2020). The genome of the sugarcane modern cultivar is derived from a few interspecific hybrids between a decaploid *S. officinarum* and octoploid *S. spontaneum*, resulting in more polyploid chromosomes ranging from 100 to 120 (Garsmeur et al., 2018; Piperidis and D'Hont, 2020; Healey et al., 2024).

In addition, the sugarcane genome contains a high frequency of gene duplication, transposable elements, and uncharacterized intergenic/intronic sequences that make sugarcane more complex than all other crops (de Setta et al., 2014). Generally, sugarcane is vegetatively propagated from axillary buds, but sugarcane will flower and produce few seeds with poor fertility and short longevity when cultivated in certain latitudes (Pierre et al., 2015). The complex genome, poor seed fertility, and high polyploidy–aneuploidy chromosome of sugarcane are bottlenecks for generating elite cultivars using conventional breeding techniques. Introducing desirable alleles into a sugarcane elite cultivar by cross-breeding will be laborious work, take many years, and require a very long duration of breeding.

The recent development in genomics has led to the advanced breeding approach, a breakthrough that allows the modification of the DNA sequence of plant cells. This advancement is particularly significant in sugarcane breeding, as it overcomes the challenges posed by seed fertility and the polyploidy level of the sugarcane chromosome. The advanced breeding approach, such as genome editing, cisgenesis/intragenesis, and RNA interference, was developed based on advances in molecular genetics, molecular biology, and plant tissue culture. These techniques have accelerated the breeding process and allowed for the development of sugarcane cultivars with traits specifically beneficial for bioethanol production, such as high sucrose content and increased biomass yield. Later, we review the progress of the advanced breeding approach in developing sugarcane cultivars in the context of bioethanol production.

4.2 SUGARCANE TISSUE CULTURE

Plant tissue culture techniques are a prominent biotechnological tool widely used in plant science research, from basic studies to commercial applications. Plant tissue culture is a process that uses plant material (plant cells, tissues, organs, seeds, or other plant parts) in a growth medium to grow new plantlets. The initial plant material is cultivated and developed in a specific and tightly controlled environment. A significant development in plant tissue culture is the regulation of morphogenesis through the use of various levels and combinations of growth regulators (Skoog and Miller, 1957). This enables the regeneration of entire plants, providing the opportunity to use *in vitro* systems for studying basic components of cell growth and differentiation, as well as for various applications of tissue culture.

Different approaches have been developed in plant tissue culture, including the use of meristem culture to obtain virus-free plants (Salokhe, 2021), the demonstration of totipotency in haploid cells or gametophytes, which allows for faster generation of isogenic lines essential for plant breeding programs (Guha and Maheswari, 1966), and the protection of hybrid embryos to get over plant species sexual incompatibility (Raghavan, 2003). The process of producing protoplasts involves enzymatic degradation of the cell walls of plant cells, followed by fusion of these naked cells to eliminate sexual incompatibility between different plant species.

This results in the production of intraspecific or interspecific somatic hybrids (Fathi and Jahani, 2012). Another application of plant tissue culture is to produce secondary elements by plant organ (Turi et al., 2014).

The availability of genetic diversity is essential in breeding programs to promote production sustainability. Breeders of plants have made extensive use of genetic variation from different genes to increase genetic diversity. Genetic diversity observed in cells and callus tissue culture is somaclonal variation that can lead to genetic or epigenetic changes and bring out favorable agronomic or production traits in breeding. Selection of resistant cells as a result of controlled selective stress will produce thousands or millions of cells that are part of the callus or cell suspension after propagation in culture. The recovered resistant cells can regenerate the entire resistant plant when grown optimally in culture medium. Consequently, it is feasible to produce plants that are resistant to drought, salinity, cold, or biotic stress, which can increase crop yields (Verma et al., 2024a, 2024b, 2024c).

Preservation of genetic diversity is critical in breeding programs to ensure sustainable production. Genetic manipulation of traits of interest can be achieved through the use of physical and chemical mutagens, as well as agents of epigenetic like histone deacetylase inhibitors and DNA demethylases. These agents are often used in combination with *in vitro* techniques to induce genetic rearrangement and epigenetic reprogramming through mutation induction, DNA and histone methylation, and acetylation of histone (Bridgen et al., 2018). According to Lu et al. (2024), sugarcane breeding entered the 2.0 era in the early 20th century, which was characterized by crossing applications, quantitative genetic and statistical techniques to choose better cultivars or plant species according to Mendel's theory of plant genetics. In this era, mutation treatment by chemical mutagenesis or radiation on seeds or shoots was used to obtain sugarcane varieties with better characteristics by selecting superior mutants.

Plants that are homozygous or isogenic are valuable resources for breeding initiatives because they are used as parent strains to produce hybrid seed that will produce high yields when grown. The development of isogenic strains using traditional breeding techniques alone can require 5–10 fertilization cycles. Today, plant tissue culture has advanced and is capable of producing haploid and double haploid plants through microspore and anther culture. By microspore or anther culture, it develops isogenic strains that can be dramatically reduced because haploid plants may be redeveloped in only one culture cycle and then doubled by colchicine treatment to obtain double haploid plants with a fixed set of homozygous chromosomes (Alvarez and Guzmán, 2018). Anther or microspore culture can also be used to improve the characteristics of hybrid plants developed by parental crosses and traditional strategies.

It is possible to recover hybrid plants from sexually compatible species by using *in vitro* culture embryo protection. Following the cross-pollination of two distinct species, hybrid embryo development occurs, but the endosperm does not always accompany the entire seed development process, and at some stage the hybrid embryo fails; it is then that the embryo may be preserved and cultured for continued development (Li et al., 2018; Akmal, 2021). By applying physicochemical techniques to combine protoplasts from two distinct sources, somatic hybridization can also be used to generate intra- or interspecific hybrids from sexually incompatible plant species. In order to generate new hybrid plants, the hybrid cells are cultivated.

4.3 SUGARCANE GENETIC TRANSFORMATION

Perhaps the most significant advancement in plant tissue culture is the advancement and implementation of genetic transformation approaches through *Agrobacterium tumefaciens* infection and particle bombardment, which enables genetic manipulation of plant species. New technologies, such as genome editing in addition to culture of tissues and contamination of *Agrobacterium tumefaciens*, offer promising alternatives for highly specific genetic manipulations of agronomic or industrial interest in crop plants. The use of tissue culture techniques has been essential in introducing genetic information into plant cells. This has allowed for the production of proteins, antibodies, vaccines, and other medicinal materials, as well as beneficial traits for higher productivity through the use of transgenic plants (Argaye, 2021; Krishna et al., 2022; Jagadesh et al., 2024). The emergence of improved and novel agricultural products that meet the growing demand in productive systems worldwide has been made possible by the development of genetic engineering and molecular biology techniques during the second half of the last century (Nerkar et al., 2018). The most promising alternatives for genetically manipulating traits of interest are *Agrobacterium* transformation, advanced plant tissue culture, and genome editing approaches (Bridgen et al., 2018; Borisjuk et al., 2019).

The initial report of successful sugarcane genetic transformation was made by Chen et al. (1987) through protoplast transformation by polyethylene glycol-induced direct uptake of DNA. After Chen's successful achievement, Rathus and Birch (1992) reported the steady development of callus from electroporated sugarcane protoplasts. Subsequently, Bower and Birch (1992) produced transgenic sugarcane plants via microprojectile bombardment of suspension culture cells and embryogenic callus. The genetic transformation of sugarcane using *A. tumefaciens* was first reported by Arencibia et al. (1998). They optimized this biological system for sugarcane callus by using the reporter *gus* gene. Shortly thereafter, Enríquez-Obregón et al. (1998) reported glufosinate-resistant sugarcane generations produced by *A. tumefaciens*-mediated transformation of the *bar* gene encoding the phosphinothricin acetyltransferase from the broad-spectrum herbicide ammonium glufosinate. Currently, the two more commonly applied techniques for sugarcane transformation are biolistic or *Agrobacterium*-based (agrotransformation) embryogenic callus induced from shoot tip or spindle leaf (Nerkar et al., 2018).

Efficient genetic transformation protocols are required for biotechnological sugarcane breeding. These protocols should allow for the insertion of few or single copies of transgenes and ensure stable high-level expression of constitutive or tissue-specific transgenes during different seasons in the field. *Agrobacterium*-mediated transformation (AMT) is a more suitable approach than biolistic transformation because it is simple and easy to perform, allowing for single- or low-copy insertion events with minimal rearrangement. The process of AMT involves three stages: (1) establishing embryogenic callus culture from shoot tip or spindle leaf, (2) genetically transforming embryogenic callus using *A. tumefaciens*, and (3) regenerating callus, selecting and improving transformed cells, and recovering plants. Callus induction and regeneration are time-consuming processes that involve repeated culture activities. The process is also variety-dependent, which limits transformation efficiency (Basso et al., 2017).

Initially, the success of genetic transformation using the *A. tumefaciens* system was limited due to its strong dependence on genotype and cocultivation conditions, such as explants, strains, target genes, culture conditions, and media. Although from the second decade of the 2000s, efforts have been performed to enhance this technique, and a variety of transgenic sugarcane events have been produced using *A. tumefaciens*-mediated transformation (Budeguer et al., 2021). *A. tumefaciens* system method offers significant advantages for commercial purposes, such as reducing copy numbers and presenting fewer problems with the emphasis and instability of the transgene (Enríquez-Obregón et al., 1998; Budeguer et al., 2021). Although the *A. tumefaciens* system is more favorable due to its low copy number in commercial varieties of transformations, Jackson et al. (2014) state that the efficiency of sugarcane transformation using the biolistic method is better (Budeguer et al., 2021).

Limitations of biological methods can be overcome through direct delivery systems, such as "biolistics," which involves shooting foreign DNA into plant cells. The effectiveness of biolistic transformation in producing transgenic sugarcane has been demonstrated by numerous examples of transgenic sugarcane with commercially valuable traits that can be applied to a wide range of genotypes (Nerkar et al., 2018). The success of this strategy depends on several advantages, including rapid gene transfer with high efficacy in specific and nonspecific tissues, no host restrictions, and no need for vectors (Radhesh Krishnan and Mohan, 2017). However, there are also some main disadvantages to consider, such as the high probability of integrating multiple copies of the transgene into the genome and the long and complicated tissue culture procedure. This procedure can lead to genome changes, genetic rearrangements, or transgene silencing. The outcome of transgenic events with particle bombardment is influenced by several factors, including the nature of the explant, the volume and integrity of DNA, the particle size of gold/tungsten, degree of pressure, and the distance between the bombardment and the target tissue (Table 4.1).

Before deciding on a method for genetic transformation experiments in sugarcane, it is crucial to determine the genotype's susceptibility to *A. tumefaciens* transformation and evaluate the efficiency of both methods. To compare the transformation efficiency between the two methods, an integrated efficiency parameter must be established. The quantity of plant material used as explants (input) and the quantity of transgenic plants (independent events) that are regenerated following the selection step (output) should be taken into account by this parameter. Determining transformation efficiency solely based on the number of plants regeneration after the selection procedure is unreliable. High selection pressure can lead to lower estimation, while less stringent selection can result in false positives.

TABLE 4.1 Transformation Method and Efficiency Rate from Biolistic and *Agrobacterium*-Mediated in Sugarcane During the Last Decades.

Sugarcane Genotype	Gene of Interest	Method of Transformation	Explants	Transgene No. of Copy	Transformation Efficiency	References
Q117	*Luciferase* reporter gene	*Agrobacterium*-mediated	Embryogenic callus	1–4	1%	Jackson et al. (2013)
Co 62175, Co 6304, Co 8021, Co 86032, Co 6907	*GUS* reporter gene	*Agrobacterium*-mediated	Nodal cutting (from 6-month-old plants)	1–3	32.6%	Mayavan et al. (2015)
RA 87-3	*EPSPS* gene	Biolistic	Embryogenic callus	6–9	2.1%	Noguera et al. (2015)
ROC 22	*Bar* gene	*Agrobacterium*-mediated	Embryogenic callus	1–5	11%	Wang et al. (2017a)
Badila	*CP SCMV*	Biolistic	Embryogenic callus	1–3	0.58%	Yao et al. (2017)
ROC 22	*Cry1Ab-EPSPS*	*Agrobacterium*-mediated	Embryogenic callus	0–7	7.3%	Wang et al. (2017a)
SPF-234, NSG-311	*shRNA CP SCMV*	Biolistic	Embryogenic callus	1	29%	Aslam et al. (2018)
Saccharum spp. hybrids	*Bar* gene	Biolistic	Leaf roll disc, embryogenic callus	1–8	39%	Ramasamy et al. (2018)
Co 86032	*GUS* reporter gene	*Agrobacterium*-mediated	Embryogenic callus	1–3	35.8%	Sathish et al. (2020)

Only two review articles have conducted direct efficiency in sugarcane comparisons between *A. tumefaciens* and biolistic transformation. A review by Jackson et al. (2013) stated that the biolistic method produces better efficiency than the biological method using *A. tumefaciens*. On the other hand, a second review by Wu et al. (2015) found that no significant variation in transformation efficiency between the two methods. Taken together, these results indicate that there is less or no difference and that both are equally likely to produce transformants suitable for scientific experimentation or industrial release. A direct comparison of AMT and particle bombardment applied whole plasmids and minimally truncated cassettes (MC) for transformation efficiency, complexity of transgene integration, and transgene expression was conducted in sugarcane plants according to Jackson et al. (2013), applied gene selectable markers and luciferase reporters. It monitored that, predicting a preference for low transformant copies with high expression, the preferred method is particle bombardment using MC at low DNA concentrations (6.6 ng per injection in our example) or AMT (in amenable genotypes). Both methods produced multiple copies of a single transformant with luciferase reporter gene activity of about 108 units/μg of the cellular protein, in a population size of transgenic sugarcane plants that is easy to produce (Table 4.1).

For breeding purposes, it is advisable to obtain low copy number transgenic events, which is also a significant strategy when applying at genetic level for commercial release. Numerous transformation events in sugarcane that were produced by *A. tumefaciens* have been demonstrated to be single transgene events, suggesting that a significant proportion of single gene fusions were produced using this approach. More reports of *A. tumefaciens*-mediated transformation have been analyzed by Southern hybridization on a limited number of plants. However, few studies have been performed in which a large number of transgenic aspects have been observed by Southern blot for *A. tumefaciens* and the transformation of biolistic. In 146 *A. tumefaciens*-transformed events, Mayavan et al. (2015) studied the transgene copy number and reported that 25% carried a single copy and 75% carried two or more copies. In an associated investigation, 15% of single-copy events and 75% of plants with 2–7 copies were identified after 33 transgenic events generated by *A. tumefaciens*-mediated transformation (Wang et al., 2017b). More recently, 55 transgenic plants produced by biolistic transformation were assessed by Ramasamy et al. (2018); of these, 30% had single-copy events and 70% had 2–7 copies.

It suggests that there is little difference between the two methods and that single insertion events are common in both techniques.

The level of transgene expression is a crucial factor in the success of generating transgenic events and introducing desired traits. The success of this process relies on various factors, some of which can be controlled by the researcher, such as promoter selection, gene nucleotide optimization, and codon usage. While other important variables, such as insertion location in gene silencing and plant genome, cannot be manipulated. Several studies have shown that in sugarcane, there is little to no correlation between the copy number of the inserted transgene and its expression level (Jackson et al., 2013; Joyce et al., 2014; Wu et al., 2015). This suggests that the main factor determining expression is correlated with the insertion site. Significant increase of more than twofold in nptII expression in sugarcane plants that were transformed using expression cassettes flanked by both insulator sequences, exogenous border sequence, and tobacco budworm sequence (Zhao et al., 2019). By inserting a Ubiquitin promoter next to a lox76 site, the researcher intended to define a region in the sugarcane genome for transgenic introduction, also known as gene stacking. It supports site-specific integration of promoterless construction of selectable markers with additional transgenes into lox76 sites to enable optimum and steady levels of gene expression and assure industial release through defined transgene integration sites (Zhao et al., 2019).

Obstacles to the nonexpression of inserted genes in transgenes (gene silencing) are often encountered in transgenic sugarcane. To overcome this barrier, several analyses have been conducted, including the use of strong promoters such as maize ubiquitin followed by rice ubiquitin (Nerkar et al., 2018). Sugarcane transgenes tested for field performance to date generally use maize Ubiquitin promoter. Jackson et al. (2014) conducted an analysis of transgene in sugarcane. They concluded that removing sequences involved in RNA instability, rare codons, and unwanted structural features, such as repetitive sequences, notably decreased silencing in the transformed sugarcane plants. To reduce gene silencing in transgenic events, another strategy is to incorporate post-transcriptional gene silencing suppressors, as suggested by Gao et al. (2013).

In summary, the evidence suggests that the industrially agreed transgenic sugarcane plants may be developed using either *A. tumefaciens* (in amenable genotypes) or biolistic-mediated transformation approaches. Both applications need the development of large numbers of transgenic events to identify low-copy number transgenic events that express the transgene at levels required for specific traits, while keeping the rest of the genome practically unchanged.

4.4 THE SUGARCANE GENOME

The modern sugarcane ($2n$ = 100–120) is developed from interspecific hybridization between *S. officinarum* ($2n$ = 80) and *S. spontaneum*. The basic chromosome number of *S. officinarum* is $x = 10$, while *S. spontaneum* has a varying chromosome number from x = 8–10. The *S. spontaneum* $x = 8$ results from the double-step rearrangement of $x = 10$ to $x = 9$ and $x = 8$, in which three chromosomes were rearranged into two chromosomes in each step. However, *S. spontaneum* with basic chromosome $x = 8$ has contributed to the development of the modern sugarcane cultivar (Piperidis and D'Hont, 2020). Crossing two species with different chromosome numbers creates a hybrid chromosome structure that differs from the parental species, in which modern sugarcane contains intact chromosomes inherited from the parental, interspecific recombinant chromosome, and inter-chromosome translocation products. The modern sugarcane cultivar has mainly 12 copies of chromosomes, where variations in the chromosome copies occur in interspecific recombinant chromosomes, and chromosomes originating from *S. spontaneum* (Piperidis and D'Hont, 2020).

The availability of a sugarcane reference genome is urgently required to assist the development of new cultivars and gene functional studies in sugarcane. Because grasses have a conserved organization structure, the BAC sequence of genomic sugarcane was aligned to the sorghum reference genome to obtain the first sugarcane reference genome. Sugarcane and sorghum are highly syntenic, indicating that the order of the DNA genome sequence is conserved in their chromosomes. This study has predicted that the sugarcane genome contains 25,316 protein-coding genes, and 17% of the proteins show no colinearity with their ortholog in sorghum (Garsmeur et al., 2018).

Significant progress has been made in developing the reference polyploidy genome of *S. spontaneum* and modern sugarcane cultivar based on the advancement of whole genome sequencing technology. The reference genome assembly of *S. spontaneum* is 2.7 Gb in size and covers 35,754 annotated gene models. The size of the polyploidy genome of the modern sugarcane cultivar R570 is 5.04 Gb, and is composed of 8.7 billion bases, containing 194,593 predicted protein-coding genes covering a complete representation sequence across 12 chromosome copies in the sugarcane genome (Healey et al., 2024). In addition, the reference genomes of *Erianthus fulvus* and *Erianthus rufipilus*, sugarcane relative species, are available online (Kui et al., 2023; Wang et al., 2023). The sugarcane reference genome provides an essential step in genomic and transcriptomic studies and assist the development of new cultivar using biotechnological approaches.

4.5 ADVANCE BREEDING APPROACH

Two factors that ensure biofuel sustainability are total biomass production and biomass conversion efficiency to fuel (Furtado et al., 2014). Both factors need to be optimized to maximize the energy generated from sugarcane. Soluble sugar and lignocellulose are the main components of biomass in sugarcane. Because sugar is the main product of sugarcane, sucrose content is the most essential trait and becomes the main subject of breeding programs for sugarcane variety improvement. Maximizing the sucrose content will significantly enhance the economic value of sugarcane as a biofuel industrial crop.

The enzymes sucrose phosphate synthase (SPS), sucrose synthase (SuSy), and invertase regulate the metabolism of sucrose in sugarcane. SPS catalyzes the transfer of the glycosyl group from uridine diphosphate-glucose to D-fructose 6-phosphate to form sucrose-6-phosphate, in which the phosphate group is then hydrolyzed by sucrose phosphate phosphatase to sucrose (Huber and Huber, 1996). SPS activity is positively correlated with the accumulation of sucrose rather than hexose sugars in the sugarcane stem (Verma et al., 2011), indicating that SPS is a critical enzyme in the biosynthesis and accumulation of sucrose. Because the sugarcane yield is determined by its sucrose content, the SPS activity is highly correlated to sugarcane productivity. SuSy catalyzes the reversible enzymatic reaction from sucrose to UDP-glucose and fructose. SuSy activity correlates positively with the accumulation of hexose sugars rather than sucrose, and its activity is higher in immature internodes than in mature internodes (Verma et al., 2011). SuSy tends to degrade sucrose into hexose sugar to provide energy for developing young tissues in sugarcane. Invertase catalyzes the breakdown of sucrose into glucose and fructose. The invertase is classified into neutral invertase (NI), active in the neutral to slightly alkaline range, and soluble acid invertase (SAI), localized at the vacuole and cell wall, active at acidic conditions (Sturm, 1999). Sucrose content is negatively correlated with SAI, NI, and SuSy activity, but only SAI has a significant negative correlation to sucrose content in sugarcane stems (Pan et al., 2009). As a result, the sucrose content in sugarcane stems is regulated by different sucrose synthesis and breakdown activities catalyzed by SPS and SAI, respectively (Zhu et al., 1997).

Lignocellulose, the main component of the cell wall, is a complex structure constituted by lignin, cellulose, hemicellulose, hydroxycinnamic acids, and pectin. Lignin is a three-dimensional biopolymer composed of three kinds of monomeric combinations of 4-hydroxyphenylpropanoids or monolignol, namely ρ-hydroxyphenyl (H), guaiacyl (G), and syringyl (S), known as monolignol. Phenylalanine or tyrosine, the precursor of monolignol biosynthesis, is converted to ρ-coumaric acid by phenylalanine ammonia-lyase (PAL) and tyrosine ammonia-lyase. The ρ-coumaric acid is then converted to caffeic acid and ferulic acid in a stepwise reaction catalyzed by 4-hydroxycinnamate 3-hydroxylase (C3H) and caffeic acid 3-*O*-methyltransferase (COMT). In the same way, a coupling reaction of ferulate 5-hydroxylase (F5H) and COMT enzymes converts ferulic acid to 5-hydroxy ferulic acid and sinapic acid. Subsequently, 4-coumarate:CoA ligase (4CL) catalyzes the joining of CoA to ρ-coumaric acid, ferulic acid, and sinapic acid to form ρ-coumaroyl CoA, feruloyl CoA, and Sinapoyl CoA. The ρ-coumaroyl CoA, feruloyl CoA, and Sinapoyl CoA undergo reduction catalyzed by cinnamoyl-CoA reductase (CCR), forming ρ-coumaroyl aldehyde, coniferyl aldehyde, and sinapyl aldehyde. Ultimately, cinnamyl alcohol dehydrogenase (CAD) catalyzed the oxidation reaction of ρ-coumaroyl aldehyde, coniferyl aldehyde, and sinapyl aldehyde to form ρ-coumaroyl alcohol (ρ-hydroxyphenol/H), coniferyl alcohol (guaiacyl/G), and sinapyl alcohol (syringyl/S) (Furtado et al., 2014; Vanholme et al., 2010). The G and S monolignol units are the main components of monocot and dicot plants lignin, but lignin from monocot accumulates higher H unit compared to dicot plant (Poovaiah et al., 2014). Lignin plays as intramolecular adhesive and provide rigidity to the cell walls structure in order to protect cells against pathogen attacks (Lee et al., 2019). High lignification of cell walls in xylem tissue reduces water permeability, enabling water transport over vertical distances in plants (Emonet and Hay, 2022) (Figure 4.1).

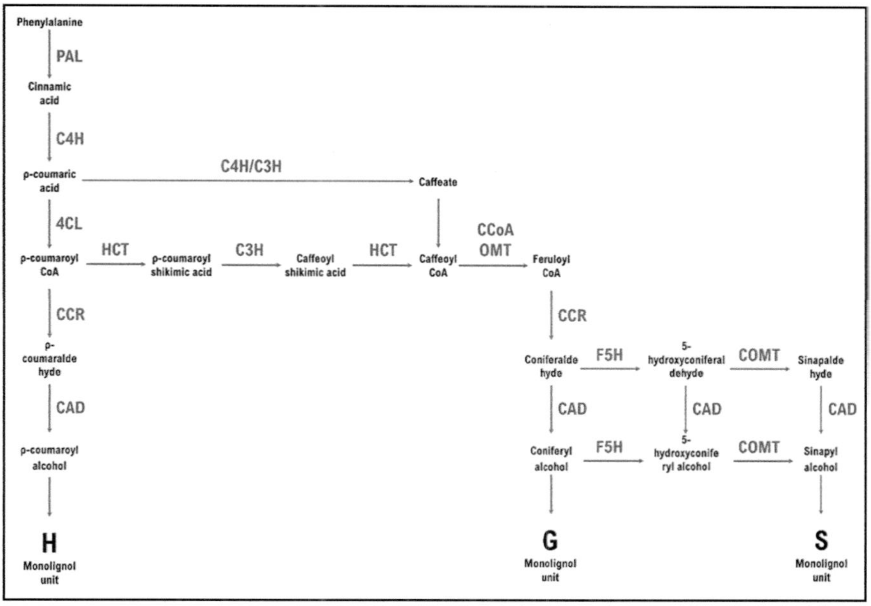

FIGURE 4.1 Lignin biosynthesis pathway.

Notes: PAL, phenylalanine ammonia-lyase; C4H, cinnamate 4-hydroxylase; CSE, caffeoyl shikimate esterase; 4CL, 4-coumarate:CoA ligase; C3H, *p*-coumarate 3-hydroxylase; HCT, *p*-hydroxycinnamoyl-CoA:quinate/shikimate *p*-hydroxycinnamoyltransferase; CCoAOMT, caffeoyl-CoA *O*-methyltransferase; CCR, cinnamoyl-CoA reductase; F5H, ferulate 5-hydroxylase; COMT, caffeic acid *O*-methyltransferase; and CAD, cinnamyl alcohol dehydrogenase.

Source: Adapted from Poovaiah et al. (2014).

Second-generation biofuel production favors lignocellulosic-based material to avoid competition with food or feed resources. However, the cellulose in the cell wall is tough to convert to fermentable sugar due to its crystalline structure and shielding by lignin (Marriott et al., 2016). The lignin prevents the access of cellulose hydrolytic enzymes during enzymatic decomposition processes and microbial fermentation (Vanholme et al., 2010). The derivatives of lignin produced from its degradation, such as syringyl lignin, inhibit yeast growth during the fermentation process and negatively affect the growth of other microorganisms associated with bioprocessing (Keating et al., 2006). Moreover, lignin binds tightly to cellulase enzymes, inhibiting cellulase accessibility to release sugars from cellulose (Saini et al., 2016). Therefore, reducing lignin biosynthesis in sugarcane is urgently required to increase biomass conversion efficiency and reduce the production cost of lignocellulosic-based bioethanol.

Besides bioethanol production, biodiesel from renewable feedstock is also expected to provide environmental benefits. Biodiesel is generated from the lipid or triacylglycerol (TAG) of oil-seed crops, such as palm, rapeseed, sunflower, and safflower, which are currently the primary sources of biodiesel. TAG has more than twice the energy density of carbohydrates, thus becoming a suitable compound for long-term energy storage (Ohlrogge and Chapman, 2011). Expanding the cultivation of oil-seed crops to match the requirement of bioenergy could compete with food and feed supply in terms of using arable land use. Since vegetative plant tissues can synthesize and store lipids, increasing lipid accumulation in that tissue has emerged the new idea to provide sources for biodiesel production (Ohlrogge and Chapman, 2011). Plants transform light energy into chemical energy during the process of photosynthesis, and use the energy to synthesize triose phosphate in the chloroplast. Triose phosphate is the precursor for synthesizing lipids, starch, and protein in the biomass of vegetative tissues (Xu and Shanklin, 2016). Because sugarcane has a high capacity for photosynthesis and biomass accumulation, it becomes an ideal feedstock for producing lipids and TAG in its vegetative tissues.

Plants synthesize and store lipids in the form of TAG through the collaboration between the chloroplast and endoplasmic reticulum. Fatty acids are synthesized in the chloroplast, which are then assembled and stored in TAG in the ER. Increasing the lipid content in vegetative tissue can be achieved by manipulating the genes expression associated in the synthesis of lipids and storage in both the chloroplast and the ER. Vanhercke et al. (2014) describe a "push-pull-protect" strategy to achieve industrial relevance of storage lipid content in vegetative tissue. The strategy is applied by increasing the flux of carbon conversion into fatty acid (push), enhancing TAG assembly (pull), and suppressing lipid turnover (protect) through multigene engineering. Increased conversion of carbon to fatty acid in the chloroplast has been achieved by overexpressing fatty acid biosynthetic genes (*ACCase, acetyl-CoA carboxylase*) (Klaus et al., 2004), overexpressing transcription factors regulating fatty acid biosynthesis (*LEC1*, *LEC2*, *LEAFY COTYLEDON*; *WRI1*, *WRINKLED 1*) (Mu et al., 2008; Andrianov et al., 2010; Yang et al., 2015), and repressing starch synthesis gene regulators (*AGPase*, *ADP glucose pyrophosphorylase*). Enhancing TAG accumulation in the ER is associated with the esterification of DAG (diacylglycerol) to TAG catalyzed by diacylglycerol:acyl-CoA acyltransferase (DGAT) and monoacylglycerol acyltransferase 1 enzymes (Andrianov et al., 2010; Petrie et al., 2012). The storage TAG is protected by overexpression of *Oleosin* (*OLE*), a structural protein protecting lipid droplets, repressing the *peroxisomal ABC transporter 1* gene expression that transfers lipid to peroxisome for β-oxidation, and repressing *sugar-dependent 1* (*SDP1*) gene expression that hydrolyzes storage TAG (Kunz et al., 2009; Winichayakul et al., 2013).

Several advanced breeding approaches, such as gene overexpression, RNA interference, and genome editing, have been applied to optimize sugarcane traits to increase biomass production and biomass conversion efficiency to biofuel. Sugar content has been improved by increasing the expression of sugar biosynthetic genes, repressing the expression of sugar-degrading enzymes, and engineering a pathway to increase sucrose production flux in sugarcane. Several efforts have been made to break the recalcitrant of lignocellulose by reducing lignin content or engineering the S/G ratio of lignin to improve lignocellulose saccharification (Table 4.2).

4.5.1 GENE OVEREXPRESSION

Gene overexpression is a technique for obtaining high-level expression of a gene using a strong promoter or upstream activation sequences to obtain a particular phenotype or to study gene function. Because SPS plays a critical role in the synthesis and accumulation of sucrose in sugarcane, enhancing the SPS activity by gene overexpression has gained much interest in improving sucrose content in sugarcane. Overexpression of sugarcane *SPS* gene (*SoSPS*) regulated by 35S promoter increases sucrose content in sugarcane leaves, followed by enhancement in glucose and fructose levels because of the increasing activity SAI in transgenic sugar-cane leaves. These transgenic plants showed a significant increase in plant height and tiller number, indicating that glucose and fructose are utilized as energy and carbon sources for plant growth (Anur et al., 2020). In the field experiment, transgenic lines overexpressing *SoSPS* have higher tiller number, plant height, and Brix and Pol percentages than nontransgenic sugarcane (Suherman et al., 2022). In a similar work, overexpression of *SoSPS* under the regulation of maize ubiquitin promoter in *Brachypodium distachyon*, a monocot plant model, not only increases plant height in the developmental stage but also increases biomass up to 52% compared to

TABLE 4.2 Advance Breeding Approach and Gene Target for Increasing Sucrose, Biomass, and Lipid Content in Sugarcane

Advance Breeding Approach	Gene Target	Application Prospective	Transformation Method	References
Overexpression	Sugarcane sucrose-phosphate synthase gene (*SoSPS*)	Enhancing sucrose content and biomass production	*Agrobacterium*-mediated transformation	Anur et al. (2020)
	Sugarcane sucrose-phosphate synthase gene (*SoSPS*)	Enhancing sucrose content and biomass production	Field evaluation studies	Suherman et al. (2022)
	Sugarcane sucrose-phosphate synthase gene (*SoSPS*)	Enhancing sucrose content and biomass production	*Agrobacterium*-mediated transformation	Falter and Voigt (2016a)
	Sugarcane ferulate-5-hydroxylase/ Coniferaldehyde 5-Hydroxylase (F5H)	Reducing lignin content to improves lignocellulose saccharification	*Agrobacterium*-mediated transformation	Portilla Llerena et al. (2024)
	Sugarcane transcription factor SHINE/wax inducer (*ShSHN1*)	Reducing lignin content to improves lignocellulose saccharification	*Agrobacterium*-mediated transformation	Martins et al. (2018)
RNAi	Soluble acid invertase (*SCINVm*)	Enhancing sucrose accumulations	Biolistic gene transfer	Ma et al. (2000)
	Neutral invertase (*NI*)	Increasing sucrose to hexose ratio	*Agrobacterium*-mediated transformation	Rossouw et al. (2007)
	Neutral invertase (*NI*)	Increasing sucrose to hexose ratio	Biolistic gene transfer	Rossouw et al. (2010)
	Caffeic acid *O*-methyltransferase (*COMT*)	Reducing lignin content to improves cell wall saccharification	Biolistic gene transfer	Jung et al. (2012)
	Caffeic acid *O*-methyltransferase (*COMT*)	Reducing lignin content to improves cell wall saccharification	Field evaluation studies	Jung et al. (2013)
	4-coumarate:coenzyme A ligase (*Sh4CL1 and Sh4CL2*)	Reducing lignin content to improves cell wall saccharification	Biolistic gene transfer	Jung et al. (2016)
	Caffeoyl-CoA *O*-methyltransferase (*CCoAOMT*), ferulate 5-hydroxylase (*F5H*) Caffeic acid *O*-methyltransferase (*COMT*)	Reducing lignin content to improves cell wall saccharification	Biolistic gene transfer	Bewg et al. (2016)

TABLE 4.2 *(Continued)*

Advance Breeding Approach	Gene Target	Application Prospective	Transformation Method	References
	BAHD acyl-coenzyme A (*CoA*) transferase	Reducing lignin content to improves cell wall saccharification	*Agrobacterium*-mediated transformation	de Souza et al. (2019)
	Fructose 6-phosphate 1-phosphotransferase (*PFP*)	Enhancing sucrose accumulations	Biolistic gene transfer	Groenewald and Botha (2008)
	Fructose 6-phosphate 1-phosphotransferase (*PFP*)	Enhancing sucrose accumulations	Biolistic gene transfer	van der Merwe et al. (2010)
Gene overexpression and RNAi	WRINKLED1 (*WRI1*) Diacylglycerol acyltransferase1-2 (*DGAT1-2*) Oleosin1 (*OLE1*) ADP-glucose pyrophosphorylase (*AGPase*) Peroxisomal ABC transporter 1 (*PAX1*)	Elevating lipid content	Biolistic gene transfer	Zale et al. (2016)
	WRINKLED1 (*WRI1*) Diacylglycerol acyltransferase1-2 (*DGAT1-2*) Oleosin1 (*OLE1*) ADP-glucose pyrophosphorylase (*AGPase*)	Elevating lipid content	Field evaluation studies	Kannan et al. (2022)
	Diacylglycerol acyltransferase1-2 (DGAT1-2) Oleosin1 (*OLE1*) Sugar-dependent 1 (*SDP1*) Trigalactosyl diacylglycerol 1 (*TGD1*)	Elevating lipid content	Biolistic gene transfer	Luo et al. (2022)
Genome editing	Caffeic acid *O*-methyltransferase (*COMT*)	Reducing lignin content to improves cell wall saccharification	*Agrobacterium* and *Biolistic* gene transfer	Jung and Altpeter (2016)
	Caffeic acid *O*-methyltransferase (*COMT*)	Reducing lignin content to improves cell wall saccharification	Field evaluation studies	Kannan et al. (2018)
	LIM domain transcription factor *(LIM)*	Reducing lignin content to improves cell wall saccharification	*Agrobacterium* and *Biolistic* gene transfer	Laksana et al. (2024)

nontransgenic at a fully senesced plant (Falter and Voigt, 2016b). Therefore, *SPS* gene overexpression is a valuable strategy in advanced breeding approaches for improving sugarcane yield and biomass.

Cell wall lignocellulose is recalcitrant to physical and enzymatic deconstruction to release its glucose because lignin prevents the cellulase enzyme from accessing cellulose microfibril in the lignocellulose complex. So, reducing lignin content is a reasonable strategy for increasing lignocellulose digestibility. The composition or ratio of monolignol in lignin affects biomass digestibility, in which the S/G ratio positively correlates with biomass digestibility in some plant species (Huntley et al., 2003). The biosynthesis of the S unit involves the F5H and COMT enzymes; F5H converts coniferaldehyde/coniferyl alcohol to 5-hydroxyconiferaldehyde/5-hydroxyconiferyl alcohol, and COMT catalyzes the conversion of 5-hydroxyconiferaldehyde/5-hydroxyconiferyl alcohol to Sinapaldehyde/Sinapyl alcohol. In poplar, the biomass digestibility increases in the S-rich lignocellulose substrate derived from the F5H overexpression plant (Huntley et al., 2003; Cai et al., 2016).

Sugarcane *F5H* (*SoF5H*) overexpression controlled by *C4H* (*cinnamate 4-hydroxylase*) rice promoter significantly increased the S/G ratio and biomass digestibility in transgenic sugarcane leaves, but the lignin content in the same organ was relatively unchanged. Interestingly, the sugarcane transgenic stem's S/G ratio and sucrose content were not altered in the transgenic sugarcane. Thus, sugarcane straw can be used as a source of biomass for the second generation of bioethanol (Portilla Llerena et al., 2024). Overexpression of sugarcane *SHN* (*ShSHN*), a member of the *ethylene responsive factor* (*ERF*)/*APETALA2* transcription factor that regulates the formation of the secondary cell wall, has increased biomass production, decreased lignin concentration, and changed the composition of lignin, which led to increasing saccharification efficiency of up to 53% in transgenic rice. *ShSHN* overexpression changed the expression pattern of secondary cell wall biosynthesis genes, including lignin biosynthetic gene suppression (Martins et al., 2018). Using the same strategy, cisgenic overexpression of *ShSHN* in sugarcane may also improve biomass and saccharification efficiency.

4.5.2 RNA INTERFERENCE

RNA interference is a gene regulatory mechanism that limits gene expression by employing a complex of small interfering RNA (siRNA) and small protein effectors by a sequence-specific RNA degradation process. The RNAi mechanism is induced by a double-strand RNA chopped by endoribonuclease Dicer to 21 or 22 nucleotide siRNA. Subsequently, one strand of siRNA is incorporated into the protein effector, argonaute (AGO), to form an RNA-induced silencing complex (RISC). In this complex, siRNA guides AGO to a specific mRNA target by sequence-specific base pairing, which then induces RNA cleavage or translation inhibition, leading to mRNA decay (Sugiharto et al., 2022). The formation of dsRNA initiates the RNAi mechanism, which can be introduced to the plant using cosuppression, sense–antisense construct, and self-complementary hairpin. The cosuppression construct expresses genes in an antisense orientation to induce dsRNA formation with its endogenous sense transcript. Sense–antisense construct expresses genes in both sense and antisense orientation under the control of separate promoters to induce dsRNA inside the cell (Waterhouse et al., 1998). Cosuppression or sense–antisense construct usually generates low silencing efficiency in plants (Stoutjesdijk et al., 2002). The self-complementary hairpin construct, in which sense and antisense are separated by intron, is expressed under the control of a constitutive promoter, generates high silencing efficiency in the plant (Wesley et al., 2001).

Pyrophosphate fructose 6-phosphate 1-phosphotransferase (PFP) catalyzes the reversible reaction of fructose 6-phosphate (F6P) and pyrophosphate to fructose 1,6-bisphosphate (F1,6P2) and inorganic phosphate (Pi) (Carnal and Black, 1979). PFP is involved in glycolysis reaction by partially participating in the reversible conversion of triose phosphate and hexose phosphate pool. PFP activity is negatively correlated with sucrose accumulation and positively correlated with respiration in the internodal tissue of sugarcane (Whittaker and Botha, 1999). The suppression of PFP activity in sugarcane using the antisense construct of PFP-β controlled by maize ubiquitin promoter has induced the accumulation of sucrose in young internode, increased hexose phosphate to triose phosphate ratio and promoted fiber biosynthesis in both young and mature internode without altered plant morphological structure (Groenewald and Botha, 2008; van der Merwe et al., 2010), suggesting that lowering PFP activity may promote carbon flux to accumulate sucrose and biomass rather than glycolysis and respiration.

Because sucrose content in sugarcane is regulated by the activity of SPS and invertase, suppression of invertase expression by RNAi is one strategy that may suppress sucrose degradation flux and increase sucrose accumulation. Ma et al. (2000) generated an antisense construct of a 1-Kb *S. robustum* SAI (*SrSAI*) cDNA sequence controlled by an ubiquitin promoter to downregulate invertase activity in the sugarcane callus culture. The callus carrying the antisense construct showed moderate invertase activity and increased sucrose accumulation up to twofold. Rossouw et al. (2007) generated a sugarcane cell line transformed with antisense of neutral invertase to induce downregulation of NI expression by RNAi mechanism. The transgenic cell line has a higher sucrose-to-hexose ratio but showed growth retardation because the reduction of hexose availability for respiration may impair cell growth. Transgenic sugarcane plants were then generated from cell lines with reduced NI activity. The transgenic plant showed decreased hexose level and increased sucrose concentration, followed by increased SuSy activity in compensation of NI activity reduction. However, the respiration rate in transgenic plants was still significantly lower compared to untransformed plants, indicating that sucrose accumulation flux in transgenic plants is higher than its degradation (Rossouw et al., 2010).

Because fermentable sugar yield is negatively correlated with lignin content in biomass destruction processes, reducing lignin content and modifying lignin composition is an attractive strategy in developing sugarcane cultivars for biofuel feedstock. To avoid adverse effects on sugarcane growth and development, the engineering of lignin generally targets the gene in the terminal step of monolignol biosynthesis. Jung et al. (2012) employed the RNAi approach to reduce lignin content in sugarcane by targeting *COMT*, a gene responsible for the final step of S unit biosynthesis. Reduction of *COMT* expression 67–97% in the transgenic sugarcane led to a reduction of lignin 3.9–13.7%, and its lignocellulose degradation processes yield fermentable sugar up to 29% and 34% with or without acid pretreatments, respectively. Field evaluation showed that transcript suppression of *COMT* is stable in the field, leading to the suppression of lignin content from 6% to 12% with no variation in crop production and agronomic performance and increased saccharification efficiency up to 32% compared to the wild type plant (Jung et al., 2013). RNAi targeting sugarcane *4CL* (*Sh4CL1*), one of the key genes responsible for monolignols biosynthesis, reduced lignin content up to 16.5% without altered biomass yield, and lignocellulose saccharification efficiency improved by a range of 52–76% compared to wild type (Jung et al., 2016). Bewg et al. (2016) employed RNAi to individually reduce the expression *of F5H* and

COMT, which are responsible for lignin biosynthesis in the S branch. RNAi suppression *of F5H* and *COMT* has reduced the S/G ratio and increased fermentable sugar during lignocellulose enzymatic hydrolysis processes. Downregulation of one gene in monolignol biosynthesis reduces lignin content and improves the saccharification efficiency of sugarcane bagasse and straw, which may contribute to economically enhancing the production of the second generation of bioethanol.

4.5.3 COMBINATION OF GENE OVEREXPRESSION AND RNA INTERFERENCE

Metabolic engineering generally involves multiple target genes and various genetic engineering approaches to obtain the desired product from the plant. Zale et al. (2016) reported that metabolic engineering for accumulating TAG in sugarcane vegetative tissue involves five gene targets arranged in three plasmids. The first plasmid carrying overexpression cassette of sugarcane codon-optimized *AtWRI* and *nptII* controlled by *OsUbi* with its first exon and intron and *CaMV35S* promoter with maize HSP70 first intron, respectively. The second plasmid carrying an overexpression cassette of codon-optimized *ZmDGAT1* and *AtOLE1* subcloned under the control of maize ubiquitin with the first intron and *CaMV35S* promoter with maize *HSP70* first intron, respectively. The third plasmid is RNAi construct with dual suppression to *ShAGPase* and *ShPAX1*, carrying a hybrid sequence of 200 bp conserved DNA sequence of each target gene, followed by *Pn4CL* intron and respective antisense of hybrid sequence. The gene overexpression and RNAi suppression cassettes were released from plasmids and transferred to sugarcane using the biolistic method. The transgenic sugarcane line carrying the cassettes has accumulated TAG ranging from 43 to 95-fold or 0.9–1.9% of the weight. The *AtWRI*, *ZmDGAT1*, *AtOLE1* overexpression, and *ShAGPase* and *ShPAX1* RNAi suppression were stable during two years of field evaluation, resulting in TAG uptake nearly 4% plant dry leaves, which is 70-fold higher than wild type plants (Kannan et al., 2022). In a similar approach, Luo et al. (2022) demonstrated that metabolic engineering for increasing TAG accumulation in sugarcane vegetative tissue involves overexpression of *ZmDGAT1* and *SiCysOLE1* and suppressing *TGD1* and *SDP1* expression. Overexpression cassette of *ZmDGAT1* and *SiCysOLE1* constructed in the first plasmid subcloned under maize ubiquitin promoter and sorghum ubiquitin promoter, respectively. The RNAi suppression cassette was constructed from 244 bp of *TGD1* or 278 bp of *SDP1* arranged in sense and anti-sense orientation separated by 94 bp of *Pn4CL* intron, in which each RNAi cassette was controlled by 35SCaMV promoter. Transgenic sugarcane carrying the overexpression and suppression cassette accumulated TAG up to 1.52% of dry weight or increased up to 30-fold compared to nontransgenic plants. Both experimental reports provide strong evidence that sugarcane is an ideal crop for optimizing and producing lipids in vegetative tissue.

4.5.4 GENOME EDITING

The genome editing approach employs an effector component that recognizes specific DNA sequences and endonucleases that cleave sequence DNA in the genome, producing double-strand breaks (DSB), which are repaired by the cellular mechanism producing nucleotides deletion, insertion, or substitution. Four types of nucleases, namely meganucleases, zinc finger nucleases, transcription activator-like effector nucleases (TALENs), and clustered regularly interspaced short palindromic repeats (CRISPR)/Cas9 system have been utilized in plant genome editing. TALEN and CRISPR/Cas9, known for their simplicity, robustness, and cost-effectively, have been applied successfully in sugarcane.

TALEN is developed from type III transcription activator-like effector (TALE) protein from plant pathogenic bacteria *Xanthomonas* that is secreted to the host plant, enters the nucleus, and binds the promoter of the host gene to activate transcription in order to facilitate pathogenic bacterial colonization (Boch and Bonas, 2010; Bogdanove et al., 2010). TALEN has two domains, that is, a customizable DNA-binding domain at the N-terminal and a nonspecific *FokI* nuclease domain at the C-terminal. The customizable DNA-binding domain comprises an array of TALE protein flanking by additional domains at N- and C terminals. TALE, composed of 33–35 amino acids, binds to a single nucleotide determined by the hypervariable amino acid at positions 12th and 13th (Deng et al., 2012). The hypervariable amino acid residues NI, NG, NN, and HD in TALE recognize nucleotides A, T, G, and C, respectively (Boch et al., 2009). TALEN works in the dimer to recognize the DNA target, in which the left and right monomer bindings enable *FokI* dimerization to induce DSB.

TALEN-mediated genome editing has successfully generated sugarcane lines with reduced lignin content, modified cell wall composition, and improved saccharification. Jung and Altpeter (2016) reported a targeted mutagenesis of *COMT* mediated by TALEN to reduce lignin content in sugarcane. TALEN and spacer were designed with 16 and 17 bp, respectively, to target conserved sequences in the first exon of the *COMT* gene and its homo(eo)log. The TALEN expression cassette was transferred to the sugarcane embryonic callus by biolistic or *Agrobacterium*-mediated transformation. The mutant of *COMT* displayed 29–32% lignin reduction, decreased S subunit level, and enhanced hemicellulose level (Kannan et al., 2018). Under field-grown evaluation, the *COMT* mutant showed agronomic performance similar to the WT, and the mutant line displayed lignin reduction up to 19.7%, S/G ratio reduction, and upgraded saccharification capacity nearly 44% according to Kannan et al. (2018).

CRISPR/Cas9 is developed from the adaptive immune system of prokaryotes against invading viruses (Hille et al., 2018). CRISPR/Cas9 genome editing comprises a Cas9 protein and a single-guide RNA (sgRNA). Cas9, an endonuclease derived from bacteria, scans the proto-adjacent motif (PAM) along the DNA sequence and cleaves the site of the target. The sgRNA identifies complementary target sequences located upstream of the PAM, where DSB occurs. CRISPR/Cas9 can target diverse sequences in the plant genome by modifying the 20 bp spacer sequence at the 5′ end of gRNA and considering the PAM motif in the target sequence. The nuclear localization signal (NLS) is fused to the Cas9 to direct the protein into the DNA sequence target in the nucleus. The constitutive U6 and U3 promoters from RNA polymerase III control the expression of sgRNA to facilitate the transcript start with G or A, respectively. The spacer of sgRNA should follow a consensus sequence $5'\text{-G(N)}_{19\text{–}22}$ or $5'\text{-A(N)}_{19\text{–}22}$, where the first G or A may or may not pair with the target sequence (Belhaj et al., 2013).

Laksana et al. (2024) reported a method to reduce lignin content in sugarcane using CRISPR/Ca9-mediated genome editing by targeting *SoLIM*, a transcription factor containing LIM domain associated with lignin production in the secondary cell wall. The sgRNA was designed to target exon two and exon five of *SoLIM.* The CRISPR/Cas9 and sgRNA expression cassette was transformed into sugarcane calli by *Agrobacterium*-mediated transformation. Based on sequencing results, CRISPR/Cas9-mediated genome editing of *SoLIM* produced several mutant types with a combination of base insertion, substitution, and deletion. All types of mutations of *SoLIM* reduced the expression of *PAL*, *C4H*, and *CAD*, genes responsible for lignin biosynthesis. The edited lines show a 9.74–51.46% lignin reduction compared to the wild type (Laksana et al., 2024).

KEYWORDS

- **breeding strategies**
- **genetic transformation**
- **lignocellulose**
- **genome editing**
- **RNA interference**
- **tissue culture**
- ***Saccharum* spp.**

REFERENCES

Aditiya HB, Mahlia TMI, Chong WT, Nur H, Sebayang AH. (2016). Second generation bioethanol production: A critical review. *Renew. Sustain. Energy Rev.* 66, 631–653. https://doi.org/10.1016/j.rser.2016.07.015

Akmal M. (2021). Embryo culture and embryo rescue in brassica. In: Islam AKMA, Hossain MA, Islam AKMM (Eds.), *Brassica Breeding and Biotechnology*. IntechOpen; Chpt. 4. https://doi.org/10.5772/intechopen.96058

Alvarez JB, Guzmán C. (2018). Interspecific and intergeneric hybridization as a source of variation for wheat grain quality improvement. *Theor. Appl. Genet.* 131(2), 225–251. https://doi.org/10.1007/s00122-017-3042-x

Anas M, Verma KK, Haq IU, Naeem M, Li Q, Liao F, Xiang YX, Li YR. (2021a). Characterization of exotic, native and wild-type genotypes of sugarcane (*Saccharum* spp. hybrids) for internal nitrogen use efficiency under different nitrogen levels. *Sugar Tech*. 23, 1258–1267. https://doi.org/10.1007/s12355-021-01014-

Anas M, Verma KK, Riaz M, Qiang L, Liao F, Liu Y, Li YR. (2021b). Physio-morphological and biochemical mechanism of nitrogen use efficiency in sugarcane (*Saccharum* spp.) genotypes under different growth stages and nitrogen levels. *J. Plant Interact.* 16, 332–343.

Andrianov V, Borisjuk N, Pogrebnyak N, Brinker A, Dixon J, Spitsin S, Flynn J, Matyszczuk P, Andryszak K, Laurelli M, Golovkin M, Koprowski H. (2010). Tobacco as a production platform for biofuel: Overexpression of Arabidopsis DGAT and LEC2 genes increases accumulation and shifts the composition of lipids in green biomass. *Plant Biotechnol. J.* 8(3), 277–287. https://doi.org/10.1111/j.1467-7652.2009.00458.x

Anur RM, Mufithah N, Sawitri WD, Sakakibara H, Sugiharto B. (2020). Overexpression of sucrose phosphate synthase enhanced sucrose content and biomass production in transgenic sugarcane. *Plants* 9(2), 200. https://doi.org/10.3390/plants9020200

Arencibia AD, Carmona ER, Tellez P, Chan M-T, Yu S-M, Trujillo LE, Oramas P. (1998). An efficient protocol for sugarcane (*Saccharum* spp. L.) transformation mediated by Agrobacterium tumefaciens. *Transgenic Res.* 7(3), 213–222. https://doi.org/10.1023/A:1008845114531

Argaye S. (2021). Development and the role of tissue culture in plant breeding: A review. *Int. J. Res. Stud. Agric. Sci.* 7(5), 27–36.

Aslam U, Tabassum B, Nasir IA, Khan A, Husnain T. (2018). A virus-derived short hairpin RNA confers resistance against sugarcane mosaic virus in transgenic sugarcane. *Transgenic Res.* 27(2), 203–210. https://doi.org/10.1007/s11248-018-0066-1

Balat M, Balat H. (2009). Recent trends in global production and utilization of bio-ethanol fuel. *Appl. Energy* 86(11), 2273–2282. https://doi.org/10.1016/j.apenergy.2009.03.015

Basso MF, da Cunha BADB, Ribeiro AP, Martins PK, de Souza WR, de Oliveira NG, Nakayama TJ, Augusto das Chagas Noqueli Casari R, Santiago TR, Vinecky F, Cançado LJ, de Sousa CAF, de Oliveira PA, de Souza SACD, Cançado GM de A, Kobayashi AK, Molinari HBC. (2017). Improved genetic transformation of sugarcane (*Saccharum* spp.) Embryogenic callus mediated by agrobacterium tumefaciens. *Curr. Protocols Plant Biol.* 2(3), 221–239. https://doi.org/10.1002/cppb.20055

Belhaj K, Chaparro-Garcia A, Kamoun S, Nekrasov V. (2013). Plant genome editing made easy: Targeted mutagenesis in model and crop plants using the CRISPR/Cas system. *Plant Methods* 9(1), 39. https://doi.org/10.1186/1746-4811-9-39

Bewg WP, Poovaiah C, Lan W, Ralph J, Coleman HD. (2016). RNAi downregulation of three key lignin genes in sugarcane improves glucose release without reduction in sugar production. *Biotechnol. Biofuels* 9(1), 270. https://doi.org/10.1186/s13068-016-0683-y

Bhatt R, Verma KK, Jamwal NS. (2023a). Can paddy-compost eliminate N fertilization for climate-smart sugarcane? *Sugar Tech.* 25, 1036–1046.https://doi.org/10.1007/s12355-023-01270-3

Bhatt R, Verma KK, Kumar R, Sanghera GS. (2023b). Foliar application of potassium salt of active phosphorus (PSAP) mitigates insect pests and improves yield along with sugarcane quality in response to agroclimatic conditions of Punjab. *Sugar Tech.* 25(3), 660–669.

Boch J, Bonas U. (2010). Xanthomonas AvrBs3 family-type III effectors: Discovery and function. *Ann. Rev. Phytopathol.* 48(1), 419–436. https://doi.org/10.1146/annurev-phyto-080508-081936

Boch J, Scholze H, Schornack S, Landgraf A, Hahn S, Kay S, Lahaye T, Nickstadt A, Bonas U. (2009). Breaking the code of DNA binding specificity of TAL-type III effectors. *Science* 326(5959), 1509–1512. https://doi.org/10.1126/science.1178811

Bogdanove AJ, Schornack S, Lahaye T. (2010). TAL effectors: Finding plant genes for disease and defense. *Curr. Opin. Plant Biol.* 13(4), 394–401. https://doi.org/10.1016/j.pbi.2010.04.010

Borisjuk N, Kishchenko O, Eliby S, Schramm C, Anderson P, Jatayev S, Kurishbayev A, Shavrukov Y. (2019). Genetic modification for wheat improvement: From transgenesis to genome editing. *BioMed Res. Int.* 2019, 6216304. https://doi.org/10.1155/2019/6216304

Bosabalidis AM, Evert RF, Russin WA. (1994). Ontogeny of the vascular bundles and contiguous tissues in the maize leaf blade. *Am. J. Bot.* 81(6), 745–752. https://doi.org/10.2307/2445653

Bower R, Birch RG. (1992). Transgenic sugarcane plants via microprojectile bombardment. *Plant J.* 2(3), 409–416. https://doi.org/10.1111/j.1365-313X.1992.00409.x

Bräutigam A, Gowik U. (2016). Photorespiration connects C3 and C4 photosynthesis. *J. Exp. Bot.* 67(10), 2953–2962. https://doi.org/10.1093/jxb/erw056

Bridgen MP, Van Houtven W, Eeckhaut T. (2018). Plant tissue culture techniques for breeding. In: Van Huylenbroeck J (Ed.), *Ornamental Crops*. Springer International Publishing; pp. 127–144. https://doi.org/10.1007/978-3-319-90698-0_6

Budeguer F, Enrique R, Perera MF, Racedo J, Castagnaro AP, Noguera AS, Welin B. (2021). Genetic transformation of sugarcane, current status and future prospects. *Front. Plant Sci.* 12, 768609. https://www.frontiersin.org/articles/10.3389/fpls.2021.768609

Byrt CS, Grof CPL, Furbank RT. (2011). C4 plants as biofuel feedstocks: Optimising biomass production and feedstock quality from a lignocellulosic perspective free access. *J. Integr. Plant Biol.* 53(2), 120–135. https://doi.org/10.1111/j.1744-7909.2010.01023.x

Cai Y, Zhang K, Kim H, Hou G, Zhang X, Yang H, Feng H, Miller L, Ralph J, Liu C-J. (2016). Enhancing digestibility and ethanol yield of Populus wood via expression of an engineered monolignol 4-O-methyltransferase. *Nat. Commun.* 7(1), 11989. https://doi.org/10.1038/ncomms11989

Carnal NW, Black CC. (1979). Pyrophosphate-dependent 6-phosphofructokinase, a new glycolytic enzyme in pineapple leaves. *Biochem. Biophys. Res. Commun.* 86(1), 20–26. https://doi.org/10.1016/0006-291X(79)90376-0

Chaudhary P, Xu M, Ahamad L, Chaudhary A, Kumar G, Adeleke BS, Verma KK, Hu D-M, Siric I, Kumar P, Popescu SM, Fayssal SA. (2023). Application of synthetic consortia for improvement of soil fertility, pollution remediation, and agricultural productivity: A review. *Agronomy* 13, 643. https://doi.org/10.3390/agronomy13030643.

Chen GL, Zheng FJ, Lin B, Lao SB, He J, Huang Z, Zeng Y, Sun J, Verma KK. (2020). Phenolic and volatile compounds in the production of sugarcane vinegar. *ACS Omega* 5, 30587–30595.

Chen G-L, Zheng F-J, Lin B, Yang Y-X, Fang X-C, Verma KK, Yang L-F. (2023). Vinegar: A potential source of healthy and functional food with special reference to sugarcane vinegar. *Front. Nutr.* 10, 1145862. https://doi.org/10.3389/fnut.2023.1145862

Chen WH, Gartland KMA, Davey MR, Sotak R, Gartland JS, Mulligan BJ, Power JB, Cocking EC. (1987). Transformation of sugarcane protoplasts by direct uptake of a selectable chimaeric gene. *Plant Cell Rep.* 6(4), 297–301. https://doi.org/10.1007/BF00272003

D'Hont A, Paulet F, Glaszmann JC. (2002). Oligoclonal interspecific origin of 'North Indian' and 'Chinese' sugarcanes. *Chromosome Res.* 10(3), 253–262. https://doi.org/10.1023/A:1015204424287

Day C, Day G. (2017). Climate change, fossil fuel prices and depletion: The rationale for a falling export tax. *Economic Model.* 63, 153–160. https://doi.org/10.1016/j.econmod.2017.01.006

de Setta N, Monteiro-Vitorello CB, Metcalfe CJ, Cruz GMQ, Del Bem LE, Vicentini R, Nogueira FTS, Campos RA, Nunes SL, Turrini PCG, Vieira AP, Ochoa Cruz EA, Corrêa TCS, Hotta CT, de Mello Varani A, Vautrin S, da Trindade AS, de Mendonça Vilela M, Lembke CG, Sato PM, de Andrade RF, Nishiyama MY Jr, Cardoso-Silva CB, Scortecci KC, Garcia AA, Carneiro MS, Kim C, Paterson AH, Bergès H, D'Hont A, de Souza AP, Souza GM, Vincentz M, Kitajima JP, Van Sluys M-A. (2014). Building the sugarcane genome for biotechnology and identifying evolutionary trends. *BMC Genom.* 15(1), 540. https://doi.org/10.1186/1471-2164-15-540

de Souza WR, Pacheco TF, Duarte KE, Sampaio BL, de Oliveira Molinari PA, Martins PK, Santiago TR, Formighieri EF, Vinecky F, Ribeiro AP, da Cunha BADB, Kobayashi AK, Mitchell RAC, de Sousa Rodrigues Gambetta D, Molinari HBC. (2019). Silencing of a BAHD acyltransferase in sugarcane increases biomass digestibility. *Biotechnol. Biofuels* 12(1), 111. https://doi.org/10.1186/s13068-019-1450-7

Deng ZN, Huang HR, Huang CM, Verma KK, Luo HB, Jiang SL, Li YJ, Cao HQ, Wu KC, Xu L. (2023). Short-term limited water irrigation influences the photosynthetic pigments, enzymatic and non-enzymatic activities in *Saccharum spontaneum* L. at vegetative stage. *Sugar Tech.* 25, 110–118. https://doi.org/10.1007/s12355-022-01175-7

Deng D, Yan C, Pan X, Mahfouz M, Wang J, Zhu J-K, Shi Y, Yan N. (2012). Structural basis for sequence-specific recognition of DNA by TAL effectors. *Science* 335(6069), 720–723. https://doi.org/10.1126/science.1215670

Dida G. (2024). Biotechnology towards energy crops. In: Kumar N (Ed.), *Industrial Crop Plants*. Springer Nature Singapore; pp. 227–253. https://doi.org/10.1007/978-981-97-1003-4_9

Emonet A, Hay A. (2022). Development and diversity of lignin patterns. *Plant Physiol.* 190(1), 31–43. https://doi.org/10.1093/plphys/kiac261

Enríquez-Obregón GA, Vázquez-Padrón RI, Prieto-Samsonov DL, De la Riva GA, Selman-Housein G. (1998). Herbicide-resistant sugarcane (*Saccharum officinarum* L.) plants by Agrobacterium-mediated transformation. *Planta* 206(1), 20–27. https://doi.org/10.1007/s004250050369

Falter C, Voigt CA. (2016a). Improving biomass production and saccharification in *Brachypodium distachyon* through overexpression of a sucrose-phosphate synthase from sugarcane. *J. Plant Biochem. Biotechnol.* 25(3), 311–318. https://doi.org/10.1007/s13562-015-0343-5

Falter C, Voigt CA. (2016b). Improving biomass production and saccharification in *Brachypodium distachyon* through overexpression of a sucrose-phosphate synthase from sugarcane. *J. Plant Biochem. Biotechnol.* 25(3), 311–318. https://doi.org/10.1007/s13562-015-0343-5

Fathi H, Jahani U. (2012). Review of embryo culture in fruit trees. *Ann. Biol. Res.* 3, 4276–4281. https://api.semanticscholar.org/CorpusID:27919830

Furtado A, Lupoi JS, Hoang NV, Healey A, Singh S, Simmons BA, Henry RJ. (2014). Modifying plants for biofuel and biomaterial production. *Plant Biotechnol. J.* 12(9), 1246–1258. https://doi.org/10.1111/pbi.12300

Gao S-J, Damaj MB, Park J-W, Beyene G, Buenrostro-Nava MT, Molina J, Wang X, Ciomperlik JJ, Manabayeva SA, Alvarado VY, Rathore KS, Scholthof HB, Mirkov TE. (2013). Enhanced transgene expression in sugarcane by co-expression of virus-encoded

RNA silencing suppressors. *PLoS One* 8(6), e66046-. https://doi.org/10.1371/journal.pone.0066046

Garsmeur O, Droc G, Antonise R, Grimwood J, Potier B, Aitken K, Jenkins J, Martin G, Charron C, Hervouet C, Costet L, Yahiaoui N, Healey A, Sims D, Cherukuri Y, Sreedasyam A, Kilian A, Chan A, Van Sluys M.-A, Swaminathan K, Town C, Bergès H, Simmons B, Glaszmann JC, van der Vossen E, Henry R, Schmutz J, D'Hont A. (2018). A mosaic monoploid reference sequence for the highly complex genome of sugarcane. *Nat. Commun.* 9(1), 2638. https://doi.org/10.1038/s41467-018-05051-5

Goldemberg J, Guardabassi P. (2010). The potential for first-generation ethanol production from sugarcane. *Biofuels, Bioprod. Bioref.* 4(1), 17–24. https://doi.org/10.1002/bbb.186

Groenewald J-H, Botha FC. (2008). Down-regulation of pyrophosphate: Fructose 6-phosphate 1-phosphotransferase (PFP) activity in sugarcane enhances sucrose accumulation in immature internodes. *Transgenic Res.* 17(1), 85–92. https://doi.org/10.1007/s11248-007-9079-x

Guha S, Maheswari SC. (1966). Cell division and differentiation of embryos in the pollen grains of datura in vitro. *Nature* 212(5057), 97–98. https://doi.org/10.1038/212097a0

Guo D-J, Li D-P, Chen Z-D, Yang B, Verma KK, Singh RK, Singh P, Khan Q, Sharma A, Qin Y, Zhang B-Q, Song X-P, Li Y-R. (2023). Effect of endophytic diazotroph *Enterobacter roggenkampii* ED5 on nitrogen-metabolism-related microecology in the sugarcane rhizosphere at different nitrogen levels. *Front. Microbiol.* 14, 1132016. https://doi.org/10.3389/fmicb.2023.1132016

Guo D-J, Li D-P, Singh RK, Singh P, Sharma A, Verma KK, Qin Y, Khan Q, Lu Z, Malviya MK, Song X-P, Xing Y-X and Li Y-R. (2021). Differential protein expression analysis of two sugarcane varieties in response to diazotrophic plant growth-promoting endophyte *Enterobacter roggenkampii* ED5. *Front. Plant Sci.* 12, 727741. https://doi.org/10.3389/fpls.2021.727741

Guo DJ, Li DP, Singh RK, Singh P, Verma KK, Sharma A, Qin Y, Khan Q, Song XP, Malviya MK, Xing YX, Li YR. (2022). Comparative transcriptome analysis of two sugarcane varieties in response to diazitrophic plant growth promoting endophyte *Enterobacter roggenkampii* ED5. *J. Plant Interact.* 17(1), 75–84.

Hailing P, Simms-Borre P. (2008). Overview of lignocellulosic feedstock conversion into ethanol-focus on sugarcane bagasse. *Int. Sugar J.* 110(1311), 191–194.

Healey AL, Garsmeur O, Lovell JT, Shengquiang S, Sreedasyam A, Jenkins J, Plott CB, Piperidis N, Pompidor N, Llaca V, Metcalfe CJ, Doležel J, Cápal P, Carlson JW, Hoarau JY, Hervouet C, Zini C, Dievart A, Lipzen A, Williams M, Boston LB, Webber J, Keymanesh K, Tejomurthula S, Rajasekar S, Suchecki R, Furtado A, May G, Parakkal P, Simmons BA, Barry K, Henry RJ, Grimwood J, Aitken KS, Schmutz J, D'Hont A. (2024). The complex polyploid genome architecture of sugarcane. *Nature* 628(8009), 804–810. https://doi.org/10.1038/s41586-024-07231-4

Hille F, Richter H, Wong SP, Bratovič M, Ressel S, Charpentier E. (2018). The biology of CRISPR-Cas: Backward and forward. *Cell* 172(6), 1239–1259. https://doi.org/10.1016/j.cell.2017.11.032

Huang CM, Wu KC, Verma KK, Luo HB, Deng ZN, Shi GY, Cao HQ, Wu XJ, Ye LP, Wei YW, Xu L. (2024). Isolation and expression analysis of cellulose synthase 3 (*Ces3*) genes from sugarcane (*Saccharum officinarum* L.). *Chem. Biol. Technol. Agric.* 11, 71. https://doi.org/10.1186/s40538-024-00597-7

Huang HR, Gao YJ, Malviya MK, Verma KK, Solanki MK, Huang YX, Li X, Deng Y, Yan J, Tang SY, Wang LW, Xu L. (2022). Genetic diversity analysis of sugarcane (*Saccharum* spp. Hybrids) among high-sucrose clones of GT series and commonly used parents by using microsatellite markers in Guangxi, China. *Sugar Tech.* 24, 397–407. https://doi.org/10.1007/s12355-021-01033-y

Huber SC, Huber JL. (1996). Role and regulation of sucrose-phosphate synthase in higher plants. *Ann. Rev. Plant Biol.* 47, 431–444. https://doi.org/10.1146/annurev.arplant.47.1.431

Huntley SK, Ellis D, Gilbert M, Chapple C, Mansfield SD. (2003). Significant increases in pulping efficiency in C4H-F5H-transformed poplars: Improved chemical savings and reduced environmental toxins. *J. Agric. Food Chem.* 51(21), 6178–6183. https://doi.org/10.1021/jf034320o

Irvine JE. (1978). Identification of cold tolerance in Saccharum and related genera through refrigerated freeze screening. *Proc. Int. Soc. Sugar Cane Technol.* 16, 147–156.

Jackson MA, Anderson DJ, Birch RG. (2013). Comparison of Agrobacterium and particle bombardment using whole plasmid or minimal cassette for production of high-expressing, low-copy transgenic plants. *Transgenic Res.* 22(1), 143–151. https://doi.org/10.1007/s11248-012-9639-6

Jackson MA, Sternes PR, Mudge SR, Graham MW, Birch RG. (2014). Design rules for efficient transgene expression in plants. *Plant Biotechnol. J.* 12(7), 925–933. https://doi.org/10.1111/pbi.12197

Jagadesh M, Dash M, Kumari A, Singh SK, Verma KK, Kumar P, Bhatt R, Sharma SK. (2024). Revealing the hidden world of soil microbes: Metagenomic insights into plant, bacteria, and fungi interactions for sustainable agriculture and ecosystem restoration. *Microbiol. Res.* 127764. https://doi.org/10.1016/j.micres.2024.127764

Joshi A, Verma KK, Rajput VD, Arora J. (2023). Metabolic engineering of microorganisms in advancing biofuel production. In: *Advances in Lignocellulosic Biofuel Production Systems*; pp. 181–202. https://doi.org/10.1016/B978-0-323-91192-4.00016-X

Joshi A, Verma KK, Rajput VD, Minkina T, Arora J. (2022). Recent advances in metabolic engineering of microorganisms for advancing lignocellulose-derived biofuels. *Bioengineered* 13(4), 8135–8163. https://doi.org/10.1080/21655979.2022.2051856.

Joyce P, Hermann S, O'Connell A, Dinh Q, Shumbe L, Lakshmanan P. (2014). Field performance of transgenic sugarcane produced using Agrobacterium and biolistics methods. *Plant Biotechnol. J.* 12(4), 411–424. https://doi.org/10.1111/pbi.12148

Jung JH, Altpeter F. (2016). TALEN mediated targeted mutagenesis of the caffeic acid O-methyltransferase in highly polyploid sugarcane improves cell wall composition for production of bioethanol. *Plant Mol. Biol.* 92(1), 131–142. https://doi.org/10.1007/s11103-016-0499-y

Jung JH, Fouad WM, Vermerris W, Gallo M, Altpeter F. (2012). RNAi suppression of lignin biosynthesis in sugarcane reduces recalcitrance for biofuel production from lignocellulosic biomass. *Plant Biotechnol. J.* 10(9), 1067–1076. https://doi.org/10.1111/j.1467-7652.2012.00734.x

Jung JH, Kannan B, Dermawan H, Moxley GW, Altpeter F. (2016). Precision breeding for RNAi suppression of a major 4-coumarate:coenzyme A ligase gene improves cell wall saccharification from field grown sugarcane. *Plant Mol. Biol.* 92(4), 505–517. https://doi.org/10.1007/s11103-016-0527-y

Jung JH, Vermerris W, Gallo M, Fedenko JR, Erickson JE, Altpeter F. (2013). RNA interference suppression of lignin biosynthesis increases fermentable sugar yields for biofuel production from field-grown sugarcane. *Plant Biotechnol. J.* 11(6), 709–716. https://doi.org/10.1111/pbi.12061

Kannan B, Jung JH, Moxley GW, Lee S-M, Altpeter F. (2018). TALEN-mediated targeted mutagenesis of more than 100 COMT copies/alleles in highly polyploid sugarcane improves saccharification efficiency without compromising biomass yield. *Plant Biotechnol. J.* 16(4), 856–866. https://doi.org/10.1111/pbi.12833

Kannan B, Liu H, Shanklin J, Altpeter F. (2022). Towards oilcane: Preliminary field evaluation of metabolically engineered sugarcane with hyper-accumulation of triacylglycerol in vegetative tissues. *Mol. Breed.* 42(10), 64. https://doi.org/10.1007/s11032-022-01333-5

Keating JD, Panganiban C, Mansfield SD. (2006). Tolerance and adaptation of ethanologenic yeasts to lignocellulosic inhibitory compounds. *Biotechnol. Bioeng.* 93(6), 1196–1206. https://doi.org/10.1002/bit.20838

Klaus D, Ohlrogge JB, Neuhaus HE, Dörmann P. (2004). Increased fatty acid production in potato by engineering of acetyl-CoA carboxylase. *Planta* 219(3), 389–396. https://doi.org/10.1007/s00425-004-1236-3

Krishna D, Sachan HK, Verma KK. (2022). Abiotic stresses in plants: An overview. In: *Understanding Abiotic Stresses.* Nova Science Publishers, Inc.: New York.

Kui L, Majeed A, Wang X, Yang Z, Chen J, He L, Di Y, Li X, Qian Z, Jiao Y, Wang G, Liu L, Xu R, Gu S, Yang Q, Chen S, Lou H, Meng Y, Xie L, Xu F, Shen Q, Singh A, Gruber K, Pan Y, Hao T, Dong Y, Li F. (2023). A chromosome-level genome assembly for *Erianthus fulvus* provides insights into its biofuel potential and facilitates breeding for improvement of sugarcane. *Plant Commun.* 4(4), 100562. https://doi.org/10.1016/j.xplc.2023.100562

Kumari A, Dash M, Singh SK, Jagadesh M, Mathpal B, Mishra PK, Pandey SK, Verma KK. (2023a). Soil microbes: A natural solution for mitigating the impact of climate change. *Environ. Monit. Assess.* 195, 1436. https://doi.org/10.1007/s10661-023-11988-y

Kumari A, Lakshmi GA, Krishna GK, Patni B, Prakash S, Bhattacharyya M, Singh SK, Verma KK. (2022). Climate change and its impact on crops: A comprehensive investigation for sustainable agriculture. *Agronomy* 12, 3008. https://doi.org/10.3390/agronomy12123008

Kumari A, Singh SK, Mathpal B, Verma KK, Garg VK, Bhattacharyya M, Bhatt R. (2023b). The biosynthesis, mechanism of action, and physiological functions of melatonin in horticultural plants: A review. *Horticulturae* 9, 913. https://doi.org/10.3390/horticulturae9080913

Kunz H-H, Scharnewski M, Feussner K, Feussner I, Flügge U-I, Fulda M, Gierth M. (2009). The ABC transporter PXA1 and peroxisomal β-oxidation are vital for metabolism in mature leaves of arabidopsis during extended darkness. *Plant Cell* 21(9), 2733–2749. https://doi.org/10.1105/tpc.108.064857

Laksana C, Sophiphun O, Chanprame S. (2024). Lignin reduction in sugarcane by performing CRISPR/Cas9 site-direct mutation of SoLIM transcription factor. *Plant Sci.* 340, 111987. https://doi.org/10.1016/j.plantsci.2024.111987

Lee M, Jeon HS, Kim SH, Chung JH, Roppolo D, Lee H, Cho HJ, Tobimatsu Y, Ralph J, Park OK. (2019). Lignin-based barrier restricts pathogens to the infection site and confers resistance in plants. *EMBO J.* 38(23), e101948. https://doi.org/10.15252/embj.2019101948

Lee TSG, Bressan EA. (2006). *The Potential of Ethanol Production from Sugarcane in Brazil.*

Li A-M, Liao F, Wang M, Chen Z-L, Qin C-X, Huang R-Q, Verma KK, Li Y-R, Que Y-X, Pan Y-Q, et al. (2023). Transcriptomic and proteomic landscape of sugarcane response to biotic and abiotic stressors. *Int. J. Mol. Sci.* 24, 8913. https://doi.org/10.3390/ijms24108913

Li Q, Chen Y, Yue F, Qian W, Song H. (2018). Microspore culture reveals high fitness of B. napus-like gametes in an interspecific hybrid between Brassica napus and B. oleracea. *PLOS ONE* 13(3), e0193548-. https://doi.org/10.1371/journal.pone.0193548

Lu G, Liu P, Wu Q, Zhang S, Zhao P, Zhang Y, Que Y. (2024). Sugarcane breeding: A fantastic past and promising future driven by technology and methods. *Front. Plant Sci.* 15. https://www.frontiersin.org/journals/plant-science/articles/10.3389/fpls.2024.1375934

Luo G, Cao VD, Kannan B, Liu H, Shanklin J, Altpeter F. (2022). Metabolic engineering of energycane to hyperaccumulate lipids in vegetative biomass. *BMC Biotechnol.* 22(1), 24. https://doi.org/10.1186/s12896-022-00753-7

Ma H, Albert HH, Paull R, Moore PH. (2000). Metabolic engineering of invertase activities in different subcellular compartments affects sucrose accumulation in sugarcane cells. *Funct. Plant Biol.* 27(11), 1021–1030. https://doi.org/10.1071/PP00029

Malviya MK, Li C-N, Lakshmanan P, Solanki MK, Wang Z, Solanki AC, Nong Q, Verma KK, Singh RK, Singh P, Sharma A, Guo D-J, Dessoky ES, Song X-P, Li Y-R. (2022). High-throughput sequencing-based analysis of rhizosphere and diazotrophic bacterial diversity among wild progenitor and closely related species of sugarcane (*Saccharum* spp. inter-specific hybrids). *Front. Plant Sci.* 13, 829337. *https://doi.org/10.3389/fpls.2022.829337.*

Malviya MK, Solanki MK, Li CN, Wang Z, Zeng Y, Verma KK, Singh RK, Singh P, Huang HR, Yang LT, Song XP, Li YR. (2021). Sugarcane-legume intercropping can enrich the soil microbiome and plant growth. *Front. Sustain. Food Syst.* 5, 606595. https://doi.org/10.3389/fsufs.2021.606595.

Marriott PE, Gómez LD, McQueen-Mason SJ. (2016). Unlocking the potential of lignocellulosic biomass through plant science. *New Phytologist.* 209(4), 1366–1381. https://doi.org/10.1111/nph.13684

Martins APB, dos S. Brito M, Mayer JLS, Llerena JPP, Oliveira JF, Takahashi NG, Carlin SD, Borges DNAF, Andrade LM, Peixoto-Júnior RF, Goldman, MH de S, Mazzafera P, Creste S, Nobile PM. (2018). Ectopic expression of sugarcane SHINE changes cell wall and improves biomass in rice. *Biomass and Bioenergy* 119, 322–334. https://doi.org/10.1016/j.biombioe.2018.09.036

Mayavan S, Subramanyam K, Jaganath B, Sathish D, Manickavasagam M, Ganapathi A. (2015). Agrobacterium-mediated in planta genetic transformation of sugarcane setts. *Plant Cell Rep.* 34(10), 1835–1848. https://doi.org/10.1007/s00299-015-1831-8

Moodley P. (2021). Sustainable biofuels: Opportunities and challenges. In: Ray RC (Ed.), *Sustainable Biofuels*. Academic Press; pp. 1–20. https://doi.org/10.1016/B978-0-12-820297-5.00003-7

Mu J, Tan H, Zheng Q, Fu F, Liang Y, Zhang J, Yang X, Wang T, Chong K, Wang X-J, Zuo J. (2008). LEAFY COTYLEDON1 is a key regulator of fatty acid biosynthesis in arabidopsis. *Plant Physiol.* 148(2), 1042–1054. https://doi.org/10.1104/pp.108.126342

Nerkar G, Thorat A, Sheelavantmath S, Kassa HB, Devarumath R. (2018). Genetic transformation of sugarcane and field performance of transgenic sugarcane. In: Gosal SS, Wani SH (Eds.), *Biotechnologies of Crop Improvement, Volume 2: Transgenic Approaches*. Springer International Publishing; pp. 207–226. https://doi.org/10.1007/978-3-319-90650-8_9

Noguera A, Enrique R, Perera MF, Ostengo S, Racedo J, Costilla D, Zossi S, Cuenya MI, Filippone MP, Welin B, Castagnaro AP. (2015). Genetic characterization and field evaluation to recover parental phenotype in transgenic sugarcane: A step toward commercial release. *Mol. Breed.* 35(5), 115. https://doi.org/10.1007/s11032-015-0300-y

Ohlrogge J, Chapman K. (2011). The seeds of green energy: Expanding the contribution of plant oils as biofuels. *Biochemist* 33(2), 34–38. https://doi.org/10.1042/BIO03302034

Oliveira MA, Verma KK, Bhatt R, Oliveira TBA. (2022). Impact of green and organic fertilizers on soil fertility and sugarcane productivity. In: Verma KK et al. (Eds.), *Agro-industrial Perspectives on Sugarcane Production under Environmental Stress*. https://doi.org/10.1007/978-981-19-3955-6_11.

Pan Y-Q, Luo H-L, Li Y-R. (2009). Soluble acid invertase and sucrose phosphate synthase: Key enzymes in regulating sucrose accumulation in sugarcane stalk. *Sugar Tech.* 11(1), 28–33. https://doi.org/10.1007/s12355-009-0005-9

Pereira SC, Maehara L, Machado CMM, Farinas CS. (2015). 2G ethanol from the whole sugarcane lignocellulosic biomass. *Biotechnol. Biofuels* 8(1), 44. https://doi.org/10.1186/s13068-015-0224-0

Petrie JR, Vanhercke T, Shrestha P, El Tahchy A, White A, Zhou X-R, Liu Q, Mansour MP, Nichols PD, Singh SP. (2012). Recruiting a new substrate for triacylglycerol synthesis in plants: The monoacylglycerol acyltransferase pathway. *PLOS ONE* 7(4), e35214-. https://doi.org/10.1371/journal.pone.0035214

Pierre JS, Perroux J, Whan A, Rae AL, Bonnett GD. (2015). Poor fertility, short longevity, and low abundance in the soil seed bank limit volunteer sugarcane from seed. *Front. Bioeng. Biotechnol.* 3. https://www.frontiersin.org/articles/10.3389/fbioe.2015.00083

Piperidis N, D'Hont A. (2020). Sugarcane genome architecture decrypted with chromosome-specific oligo probes. *Plant J.* 103(6), 2039–2051. https://doi.org/10.1111/tpj.14881

Poovaiah CR, Nageswara-Rao M, Soneji JR, Baxter HL, Stewart Jr CN. (2014). Altered lignin biosynthesis using biotechnology to improve lignocellulosic biofuel feedstocks. *Plant Biotechnol. J.* 12(9), 1163–1173. https://doi.org/10.1111/pbi.12225

Portilla Llerena JP, Kiyota E, dos Santos FRC, Garcia JC, de Lima RF, Mayer JLS, dos Santos Brito M, Mazzafera P, Creste S, Nobile PM. (2024). ShF5H1 overexpression increases syringyl lignin and improves saccharification in sugarcane leaves. *GM Crops Food* 15(1), 67–84. https://doi.org/10.1080/21645698.2024.2325181

Radhesh Krishnan S, Mohan C. (2017). Methods of sugarcane transformation. In: Mohan C (Ed.), *Sugarcane Biotechnology: Challenges and Prospects*. Springer International Publishing; pp. 51–60. https://doi.org/10.1007/978-3-319-58946-6_4

Raghavan V. (2003). One hundred years of zygotic embryo culture investigations. *In Vitro Cell. Dev. Biol. - Plant* 39(5), 437–442. https://doi.org/10.1079/IVP2003436

Ramasamy M, Mora V, Damaj MB, Padilla CS, Ramos N, Rossi D, Solís-Gracia N, Vargas-Bautista C, Irigoyen S, DaSilva JA, Mirkov TE, Mandadi KK. (2018). A biolistic-based genetic transformation system applicable to a broad-range of sugarcane and energycane varieties. *GM Crops Food* 9(4), 211–227. https://doi.org/10.1080/21645698.2018.1553836

Rathus C, Birch RG. (1992). Stable transformation of callus from electroporated sugarcane protoplasts. *Plant Sci.* 82(1), 81–89. https://doi.org/10.1016/0168-9452(92)90010-J

Renouf MA, Wegener MK, Nielsen LK. (2008). An environmental life cycle assessment comparing Australian sugarcane with US corn and UK sugar beet as producers of sugars for fermentation. *Biomass Bioenerg.* 32(12), 1144–1155. https://doi.org/10.1016/j.biombioe.2008.02.012

Rossouw D, Bosch S, Kossmann J, Botha FC, Groenewald J-H. (2007). Downregulation of neutral invertase activity in sugarcane cell suspension cultures leads to a reduction in respiration and growth and an increase in sucrose accumulation. *Funct. Plant Biol.* 34(6), 490–498. https://doi.org/10.1071/FP06214

Rossouw D, Kossmann J, Botha FC, Groenewald J-H. (2010). Reduced neutral invertase activity in the culm tissues of transgenic sugarcane plants results in a decrease in respiration and sucrose cycling and an increase in the sucrose to hexose ratio. *Funct. Plant Biol.* 37(1), 22–31. https://doi.org/10.1071/FP08210

Saini JK, Patel AK, Adsul M, Singhania RR. (2016). Cellulase adsorption on lignin: A roadblock for economic hydrolysis of biomass. *Renew. Energy* 98, 29–42. https://doi.org/10.1016/j.renene.2016.03.089

Salokhe S. (2021). Development of an efficient protocol for production of healthy sugarcane seed cane through Meristem culture. *J. Agric. Food Res.* 4, 100126. https://doi.org/10.1016/j.jafr.2021.100126

Sánchez ÓJ, Cardona CA. (2008). Trends in biotechnological production of fuel ethanol from different feedstocks. *Bioresour. Technol.* 99(13), 5270–5295. https://doi.org/10.1016/j.biortech.2007.11.013

Sathish D, Theboral J, Vasudevan V, Pavan G, Ajithan C, Appunu C, Manickavasagam M. (2020). Exogenous polyamines enhance somatic embryogenesis and Agrobacterium tumefaciens-mediated transformation efficiency in sugarcane (*Saccharum* spp. hybrid). *In Vitro Cell. Dev. Biol. - Plant* 56(1), 29–40. https://doi.org/10.1007/s11627-019-10022-6

Schlüter U, Weber APM. (2020). Regulation and evolution of C4 photosynthesis. *Ann. Rev. Plant Biol.* 71, 183–215. https://doi.org/10.1146/annurev-arplant-042916-040915

Sharma A, Singh RK, Singh P, Vaishnav A, Guo DJ, Verma KK, Li DP, Song XP, Malviya MK, Khan N, Lakshmanan P, Li YR. (2021). Insights into the bacterial and nitric oxide-induced salt tolerance in sugarcane and their growth-promoting abilities. *Microorganisms* 9, 2203. https://doi.org/10.3390/microorganisms9112203.

Sharma A, Song X-P , Singh RK, Vaishnav A, Gupta S, Singh P, Guo D-J, Verma KK, Li Y-R. (2022). Impact of carbendazim on cellular growth, defence system and plant growth promoting traits of *Priestia megaterium* ANCB-12 isolated from sugarcane rhizosphere. *Front. Microbiol.* 13, 1005942. doi:10.3389/fmicb.2022.1005942

Singh P, Singh RK, Li HB, Guo DJ, Sharma A, Verma KK, Solanki MK, Upadhyay SK, Lakshmanan P, Yang LT, Li YR. (2023). Nitrogen fixation and phytohormone stimulation of sugarcane plant through plant growth promoting diazotrophic *Pseudomonas*.

Biotechnol. Genetic Eng. Rev. 40(1), 15–35. https://doi.org/10.1080/02648725.2023.2177814

Singh P, Singh RK, Song QQ, Li HB, Guo DJ, Malviya MK, Verma KK, Song XP, Lakshmanan P, Yang LT, Li YR. (2021). Comparative analysis of protein and differential responses of defense-related gene and enzyme activity reveals the long-term molecular responses of sugarcane inoculated with *Sporisorium scitamineum*. *J. Plant Interact.* 16(1), 12–29.

Singh RK, Singh P, Li HB, Song QQ, Guo DJ, Solanki MK, Verma KK, Malviya MK, Song XP, Lakshmanan P, Yang LT, Li YR. (2020). Diversity of nitrogen-fixing rhizobacteria associated with sugarcane: A comprehensive study of plant-microbe interactions for growth enhancement in *Saccharum* spp. *BMC Plant Biol.* 20, 220. https://doi.org/10.1186/s12870-020-02400-9.

Skoog F, Miller CO. (1957). Chemical regulation of growth and organ formation in plant tissues cultured in vitro. *Symposia Soc. Exp. Biol.* 11, 118–130. http://europepmc.org/abstract/MED/13486467

Solanki MK, Verma KK, Dastogeer KMG, Mora-Poblete F, Mundra S. (2023). Microbial resilience in plant nutrient management towards sustainable farming. *Front. Microbiol.* 14, 1280811. https://doi.org/10.3389/fmicb.2023.1280811

Somerville C, Youngs H, Taylor C, Davis SC, Long SP. (2010). Feedstocks for lignocellulosic biofuels. *Science* 329(5993), 790–792. https://doi.org/10.1126/science.1189268

Song X-P, Mo F, Verma KK, Wei J, Zhang X, Yang L, Li YR. (2019). Effect of sugarcane smut (*Ustilago scitaminea* Syd.) on ultrastructure and biochemical indices of sugarcane. *Biomed. J. Sci. Tech. Res.* 002950. https://doi.org/10.26717/BJSTR.2019.17.002950.

Song XP, Mo F, Yan M, Zhang X, Zhang B, Huang X, Huang D, Pan Y, Verma KK, Li YR. (2022). Effect of smut infection on the photosynthetic physiological characteristics and related defense enzymes of sugarcane. *Life* 12, 1201. https://doi.org/10.3390/life12081201

Song XP, Verma KK, Tian DD, Zhang XQ, Liang YJ, Huang X, Li CN, Li YR. (2021). Exploration of silicon functions to integrate with biotic stress tolerance and crop improvement. *Biol. Res.* 54, 19. https://doi.org/10.1186/s40659-021-00344-4

Spreitzer RJ, Salvucci ME. (2002). RUBISCO: Structure, regulatory interactions, and possibilities for a better enzyme. *Ann. Rev. Plant Biol.* 53, 449–475. https://doi.org/10.1146/annurev.arplant.53.100301.135233

Stoutjesdijk PA, Singh SP, Liu Q, Hurlstone CJ, Waterhouse PA, Green AG. (2002). hpRNA-mediated targeting of the arabidopsis FAD2 gene gives highly efficient and stable silencing. *Plant Physiol.* 129(4), 1723–1731. https://doi.org/10.1104/pp.006353

Sturm A. (1999). Invertases. Primary structures, functions, and roles in plant development and sucrose partitioning. *Plant Physiol.* 121(1), 1–8. https://doi.org/10.1104/pp.121.1.1

Sugiharto B, Harmoko R, Sawitri WD. (2022). Biotechnological approaches to improve sugarcane quality and quantum under environmental stresses. In: Verma KK, Song X-P, Rajput VD, Solomon S, Li Y-R, Rao GP (Eds.), *Agro-industrial Perspectives on Sugarcane Production under Environmental Stress*. Springer Nature Singapore; pp. 267–300. https://doi.org/10.1007/978-981-19-3955-6_14

Sugiyama T, Mizuno M, Hayashi M. (1984). Partitioning of nitrogen among ribulose-1,5-bisphosphate carboxylase/oxygenase, phosphoenolpyruvate carboxylase, and pyruvate orthophosphate dikinase as related to biomass productivity in maize seedlings 1. *Plant Physiol.* 75(3), 665–669. https://doi.org/10.1104/pp.75.3.665

Suherman, Wijayanto SI, Anur RM, Neliana IR, Dewanti P, Sugiharto B. (2022). Field evaluation on growth and productivity of the transgenic sugarcane lines overexpressing sucrose-phosphate synthase. *Sugar Tech.* https://doi.org/10.1007/s12355-022-01121-7

Taylor SH, Hulme SP, Rees M, Ripley BS, Ian Woodward F, Osborne CP. (2010). Ecophysiological traits in C3 and C4 grasses: A phylogenetically controlled screening experiment. *New Phytologist.* 185(3), 780–791. https://doi.org/10.1111/j.1469-8137.2009.03102.x

Tian D, Qin L, Verma KK, Wei L, Li J, Li B, Zhou W, He Z, Wei D, Huang S, Long S, Huang Q, Li C, Wei S. (2023). Transcriptomic and metabolomic differences between banana varieties which are resistant or susceptible to *Fusarium wilt. PeerJ* 11, e16549. https://doi.org/10.7717/peerj.16549

Turi CE, Axwik KE, Murch SJ. (2014). In vitro conservation, phytochemistry, and medicinal activity of *Artemisia tridentata* Nutt.: Metabolomics as a hypothesis-generating tool for plant tissue culture. *Plant Growth Regul.* 74(3), 239–250. https://doi.org/10.1007/s10725-014-9915-y

van der Merwe MJ, Groenewald J-H, Stitt M, Kossmann J, Botha FC. (2010). Downregulation of pyrophosphate: D-fructose-6-phosphate 1-phosphotransferase activity in sugarcane culms enhances sucrose accumulation due to elevated hexose-phosphate levels. *Planta* 231(3), 595–608. https://doi.org/10.1007/s00425-009-1069-1

Vanhercke T, El Tahchy A, Liu Q, Zhou X-R, Shrestha P, Divi UK, Ral J-P, Mansour MP, Nichols PD, James CN, Horn PJ, Chapman KD, Beaudoin F, Ruiz-López N, Larkin PJ, de Feyter RC, Singh SP, Petrie JR. (2014). Metabolic engineering of biomass for high energy density: Oilseed-like triacylglycerol yields from plant leaves. *Plant Biotechnol. J.* 12(2), 231–239. https://doi.org/10.1111/pbi.12131

Vanholme R, Demedts B, Morreel K, Ralph J, Boerjan W. (2010). Lignin biosynthesis and structure. *Plant Physiol.* 153(3), 895–905. https://doi.org/10.1104/pp.110.155119

Verma KK, Anas M, Chen Z, Rajput VD, Malviya MK, Verma CL, Singh RK, Singh P, Song XP, Li YR. (2020f). Silicon supply improves leaf gas exchange, antioxidant defense system and growth in sugarcane responsive to water limitation. *Plants* 9, 1032, https://doi.org/10.3390/plants9081032.

Verma KK, Joshi A, Song X-P, Liang Q, Xu L, Huang H-r, Wu K-C, Seth CS, Arora J and Li Y-R. (2024a). Regulatory mechanisms of plant rhizobacteria on plants to the adaptation of adverse agroclimatic variables. *Front. Plant Sci.* 15, 1377793. https://doi.org/10.3389/fpls.2024.1377793

Verma KK, Joshi A, Song X-P, Singh S, Kumari A, Arora J, Singh SK, Solanki MK, Seth CS, Li Y-R. (2024b). Synergistic interactions of nanoparticles and plant growth promoting rhizobacteria enhancing soil-plant systems: A multigenerational perspective. *Front. Plant Sci.* 15, 1376214. https://doi.org/10.3389/fpls.2024.1376214

Verma KK, Li DM, Singh M, Rajput VD, Malviya MK, Minkina T, Singh RK, Singh P, Song XP, Li YR. (2020a). Interactive role of silicon and plant-rhizobacteria mitigating abiotic stresses: A new approach for sustainable agriculture and climate change. *Plants* 9, 1055. https://doi.org/10.3390/plants9091055.

Verma KK, Liu X-H, Wu K-C, Singh RK, Song QQ, Malviya MK, Song X-P, Singh P, Verma CL, Li YR. (2020e). The impact of silicon on photosynthetic and biochemical responses of sugarcane under different soil moisture levels. *Silicon* 12, 1355–1367.

Verma KK, Rajput VD, Singh M, Verma CL, Minkina T. (2022a). Influence of waterlogging on plant physiology and molecular biology. In: *Understanding Abiotic Stresses.* Nova Science Publishers, Inc.: New York.

Verma KK, Siddiqui N, Singh M. (2022b). Understanding the plant physiological and molecular adaptation strategies to soil salinity. In: *Understanding Abiotic Stresses.* Nova Science Publishers, Inc.: New York. https://doi.org/ 10.52305/UBUJ4024

Verma KK, Singh P, Song X-P, Malviya MK, Singh RK, Chen G-L, Solomon S, Li YR. (2020b). Mitigating climate change for sugarcane improvement: Role of silicon in alleviating abiotic stresses. *Sugar Tech.* 22(5), 741–749. https://doi.org/10.1007/s12355-020-00831-0.

Verma KK, Singh RK, Song QQ, Singh P, Zhang B-Q, Song X-P, Chen G-L, Li YR. (2019a). Silicon alleviates drought stress of sugarcane plants by improving antioxidant responses. *Biomed. J. Sci. Tech. Res.* 002957. https://doi.org/10.26717/BJSTR.2019.17.002957

Verma KK, Song X, Degu HD, Guo D, Joshi A, Huang H, Xu L, Singh M, Huang D, Rajput VD, Li YR. (2023a). Recent advances in nitrogen and nano-nitrogen fertilizers for sustainable crop production: A mini-review. *Chem. Biol. Technolog. Agricult.* 10, 1–14. https://doi.org/10.1186/s40538-023-00488-3

Verma KK, Song X, Verma CL, Huang H, Singh M, Xu L, Wu J, Khan ZH, Zhang B, Li Y. (2023b). Mathematical modeling of climate and fluoride effects on sugarcane photosynthesis with silicon nanoparticles. *Plant Physiol. Biochem.* 204, 108089. https://doi.org/10.1016/j.plaphy.2023.108089

Verma KK, Song XP, Budeguer F, Nikpay A, Enrique R, Singh M, Zhang B, Wu, JM, Li YR. (2022c). Genetic engineering: An efficient approach to mitigating biotic and abiotic stresses in sugarcane cultivation. *Plant Signal. Behav.* 17(1), e2108253.

Verma KK, Song XP, Joshi A, Rajput VD, Singh M, Sharma A, Singh RK, Li DM, Arora J, Minkina T, Li YR. (2022d). Nanofertilizer possibilities for healthy soil, water and food in future: An overview. *Front. Plant Sci.* 13, 865048. *https://doi.org/10.3389/fpls.2022.865048*

Verma KK, Song XP, Joshi A, Tian DD, Rajput VD, Singh M, Arora J, Minkina TM, Li YR. (2022e). Recent trends in nano-fertilizer for advancing sustainable agriculture under the era of climate change to ensure food security for future generations. *Nanomaterials* 12, 173. https://doi.org/10.3390/nano12010173.

Verma KK, Song XP, Kumari A, Jagadesh M, Singh SK, Bhatt R, Singh M, Seth CS, Li YR. (2024c). Climate change adaptation: Challenges for agricultural sustainability. *Plant, Cell Environ.* https://doi.org/10.1111/pce.15078

Verma KK, Song XP, Li DM, Singh M, Wu JM, Singh RK, Sharma A, Zhang BQ, LI YR. (2022f). Silicon and soil microorganisms improves rhizospheric soil health with bacterial community, plant growth, performance and yield. *Plant Signal. Behav.* 17(1), e2104004.

Verma KK, Song XP, Lin B, Guo DJ, Singh M, Rajput VD, Singh RK, Singh P, Sharma A, Malviya MK, Chen GL, Li YR. (2022g). Silicon induced drought tolerance in crop plants: Physiological adaptation strategies. *Silicon* 14(6), 2473–2487. https://doi.org/10.1007/s12633-021-01071-x

Verma KK, Song XP, Rajput VD, Boldyreva V, Zhang BQ, Minkina T, Li YR. (2022h). Morpho-physiological, biochemical and ultrastructural modifications on sugarcane to prolonged water deficit. In: Verma KK et al. (Eds.), *Agro-industrial Perspectives on Sugarcane Production under Environmental Stress*. https://doi.org/10.1007/978-981-19-3955-6_11.

Verma KK, Song XP, Rajput VD, Solomon S, Li YR, Rao GP. (2022i). *Agro-industrial Perspectives on Sugarcane Production under Environmental Stress*. Springer Nature Singapore, https://doi.org/10.1007/978-981-19-3955-6

Verma KK, Song XP, Singh M, Tian DD, Rajput VD, Minkina T, Li YR. (2023c). Association of silicon and soil microorganisms induces stress mitigation, increasing plant productivity. In: de Mello Prado R (Ed.), *Benefits of Silicon in the Nutrition of Plants*; pp. 299–328. https://doi.org/10.1007/978-3-031-26673-7_17

Verma KK, Song XP, Tian DD, Guo DJ, Chen ZL, Zhong CS, Nikpay A, Singh M, Rajput VD, Singh RK, Minkina T, Li YR. (2021a). Influence of silicon on biocontrol strategies to manage biotic stress for crop protection, performance and improvement. *Plants* 10, 2163. https://doi.org/10.3390/plants10102163

Verma KK, Song XP, Tian DD, Singh M, Verma CL, Rajput VD, Singh RK, Sharma A, Singh P, Malviya MK, Li YR. (2021b). Investigation of defensive role of silicon during drought stress induced by irrigation capacity in sugarcane: Physiological and biochemical characteristics. *ACS Omega* 6, 19811–19821.

Verma KK, Song XP, Verma CL, Chen ZL, Rajput VD, Wu KC, Liao F, Chen GL, Li YR. (2021c). Functional relationship between photosynthetic leaf gas exchange in response to silicon application and water stress mitigation in sugarcane. *Biol. Res.* 54, 15. https://doi.org/10.1186/s40659-021-00338-2

Verma KK, Song XP, Verma CL, Malviya MK, Guo DJ, Rajput VD, Sharma A, Wei KJ, Chen GL, Solomon S, Li YR. (2021d). Predication of photosynthetic leaf gas exchange of sugarcane (*Saccharum* spp.) leaves in response to leaf positions to foliar spray of potassium salt of active phosphorus under limited water irrigation. *ACS Omega* 6, 2396–2409.

Verma KK, Song X-P, Xu L, Huang H-R, Liang Q, Seth CS, Li Y-R. (2023d). Nano-microplastic and agro-ecosystems: A mini-review. *Front. Plant Sci.* 14, 1283852. https://doi.org/10.3389/fpls.2023.1283852

Verma KK, Song XP, Yadav G, Degu HD, Parvaiz A, Singh M, Huang HR, Mustafa G, Xu L, Li YR. (2022j). Impact of agroclimatic variables on proteo-genomics in sugarcane (*Saccharum* spp.) plant productivity. *ACS Omega* 7, 22997–23008.

Verma KK, Song XP, Zeng Y, Guo DJ, Sing M, Rajput VD, Malviya MK, Wei KJ, Sharma A, Li DP, Chen GL, Li YR. (2021e). Foliar application of silicon boosts growth, photosynthetic leaf gas exchange, antioxidative response and resistance to limited water irrigation in sugarcane (*Saccharum officinarum* L.). *Plant Physiol. Biochem.* 166, 582–592.

Verma KK, Song XP,Zeng Y, Li DM, Guo DJ, Rajput VD, Chen GL, Barakhov A, Minkina TM, Li YR. (2020c). Characteristics and correlation of leaf stomata and its relationship with photosynthesis on *Saccharum* spp. under different irrigation and silicon application. *ACS Omega* 5, 24145–24153.

Verma KK, Vatsal S, Singh M, Rajput VD, Minkina T. (2022k). Coping with water stress: Effects and adaptive mechanisms to plants. In: *Understanding Abiotic Stresses*. Nova Science Publishers, Inc.: New York.

Verma KK, Wu KC, Verma CL, Li DM, Malviya MK, Singh RK, Singh P, Chen GL, Song XP, Li YR. (2020d). Developing mathematical model for diurnal dynamics on photosynthetic responses in sugarcane responsive to different irrigation and silicon application. *PeerJ* 8, e10154. https://doi.org/10.7717/peerj.10154.

Verma KK, Wu K-C, Singh P, Malviya MK, Singh RK, Song X-P, Li YR. (2019b). The protective role of silicon in sugarcane under water stress: Photosynthesis and antioxidant enzymes. *Biomed. J. Sci. Tech. Res.* 002685. https://doi.org/0.26717/BJSTR.2019.15.002685

Verma AK, Upadhyay SK, Verma PC, Solomon S, Singh SB. (2011). Functional analysis of sucrose phosphate synthase (SPS) and sucrose synthase (SS) in sugarcane (Saccharum) cultivars. *Plant Biol.* 13(2), 325–332. https://doi.org/10.1111/j.1438-8677.2010.00379.x

Waclawovsky AJ, Sato PM, Lembke CG, Moore PH, Souza GM. (2010). Sugarcane for bioenergy production: An assessment of yield and regulation of sucrose content. *Plant Biotechnol. J.* 8(3), 263–276. https://doi.org/10.1111/j.1467-7652.2009.00491.x

Wang T, Wang B, Hua X, Tang H, Zhang Z, Gao R, Qi Y, Zhang Q, Wang G, Yu Z, Huang Y, Zhang Z, Mei J, Wang Y, Zhang Y, Li Y, Meng X, Wang Y, Pan H, Chen S, Li Z, Shi H, Liu X, Deng Z, Chen B, Zhang M, Gu L, Wang J, Ming R, Yao W, Zhang J. (2023). A complete gap-free diploid genome in Saccharum complex and the genomic footprints of evolution in the highly polyploid Saccharum genus. *Nat. Plants* 9(4), 554–571. https://doi.org/10.1038/s41477-023-01378-0

Wang WZ, Yang BP, Feng CL, Wang JG, Xiong GR, Zhao TT, Zhang SZ. (2017a). Efficient sugarcane transformation via bar gene selection. *Trop. Plant Biol.* 10(2), 77–85. https://doi.org/10.1007/s12042-017-9186-7

Wang WZ, Yang BP, Feng XY, Cao ZY, Feng CL, Wang JG, Xiong GR, Shen LB, Zeng J, Zhao TT, Zhang SZ. (2017b). Development and characterization of transgenic sugarcane with insect resistance and herbicide tolerance. *Front. Plant Sci.* 8, 1535. https://doi.org/10.3389/fpls.2017.01535

Waterhouse PM, Graham MW, Wang M-B. (1998). Virus resistance and gene silencing in plants can be induced by simultaneous expression of sense and antisense RNA. *Proc. Natl. Acad. Sci.* 95(23), 13959–13964. https://doi.org/10.1073/pnas.95.23.13959

Wesley SV, Helliwell CA, Smith NA, Wang M, Rouse DT, Liu Q, Gooding PS, Singh SP, Abbott D, Stoutjesdijk PA, Robinson SP, Gleave AP, Green AG, Waterhouse PM. (2001). Construct design for efficient, effective and high-throughput gene silencing in plants. *Plant J.* 27(6), 581–590. https://doi.org/10.1046/j.1365-313X.2001.01105.x

Whittaker A, Botha FC. (1997). Carbon partitioning during sucrose accumulation in sugarcane internodal tissue. *Plant Physiol.* 115(4), 1651–1659. https://doi.org/10.1104/pp.115.4.1651

Whittaker A, Botha FC. (1999). Pyrophosphate: d-fructose-6-phosphate 1-phosphotransferase activity patterns in relation to sucrose storage across sugarcane varieties. *Physiol. Plant.* 107(4), 379–386. https://doi.org/10.1034/j.1399-3054.1999.100402.x

Winichayakul S, Scott RW, Roldan M, Hatier J-HB, Livingston S, Cookson R, Curran AC, Roberts NJ. (2013). In vivo packaging of triacylglycerols enhances arabidopsis leaf biomass and energy density. *Plant Physiol.* 162(2), 626–639. https://doi.org/10.1104/pp.113.216820

Wu K-C, Huang C-M, Verma KK, Deng Z-N, Huang H-R, Pang T, Cao H-Q, Luo H-B, Jiang S-L, Xu L. (2022). Transcriptomic responses of *Saccharum spontaneum* roots in response to polyethylene glycol-6000 stimulated drought stress. *Front. Plant Sci.* 13, 992755. https://doi.org/10.3389/fpls.2022.992755

Wu H, Awan FS, Vilarinho A, Zeng Q, Kannan B, Phipps T, McCuiston J, Wang W, Caffall K, Altpeter F. (2015). Transgene integration complexity and expression stability following biolistic or Agrobacterium-mediated transformation of sugarcane. *In Vitro Cell. Dev. Biol. Plant* 51(6), 603–611. https://doi.org/10.1007/s11627-015-9710-0

Xu L, Deng Z-N, Wu K-C, Malviya MK, Solanki MK, Verma KK, Pang T, Li Y-J, Liu X-Y, Kashyap BK, Dessoky ES, Wang W-Z and Huang H-R. (2022). Transcriptome analysis reveals a gene expression pattern that contributes to sugarcane bud propagation induced by indole-3-butyric acid. *Front. Plant Sci.* 13, 852886. https://doi.org/10.3389/fpls.2022.852886

Xu C, Shanklin J. (2016). Triacylglycerol metabolism, function, and accumulation in plant vegetative tissues. *Ann. Rev. Plant Biol.* 67, 179–206. https://doi.org/10.1146/annurev-arplant-043015-111641

Yadav P, Priyanka P, Kumar D, Yadav A, Yadav K. (2019). Bioenergy crops: Recent advances and future outlook. In: Rastegari AA, Yadav AN, Gupta A (Eds.), *Prospects of Renewable Bioprocessing in Future Energy Systems*. Springer International Publishing; pp. 315–335. https://doi.org/10.1007/978-3-030-14463-0_12

Yang Y, Munz J, Cass C, Zienkiewicz A, Kong Q, Ma W, Sanjaya, Sedbrook J, Benning C. (2015). Ectopic expression of WRINKLED1 affects fatty acid homeostasis in *Brachypodium distachyon* vegetative tissues. *Plant Physiol.* 169(3), 1836–1847. https://doi.org/10.1104/pp.15.01236

Yao W, Ruan M, Qin L, Yang C, Chen R, Chen B, Zhang M. (2017). Field performance of transgenic sugarcane lines resistant to sugarcane mosaic virus. *Front. Plant Sci.* 8, 104. https://doi.org/10.3389/fpls.2017.00104

Yuan JS, Tiller KH, Al-Ahmad H, Stewart NR, Stewart CN. (2008). Plants to power: Bioenergy to fuel the future. *Trends Plant Sci.* 13(8), 421–429. https://doi.org/10.1016/j.tplants.2008.06.001

Zale J, Jung JH, Kim JY, Pathak B, Karan R, Liu H, Chen X, Wu H, Candreva J, Zhai Z, Shanklin J, Altpeter F. (2016). Metabolic engineering of sugarcane to accumulate energy-dense triacylglycerols in vegetative biomass. *Plant Biotechnol. J.* 14(2), 661–669. https://doi.org/10.1111/pbi.12411

Zhao Y, Kim JY, Karan R, Jung JH, Pathak B, Williamson B, Kannan B, Wang D, Fan C, Yu W, Dong S, Srivastava V, Altpeter F. (2019). Generation of a selectable marker free, highly expressed single copy locus as landing pad for transgene stacking in sugarcane. *Plant Mol. Biol.* 100(3), 247–263. https://doi.org/10.1007/s11103-019-00856-4

Zheng F-J, Lin B, Yang Y-X, Fang X-C, Verma KK, Chen G-L. (2023). Efficacy and functionality of sugarcane original vinegar on mice. *Front. Microbiol.* 14, 1224666. https://doi.org/10.3389/fmicb.2023.1224666

Zhu X-G, Long SP, Ort DR. (2010). Improving photosynthetic efficiency for greater yield. *Ann. Rev. Plant Biol.* 61, 235–261. https://doi.org/10.1146/annurev-arplant-042809-112206

Zhu YJ, Komor E, Moore PH. (1997). Sucrose accumulation in the sugarcane stem is regulated by the difference between the activities of soluble acid invertase and sucrose phosphate synthase. *Plant Physiol.* 115(2), 609–616. https://doi.org/10.1104/pp.115.2.609

CHAPTER 5

Germplasm Diversity and Related Genera for Use in Sugarcane Betterment

SHAHID AFGHAN,[1*] MUHAMMAD EHSAN KHAN,[1] KRISHAN K. VERMA,[2] and AMIN NIKPAY[3]

[1]*Sugarcane Research and Development Board, Ayub Agricultural Research Institute, Faisalabad, Punjab, Pakistan*

[2]*Sugarcane Research Institute, Guangxi Academy of Agricultural Sciences/Key Laboratory of Sugarcane Biotechnology and Genetic Improvement (Guangxi)/Guangxi Key Laboratory of Sugarcane Genetic Improvement, Ministry of Agriculture and Rural Affairs, Nanning, Guangxi, China*

[3]*Department of Plant Protection, Sugarcane & By-products Development Company, Salman Farsi Agro-Industry, Ahwaz, Iran*

*Corresponding author

ABSTRACT

The significance of sugarcane germplasm, genetic material vital for breeding and preserving varieties is immense for global agriculture and the economy. Leading producers like Brazil, India, China, Thailand, Pakistan, Mexico, and Australia have actively collected and assessed diverse germplasm since the 1950s. This chapter involves utilizing noble and wild species, synthesizing desired traits, refining selection techniques, and advancing molecular genetics. International exchange of elite clones remains pivotal. Brazil has developed over 200 high-yield, resilient cultivars and pioneered transgenic clones for drought and pest resistance, though commercial use is pending. India operates a national network

Sugarcane Transformation: Biofuels, Biotechnology, and Beyond.
Krishan K. Verma, Xiu-Peng Song, Hai-Rong Huang, Aradhna Kumari, and Yang-Rui Li (Eds.)

DOI: 10.1201/9781779641571-5

of breeding centers, maintaining a core collection of 400 accessions. China preserves over 3300 accessions and has successfully introgressed genes from wild relatives like *Saccharum spontaneum*, whose genome was recently sequenced, providing a valuable genomic resource. In Pakistan, breeding relies heavily on imported germplasm due to climatic constraints, resulting in 36 commercially released varieties. A landmark global achievement was the 2024 complete genome sequencing of the R570 cultivar, unlocking new potential for improving yield, ratoon ability, and stress resistance. This chapter summarizes the status of sugarcane germplasm in major producing countries, highlighting its critical role in crop enhancement.

5.1 INTRODUCTION

The genetic variety of sugarcane germplasm among several *Saccharum* species and related taxa provides an extensive gene pool from which new varieties of any trait can be developed. Most of the countries engaged in sugarcane improvement have a big collection of sugarcane clones. Sugarcane betterment became a professionally guided and scientifically endeavor in 2024 when 8.70 billion decoded base pairs make up the full sugarcane genome of R570 that is 20 times more than the rice reference genome and three times more than the human genome (Healey et al., 2024). The initial clones for variety development globally were developed at this time by researchers in India, China, Brazil, USA, Réunion, Thailand, Japan, and Indonesia through the generation of interspecific hybrids between *Saccharum officinarum* and *Saccharum spontaneum*. Interspecific crosses have since served as the cornerstone for the growth of the modern sugarcane industry worldwide.

Changes in the environment, human activity, and the crop's increasing significance for the manufacture of sugar, ethanol, and energy from its biomass all contribute to the need for systematic research into the preservation, characterization, and use of accessions from every genera in the *Saccharum* complex. It is clear that several workers have made extensive use of the genetic resources, although the number of ancestral clones from *S. officinarum, S. spontaneum*, and *Saccharum barberi* that have been successful in producing released varieties is limited.

It is exciting to learn that a large number of hybrids produced from different species combinations—many of which are unique creations—have succeeded in producing clones that are either commercially or nearly economically viable. Through conventional breeding, genomic breeding, and gene harvesting, all associated activities related to collection, preservation, characterization, documentation, and exploitation should strive toward better systematization. With the help of excellent molecular, genomic, and bioinformatics technologies, the genetic diversity found in the germplasm provides chances for developing sugarcane varieties that may significantly increase yield in any setting and produce a variety of goods in the years to come (Sharma et al., 2021; Guo et al., 2021, 2022, 2023; Huang et al., 2024). Germplasm diversity and related genera are essential for sugarcane improvement through breeding and genetic research. Sugarcane (*S. officinarum*) is a complex hybrid crop, and its genetic base is narrow due to its evolution from a few ancestral species.

To broaden its genetic diversity, breeders and researchers explore related genera and wild species for desirable traits such as disease resistance, pest-insect tolerance, high/low-temperature/drought tolerance, ratoonability, and increased sugar content. Some related genera and species used in sugarcane improvement include:

- *S. spontaneum* (Wild sugarcane)
- *Saccharum robustum* (Robust cane)
- *S. barberi* (Indian sugarcane)
- *Saccharum sinense* (Chinese sugarcane)
- *Erianthus arundinaceus* (Wild cane)
- *Sorghum* spp. (Sorghum)
- *Zea mays* (Maize)
- *Miscanthus* spp. (Miscanthus)

These species and genera offer valuable genetic resources for sugarcane breeding programs, enabling the development of more resilient, productive, and sustainable sugarcane varieties. Some of the desirable traits and characteristics targeted through germplasm diversity and related genera include:

- Disease resistance (e.g., red rot, yellow spot)
- Drought tolerance

- Improved sugar content and yield
- Increased resistance to pests and nematodes
- Enhanced adaptability to diverse environments
- Improved ratooning ability (regrowth after harvest)

By exploring and utilizing the genetic diversity within sugarcane and related genera, researchers and breeders can develop more robust and sustainable sugarcane varieties, addressing the challenges of climate change (Kumari et al., 2023a, 2023b; Chaudhary et al., 2023; Jagadesh et al., 2024), disease, and water scarcity, while meeting the increasing demand for sugar and bioenergy (Singh et al., 2020; Chen et al., 2020, 2023; Solanki et al., 2023; Zheng et al., 2023; Bhatt et al., 2023a, 2023b). Sugarcane (*Saccharum* spp.) is a crop of significant economic importance, and its improvement is crucial for meeting the increasing demand for sugar and biofuels (Verma et al., 2019a, 2019b; 2020a, 2020b; 2022a, 2022b, 2022c; 2023a, 2023b; Joshi et al., 2022, 2023). Germplasm diversity is essential for crop improvement, and sugarcane has a rich genetic resource base (Verma et al., 2023c, 2023d; Deng et al., 2023).

This chapter explores the germplasm diversity of sugarcane and related genera, their utilization in sugarcane betterment, and the potential for future improvements along with germplasm status of leading sugarcane-producing countries of the world (Verma et al., 2020c, 2020d; Anas et al., 2021a, 2021b; Wu et al., 2022; Xu et al., 2022; Li et al., 2023). The finding of DNA markers will be crucial for the development of cultivars that are resistant to both biotic and abiotic stress. The introduction of new cane varieties will increase production capacity and reduce costs for the production of sugar and other important economic products. The probability of varietal failure owing to environmental pressures such as drought, frost, and salinity hazards is mitigated by the evolution of new site-specific cultivars. Innovative varieties that maximize the use of fertilizer, irrigation resources, and harvesting would benefit growers economically. The growers' level of living will rise as a result of their economic advantages. Influence of diseases and pest-tolerant varieties would create friendly environment for sustainable and profitable production (Song et al., 2019, 2021, 2022; Singh et al., 2023).

5.2 ORIGIN OF SUGARCANE

Sugarcane, originally native to tropical regions, is now cultivated in both tropical and subtropical areas, including climates previously considered unsuitable for its growth (Verma et al., 2021a, 2021b; 2022c, d, 2022e; Sharma et al., 2022). This widespread cultivation is possible due to morphological and genetic modifications resulting from interbreeding among various species. Through this process, cane varieties resistant to diverse growing conditions have been developed (Oliveira et al., 2022; Huang et al., 2022). Understanding the origin and botanical classification of sugarcane is crucial for harnessing specific genomes to enhance particular traits. Sugarcane, a perennial grass, consists of cultivated varieties that are hybrids of different species within the genus *Saccharum*. This genus traces its ancestry back to specific family predecessors.

Family	:	Poaceae (Gramineae)
Subfamily	:	Panicoideae
Tribe	:	Andropogoneae
Subtribe	:	Saccharineae
Group	:	Saccharastrae
Genus	:	*Saccharum*

Source: Malik (2018).

5.2.1 GERMPLASM DIVERSITY IN SUGARCANE

Sugarcane is a complex crop with a wide range of genetic diversity (Malviya et al., 2021, 2022). Within the tribe Andropogoneae, aside from the genus *Saccharum*, there are eight other genera: *Miscanthus*, *Erianthus, Narenga, Sclerostachya, Miscanthidium, Eccoilopus, Erichrysis*, and *Imperata*. Due to morphological similarities, cross-ability, breeding behavior, and evolutionary relationships, cane breeders primarily focus on the genera *Saccharum*, *Erianthus*, *Narenga*, *Miscanthus*, and *Sclerostachya*, collectively termed the *Saccharum* complex (Mukerjee 1957) as shown in Table 5.1. Additionally, the genus Ripidium was later identified as a section of Erianthus.

TABLE 5.1 *Saccharum* Species and the Related Genera Utilized for the Transfer of Species Traits

Species/Genera	Characters
S. officinarum	Sucrose, stalk weight
S. bareri	Sucrose, resistance to drought, cold, salinity, red rot
S. sinense	Sucrose, resistance to drought, cold, salinity, red rot
S. robustum	Yield, fiber, water logging resistance
S. spontaneum	Tillering, vigor, yield, resistance to red rot, smut, rust, pests, drought, waterlogging.
Saccharum edule	Sucrose, stalk weight, resistance to drought
Erianthus spp.	High biomass production, fiber, pest resistance, drought, waterlogging, salinity.
Miscanthus	Cold tolerance, diseases resistance
Neranga	High tillering, earliness, water logging resistance
Sclerostachya	High tillering, earliness, water logging resistance

Source: Adapted with permission from Malik (2018).

The breeding strategies of various sugarcane-producing countries aim to expand genetic diversity by crossing and backcrossing existing hybrids with wild relatives (Legendre 1989). Besides utilizing *S. spontaneum* and *S. robustum*, incorporating *Erianthus* species is crucial for enhancing productivity and adaptability (Hogarth et al., 1997). Other strategies include interspecific enhancement of individual *Saccharum* species before interspecific hybridization (Nair et al., 1998) and diversifying the cytoplasm of new hybrids using *S. barberi*, *S. robustum*, and *Erianthus* as female parents. The goal is to develop novel varieties as complex hybrids of interspecific and intergeneric origin, with Sclerostachya also contributing to hybrid development (Legendre, 1989).

In most hybridization programs, using *Erianthus* has proven effective in generating varieties with high biomass production and resilience to pests and diseases (Nair, 2005; Singh et al., 2021). The high polyploidy of *Saccharum* species poses challenges for precise analysis of their relationships using nuclear markers. Genomic *in situ* hybridization studies have revealed that *S. barberi* and *S. sinense* originated from interspecific crosses between *S. officinarum* and *S. spontaneum*. The species S. officinarum, *S. barberi*, and S. *sinense* represent the sweet forms resulting from introgression within the *Saccharum* complex (Daniels et al., 1975). According to Daniels et al. (1975), the Indo-Burma-China region, along with *S. spontaneum, Erianthus sect.*

Ripidium, and *Sclerostachya*, forms the center of the *Saccharum* complex. Naidu and Srenivasan (1987) suggest that there are two major centers of diversity for sugarcane: New Guinea and adjoining islands, along with the chain of Indonesia for *S. officinarum and S. robustum*, and northern India and the region bordering China and Burma for *S. spontaneum* and related grasses such as *Erianthus, Sclerostachya*, and *Narenga*.

5.2.1.1 *SACCHARUM OFFICINARUM* L. (NOBILITY CANE) (2N = 80)

The species is primarily utilized for sugar production and is commonly referred to as noble cane. Generally, the species boasts thick, soft, and juicy stems with low fiber content, resulting in high sucrose levels. They typically exhibit poor tillering, shallow root systems, broad leaves, and a tendency to self-trash. While lacking resistance and hardiness, they possess high individual stalk weights and yield substantial cane and sugar under optimal conditions, making them ideal for milling in the sugar industry (Verma et al., 2020e, 2020f; 2021c; 2022f, 2022g, 2022h). Due to their soft and sweet nature, certain varieties are highly enjoyed as chewing cane (Figure 5.1).

FIGURE 5.1 *Saccharum officinarum* L.

As a group, they are susceptible to many serious cane diseases, although certain strains exhibit resistance to smut, Fiji disease, and downy mildew. Recognized as donors of sucrose genes, they are extensively utilized as premier breeding stock. While selfing may result in inbreeding depression, significant improvement has been achieved through intervariety crosses. In interspecific crosses with *S. spontaneum* and *S. barberi*, the species typically transmits "2*n*" gametes, but "*n*" gametes in interspecific crosses. Generally considered to possess $2n$ = 80 chromosomes (Vijayalakshmi, 1967), this species has been described by some as an octaploid or of even higher ploidy (Roche and Daniel, 1987). However, Jagathesan and Ratnambal (1967) indicated that its ploidy is not straightforward; certain varieties exhibit a variable number of chromosomes ranging from 68 to 99. Researchers have two main hypotheses: (1) that it evolved from *S. robustum* through long-term natural selection in New Guinea (Grassl, 1977), and (2) that it arose through introgression from *S. spontaneum*, *Miscanthus*, *Erianthus*, and *E. arundinaceus*, later migrating to regions of diversity where it adapted (Roche and Daniel, 1987).

5.2.1.2 SACCHARUM ROBUSTUM BRANDES AND JESW. EX GRASSL: (2N = 60, 80)

The stalk of this species is thick, tall, and erect, with some clones reaching heights of up to 10 m. The stalks tend to be thicker near the nodes, displaying swollen growth rings. They are characterized by their hardness, toughness, and high fiber content, with a pithy center containing minimal juice and sugar. Certain varieties are noted for their high Brix levels (Chandran et al., 2020). The species is primarily found in New Guinea, Indonesia, and partially in the Philippines. Geographically, it is the closest wild relative of *S. officinarum*. According to Daniel and Roche (1987), the species exhibits a highly diverse population derived from introgression of *S. spontaneum* with other genera such as *Erianthus*, *Arundinaceum*, and *Miscanthus* in New Guinea. Various forms of *S. robustum* have been utilized in breeding efforts, with varying degrees of success in regions such as Barbados, Queensland, Florida, and Hawaii (Babu, 1990). Notably, the varieties CP 50-28, CP 55-30, and CP 70-1527 are descendants of *robustum* hybrids.

5.2.1.3 SACCHARUM BARBERI JESW (2N = 111–120)

The species originates from northern India and is commonly referred to as Indian canes, named after the renowned breeder C. A. Barber.

Barber suggested that the species emerged from *S. spontaneum* through mutation and selection. Scientist proposed that it originated from natural hybridization between *S. officinarum* and *S. spontaneum* in the regions of West Bengal, Bihar, and Orissa. Extensive studies by Barber (1916, 1917, and 1922) categorized the species into five groups. These groups, as reviewed by Babu (1990), include *S. barberi*, *Mungo*, *Narenga*, *Saretha*, *Katha*, *Sunnabile*, and *Dhaulu* (Figure 5.2).

FIGURE 5.2 *Saccharum barberi.*

Source: The Agri times. https://theagritimes.blogspot.com/2016/06/saccharum-barberi.html

Generally, the canes are small-barreled, hard, and high in fiber content, with vigorous root systems. The juice from the Katha variety was particularly enjoyed as a beverage. These canes exhibit tolerance to extreme temperatures, drought, and waterlogging, along with immunity to sereh disease. Certain varieties display resistance to mosaic virus, although as a whole, they lack consistent resistance. Flowering in these species tends to be sparse and late, necessitating the identification of specific clones for flowering. Flowers exhibit a wide range of fertility, from normal seed setting in the Katha group to complete sterility in the Mungo group (Ethirajan, 1987). Male fertile varieties have been identified and utilized for interspecific hybrids. For instance, the variety Co 213 resulted from a cross between Black cheribon (*S. officinarum*) and chunnee (*S. barberi*).

5.2.1.4 SACCHARUM SINENSE (2N = 116–118)

The species encompasses Chinese and Japanese canes, characterized by their hard, thin, fibrous stalks, and tall stature with relatively broader leaves, primarily used as fodder. While the stalks are typically small-barreled, the Uba variety stands out for its thick, soft, and sweet characteristics, cultivated in China since prehistoric times. Although Uba boasts satisfactory sucrose content, its yield is relatively low. In China, it is primarily utilized for chewing and, to a lesser extent, for crude sugar production (Figure 5.3).

FIGURE 5.3 *Saccharum sinense.*

Source: farkomer. https://creativecommons.org/licenses/by/2.0/

Scientist also proposed that the species originated as a natural hybrid between indigenous forms of *S. spontaneum* and *S. officinarum*. Conversely, the species emerged in the China–Japan area through introgression between *S. officinarum* and tetra-ploid Miscanthus. Sreenivasan et al. (1987) theorized that various members of the *Saccharum* complex, including *Erianthus*, *Miscanthus*, *Sclerostachya*, and *S. officinarum*, contributed to introgression at different levels of ploidy in these species. Both *S. barberi* and *S. sinense* play significant roles in cane breeding due to their desirable traits such as a heavy, deep root system, and drought tolerance. Commercial varieties such as PoJ 213, PoJ 234, H32-1063, and Co213 are outcomes of their crosses. However, their utility in breeding is constrained by poor flowering and low fertility, likely linked to aneuploidy and irregular meiosis resulting from their hybrid origin (Sreenivasan and Jalaji, 1985).

5.2.1.5 SACCHARUM SPONTANEUM L. (2N = 40–128)

This species demonstrates high adaptability and resilience, exhibiting resistance to diseases, cold, drought, and other challenging growing conditions (Xu et al., 2022). Consequently, it is extensively utilized in interbreeding with *S. officinarum* to cultivate adaptable and robust germplasm suitable for diverse environments. Within this species, a diverse range of wild canes exists, spanning from dwarf bushy types with narrow leaves and thin stalks to erect specimens with broad leaves and thick, tall stems. Stalks may exhibit hardness, pithiness, or hollow centers. Notably, the species demonstrates prolific tillering and possesses a vigorous root system. While some varieties exhibit low sugar content and minimal juice quality, certain species display superior juice quality (Vijayalakshmi, 1967). Clones are immune to sereh and mosaic viruses and resistant to red rot. Additionally, they exhibit resilience to drought, cold, and salinity. Through its contributions of vigor, hardness, tillering capacity, and resistance to environmental stresses, diseases, and pests, this species enhances the adaptability and robustness of commercial varieties (Naidu and Sreenivasan, 1979).

Its habitat extends from Japan and Indonesia/New Guinea through the Indo-Pakistan subcontinent to the Mediterranean and Africa. India is recognized as both the center of origin and the center of diversity of *S. spontaneum* (Babu, 1980). It is widely acknowledged that this species originated in northern India and evolved through introgression among members of the *Saccharum* complex (Daniel et al., 1975).

It is rich in genetic variability with reference to morphological adaptation and taxonomic differences in different habitats of riverbeds, sea coasts, and forests. In their wild form, tall and thick types have been used as shelter, wind break, and fencing material of gardens (Mukerjee, 1957). Varieties with thin stalks and vigorous tillering are used as fodder for animals in famine days. The species has been extensively used in nobalization in *S. officinarum and S. robustum*. In India, the first product of nobalization was Co 205 (*Vellai* × *S. spontaneum*). Some try-specific crosses among (*S. officinarum* × *S. barberi*) × *S. spontaneum* have been made to develop useful commercial canes in India.

5.2.1.6 SACCHARUM EDULE HASSK (2N = 60–80)

It is a traditional Melanesian vegetable grown for its edible inflorescence in villages of New Guinea and Fiji (Daniels, 1967). With pubescent leaves, swollen and aborted inflorescence, and sterility, it holds limited importance for breeding.

Artschwager and Brandes (1958) proposed it as a mutant form of *S. robustum*, also suggested its origin from introgression of *S. robustum*. Currently, it is seen as a product of introgression involving *S. officinarum* or *S. robustum* with other genera (Roche and Daniels, 1987). The discussion highlights six sugarcane species and related genera such as *S. erianthus*, *S. miscanthus*, *S. narenga*, and *S. sclerestachya*, forming essential genetic resources known as the *Saccharum* complex (Mukerjee, 1957; Daniels et al.,, 1975). Though modern cane varieties are hybrids replacing pure species, the basic sources remain preserved in gene pools, conserved in two global collections such as in USA and India. A standardized method for *in vitro* storage at Kannur, India ensures preservation without significant changes.

5.3 BREEDING OBJECTIVES RELATED TO GERMPLASM

Sugarcane breeding aims to develop varieties with improved agronomic traits and resilience to biotic and abiotic stresses (Verma et al., 2021d, 2021e; 2022i, 2022j, 2022k). Germplasm collections provide breeders with a diverse pool of genetic material to introduce desirable traits into breeding populations through hybridization and selection.

5.3.1 ADAPTATION TO DIFFERENT ENVIRONMENTS

Sugarcane germplasm from different regions exhibits adaptability to various agro-climatic conditions, including temperature extremes, water availability, and soil types (Kumari et al., 2022; Krishna et al., 2022; Tian et al., 2023). By tapping into this diversity, breeders can develop varieties tailored to specific environments, thus expanding sugarcane cultivation to new regions.

5.3.2 DISEASE AND PEST RESISTANCE

Wild relatives of sugarcane often possess natural resistance to pests and diseases that affect cultivated varieties. Germplasm collections house these resistant traits, which can be introgressed into commercial cultivars to enhance their resistance to pathogens and pests, reducing the reliance on chemical inputs for pest management.

5.3.3 QUALITY IMPROVEMENT

Germplasm resources contain genetic variability for sugar content, fiber quality, juice purity, and other quality-related traits. Breeding programs leverage this diversity to develop varieties with improved sugar yield and quality, meeting the demands of sugar and ethanol industries.

5.3.4 STRESS TOLERANCE

Climate change and environmental stresses pose significant challenges to sugarcane production (Verma et al., 2024a, 2024b, 2024c). Germplasm collections offer a repository of genetic resources with traits conferring tolerance to drought, salinity, waterlogging, and other environmental stresses, enabling the development of resilient varieties.

5.3.5 HYBRID VIGOR

Hybridization between different germplasm sources can result in hybrid vigor or heterotic, leading to superior performance in yield, stress tolerance, and other traits compared with the parental lines. Germplasm collections facilitate the creation of diverse hybrid combinations to exploit hybrid vigor in sugarcane breeding.

5.3.6 LONG-TERM CONSERVATION

Germplasm collections serve as a safeguard against genetic erosion and loss of biodiversity. Through systematic conservation efforts, valuable genetic resources are preserved for future breeding endeavors, ensuring the sustainability of sugarcane cultivation (Nair 2005). Sugarcane stands as a pivotal global food and energy crop, contributing to 80% of the world's sugar and 60% of bioethanol production (Yang et al., 1980). Notably, substantial financial contributions from the governmental and commercial sectors during this era bolstered the import of germplasm and the inception of new breeding programs. The generation of genetic variability hinges on crossing heterozygous individuals and selecting superior segregating populations. Across different countries, sugarcane breeding programs rely on diverse germplasm collections, each facing the common challenge of synchronizing flowering among various genotypes. While staggered planting offers a partial solution, it proves inefficient in many cases (Naidu and Sreenivasan, 1987). There are numerous sugarcane germplasm collections located in various countries around the world. Some of the countries with significant sugarcane germplasm collections and breeding programs are discussed below.

5.4 STATUS OF SUGARCANE GERMPLASM IN BRAZIL

The largest sugarcane producer and a leading supplier of sugar and ethanol, Brazil employs a strategic approach to consistently rejuvenate its germplasm collection through recurrent selection processes. This involves annually incorporating elite clones that excel in key traits during advanced selection stages, alongside introducing clones from both domestic breeding institutions and international sources (Cursi et al., 2019). Brazil's germplasm collection and breeding programs have been active since the 1960s, selecting germplasm according to the growing needs and breeding objectives. The main breeding objectives include high cane and sugar yield, sugar content, disease resistance, sprouting, ratooning ability, adaptability, growth rate, synchronized flowering, high fiber content, energy canes, and resistance to abiotic stresses (Cursi et al., 2019).

Brazil has imported hundreds of basic germplasm accessions from the world collection in Miami to use in introgression and development of energy canes. Since then, various introgression aspects with wild species, like *S. spontaneum*, focusing on selecting resistance plants with optimum fiber and low sucrose content but high biomass productivity. Currently, there are five germplasm collection centers located in northeastern Brazil, which have suitable climatic conditions for flowering and fertile pollen production. These two collections serve the RIDESA (RB) programs, one serves the CTC program, one serves the IAC program, and one serves the BioVertis/GranBio program, the latter being exclusively dedicated to energy-cane breeding (Cursi et al., 2021). Almost 5000 accessions were available as germplasm collection in these stations of Brazil (Table 5.2).

TABLE 5.2 List of Germplasm Collection of the Flowering and Crossing Stations at Brazil

Species of Hybrids	Serra do Quro	Devaneio (RIDESA)	BioVertis/ GranBio
S. officinarum	44	6	53
S. sinense	5	2	10
S. robustum	9	4	26
S. spontaneum	39	6	166
S. barberi	6	5	18
S. edule	1	1	1
Erianthus ssp.	6	1	33
Miscanthus	2	–	13
Unknown species	159	1	–
Hybrids/others	2794	970	480
Total	3065	996	800

Source: Adapted with permission from Cursi et al. (2021).

In recent years, RIDESA has been actively engaged in variety exchanges with several sugarcane-producing countries, including Australia, Argentina, Colombia, Pakistan, and Reunion Island. Additionally, new agreements have been established with China, Ecuador, South Africa, Pakistan, and the Philippines, resulting in the exchange of approximately 60 varieties annually. These initiatives primarily aim to introduce new sources of variability for traits of agricultural and industrial significance, and tolerance to pathogenic diseases (Cursi et al., 2019). Genetically, Brazil continually expands its germplasm collections to incorporate new alleles or increase the frequency of favorable ones, facilitating ongoing variety improvement.

Unproductive parents are removed based on their performance in family trials and their extensive use in breeding programs. Parental evaluation, particularly at universities involved in RIDESA, relies on family test results and estimates breeding values through mixed models.

Various crossing methods are employed to produce progenies for the selection of superior individuals, including biparental crosses, poly-crosses, and free pollination, the latter being feasible due to abundant viable flowers. RIDESA conducts approximately 5000 crosses annually, with proportions varying across breeding stations and programs. The correct selection of parents is a significant challenge for breeders, given the vast number of potential combinations, underscoring the importance of each institution's unique strategy for genetic progress.

5.5 STATUS OF SUGARCANE GERMPLASM IN INDIA

Sugarcane breeding in India has contributed fundamentally to sugarcane improvement worldwide. India is one of the principal centers of origin of the *Saccharum* germplasm complex (*Saccharum*, *Erianthus*, *Sclerostachya*, *Narenga*, *Miscanthus*) and clones bred in India are ancestral to all commercial sugarcane varieties worldwide. Sugarcane has been cultivated in India since 5000 BC, but its genetic enhancement through breeding began in the nineteenth century (Chakravarthi et al., 2015, 2016). A significant milestone in yield enhancement occurred in 1918 with the introduction of first interspecific Co 205 hybrid cultivar. This cultivar superseded previous varieties, establishing interspecific hybrids as the cornerstone of future sugarcane improvement efforts. Tri-specific hybrids, known as Coimbatore canes (Co canes), [(*S. officinarum* × *S. spontaneum*) × *S. barberi*] developed by the Indian Council of Agriculture Research-Sugarcane Breeding Institute, Coimbatore, India, gained popularity in India and several other countries as both varieties and parent cultivars.

Enhanced sugarcane varieties have made significant contributions to increasing yields from 1930 to 2021–2022 (30.9–82.6 t/ha), and sugar recovery rates have increased about 9–11% in the same time. Currently, a national network of sugarcane varietal improvement programs, comprising 24 sugarcane centers covering the entire country, is operational. This program relies on seedlings derived from an annual crossing program at the National Hybridization Garden, ICAR-SBI, Coimbatore (Balasundaram, 1995).

Varietal testing, managed by 36 centers, has identified 124 varieties through the All India Coordinated Research Program (Sugarcane) tailored for five agro-climatic zones since 1982. From 2000 to 2022, 72 varieties were developed for industrial purposes. However, the current varieties, Co 0238, cultivated across 2.77 mha, and Co 86032, covering 0.89 mha, are the two predominant cultivars in India (Bakshi et al., 2021; Philip et al., 2013). Micropropagation has been instrumental in rapid multiplication and enhanced adoption efficiency of new cultivars. Additionally, genetic improvement through conventional breeding is now complemented by molecular research aimed at enhancing biotic and abiotic stress tolerance, particularly against red rot and drought, as well as biomass production, positioning sugarcane as a significant source of both sugar and renewable energy (Bakshi et al., 2021).

5.5.1 ESTABLISHMENT OF SUGARCANE RESEARCH STATIONS

A flowchart of the crossing and selection stages in sugarcane breeding generally followed in India is shown in Table 5.3.

Genetic engineering of sugarcane began at ICAR-SBI, focusing initially on developing sugarcane resistant to borer pests using Cry proteins (*Cry 1 Ab*) and proteinase inhibitors. Later efforts included creating transgenic varieties with genes coding for antimicrobial peptides, chitinase, and glucanase to combat red rot (Anon 2007). A limited field trial of insect-resistant transgenic sugarcane at ICAR-SBI, Coimbatore during 2011–2012 showed no variation in yield or quality due to transgenesis (Subramonian et al., 2012). The trial, however, did not establish a differential response of transgenics because of the low borer incidence in the habitat. Despite this, a significant negative relationship between dead heart levels and leaf Cry content indicated the toxin's effectiveness against borer activity even with moderate pest activity. The study also suggested that Cry1Ab was not antagonistic to internode borer.

ICAR-SBI Coimbatore also developed transgenic sugarcane overexpressing pea DNA helicase (PDH45), heat shock protein from *Erianthus* spp. (*EaHSP70*), and *EaDREB2* genes, though these have not yet been field tested. Recently, genome editing for both biotic and abiotic stress tolerance and the production of value-added products in sugarcane has begun. ICAR-SBI was the first in India to produce transgenic sugarcane tolerant

TABLE 5.3 Flowchart of the Selection criteria and Number of Clones Selected in India

Year	Activity	Selection Criteria and Number of Clones Selected
I	Choice of parents and hybridization (800–1000 crosses)	Parents are selected based on pedigree, molecular diversity, combining ability, and registered genetic stock
II	Seedling raising and transplanting in ground nursery (200,000–500,000 seedlings)	Sowing in March to August, transplanting in June–December
III	Ratooning ground nursery selection and planting in I clonal trial (30,000–40,000 clones)	Brix, number of canes/stool, flowering, cane diameter, cane height, cane morphology (20% of clones selected)
IV	Selection in I clonal trial	Brix, number of canes/stool, flowering, cane yield, parameters (25% of clonal selected)
V	Clonal trails	One row trials in August RCBD/RDB (25% of clones selected)
VI	Prezonal varietal trials (PZVT) multiplication	250 clones—two rows trials; juice quality analysis at 240, 300, and 360 days post planting. screening for red rot, crop stand, flowering
VII	PZVT trials/final clonal trials (RBD)	60–100 clones—yield, quality evaluation, red rot screening by plug and nodal inoculation smut screening, ethanol and fiber content, PoL% cane
VIII	Selection elite clones for multiplication testing	20–25 Co canes, 30–40 Co-allied, 2–5 other center selections
IX, X	Multiplication and exchange of AICRP entries	60–75 to AICRPS trials (5–25 clones for the five agro-climatic zones multiplication for 2 years
XI	Initial varietal trials	1 year in RBDa
XI	Advanced varietal trials	Two plant crops and one ratoon crop to select the best clones
XIII	Varietal release	Within 2 years of AICRP multilocational testing

[a]RBD—randomized block design.

Source: Adapted with permission from Bakshi et al. (2021).

to herbicides using the biolistic transformation method (Subramonian et al., 1999). This was followed by the development of herbicide BASTA-resistant sugarcane through *Agrobacterium*-mediated transformation (Manickavasagam et al., 2004). This transformation method was further refined by Sathish et al. (2020). Additionally, research characterized several promoters for transgene expression, including the constitutive promoter Port Ubi 2.3 from *Porteresia coarctata* (Philip et al., 2013), a stress-inducible promoter of the MYB transcription factor from *Saccharum* (Prabu and Prasad 2012), and a PR10 promoter from *E. arundinaceus* (Chakravarthi et al., 2015, 2016).

5.6 STATUS OF SUGARCANE GERMPLASM IN CHINA

Sugarcane is cultivated in nonirrigated regions in China, primarily on sloping, steep land, and other areas with comparatively weaker conditions for agricultural production. The area of sugarcane declined to from last 10 years from 1.81 to 1.16 mha. The yield of sugarcane ranged from 7.85 to 12.57 MTs with an average of 9.87 MTs per crushing season. Currently, six main research centers were working on sugarcane breeding and germplasm maintains as shown in Table 5.4.

TABLE 5.4 Name of Institutes Having Germplasm Collection in China

Name of Institute	Germplasm Collection
Institute of Nanfan & Seed Industry, Guangdong Academy of Science, China (INSI-GAS)	2400
Yunnan Sugarcane Research Institute, Yunnan Academy of Agricultural Sciences (YSRI-YAAS)	100
Sugarcane Research Institute, Guangxi Academy of Agricultural Sciences (SRI-GXAAS)	300
Fujian Agriculture and Forestry University	200
Guangxi University, Nanning, Guangxi	200
Institute of Tropical Bioscience and Biotechnology, Chinese Academy of Tropical Agricultural Sciences (ITBB-CATAS)	100
Total	3300

Source: Adapted with permission from Yongwen et al. (2021).

China has made significant advances in variety selection methods, biotechnology, and collecting and evaluating of sugarcane germplasm. 3,300 *Saccharum* and allied species are kept at the China National Nursery of Sugarcane Germplasm Resources (CNNSGR) in Kunming, Yunnan province. Through intra and interspecific crossings between *Saccharum* spp. and related species, genes from wild germplasm, including *S. spontaneum*, *N. porphyrocoma*, *E. rockii*, and *E. fulvus*, have been successfully introgressed. Important worldwide sugarcane breeding programs still have elite imported clones that are useful as parental material for developing new varieties. Different clones from innovative crosses at the Hainan Sugarcane Breeding Station (HSBS) with native Chinese *S. spontaneum* germplasm have become significant parents in the Chinese breeding program, leading to the release of 37 commercial cultivars. Additionally, transgenic sugarcane clones resistant to insects and diseases, and tolerant to drought and low potassium, have been developed, although their commercial use remains uncertain (Ahmad et al., 2018).

Current major constraints in China's sugarcane industry include low level of mechanical harvesting, inadequate ratoon ability and diseases, like smut and the sugarcane borer. Therefore, developing system of production and plant cultivars that are compatible with mechanical harvesting, susceptible to smut, and have better ratooning is now a high priority. Varieties such as POJ2725, POJ2878, CP28-11, CP29-116, CP29-320, CP34-120, Co290, and Co281 were introduced in Guangdong, Guangxi, Sichuan, and Fujian provinces, with some widely cultivated in China (Li 2000). Furthermore, around 178 F series and 28 ROC series varieties were released in Taiwan (Wu et al. 2005), with some, including F134, ROC16, and ROC22, becoming major cultivars in mainland China.

The HSBS in Hainan province, southern China, established under mainland China's administration, is responsible for germplasm collection, introduction, conservation, and innovative utilization. It produces sugarcane fuzz (seed) for all sugarcane breeding institutions in China and remains the major sugarcane crossing station, providing fuzz from about 2000 crosses each year to institutes nationwide. Besides HSBS, other sugarcane crossing stations in China generally produce seeds for the selection of breeding programs within their respective institutes. Approximately, 226 commercial varieties were released in mainland China between 1950 and 2009, that is, 192 cultivars bred in mainland China and 34 introduced cultivars (Zhang et al., 2009a, 2009b). 186 varieties have parental records. Pedigree analysis showed that 163 varieties were bred by 21 parents, accounting for 88% of the total. Additionally, 184 varieties trace back to POJ2878, and 122 were derived from F134.

TABLE 5.5 Sugarcane Germplasm in China National Nursery of Sugarcane Germplasm Resources (CNNSGR)

Genus	Species	CNNSGR
Saccharum	*S. officinarum*	59
	S. barberi	6
	S. sinense	22
	S. robustum	6
	S. spontaneum	905
	Foreign hybrid	767
	Domestic hybrid	1085
Erianthus	*E. arundinaceus*	301
	E. fulvus	64
	E. rockii	61
Narenga	*N. porphyrocoma*	11
Miscanthus	*M. floridulus*	2
	M. sinensis	58
Imperata	*I. cylindrical*	17
Pennisetum	*P. schumac*	2

Source: Adapted with permission from Yongwen et al. (2021).

Chinese researchers have conducted a significant amount of research on the progeny's chromosomal transmission and the hybridization of sugarcane cultivars with *N. porphyrocoma, E. rockii,* and *E. fulvus* (Liu et al. 2012; Lin as al. 2014, 2017; Lu et al., 2016; Chang et al., 2020; Shenetal., 2020). With the exception of *S. spontaneum* and *E. arundinaceus*, its usage in the development of commercial cultivars has been relatively restricted. With over 37 cultivars originating from the Yacheng series parents, which bred *S. spontaneum* by-crossing and backcrossing with commercial varieties of sugarcane, proved to be the most successful material. The HSBS also produced some good progeny of the sugarcane hybrid with *E. arundinaceus*, including YCE07-71, YCE05-150, and YCE06-164, some of which were utilized as parents in commercial breeding programs. CNNSGR at the Sugarcane Research Institute of Yunnan Academy of Agricultural Sciences served the country's most sugarcane germplasm resources (Figure 5.4–5.8). By following the guidelines given forth by the CMARA, international researchers can also get in communication with other sugarcane breeding organizations directly in order to introduce germplasm resources, in addition to the CNNSGR. The primary goal of research to develop pest-resistant sugarcane has been to boost the crop's resistance to borers through the introduction of transgenic Bt genes. In China, diseases have always posed a serious threat to the crop and do so even after significant breeding programs. There are 50 sugarcane diseases identified in China, the most common ones being rust, pokkah boeng, ratoon stunting disease, mosaic, and smut (Table 5.5).

FIGURE 5.4 GT series of China varieties.

FIGURE 5.5 *Saccharum aruderanium.*

FIGURE 5.6 Introgression clones.

FIGURE 5.7 Domestic sugarcane varieties at Sugarcane Research Institute, Nanning, China.

5.7 STATUS OF SUGARCANE GERMPLASM IN THAILAND

Thailand ranks as the world's fourth-largest sugarcane producer, following Brazil, India, and China, and is the second-largest sugar exporter globally, after Brazil. Over the past 5 years (2015–2020), Thailand has achieved an average sugarcane yield of around 63 tons per hectare, with an average commercial cane sugar (CCS) content of 12.41%. Currently, an estimated 330,000 sugarcane growers cultivate the crop across the country. In the

FIGURE 5.8 *Saccharum spontaneum.*

2021–2022 milling season, sugarcane production peaked at 140.9 MTs, covering 1.58 mha of cultivated land. However, production has since declined due to recurrent drought and a shift in farmers' preferences toward more lucrative crops.

Since its establishment in 1938, Thailand's sugarcane breeding efforts have continually enhanced cane productivity by introducing improved varieties, with the notable Khon Kaen 3 now dominating approximately 85% of the nation's sugarcane production. To sustainably enhance the productivity of the Thai sugar industry, ongoing breeding programs primarily target cane yield and sugar content, with increasing attention directed toward fiber yield, pest and disease resistance, adaptability to diverse environmental conditions (e.g., sandy or clay soil, drought resistance), ratooning capability, and suitability for mechanical harvesting. The incorporation and advancement of biotechnology are poised to further augment the development of desirable traits (Weerathaworn, 2006).

Currently, 800 sugarcane germplasm were available at different research stations of Thailand working on sugarcane crop improvement programs (Figure 5.9). The pivotal institutions engaged in breeding, germplasm maintenance and associated research, the historical progression and achievements of sugarcane breeding programs, the current methodologies utilized in breeding initiatives, and the prominent cultivars and germplasm collections (Chaisan et al., 2018). Additionally, it explores the historical trajectory and contemporary application of genetics and biotechnology in enhancing sugarcane cultivation (Chaiyabut et al., 2013).

FIGURE 5.9 Thailand sugarcane variety Khon Kaen 3.

5.8 STATUS OF SUGARCANE GERMPLASM IN SOUTH AFRICA

The South African Sugarcane Research Institute (SASRI) initiated its breeding program after imported cultivars demonstrated poor adaptability and susceptibility to pests and diseases. In the 1940s, SASRI began importing crosses and later established its own crossing utilities. Some of the most successful cultivars in the world, like NCo310 and NCo376—which were widely cultivated—were the result of the breeding initiative. These cultivars were chosen from crosses import from the Sugarcane Breeding Institute in Coimbatore, India (Balasundaram, 1995). At present, the program conducts four regional breeding initiatives that produce cultivars for irrigated, high-altitude, and coastal areas. Among the purposes of breeding are a high cane yield, sugar content, ratooning ability, resistance to pests and diseases, resistance to lodging, shy blossoming, and adaptability to irrigated, coastal, high-altitude, and high-altitude conditions (Amin et al., 1972). Seventy-seven cultivars have been released, some of which are cultivated in Southern Africa. A core germplasm collection that includes imported and elite clones as well as clones derived via introgression breeding supports the regional breeding initiatives.

5.9 STATUS OF SUGARCANE GERMPLASM IN AUSTRALIA

Sugarcane is a crucial crop for the economy of Australian, and plant breeding program has played a significant role in the success of the sugarcane industry (Berding et al., 1987). At present, Sugar Research Australia and Wilmar Sugar work together to manage the Australian sugarcane breeding program at six research stations (Berding et al., 2003). Crossing is primarily conducted at Meringa, while selection occurs at six different stations. The selection process involves different phages such as progeny, clonal and final assessment trials. It takes nearly 12 years from seed sowing in progeny assessment trials to the release of a commercial variety (Berding et al., 2003). Key infrastructure and initiatives to enhance the program are also in place to ensure its efficient and effective operation.

The Australian Sugarcane Breeding Program's primary goals are:

- Higher yields and profits from cane through increased sugar content and cane yield, or CCS. By 2015, targets have been established for a 0.5 unit CCS and an increase in sugar yield of more than 300 kg sugar/ha/year (Cox et al., 2005).
- Reducing losses due to diseases and pests by releasing cultivars that are resistant.
- Enhanced ratooning performance and environmental adaptability.
- The development of cultivars with acceptable levels of sugar and fiber quality and quantity.
- Keeping a variety of germplasm for cross-pollinating to produce new cultivars with different applications for sugarcane like high biomass (Berding et al., 1997).

The Australian Sugarcane Breeding Program consists of three main parts that achieve these goals:

1. An increase in population through:
    - Parental germplasm selection.
    - The introduction of innovative germplasm, including cultivars, from international breeding initiatives (such as those in the USA, Argentina, Brazil, China, Mauritius, Columbia, and South Africa). Many hybrid cultivars have been released in Australia as a result of crosses conducted with these cultivars, even though they often do not perform well in Australian conditions.

- The Australian Sugarcane Breeding Program's interstation exchange, in which each region's most promising clones are chosen and transferred to all other areas for utilization in selection trials (Berding, 1976).
- Because population improvement is a very dynamic process, it is important to source, maintain, assess, and manage the germplasm collection.

2. Development of genetic heterogeneity via cross pollination.
3. Conduct regional selection programs.

5.10 STATUS OF SUGARCANE GERMPLASM IN PAKISTAN

Pakistan is the sixth largest producer of cane and the ninth largest producer of sugar with an area of 1.25 mha, production of 88.61 MTs, and yields of 70.35 tons/ha. The research institutes working on sugarcane crop production has released 28 sugarcane varieties so far for general cultivation in Punjab including varieties such as CoL 29, CoL 54, BL 4, Triton, CP 72-2086, CP 77-400, SPF 213, CPF 237, HSF 240, CPF 247, CPF 248, CPF 249, CPF-250, CPF-251, CPF-252, and CPF-253 (Table 5.6). In Sindh, four varieties were approved for general cultivation, namely, Thatta-10, Thatta-2109, Thatta-326, and YT-55 Thatta. In KPK, four varieties were approved for general cultivation in Israr Shaheed, Abdul Qayyum, Mardan, and Gi Rehman.

TABLE 5.6 Sugarcane Varieties Developed by Research Institutes Working on Varietal Development in Pakistan

S. No.	Variety	Year of Approval	Potential Yield (tons/ha)	Sugar Recovery (%)
		Punjab		
1	CoL-29	1954	69	10.10
2	CoL-44	1954	74	8.93
3	CoL-54	1963	74	9.63
4	BL-19	1966	84	9.49
5	BL-4	1968	84	10.34
6	L 116	1973	74	10.81
7	L 118	1975	82	8.23
8	Triton	1983	84	10.10
9	BF 162	1990	89	10.35

TABLE 5.6 *(Continued)*

S. No.	Variety	Year of Approval	Potential Yield (tons/ha)	Sugar Recovery (%)
10	CP 43-33	1996	79	11.69
11	CP 72-2086	1996	84	12.35
12	CP 77-400	1996	89	11.90
13	CoJ 84	2000	89	9.80
14	SPS-213	2000	89	10.50
15	CPF-237	2000	94	12.50
16	CPF-240	2002	94	12.50
17	SPF-234	2002	99	11.60
18	SPF-245	2004	99	11.00
19	HSF-242	2006	101.1	12.50
20	CPF-243	2006	101.1	12.55
21	CPF-246	2011	104	12.15
22	CPF-247	2011	104	12.25
23	CPF-248	2013	119	12.45
24	CPF-249	2016	115	12.46
25	CPF-250	2019	110	14.1
26	CPF-251	2019	115	14.3
27	CPF-252	2019	120	15.9
28	CPF-253	2019	118	15.9
		Sindh		
29	Thatta-10	2004	150	12.15
30	Thatta-2109	2016	140	13.50
31	Thatta-326	2016	150	12.25
32	YT-55 Thatta	2018	140	12.50
		KPK		
33	Israr Shaheed	2017	90	13.40
34	Abdul Qayyum	2017	89	13.69
35	Mardan	2021	87	12.71
36	Gi Rehman	2021	85	12.46

https://parc.gov.pk/index

Source: Adapted with permission from Malik (2018).

Sugarcane is primarily cultivated in south and southwest areas of Pakistan, and it is a very limited profitable crop than others. It grows in irrigated areas. In Punjab, Rahim Yar Khan, Sargodha, and Faisalabad are major sugar-producing districts.

5.10.1 CANE VARIETY DEVELOPMENT PROGRAM IN PAKISTAN

Pakistan has a good infrastructure of research work in three cane growing provinces. Since cross-breeding facilities have not been developed, variety development program is mainly carried out through introduction of cane varieties and import of cane fuzz. The variety evolution program at the provincial research stations is summarized in this section. Imported fuzz and germplasm constitute the primary reservoir for variety selection in the province. Mass selection is conducted from the seedling population derived from fuzz and imported germplasm. These seedling clones and germplasm undergo rigorous evaluation across various selection stages to assess yield, quality, and disease tolerance. Field trials are conducted in diverse agro-climatic regions of the province to gauge varietal adaptability to different ecological conditions.

Before 1947, cane varieties and fuzz were procured from Coimbatore, India. The main variety cultivated was Co 312, with some areas allocated to CoL 5, CoL 9, and Co 313. Subsequently, during the 1950s and 1960s, cane varieties like CoL 29, CoL 44, and CoL 54, sourced from Coimbatore fuzz, were approved for commercial cultivation in Punjab. Over time, various sources of fuzz supply were explored, aiding variety selection efforts at the Sugarcane Research Station, Faisalabad (Malik et al., 1986). A cane flower station was established in the Murree Hills during 1953–1954. However, due to sporadic flowering in some varieties and challenges in obtaining desired fuzz, only two varieties, L-116 and L-118, were developed from open-pollinated fuzz in the past five decades. Consequently, the primary reliance has been on fuzz imported from foreign cane breeding stations (Malik et al., 1986).

Imports from Barbados during 1952–1953 led to the development of two commercial varieties, BL 4 (1968) and BL 19 (1967). BL 4 gained popularity in lower Sindh due to its high yield potential and sugar content (Tables 5.7 and 5.8). Similar success followed imports from Canal Point and Louisiana (USA) starting in 1980, resulting in the release of notable varieties under the name CPF. The latest releases, including CPF-250, CPF-251, CPF-252, and CPF-253, have garnered favor from both growers and the sugar industry (Malik et al., 1986). Additionally, some fuzz has been received from Brazil, leading to the development of SPF varieties. SPF 234 gained popularity in southern Punjab but is now facing challenges due to susceptibility to red rot. Local and exotic cane varieties are continuously collected and evaluated for adaptability and performance in various selection stages at the research station and field trials. Notable varieties like Triton and Co 1148 have enjoyed periods of popularity in the province, despite certain limitations (Malik et al., 1986).

Following are issues of sugarcane breeding and variety development in Pakistan:

- With no full-scale breeding of sugarcane, the sugarcane industry in Pakistan would find it difficult to sustain itself.
- Climatic conditions are not conducive for production of genetically improved and viable true seed (fuzz) of sugarcane.
- Varieties are developed through selection breeding in Pakistan from imported fuzz or imported/exchanged as introduction.
- The varieties developed through selection breeding have issue of poor performer because the good quality gene cannot be transferred to better biometric traits.
- Breeders' rights/propriety and limited genetic resources.
- Stability across environments and climate change adaptation

5.10.2 PRIORITIES OF SUGARCANE CROP IMPROVEMENT IN PAKISTAN

- Sugarcane crop improvement through R&D at sugarcane industry level
- Develop cane cultivars having stress resistance capacity and excellent ratoon ability.
- Improve productivity growth and genetic solutions for commercial cane farming.
- Accelerate innovative development of high-performing varieties—success of the industry's sugarcane crop production and protection
- Establish infrastructure and HR for high-quality sugarcane using basic, commercial breeding, and selection breeding technologies (Figure 5.10).

TABLE 5.7 Progressive Release of Cane Varieties in Cane Growing Provinces of Pakistan

Period	Varieties Evolved through Fuzz or Variety Selection					
	Punjab Province		Sindh Province		KP Province	
	Fuzz	Exotic/Local Varieties	Fuzz	Exotic/Local Varieties	Fuzz	Exotic/Local Varieties
1950–1955	CoL 29 CoL 44	–	–	Co 421	–	Co 622
1955–1960	–	–	–	–	–	CoS 425 CoS 321
1960–1965	CoL 54	–	–	Co 547, PR 1000	–	NCo 310 CP 48.101
1965–1970	BL 4, BL 19	–	–	Co L 54, Co 622	–	–
Period	PUNJAB		SINDH		NWFP	
	Fuzz	Exotic/ local varieties	Fuzz	Exotic/local varieties	Fuzz	Exotic/local varieties
1970–1975	L 116	–	–	BL 4 NCo 310 Co 997	–	CP 44-101
1975–1980	L 118	Co 975	–	Co 658 L 113 L 116	–	–
1980–1985	–	Tritron, Co 1148	–	–	IM 61	CP 51-21 CP 65-357 L 62-96
1985–1990	BF 162	–	–	–	CPM 13	Co 1321
1990–1995	–	–	–	CP 67-412	–	Mardan 92 Mardan 93 Bannu 01

TABLE 5.7 *(Continued)*

Period	Varieties Evolved through Fuzz or Variety Selection					
	Punjab Province		Sindh Province		KP Province	
	Fuzz	Exotic/Local Varieties	Fuzz	Exotic/Local Varieties	Fuzz	Exotic/Local Varieties
1995–2000	SPSG 26 SPF 213 CPF 237	CP 43-33 CP72-2086, CP 77-400	NIA-98 Th 10	BF 129, SPSG 26	JN 88-01 ABID 96 SN-1	CP 77-400
2000–2005	SPF234 HSF 240	–	Larkana 2001 NIA.2004	–	–	Mardan-01 Mardan-04 MC 421 Mardan 2005
2005–2010	HSF 242 CPF243 CPF 245	–	HS 12	–	–	–
2010–2015	CPF 246 CPF 247 CPF 248	–	Th 326 Th 2109-	–	–	–
2015–2020	CPF-249 CPF-250 CPF-251 CPF-252 CPF-253		NIA-2011			

https://parc.gov.pk/index

Source: Adapted with permission from Malik (2018).

TABLE 5.8 Import of Sugarcane Germplasm from Different Cane Growing Countries during Last 50 years (1974–2024)

Institution/Organization	Country
SRI, GXAAS, Nanning	China
CSR, BSES (SRA)	Australia
USDA-ARS CP/Ho	USA
COPERSUCAR/RIDESA	Brazil
CIRAD	Reunion
SASRI	South Africa
MSIRI	Mauritius
SRI	Sri Lanka
CARDI	Barbados
PSI	The Philippines

Source: SRDB Punjab, Pakistan 2024.

FIGURE 5.10 Sugarcane varieties: selection breeding in Pakistan.

5.10.3 APPROACHES OF SUGARCANE GERMPLASM BETTERMENT IN PAKISTAN

- Collaborative R&D work/projects at international level through public–private partnerships.
- Exchange/import of germplasm from leading sugar-producing countries of the world.
- Training of sugarcane scientists on basic, commercial and molecular breeding.
- Breeding to shorten overall cycle times of sugarcane.

- Development of varieties with traits such as high sugar yield, disease resistance, stress tolerance, and ratoon-ability.
- Large-scale multiplication of the new varieties to reduce cost of production and increase profitability for stakeholders.
- Development of better germplasm for better varieties with better seed production to obtain better crop and better yield for better profitability (Figures 5.11–5.13; Table 5.9)

5.11 STATUS OF SUGARCANE GERMPLASM IN MAURITIUS

Sugarcane was introduced to Mauritius in 1639 and gradually became the island's primary cash crop for centuries. Although efforts to improve sugarcane genetics date back to 1891, large-scale sugarcane breeding began in Mauritius in 1930 with the establishment of the Sugarcane Research Station. In 1953, the Mauritius Sugar Industry Research Institute (MSIRI) was founded, taking responsibility for all research related to the Mauritian sugar industry, with a significant focus on sugarcane breeding. Today, the MSIRI maintains a germplasm collection of 2500 accessions. Its breeding program includes a hybridization phase, where each year, 250–500 different family combinations are created in controlled glasshouses, producing 50,000–100,000 seedlings. The selection program, which spans 15 years, comprises six selection stages. Genetic studies support the program to enhance breeding and selection efficiency, aiming to produce resilient varieties with higher productivity and lower production costs.

To date, the MSIRI has released 76 sugarcane varieties. The most recent varieties, which exhibit high productivity, low flowering behavior, and adaptability to various climatic conditions, are expected to significantly contribute to the industry's sustainability (Amin et al., 1972). New challenges include effectively utilizing sugarcane as a dedicated bioenergy crop while continuing to focus on increasing sugar yield and applying new genetic and biotechnological tools in breeding and selecting superior, multipurpose cultivars for the industry (Amin et al., 1972).

5.12 FUTURE DIRECTIONS FOR SUGARCANE GERMPLASM

Future research should focus on the points mentioned below.

FIGURE 5.11 Import of sugarcane germplasm of ZZ-Series from PR China.

FIGURE 5.12 Import of sugarcane germplasm of RB-Series from Brazil.

FIGURE 5.13 CP00-1101 as introduction at SRDB Pakistan.

TABLE 5.9 Biometric Traits of Commercial and Promising Clones Developed in Pakistan (1974–2024)

Description	Commercial/ Promising Clones	Production (t/ha)	Sugar Recovery (%)	Pest Insect	Disease Reaction	Ratoon Index
Commercial	55 = 35 Punjab, 10 Sindh, 10 KPK	70–90	10–11	MT-MS	R-MR	0.7–0.8
Promising	10 = 6 Punjab, 2 Sindh, 2 KPK	90–120	11–12	T-MT	HR-R	0.9–1.0
Gene Bank	400 = ZZ, RB, CP, SP, M, CSR/Q, P, B, Co, SL, N, CIRAD	60–150	7–14	T-MS	HR-HS	0.5–1.20

5.12.1 EXPLORING UNTAPPED GENETIC RESOURCES

The exploration of untapped genetic resources in sugarcane germplasm presents significant opportunities for advancing sugarcane cultivation and production. Sugarcane, a highly complex hybrid, benefits immensely from genetic diversity, which can drive improvements in yield, disease tolerance, and adaptation to the era of climate change.

5.12.2 DEVELOPING NEW BREEDING STRATEGIES

To effectively explore and utilize sugarcane germplasm, several methodologies are employed.

5.12.2.1 GENOMIC SEQUENCING AND MARKER-ASSISTED SELECTION

Advances in genomic sequencing allow for the detailed mapping of genetic variations within germplasm collections. Marker-assisted selection (MAS) can then be used to identify and propagate desirable traits.

5.12.2.2 HYBRIDIZATION AND BACKCROSSING

Traditional breeding techniques such as hybridization and backcrossing remain crucial. By crossing high-performing cultivars with wild species or landraces, breeders can introduce and stabilize beneficial traits.

5.12.2.3 BIOTECHNOLOGICAL APPROACHES

Genetic engineering and CRISPR (clustered regularly interspaced short palindromic repeats) technology offer precise tools for incorporating specific genes into sugarcane varieties. This can expedite the development of cultivars with improved characteristics.

5.12.2.4 PHENOTYPIC SCREENING

Field trials and phenotypic screening of germplasm collections help identify individuals with superior traits. This hands-on approach ensures that selected traits perform well under real-world conditions.

5.12.3 INTEGRATING GENOMICS AND PRECISION BREEDING

The integration of genomics and precision breeding in sugarcane germplasm represents a significant advancement in agricultural science, promising to enhance sugarcane productivity, disease resistance, and adaptability. This approach leverages cutting-edge genomic tools and precise breeding techniques to optimize the genetic potential of sugarcane.

5.13 THE ROLE OF GENOMICS IN SUGARCANE BREEDING

Genomics involves the comprehensive study of the entire genome of an organism, providing detailed insights into its genetic makeup. In sugarcane breeding, genomics plays a critical role as discussed below.

5.13.1 GENOME MAPPING AND SEQUENCING

Detailed mapping and sequencing of the sugarcane genome allow breeders to identify specific genes associated with desirable traits such as high yield, pest resistance, and environmental resilience.

5.13.2 MARKER-ASSISTED SELECTION

Genomic markers linked to beneficial traits can be used to screen and select breeding candidates more efficiently. MAS accelerates the breeding process by reducing the reliance on lengthy phenotypic evaluations (Ahmad et al., 2018).

5.13.3 GENOMIC SELECTION

Genomic selection (GS) uses genome-wide markers to predict the breeding value of individuals in a population. This approach enables the selection of the best candidates for breeding programs based on their genetic potential, even before they are fully grown.

5.14 PRECISION BREEDING TECHNIQUES IN SUGARCANE

Precision breeding encompasses various techniques that allow for the targeted manipulation of an organism's genetic material. In sugarcane breeding, the following precision breeding methods are particularly impactful.

5.14.1 CRISPR/CAS9 GENE EDITING

This revolutionary technology enables precise modifications of the sugarcane genome. CRISPR/Cas9 can be used to knock out undesirable genes or introduce beneficial ones, facilitating the development of sugarcane varieties with enhanced traits.

5.14.2 TRANSGENICS

The introduction of genes from other species into the sugarcane genome can impart new traits, such as increased pest resistance or improved stress tolerance. Transgenic approaches must be carefully managed to address regulatory and public acceptance issues.

5.14.3 HYBRID BREEDING

Combining traditional breeding with genomic insights, hybrid breeding involves crossing different sugarcane varieties to produce offspring with superior traits. Genomic data guides the selection of parent plants to maximize the likelihood of desirable outcomes.

The most recent crop to possess the entire genome is sugarcane, following rice, sorghum, maize, bananas, and wheat. And not without reason, since its genome is the most complex: there are at least 12 copies of each chromosome, and a total of 114 chromosomes! This colossal work involved one of the world's leading sequencing centers, the Joint Genome Institute in the USA, and a team of 35 scientists from four countries, including from CIRAD, which was heavily involved in the strategy to decode and analyze the new genome.

5.15 CONCLUSION

Breeders have grown impatient because although traditional sugarcane breeding techniques have been successful in producing cultivars that are adaptable to new conditions and pathogens, the progress in sugar production has now peaked. It will be considerably simpler to find genes of agronomic importance due to that reference sequence and the insight it provides into the architecture of the sugarcane genome. It will speed up efforts to breed varieties that are more suited to upcoming conditions in the environment (Healey et al., 2024). Sugarcane germplasm plays a crucial role in sugarcane breeding programs by providing genetic diversity, traits for adaptation and stress tolerance, disease resistance, and quality. The genome is freely available on the sugarcane hub: https://sugarcane-genome.cirad.fr/

Its systematic conservation and utilization contribute significantly to the development of improved sugarcane varieties with enhanced productivity, sustainability, and resilience. The top sugarcane-producing countries such as Brazil, India, and China have Houses extensive germplasm collections and focuses on recurrent selection, utilizing both domestic and international clones. The breeding objectives include high yield, disease resistance, and adaptability to various stresses. The emphasis is on high cane, sugar yield, energy canes, and resilience to abiotic stresses. The introduction of high-fiber and low-sucrose varieties for biomass production is notable. Countries such as China, Pakistan, India, and Thailand face challenges in mechanized harvesting and managing diseases like red rot, smut, and sugarcane borer. Continued research in genetic engineering and biotechnology is essential for developing superior, multipurpose cultivars. The concerted global efforts in sugarcane germplasm collection and breeding have led to remarkable advancements, ensuring the development of high-yielding, disease-resistant, and climate-adaptable varieties.

The use of emerging GS technologies is very effective at addressing past problems. Greater progress could be made through a coordinated international network involving sharing data, results, and germplasm for mutual benefits rather than smaller independent efforts. The need to develop methodology for selecting sugarcane clones for dry environments, screening of elite sugarcane germplasm for developing high sugar varieties, and evaluation of cultivars in different environments is key in identifying the desirable characteristics, for example robustness and long-term evaluation of sugarcane cultivars will be helpful in monitoring cultivars deterioration and breeding achievement. Continued international collaboration, innovative breeding techniques, and addressing current challenges with development of germplasm diversity and related genera for use in sugarcane betterment will be crucial for the future success, competitiveness, and sustainability of the sugarcane industry at global level.

KEYWORDS

- **sugarcane germplasm**
- **associated cultivars**
- **varietal improvement**
- ***Saccharum* spp.**

REFERENCES

Ahmad A, Wang J-D, Pan Y-B, Sharif R, Gao S-J. (2018). Development and use of simple sequence repeats (SSRs) markers for sugarcane breeding and genetic studies. *Agronomy.* 8, 260.

Amin MH, Kassim ES, Bayaumi M, Menshawi ZA. (1972). Growth and flowering of sugarcane in relation to photoperiod and humidity. *Proc. Int. Soc. Sugarcane Tech.* 17, 348–353.

Anas M, Verma KK, Haq IU, Naeem M, Li Q, Liao F, Xiang YX, Li YR. (2021a). Characterization of exotic, native and wild-type genotypes of sugarcane (*Saccharum* spp. Hybrids) for internal nitrogen use efficiency under different nitrogen levels. *Sugar Tech.* 23, 1258–1267. https://doi.org/10.1007/s12355-021-01014-1.

Anas M, Verma KK, Riaz M, Qiang L, Liao F, Liu Y, Li YR. (2021b). Physio-morphological and biochemical mechanism of nitrogen use efficiency in sugarcane (*Saccharum* spp.) genotypes under different growth stages and nitrogen levels. *J. Plant Interact.* 16, 332–343.

Artschwager E, Brandes E.W. (1958). *Sugarcane (S. officinarum) Origin, Classification, Characterization and Description of Reproductive Clones.* U.S. Department of Agriculture, 122, p. 307.

Babu CN. (1990). *Sugarcane*. Allied Publishers Ltd.: New Delhi. India.

Bakshi R, Hemaprabha G, Singh BD, Appunu C. (2021). History and current status of sugarcane breeding, germplasm development and molecular biology in India. *Sugar Tech.* 24(1), 4–29.

Balasundaram N. (1995). *Annual Report Sugarcane Breeding Institute.* Coimbatore.

Barber CR. (1916). Classification of indigenum Indian canes. *Agric. J. India.* 8(3), 371–376.

Barber CR. (1917). Studies in Indian sugarcane no. 3. The classification of Indian canes with special reference to the Saretha and Sonabile group. *Mem. Dept. Agri. India, Bot.* 9, 133–219.

Barber CR. (1922). The classification of Indian canes. *Int. Sug. J.* 24, 18–21.

Berding N, Skinner JC, (1987). Traditional Breeding Methods. Copersucar International Sugarcane Breeding Workshop. CTC, Piracicaba, SP Brazil, pp. 269–320.

Berding N. (1976). Techniques for ripening and drying sugarcane panicles. *ISSCT Sugarcane Breeding Newsle.* 18, 17–22.

Berding N. (1995). Improving flowering for breeding; progress and prospects. *Proc. Aust. Soc. Sugarcane Tech. Conf.* 17, 162–171.

Berding N, Hogarth DM, Cox MC. (2003). Plant improvement of sugarcane. In: James G. (Eds.), *Sugarcane*, 2nd edn., Blackwell Publishing.

Berding N, Moore PH, Smith GR. (1997). Advances in breeding technology for sugarcane. In: Keating BA, Wilson JR. (Eds.). *Intensive Sugarcane Production; Meeting the Challenges Beyond* 2000, 189 220. CAB International, Wallingford, UK.

Berding N, Brain T. Roach. (1987). Germplasm collection, maintenance, and use. In: ed. Heinz DJ (Ed.), *Sugarcane Improvement Through Breeding*. Elsevier: Amsterdam, pp. 143–210.

Bhatt R, Verma KK, Jamwal NS. (2023a). Can paddy-compost eliminate N fertilization for climate-smart sugarcane? *Sugar Tech.* 25, 1036–1046. https://doi.org/10.1007/s12355-023-01270-3.

Bhatt R, Verma KK, Kumar R, Sanghera GS. (2023b). Foliar application of potassium salt of active phosphorus (PSAP) mitigates insect pests and improves yield along with sugarcane quality in response to agroclimatic conditions of Punjab. *Sugar Tech.* 25(3), 660–669.

Chaisan T, Wannarat W, Authapun J. (2018). Creation of genetic variation and selection for drought tolerance in sugarcane. In *ISSCT 12th Germplasm & Breeding Workshop.*

Chaiyabut A, Poeaim S, Distabanjong K. (2013). Genetic Relationship in *Saccharum spontaneum* and Their Hybrids by RAPD. National genetics conference 2013 (NGC2013). *Thai J. Genet.* S(1), 206–209.

Chakravarthi M, Philip A, Subramonian N. (2015). Truncated ubiquitin 50 regulatory region from *Erianthus arundinaceus* drives enhanced transgene expression in heterologous systems. *Mol. Biotechnol.* 57, 820–835.

Chakravarthi M, Syamaladevi DP, Harunipriya P, Augustine SM, Subramonian N. (2016). A novel PR10 promoter from *Erianthus arundinaceus* directs high constitutive transgene expression and is enhanced upon wounding in heterologous plant systems. *Mol. Biol. Rep.* 43, 17–30.

Chandran K, Nisha M, Gireesan PP. (2020). Characterization of progenies from polycrosses of *S. robustum* Clones *f. sanguineum*. *Sugar Tech.* 22, 379–388. https://doi.org/10.1007/s12355-019-00785-y.

Chaudhary P, Xu M, Ahamad L, Chaudhary A, Kumar G, Adeleke BS, Verma KK, Hu D-M, Siric I, Kumar P, Popescu SM, Fayssal SA. (2023). Application of synthetic consortia for improvement of soil fertility, pollution remediation, and agricultural productivity: A review. *Agronomy.* 13, 643. https://doi.org/10.3390/agronomy13030643.

Chen GL, Zheng FJ, Lin B, Lao SB, He J, Huang Z, Zeng Y, Sun J, Verma KK. (2020). Phenolic and volatile compounds in the production of sugarcane vinegar. *ACS Omega.* 5, 30587−30595.

Chen G-L, Zheng F-J, Lin B, Yang Y-X, Fang X-C, Verma KK, Yang L-F. (2023). Vinegar: A potential source of healthy and functional food with special reference to sugarcane vinegar. *Front. Nutr.* 10, 1145862. https://doi.org/10.3389/fnut.2023.1145862.

Cox MC, Stringer JK, Cervellin RJ. (2005). Productivity increases from new varieties in the Queensland sugar industry. *Proc. Aust. Soc. Sugar Cane Technol.* 25, 124–132.

Cursi, DE, Hoffmann HP, Barbosa GVS, Fernandes Junior AR, Balsalobre TWA, Bressiani JA, Gazaffi R, Chapola RG, Diniz CA, Santos JM, Carneiro MS. (2021). History and current status of sugarcane breeding, germplasm development and molecular genetics in Brazil. *Sugar Tech.* 24(1), 112–133. https://doi.org/10.1007/s12355-021-00951-1.

Cursi DE, Gazaffi R, Hoffmann HP, Roberto Giacomini Chapola, Fernandes Junior AR, Carneiro MS, Antonio Augusto F. Garcia. (2019). Genetic gain prediction by different family-selection strategies. In *Proceedings of the International Society of Sugar Cane Technologists*, ISSCT: Tucumán, pp. 693–697.

Daniels J, Roach BT. (1987). Taxonomy and evolution. In: *Sugarcane Important Through Breeding* Heinz DJ (Ed.), Elsevier Amsterdam: New York, Tokyo.

Daniels J. (1967). Sugarcane and related plant of Fiji Islands. *Proc. Int. Soc. Sugarcane Tech.* 12, 1004–1013.

Daniels JP, Smith N, Poton H, Williams CA. (1975). The origin of the genus *Saccharum* ISSCT. *Sugarcane Breeding Newslett.* 26, 24–39.

de Silva AGM, Degaspari N, Carvalho LCC, de Rodrigues PA, Baglione BM. (1989). The use of Ethepon to manage sugarcane varieties in different locations of the central-southern region of Brazil. *Proc. Int. Soc. Sugarcane Tech.* 20, 635–646.

Degroot W, George EF. (1962). Breeding of sugarcane varieties in Mauritius. *Proc. Int. Soc. Sugarcane Tech.* 11, 436–770.

Deng ZN, Huang HR, Huang CM, Verma KK, Luo HB, Jiang SL, Li YJ, Cao HQ, Wu KC, Xu L. (2023). Short-term limited water irrigation influences the photosynthetic pigments, enzymatic and non-enzymatic activities in *Saccharum spontaneum* L. At vegetative stage. *Sugar Tech.* 25, 110–118. https://doi.org/10.1007/s12355-022-01175-7.

Ethirajan AS. (1987). Sugarcane hybridization techniques. *Copersucar Int. Sugarcane Breeding Wor*kshop. CTC, Piracicaba: SP Brazil, pp. 129–147.

Guo D-J, Li D-P, Chen Z-D, Yang B, Verma KK, Singh RK, Singh P, Khan Q, Sharma A, Qin Y, Zhang B-Q, Song X-P, Li Y-R. (2023). Effect of endophytic diazotroph *Enterobacter roggenkampii* ED5 on nitrogen-metabolism-related microecology in the sugarcane rhizosphere at different nitrogen levels. *Front. Microbiol.* 14, 1132016. https://doi.org/10.3389/fmicb.2023.1132016.

Guo D-J, Li D-P, Singh RK, Singh P, Sharma A, Verma KK, Qin Y, Khan Q, Lu Z, Malviya MK, Song X-P, Xing Y-X, Li Y-R. (2021). Differential protein expression analysis of two sugarcane varieties in response to diazotrophic plant growth-promoting endophyte *Enterobacter roggenkampii* ED5. *Front. Plant Sci.* 12, 727741. https://doi.org/10.3389/fpls.2021.727741.

Guo DJ, Li DP, Singh RK, Singh P, Verma KK, Sharma A, Qin Y, Khan Q, Song XP, Malviya MK, Xing YX, Li YR. (2022). Comparative transcriptome analysis of two sugarcane varieties in response to diazitrophic plant growth promoting endophyte *Enterobacter roggenkampii* ED5. *J. Plant Interact.* 17(1), 75–84.

Healey AL, Garsmeur O, Lovell JT, Shengquiang S, Sreedasyam A, Jenkins J, Plott CB, Piperidis N, Pompidor N, Llaca V, Metcalfe CJ, Doležel J, Cápal P, Carlson JW, Hoarau JY, Hervouet C, Zini C, Dievart A, Lipzen A, Williams M, Boston LB, Webber J, Keymanesh K, Tejomurthula S, Rajasekar S, Suchecki R, Furtado A, May G, Parakkal P, Simmons BA, Barry K, Henry RJ, Grimwood J, Aitken KS, Schmutz J, D'Hont A. (2024). The complex polyploid genome architecture of sugarcane. *Nature*. 628(8009), 804–810. https://doi.org/10.1038/s41586-024-07231-4.

Hogarth DM, Cox MC, Bull JK. (1997). Sugarcane improvement: Past achievements and future prospects. In: *Crop Improvement for the 2nd Century*, pp. 29–56.

Huang CM, Wu KC, Verma KK, Luo HB, Deng ZN, Shi GY, Cao HQ, Wu XJ, Ye LP, Wei YW, Xu L. (2024). Isolation and expression analysis of cellulose synthase 3 (Ces3) genes from sugarcane (*Saccharum officinarum* L.). *Chem. Biol. Technol. Agric.* 11, 71. https://doi.org/10.1186/s40538-024-00597-7.

Huang HR, Gao YJ, Malviya MK, Verma KK, Solanki MK, Huang YX, Li X., Deng Y, Yan J, Tang SY, Wang LW, Xu L. (2022). Genetic diversity analysis of sugarcane (*Saccharum* spp. Hybrids) among high-sucrose clones of GT series and commonly used parents by using microsatellite markers in Guangxi, China. *Sugar Tech.* 24, 397–407. https://doi.org/10.1007/s12355-021-01033-y.

Jagadesh M, Dash M, Kumari A, Singh SK, Verma KK, Kumar P, Bhatt R, Sharma SK. (2024). Revealing the hidden world of soil microbes: Metagenomic insights into plant, bacteria, and fungi interactions for sustainable agriculture and ecosystem restoration. *Microbiol. Res.* 127764. https://doi.org/10.1016/j.micres.2024.127764.

Jagathesan D, Ratnambal MJ. (1967). Karyotype analysis in *Saccharum officinarum*. *Nucleus*. 12(1), 23–30.

Joshi A, Verma KK, Rajput VD, Arora J. (2023). Metabolic engineering of microorganisms in advancing biofuel production. In: *Advances in Lignocellulosic Biofuel Production Systems*. pp. 181–202. https://doi.org/10.1016/B978-0-323-91192-4.00016-X.

Joshi A, Verma KK, Rajput VD, Minkina T, Arora J. (2022). Recent advances in metabolic engineering of microorganisms for advancing lignocellulose-derived biofuels, *Bioengineered*. 13(4), 8135–8163. https://doi.org/10.1080/21655979.2022.2051856.

Krishna D, Sachan HK, Verma KK. (2022). Abiotic stresses in plants: an overview. In: *Understanding Abiotic Stresses Publisher*. Nova Science Publishers, Inc.: New York.

Kumari A, Dash M, Singh SK, Jagadesh M, Mathpal B, Mishra PK, Pandey SK, Verma KK. (2023a). Soil microbes: A natural solution for mitigating the impact of climate change. *Environ. Monit. Assess.* 195, 1436. https://doi.org/10.1007/s10661-023-11988-y.

Kumari A, Lakshmi GA, Krishna GK, Patni B, Prakash S, Bhattacharyya M, Singh SK, Verma KK. (2022). Climate change and its impact on crops: A comprehensive investigation for sustainable agriculture. *Agronomy.* 12, 3008. https://doi.org/10.3390/agronomy12123008.

Kumari A, Singh SK, Mathpal B, Verma KK, Garg VK, Bhattacharyya M, Bhatt R. (2023b). The biosynthesis, mechanism of action, and physiological functions of melatonin in horticultural plants: A review. *Horticulturae*. 9, 913. https://doi.org/10.3390/horticulturae9080913.

Legendre BL (1989). Use of germplasm for sugarcane improvement in Louisiana. *Proc. Int. Soc. Sugarcane Tech.* 20, 883–891.

Legendre BL, Burner BM. (1997). New opportunities in the development of sugarcane cultivars. Augmenting traditional approaches with molecular techniques. *Sugarcane*. 5, 4–10.

Leverington KC. (1987). Varieties and their relationship to the control of diseases and pests. *Copersucar Int. Sugarcane Breeding Workshop*. CTC, Piracicaba: SP Brazil, pp. 149–159.

Li A-M, Liao F, Wang M, Chen Z-L, Qin C-X, Huang R-Q, Verma KK, Li Y-R, Que Y-X, Pan Y-Q, et al. (2023). Transcriptomic and proteomic landscape of sugarcane response to biotic and abiotic stressors. *Int. J. Mol. Sci.* 24, 8913. https://doi.org/10.3390/ijms24108913.

Ming-Chin L. (1981). Invitro methods applied to sugarcane improvement. In: Thorpe JA. (Eds.), *Plant Tissue Culture-Methods and Application in Agriculture*. Academic Press: NY, USA, pp. 299–323.

MacColls D. (1977). Some aspects of the flowering in sugarcane in Barbados and its control in breeding programme. *Ann. Bot. (London).* 41, 191–207.

Malik KB. (2018). *Cane and Sugar Production*. Punjab Agricultural Research Board.

Malik KB, Chaudhry AG, Nayyar MM. (1986). Hybridization problems in sugarcane. In *Proceedings of Seminar on Sugarcane Production*. Faisalabad.

Malviya MK, Li C-N, Lakshmanan P, Solanki MK, Wang Z, Solanki AC, Nong Q, Verma KK, Singh RK, Singh P, Sharma A, Guo D-J, Dessoky ES, Song X-P, Li Y-R. (2022). High-throughput sequencing-based analysis of rhizosphere and diazotrophic bacterial diversity among wild progenitor and closely related species of sugarcane (*Saccharum* spp. inter-specific hybrids). *Front. Plant Sci.* 13, 829337. https://doi.org/10.3389/fpls.2022.829337.

Malviya MK, Solanki MK, Li CN, Wang Z, Zeng Y, Verma KK, Singh RK, Singh P, Huang HR, Yang LT, Song XP, Li YR. (2021). Sugarcane-legume intercropping can enrich the soil microbiome and plant growth. *Front. Sustain. Food Syst.* 5, 606595. https://doi.org/10.3389/fsufs.2021.606595.

Mangelsdorf AJ. (1956). Sugarcane breeding. In retrospect and prospect. *Proc. Int. Soc. Sugarcane Tech.* 9(1), 560–575.

Mariotti JA (1972). *On the Effectiveness of Genetic Parameters Used in the Selection of Sugarcane Population.* ISSCT Sugarcane Breed: Newsle. vol. 29, p. 8.

Menhilal AK, et a1. (2000). *Extended Summaries: International Conference on Managing Natural Resources.*

Miller JD. (1994). The philosophy of the Canal Point breeding program: Major issues in the establishment of new breeding program. *Sugar J.* 57(5), 13–15.

Miller JD, Li QW. (1995). Effect of photoperiod treatment on initiation, emergence and flowering date of elite and exotic sugarcane clones. *Sugarcane.* 6, 4–11.

Mukerjee SK. (1957). Origin and distribution of *Saccharum. Bot. Gaz.* 119, 55–61.

Nagaraju MS, et al. (2000). *Coop. Sug.* 31, 391–395.

Naidu KN, Sreenivasan TV. (1987). Conservation of sugarcane germ plasm. In: Heinz DJ (Ed.), *Sugarcane Improvement Through Breeding.* Elsevier: Amsterdam, New York, Tokyo, pp. 33–53.

Naik GR. (2001). *Sugarcane Biotechnology.* Oxford & IBH Publishing Co. Pvt. Ltd.: Delhi, Calcutta.

Nair NV. (2005). *Sugarcane Genetic Resources: Collection, Conservation and Utilization. International Training Course on Genetic Improvement of Sugarcane for Use as Food, Fuel and Fiber.* 12–26 October, 2005. Sugarcane Breeding Institute: Coimbatore.

Nair NV, Sreenivasan TV, William Jebadhas A. (1998). Intra-specific improvement in *Saccharum. Proceedings of the 1st National Plant Breeding Congress.* Indian Society of Plant Breeders, Tamil Nadu Agricultural University: Coimbatore, pp. 140–145.

Natarajan S, Basnayake J, Wei X, Lakshmanan P. (2019). High-throughput phenotyping of indirect traits for early-stage selection in sugarcane breeding. *Remote Sens.* 11, 2952.

Oliveira MA, Verma KK, Bhatt R, Oliveira TBA. (2022). Impact of green and organic fertilizers on soil fertility and sugarcane productivity. In: Verma KK, et al. (Eds.), *Agro-Industrial Perspectives on Sugarcane Production Under Environmental Stress.* https://doi.org/10.1007/978-981-19-3955-6_11.

Philip A, Syamaladevi DP, Chakravarthi M, Gopinath K, Subramonian N. (2013). 50 regulatory region of ubiquitin 2 gene from *Porteresia coarctata* makes efficient promoters for trans gene expression in monocots and dicots. *Plant Cell Rep.* 32(8), 1199–1210.

Prabu GR, Prasad DT. (2012). Functional characterization of sugarcane MYB transcription factor gene promoter (PScMY BAS1) in response to abiotic stresses and hormones. *Plant Cell Rep.* 31, 661–669.

Prasad NR, Neelamathi D, Sreenivasan TV, Tripathi BK. (1998). Storage studies in sugarcane de-fuzzed seed. *Indian Sugar.* 48(8), 577–579.

Quebedeaus JP. (1992). Breeding sugarcane in green house at LSU sugarcane center St. Gabrieal unpublished report.

Ramdoyal K, Badaloo G, Goburdhun P (2003). An international survey of sugarcane crossing programs. *Sugarcane International* (July–August): 3–11

Ramdoyal K, Badaloo G, Mangar M, Goburdhun P, Domaingue R. (2000). Optimizing seed-setting in sugarcane in relation to fertility of inflorescence and harvesting time. *Sugarcane Int*. 9–14.

RIDESA-Rede Interuniversita´ria Para o Desenvolvimento do Setor Sucroenerge´tico; 2020. Censo Varietal Brasil 2017/18. Retrieved May 16, 2024. https://www.ridesa.com.br/censo-varietal

Roach BT, Mullins RT. (1981). Testing sugarcane for water logging tolerance. *Proc. Aust. Soc. Sugarcane Tech.* 5–102.

Roche BT, Daniels J. (1987). A review of the origin and improvement of sugarcane. *Copersucar Int. Sugarcane Breeding Workshop*. CTC, Piracicaba: SP Brazil, pp. 1–33.

Santana I. (1991). Study of variability in micro-propagation of sugarcane. *Cuba Azucar*. 18(2), 74

Selvi A. (2005). *Molecular Markers and Genomics for Sugarcane Varietal Improvement: International Training Course on Genetic Improvement of Sugarcane for Use as Food, Fuel and Fibre*. Sugarcane Breeding Institute: Coimbatore.

Shankaraiah C, et al. (1999). *Coop. Sug.* 30, 965–969.

Sharma A, Singh RK, Singh P, Vaishnav A, Guo DJ, Verma KK, Li DP, Song XP, Malviya MK, Khan N, Lakshmanan P, Li YR. (2021). Insights into the bacterial and nitric oxide-induced salt tolerance in sugarcane and their growth-promoting abilities. *Microorganisms*. 9, 2203. https://doi.org/10.3390/microorganisms9112203.

Sharma A, Song X-P, Singh RK, Vaishnav A, Gupta S, Singh P, Guo D-J, Verma KK, Li Y-R. (2022). Impact of carbendazim on cellular growth, defence system and plant growth promoting traits of Priestia megaterium ANCB-12 isolated from sugarcane rhizosphere. *Front. Microbiol*. 13, 1005942. https://doi.org/:10.3389/fmicb.2022.1005942.

Shinde DB, et al. (1992). *Indian Sug*. 42(3):903–908.

Singh P, Singh RK, Li HB, Guo DJ, Sharma A, Verma KK, Solanki MK, Upadhyay SK, Lakshmanan P, Yang LT, Li YR. (2023). Nitrogen fixation and phytohormone stimulation of sugarcane plant through plant growth promoting diazotrophic Pseudomonas. *Biotechnol. Genet. Eng. Rev.* 1–21. https://doi.org/10.1080/02648725.2023.2177814.

Singh P, Singh RK, Song QQ, Li HB, Guo DJ, Malviya MK, Verma KK, Song XP, Lakshmanan P, Yang LT, Li YR. (2021). Comparative analysis of protein and differential responses of defense-related gene and enzyme activity reveals the long-term molecular responses of sugarcane inoculated with *Sporisorium scitamineum*. *J. Plant Interact.* 16(1), 12–29.

Singh RK, Singh P, Li HB, Song QQ, Guo DJ, Solanki MK, Verma KK, Malviya MK, Song XP, Lakshmanan P, Yang LT, Li YR. (2020). Diversity of nitrogen-fixing rhizobacteria associated with sugarcane: A comprehensive study of plant-microbe interactions for growth enhancement in *Saccharum* spp. *BMC Plant Biol.* 20, 220. https://doi.org/10.1186/s12870-020-02400-9.

Singh O, Rathan N. (1998). A review on low temperature stress resistance in sugarcane. *India Sugar*, XLIII (6), 421–430.

Skinner JC. (1967). Grading varieties for selection. *Proc. Int. Soc. Sugar Cane Technol.* 12, 938–949.

Skinner JC. (1972). Description of sugarcane clones-III. Botanical description. *Proc. Int. Soc. Sugarcane Tech.* 14, 124–127.

Smith B. (1977). Tassel control in drip irrigated fields at Wailuku Sugar Company. *Hawaii. Sugar Technol. Rep.* 36, 81–83.

Solanki MK, Verma KK, Dastogeer KMG, Mora-Poblete F, Mundra S. (2023). Microbial resilience in plant nutrient management towards sustainable farming. *Front. Microbiol.* 14, 1280811. https://doi.org/10.3389/fmicb.2023.1280811.

Song X-P, Mo F, Verma KK, Wei J, Zhang X, Yang L, Li YR. (2019). Effect of sugarcane smut (*Ustilago scitaminea* Syd.) on ultrastructure and biochemical indices of sugarcane. *Biomed J. Sci. Tech. Res.* 17(1), 12546–12551. https://doi.org/10.26717/BJSTR.2019.17.002950.

Song XP, Mo F, Yan M, Zhang X, Zhang B, Huang X, Huang D, Pan Y, Verma KK, Li YR. (2022). Effect of smut infection on the photosynthetic physiological characteristics and related defense enzymes of sugarcane. *Life.* 12, 1201. https://doi.org/10.3390/life12081201.

Song XP, Verma KK, Tian DD, Zhang XQ, Liang YJ, Huang X, Li CN, Li YR. (2021). Exploration of silicon functions to integrate with biotic stress tolerance and crop improvement. *Biol. Res.* 54, 19. https://doi.org/10.1186/s40659-021-00344-4.

Sreenivasan TV, Jalaji NC. (1985). In vitro sugarcane germplasm storage. *Sugarcane.* 1, 1–2.

Sreenivasan TV, Ahloowalia BS, Heinz DJ. (1987). Cytogenetics In: Heinz DJ. (Eds.), *Sugarcane Improvement Through Breeding.* Elsevier, Amsterdam: New York, and Tokyo, pp. 211–253.

Srinivas S. (1996). *Coop. Sug.* 29, 117–123.

Stevenson GC. (1965). *Flowering in Sugarcane in Genetics and Breeding of Sugarcane.* Longmans, Green and Co. Ltd.: London.

Subramonian N, Premachandran MN, Padmanaban P, Srikanth J, Nisha M. Transgenics. ICAR Sugarcane Breeding Institute, Annual Report, 30–31.

Subramonian N, Sridhar VV, Thomas G, Hemaprabha G, Tripathi BK. (1999). Expression of a transferred herbicide gene in a sugarcane cultivar. *International Society for Sugar Cane Technologists XXIII Congress.* 1, 22–26.

Sudawa S. (1987). *Annual Report Sugarcane Breeding Institute*, Coimbatore.

Tian D, Qin L, Verma KK, Wei L, Li J, Li B, Zhou W, He Z, Wei D, Huang S, Long S, Huang Q, Li C, Wei S. (2023). Transcriptomic and metabolomic differences between banana varieties which are resistant or susceptible to Fusarium wilt. *PeerJ.* 11, e16549 https://doi.org/10.7717/peerj.16549.

Torres JS, Rodriguez LA. (1996). Soil compaction management for sugarcane. *Proc. Int. Soc. Sugarcane Tech.* 22, 222–230.

Verma KK, Anas M, Chen Z, Rajput VD, Malviya MK, Verma CL, Singh RK, Singh P, Song XP, Li YR. (2020f). Silicon supply improves leaf gas exchange, antioxidant defense system and growth in sugarcane responsive to water limitation. *Plants.* 9, 1032, https://doi.org/10.3390/plants9081032.

Verma KK, Joshi A, Song X-P, Liang Q, Xu L, Huang H-R, Wu K-C, Seth CS, Arora J and Li Y-R (2024a). Regulatory mechanisms of plant rhizobacteria on plants to the adaptation of adverse agroclimatic variables. *Front. Plant Sci.* 15, 1377793. https://doi.org/10.3389/fpls.2024.1377793.

Verma KK, Joshi A, Song X-P, Singh S, Kumari A, Arora J, Singh SK, Solanki MK, Seth CS, Li Y-R. (2024b). Synergistic interactions of nanoparticles and plant growth promoting rhizobacteria enhancing soil-plant systems: A multigenerational perspective. *Front. Plant Sci.* 15, 1376214. https://doi.org/10.3389/fpls.2024.1376214.

Verma KK, Li DM, Singh M, Rajput VD, Malviya MK, Minkina T, Singh RK, Singh P, Song XP, Li YR. (2020a). Interactive role of silicon and plant-rhizobacteria mitigating abiotic stresses: A new approach for sustainable agriculture and climate change. *Plants.* 9, 1055. https://doi.org/10.3390/plants9091055.

Verma KK, Rajput VD, Singh M, Verma CL, Minkina T. (2022a). Influence of waterlogging on plant physiology and molecular biology. In: *Understanding Abiotic Stresses Publisher.* Nova Science Publishers, Inc.: New York.

Verma KK, Siddiqui N, Singh M. (2022b). Understanding the plant physiological and molecular adaptation strategies to soil salinity. In: *Understanding Abiotic Stresses Publisher*. Nova Science Publishers, Inc.: New York. https://doi.org/10.52305/UBUJ4024.

Verma KK, Singh P, Song X-P, Malviya MK, Singh RK, Chen G-L, Solomon S, Li YR. (2020b). Mitigating climate change for sugarcane improvement: Role of silicon in alleviating abiotic stresses. *Sugar Tech.* 22(5), 741–749. https://doi.org/10.1007/s12355-020-00831-0.

Verma KK, Singh RK, Song QQ, Singh P, Zhang B-Q, Song X-P, Chen G-L, Li YR. (2019a). Silicon alleviates drought stress of sugarcane plants by improving antioxidant responses. *Biomed J. Sci. Tech. Res.* 17(1), 12580–12586. https://doi.org/10.26717/BJSTR.2019.17.002957.

Verma KK, Song X, Degu HD, Guo D, Joshi A, Huang H, Xu L, Singh M, Huang D, Rajput VD, Li YR. (2023a). Recent advances in nitrogen and nano-nitrogen fertilizers for sustainable crop production: A mini-review. *Chem. Biol. Technol. Agricult.* 10, 1–14. https://doi.org/10.1186/s40538-023-00488-3.

Verma KK, Song X, Verma CL, Huang H, Singh M, Xu L, Wu J, Khan ZH, Zhang B, Li Y. (2023b). Mathematical modeling of climate and fluoride effects on sugarcane photosynthesis with silicon nanoparticles. *Plant Physiol. Biochem*. 204, 108089. https://doi.org/10.1016/j.plaphy.2023.108089.

Verma KK, Song XP, Budeguer F, Nikpay A, Enrique R, Singh M, Zhang B, Wu, JM, Li YR. (2022c). Genetic engineering: An efficient approach to mitigating biotic and abiotic stresses in sugarcane cultivation. *Plant Signal. Behav.* 17(1), e2108253.

Verma KK, Song XP, Joshi A, Rajput VD, Singh M, Sharma A, Singh RK, Li DM, Arora J, Minkina T, Li YR. (2022d). Nanofertilizer possibilities for healthy soil, water and food in future: An overview. *Front. Plant Sci.* 13, 865048. https://doi.org/10.3389/fpls.2022.865048.

Verma KK, Song XP, Joshi A, Tian DD, Rajput VD, Singh M, Arora J, Minkina TM, Li YR. (2022e). Recent trends in nano-fertilizer for advancing sustainable agriculture under the era of climate change to ensure food security for future generations. *Nanomaterials*. 12, 173. https://doi.org/10.3390/nano12010173.

Verma KK, Song XP, Kumari A, Jagadesh M, Singh SK, Bhatt R, Singh M, Seth CS, Li YR. (2024c). Climate change adaptation: Challenges for agricultural sustainability. *Plant Cell Environ.* 48(4), 2522–2533. https://doi.org/10.1111/pce.15078.

Verma KK, Song XP, Li DM, Singh M, Wu JM, Singh RK, Sharma A, Zhang BQ, LI YR. (2022f). Silicon and soil microorganisms improves rhizospheric soil health with bacterial community, plant growth, performance and yield. *Plant Signal. Behav*. 17(1), e2104004.

Verma KK, Song XP, Lin B, Guo DJ, Singh M, Rajput VD, Singh RK, Singh P, Sharma A, Malviya MK, Chen GL, Li YR. (2022g). Silicon induced drought tolerance in crop plants: Physiological adaptation strategies. *Silicon*. 14(6), 2473–2487. https://doi.org/10.1007/s12633-021-01071-x.

Verma KK, Song XP, Rajput VD, Boldyreva V, Zhang BQ, Minkina T, Li YR. (2022h). Morpho-physiological, biochemical and ultrastructural modifications on sugarcane to prolonged water deficit. In: Verma KK, et al. (Eds.), *Agro-Industrial Perspectives on Sugarcane Production Under Environmental Stress*. https://doi.org/10.1007/978-981-19-3955-6_11.

Verma KK, Song XP, Rajput VD, Solomon S, Li YR, Rao GP. (2022i). *Agro-Industrial Perspectives on Sugarcane Production under Environmental Stress*. Springer Nature: Singapore, https://doi.org/10.1007/978-981-19-3955-6.

Verma KK, Song XP, Singh M, Tian DD, Rajput VD, Minkina T, Li YR. (2023c). Association of silicon and soil microorganisms induces stress mitigation, increasing plant productivity. In: de Mello Prado R. (Eds.), *Benefits of Silicon in the Nutrition of Plants*. pp. 299–328. https://doi.org/10.1007/978-3-031-26673-7_17.

Verma KK, Song XP, Tian DD, Guo DJ, Chen ZL, Zhong CS, Nikpay A, Singh M, Rajput VD, Singh RK, Minkina T, Li YR. (2021a). Influence of silicon on biocontrol strategies to manage biotic stress for crop protection, performance and improvement. *Plants*. 10, 2163. https://doi.org/10.3390/plants10102163.

Verma KK, Song XP, Tian DD, Singh M, Verma CL, Rajput VD, Singh RK, Sharma A, Singh P, Malviya MK, Li YR. (2021b). Investigation of defensive role of silicon during drought stress induced by irrigation capacity in sugarcane: Physiological and biochemical characteristics. *ACS Omega*. 6, 19811–19821.

Verma KK, Song XP, Verma CL, Chen ZL, Rajput VD, Wu KC, Liao F, Chen GL, Li YR. (2021c). Functional relationship between photosynthetic leaf gas exchange in response to silicon application and water stress mitigation in sugarcane. *Biol. Res*. 54, 15. https://doi.org/10.1186/s40659-021-00338-2.

Verma KK, Song XP, Verma CL, Malviya MK, Guo DJ, Rajput VD, Sharma A, Wei KJ, Chen GL, Solomon S, Li YR. (2021d). Predication of photosynthetic leaf gas exchange of sugarcane (*Saccharum* spp.) leaves in response to leaf positions to foliar spray of potassium salt of active phosphorus under limited water irrigation. *ACS Omega*. 6, 2396–2409.

Verma KK, Song X-P, Xu L, Huang H-R, Liang Q, Seth CS, Li Y-R. (2023d). Nano-microplastic and agro-ecosystems: A mini-review. *Front. Plant Sci*. 14, 1283852. https://doi.org/10.3389/fpls.2023.1283852.

Verma KK, Song XP, Yadav G, Degu HD, Parvaiz A, Singh M, Huang HR, Mustafa G, Xu L, Li YR. (2022j). Impact of agroclimatic variables on proteo-genomics in sugarcane (*Saccharum* spp.) plant productivity. *ACS Omega*. 7, 22997–23008.

Verma KK, Song XP, Zeng Y, Guo DJ, Sing M, Rajput VD, Malviya MK, Wei KJ, Sharma A, Li DP, Chen GL, Li YR. (2021e). Foliar application of silicon boosts growth, photosynthetic leaf gas exchange, antioxidative response and resistance to limited

water irrigation in sugarcane (*Saccharum officinarum* L.). *Plant Physiol. Biochem.* 166, 582–592.
Verma KK, Song XP, Zeng Y, Li DM, Guo DJ, Rajput VD, Chen GL, Barakhov A, Minkina TM, Li YR. (2020c). Characteristics and correlation of leaf stomata and its relationship with photosynthesis on *Saccharum* spp. under different irrigation and silicon application. *ACS Omega.* 5, 24145–24153.
Verma KK, Vatsal S, Singh M, Rajput VD, Minkina T. (2022k). Coping with water stress: effects and adaptive mechanisms to plants. In: *Understanding Abiotic Stresses Publisher.* Nova Science Publishers, Inc.: New York.
Verma KK, Wu KC, Verma CL, Li DM, Malviya MK, Singh RK, Singh P, Chen GL, Song XP, Li YR. (2020d). Developing mathematical model for diurnal dynamics on photosynthetic responses in sugarcane responsive to different irrigation and silicon application. *PeerJ.* 8, e10154. https://doi.org/10.7717/peerj.10154.
Verma KK, Liu X-H, Wu K-C, Singh RK, Song QQ, Malviya MK, Song X-P, Singh P, Verma CL, Li YR. (2020e). The impact of silicon on photosynthetic and biochemical responses of sugarcane under different soil moisture levels. *Silicon.* 12, 1355–1367.
Verma KK, Wu K-C, Singh P, Malviya MK, Singh RK, Song X-P, Li YR. (2019b). The protective role of silicon in sugarcane under water stress: Photosynthesis and antioxidant enzymes. *Biomed J. Sci. Tech. Res.* 15(2), 1–7. https://doi.org/0.26717/BJSTR.2019.15.002685.
Vijayalakshmi U. (1967). *Taxonomy of Sugarcane.* ISSCT Sugarcane Breed: Newsletter. 20, pp. 52–64.
Wang KY. (1980). Studies on flowering of sugarcane in the South of Hainan, China. *Proc. Int. Soc. Sugarcane Tech.* 17, 1301–1306.
Weerathaworn P. (2006). Mitr Phol sugarcane research center, Thailand. *Sugar Tech.* 8, 101–103. https://doi.org/10.1007/BF02943641.
Wu K-C, Huang C-M, Verma KK, Deng Z-N, Huang H-R, Pang T, Cao H-Q, Luo H-B, Jiang S-L, Xu L. (2022). Transcriptomic responses of *Saccharum spontaneum* roots in response to polyethylene glycol-6000 stimulated drought stress. *Front. Plant Sci.* 13, 992755. https://doi.org/10.3389/fpls.2022.992755.
Xu L, Deng Z-N, Wu K-C, Malviya MK, Solanki MK, Verma KK, Pang T, Li Y-J, Liu X-Y, Kashyap BK, Dessoky ES, Wang W-Z and Huang H-R. (2022). Transcriptome analysis reveals a gene expression pattern that contributes to sugarcane bud propagation induced by indole-3-butyric acid. *Front. Plant Sci.* 13, 852886. https://doi.org/10.3389/fpls.2022.852886.
Yang S, Chen J. (1980). Germination response of sugarcane cultivars to soil moisture and temperature. *Proc. Int. Soc. Sugarcane Tech.* 17(1), 30–37.
Qi Y, Gao X, Zeng Q, Zheng Z, Wu C, Yang R, Feng X, Wu Z, Fan L, Huang Z. (2021). Sugarcane breeding, germplasm development and related molecular research in China. *Sugar Tech.* 24, 73–85. https://doi.org/10.1007/s12355-021-01055-6.
Zheng F-J, Lin B, Yang Y-X, Fang X-C, Verma KK, Chen G-L. (2023). Efficacy and functionality of sugarcane original vinegar on mice. *Front. Microbiol.* 14, 1224666. https://doi.org/10.3389/fmicb.2023.1224666.

CHAPTER 6

Advances in Molecular Markers for Upgrading Sugar Content

EID M. MEHAREB

Sugar Crops Research Institute, Agricultural Research Center, Giza, Egypt

*Corresponding author

ABSTRACT

Advances in molecular marker technologies have significantly enhanced their precision, making them vital tools in modern sugarcane breeding. These markers offer considerable advantages over traditional methods. Over the past decade, biotechnological progress has improved our understanding of the sugarcane genome, particularly the molecular features of sucrose transport and accumulation. Molecular marker strategies help elucidate sucrose metabolism, aiding in diversity studies, marker identification, and mapping. This is crucial because sugar content is a complex trait governed by multiple genes and enzymes. Genomic prediction, which combines precise breeding value estimates with shorter generation intervals, promises to accelerate genetic gains. Breeding programs can use selection trials to develop training populations for these predictions. While sugarcane's autopolyploid genome (over 80 chromosomes) is highly complex, marker technology has evolved from early hybridization-based methods to high-throughput sequencing capable of detecting numerous single-nucleotide polymorphisms (SNPs). The choice of genotyping platform from low-density, scalable methods to whole-genome approaches depends on the breeding application. Continued advancements in biotechnology and molecular biology are poised to play an increasingly important role in enhancing this vital crop.

Sugarcane Transformation: Biofuels, Biotechnology, and Beyond.
Krishan K. Verma, Xiu-Peng Song, Hai-Rong Huang, Aradhna Kumari, and Yang-Rui Li (Eds.)

DOI: 10.1201/9781779641571-6

6.1 INTRODUCTION

Sugarcane (*Saccharum* sp.), with an annual worth of over US$143 billion, is arguably the most valuable crop in the world (Verma et al., 2020a, 2020b, 2022a, 2022b). Over 100 countries cultivate sugarcane, which accounts for over 80% of the world's sucrose supply (Da Silva and Bressiani, 2005; Chen et al., 2022, 2023; Verma et al., 2023a, 2023b; Huang et al., 2024). The analysis of gene expression in sugarcane populations for sucrose concentration showed a possible overlap of sugar biosynthesis and cell wall metabolic processes and recommended that signaling and transcriptional regulators could be beneficial as molecular markers in breeding programs for transgenic plants in a sugarcane improvement program (Guo et al., 2021, 2023; Huang et al., 2022; Li et al., 2023). Responses to sucrose biosynthesis in monocots and dicots have been shown to be maintained, according to an investigation on gene expression associated with sucrose concentration. The aforementioned discovery implies the existence of significant agents in the transduction pathways of sugar signals (Oliveira et al., 2022; Song et al., 2019, 2021, 2022; Tian et al., 2023). Further research is required to clarify the potential involvement of the discovered genes in the sugarcane sucrose accumulation process, given the variety of roles that have been described for them (Papini et al., 2009; Singh et al., 2021; Verma et al., 2020c, 2022c, 2023d, 2023c).

The primary goal of most sugarcane breeding programs is to develop cultivars with improved cane and sugar yields (Anas et al., 2021a, 2021b; Verma et al., 2020d, 2020e, 2022e, 2022f, 2022g; Bhatt et al., 2023a, 2023b). Sucrose accumulation is of particular interest in sugarcane (*Saccharum* sp.) as it supplies more than 75% of world's sucrose (Zheng et al., 2023; Verma et al., 2019a, 2019b; 2023d). It is a dynamic process of a continuous cleavage and synthesis of sugar storage parenchyma tissue (Verma et al., 2021a, 2021b, 2021c; Joshi et al., 2022, 2023; Solanki et al., 2023). Still, in sugarcane, extensive research is being conducted to increase yield because the molecular mechanisms governing sucrose accumulation are still poorly understood (Singh et al., 2020, 2023; Verma et al., 2020f, 2021d, 2021e; Malviya et al., 2021; Chaudhary et al., 2023; Wu et al., 2022). By catalyzing the reversible conversion of sucrose and UDP into UDP-glucose and fructose, sucrose synthase (SuSy, UDP-glucose: D-fructose 2-a-D-glu cosyltransferase, EC 2.4.1.13) is an essential enzyme in the regulation of sucrose buildup and partitioning in the sugar biosynthesis pathway.

However, relatively little is known about the members of the SuSy gene family in sugarcane, and as a result, no conclusive research on the allelic diversity of SuSy gene families in *Saccharum* species has been published. Sugarcane cultivars are usually hybrids of 2–4 cultivars with 100–130 chromosomes. Therefore, understanding sucrose accumulation and its genetics is a very difficult and slow process (Verma et al., 2022h, 2022i, 2022j, 2022k). Due to its intricate polyploidy, genome-wide quantitative trail locus (QTL) analysis and map-based cloning are also a daunting task. Transcriptome analysis is a powerful tool to identify genes for important agronomic traits (Xu et al., 2022; Deng et al., 2023; Li et al., 2023; Jagadesh et al., 2024). It can analyze any genome and identify important genes irrespective of genome complications. One cycle in traditional sugarcane variety improvement initiatives takes an average of 12 years, ranging from hybridization to cultivar release. Since methods to improve early and indirect selection in sugarcane would have a significant influence on breeding for high sugar content and might improve selection reliability for early sugar content, molecular markers are a useful tool for breeders.

6.2 SUGARCANE TAXONOMY

Sugarcane's taxonomy and phylogeny are complex, since plants from five genera share traits and constitute the closely related "*Saccharum* complex," which is a result of interbreeding. According to Daniels and Roach (1987), the *Saccharum* complex consists of *Saccharum*, *Erianthus* section *Ripidium*, *Miscanthus* section *Diandra*, *Narenga*, and *Sclerostachya*. Due to the high levels of polyploidy (polyploids having over two sets of chromosomes) and often uneven chromosomal numbers (aneuploidy), these genera have seen numerous revisions to their taxonomic relationships, which has made taxonomy challenging (Sreenivasan et al., 1987; Daniels and Roach, 1987). According to Hodkinson et al. (2002), recent genetic study of the *Saccharum* complex genera has resulted in proposals that the taxonomy should be revised because several segments seem to be polyphyletic.

6.3 CLASSIFICATION OF MOLECULAR MARKERS AND THEIR USE IN SUGARCANE

Molecular markers are important equipment in present-day plant breeding programs (Guo et al., 2022; Malviya et al., 2022). They allow plant breeders to pick out and choose acceptable trends with extra precision, efficiency, and pace when compared with standard breeding methods. Molecular markers are genetic versions that may be detected on the DNA or protein level (Li et al., 2023; Verma et al., 2024a, 2024b, 2024c). They are used for unique genes or areas of the genome which can be related to favored trends

6.3.1 RESTRICTION FRAGMENT LENGTH POLYMORPHISMS

Among the earliest molecular markers created were restriction fragment length polymorphisms (RFLPs). They are predicated on differences in DNA sequences that particular restriction enzymes can identify. On an agarose gel, variations in the fragment lengths caused by enzyme digestion can be seen. Although labor-intensive and requiring enormous amounts of DNA, RFLPs are instructive (Williams, 1989). Correspondingly, southern hybridization is used in the creation of polymorphism that is the basis for RFLP markers (Botstein et al., 1980). Certain combinations of probes and enzymes define these markers. The first sugarcane genetic maps were created with poor and unequal coverage using RFLP markers (da Silva et al., 1993; Ming et al., 2002). Initially, variation in the polyploid sugarcane genome was only partially detected using high copy number ribosomal RFLP markers. Wood (1987) was able to differentiate between *Saccharum barberi* and *Saccharum sinense* accessions using this method, but he was unable to identify variation within a species. The ability to identify homologous single dose markers with these RFLP markers allowed linkage groups (LGs) to be converted into homology groups (HGs). Since the probes employed by Ming et al. (1998, 2002) were screened across populations, genetic map comparisons could be made by aligning different types. Furthermore, structural genomic similarities and differences were beginning to be discovered, and comparative mapping between sugarcane and sorghum could be done to a limited extent. Chromosome pairing affinities at meiosis in sugarcane were determined by an additional and intriguing application of RFLP markers. The ability to utilize probes from several related species was the main benefit of RFLP markers. Ming made use of genomic clones from sugarcane, sorghum, maize, rice, barley, and oat, as well as RFLP probes made from complementary DNA (cDNA) (Aitken, 2022).

Additionally, sugarcane polymorphism research has made specific use of sugar gene sequences. It is possible that high and low sugar clones in a population might be distinguished using expressed sequence tag (EST)-derived RFLP segment cDNA for SuSy, suggesting that these clones could be employed as indicators of sugar content (da Silva and Bressiani, 2005).

6.3.2 AMPLIFIED FRAGMENT LENGTH POLYMORPHISMS

Using different combinations of restriction enzymes and primers, amplified fragment length polymorphism (AFLP) combines DNA restriction and amplification. Autoradiography is used to visualize the variability (Vos et al., 1995; Simpson, 1997; Blears et al., 1998). This entails preselective and selective amplification, ligation of adapters to the cut end, digestion of genomic DNA with restriction enzymes (a rare cutter and a frequent cutter), separation of the fragments by electrophoresis, and autoradiography visualization.

Hoarau et al. (2001) used 295 members of a selfed population of the sugarcane variety R570 to create an AFLP genomic map. With the help of reference *Saccharum officinarum* DNA, it was possible to detect 11 recombinant LGs between *Saccharum spontaneum* and *S. officinarum*, and assign 11 LGs to each of the two species using the 887 single-dose markers on this map, which were organized into 120 LGs. Much bigger populations could be examined with more markers because to the introduction of AFLPs in sugarcane. Two major clusters within the global collection of *S. spontaneum*, corresponding accessions accumulated from the north or south of the Tropic of Cancer, were identified by Aitken et al. (2018) after analyzing 430 *S. spontaneum* and 255 variations using over 700 markers.

6.3.3 POLYMERASE CHAIN REACTION-BASED MARKERS

The polymerase chain reaction (PCR), which amplifies tiny DNA fragments repeatedly in a discriminatory way, is used by these markers. Using a heat cycler, the primers are annealed to the complementary sequence in the DNA strands, denaturing the DNA strands in the process. The primers are then extended to generate small DNA fragments of 200 bp or even up to 3.4 kb. The polymorphism can be seen by electrophoresis or other appropriate nongel-based techniques. It results from differential amplification of DNA sequences in distinct genotypes. Random primers may be used in PCR-based markers, similar to random-amplified polymorphic DNA (RAPD) (Williams et al., 1990), in which random primers are used to amp up the target DNA several times. Although it is not necessary to be aware of gene sequence information in this case, the assay's sensitivity to experimental circumstances means that it cannot be repeated. These two PCR markers were employed by Glynn et al. (2013) in stage 2 of the Canal Point sugarcane breeding and selection program to screen a sizable set of parental clones and genotypes. Indian sugarcane varieties play a role in the lineage of elite clones of other countries, as evidenced by the fact that 70% of these simple sequence repeat (SSR) markers amplified PCR products from the genomic DNA of a Louisiana variety compared with the 30% transferability of International Sugarcane Microsatellite Consortium SSR markers to US sugarcane cultivars (Pan, 2006). Interspecific hybrids (ISH) known as ISH-1 to ISH-29 were produced as a consequence of a cross between the optimum sucrose sugarcane clone, such as *S. officinarum* "Gungera" and the lower sucrose clone, like *S. spontaneum* "SES 603." The hybrids were ranked according to the amount of sucrose they contained. The four hybrids were chosen as genotypes with extremely high sucrose levels (65–100 mg/g tissue), while the genotypes with very low sucrose levels (2–25 mg/g tissue) were ISH-10, ISH-11, ISH-12, and ISH-25. Using specific primers, the DNA from the leaves of the parent clones and the progeny hybrids was amplified to find markers for sucrose concentration (Vinayak et al., 2010). Ten specific primers were analyzed: primers A and B, which identify polymorphisms in the promoter region of the SuSy-2 gene; primers AI, SS, and SPS, which were created based on the nucleotide sequences of the enzymes' respective genes; and primers MSSCIR43, MSSCIRI, SMC226CG, SMC1039CG, and SCB07, which were chosen based on their association with the sucrose accumulation process. DNA products unique to clones with low or high sucrose content were found. Only maximum sucrose genotypes, that is, Gungera, ISH-1, ISH-5, ISH-17, and ISH-23, were amplified by primers "A" and "AI" (230 and 500 bp, respectively), whereas lower sucrose genotypes, such as SES 603, ISH-10, ISH-11, ISH-12, and ISH-25, only produced a 920 bp amplified product by primer SMC226CG. Ten random decamer primers were also tested; however, the results showed no correlation between the genotypes' sucrose content and their products.

6.3.4 SIMPLE SEQUENCE REPEATS OR MICROSATELLITES OR SEQUENCE TAGGED SITES

SSRs are brief DNA sequences made up of tandem repeats with different numbers of repeats resulting in length polymorphisms. They are codominant, highly variable, and provide a wide variety of applications in plant breeding due to their prevalence in the genome (Kalia et al., 2011; Sharma et al., 2021, 2022; Meena et al., 2022). Short repeated sequences known as microsatellite markers are used to determine polymorphisms using the PCR method (Gupta et al., 1996; Powell et al., 1996). Instead of using random primers, sequence-specific primers are employed here. Therefore, in this instance, repeatability and dependability are higher. These are effective instruments for mapping investigations, marker-assisted selection, genotype differentiation, genetic diversity analysis, and seed purity assessment. In sugarcane, the introduction of microsatellite or SSR markers (Cordeiro et al., 2000) allowed for the addition of more markers, which, similar to RFLP markers, detected several homozygous alleles using a single set of primers. A capillary sequencing apparatus could also be used to automatically score SSR markers, and the accuracy of automatic genotyping was enhanced by their decreased complexity. In order to provide information on the number of LGs inside HGs, SSRs were swiftly added to AFLP genetic maps (Aitken et al., 2005). Using 65 microsatellites markers and an enhanced genetic linkage map of the sugarcane (*Saccharum* spp. hybrids) cultivar "LCP 85–384," quantitative trait loci regulating sucrose content were identified (Liu et al., 2016).

6.3.5 SINGLE-NUCLEOTIDE POLYMORPHISM

Single base pair mutations in DNA sequences are known as single-nucleotide polymorphisms (SNPs). They are the most prevalent kind of genetic variation and are easily found with a variety of techniques, including next-generation sequencing (NGS) and microarrays. SNPs are employed in genome-wide association studies (GWAS) and high-throughput genotyping (Marth et al., 1999; Batra et al., 2014). SNPs are point mutations when a single nucleotide is changed at a specific locus. They provide an endless supply of polymorphism that is beneficial for high-resolution mapping research. These can be applied in situations where genomics is well developed.

6.3.6 SINGLE-STRAND CONFORMATION POLYMORPHISM

Sugarcane, like other crops, has also been treated with single-strand conformation polymorphism derived from genomic sequences and ESTs (Fukuoka et al., 1994; Swapna et al., 2011a). Here, the amplified products undergo electrophoresis after being transformed into single strands. In this instance, polymorphism resulting from single-strand conformational variations is visualized. It is unknown exactly what sequence changes lead to the creation of distinct secondary structures, which are reflected in the shift in the gel's mobility. Sequence rearrangements, single-nucleotide insertions–deletions, substitutions, or a combination of these alterations, as shown in other plants, could be the cause of this (Fukuoka et al., 1994; Bodenes et al., 1996).

6.3.7 CLEAVED AMPLIFIED POLYMORPHIC SEQUENCES AND DERIVED CLEAVED AMPLIFIED POLYMORPHIC SEQUENCES

SNPs that add or remove restriction enzyme recognition sites serve as the foundation for the cleaved amplified polymorphic sequences and derived cleaved amplified polymorphic sequences markers. Different genotypes can be distinguished after PCR amplification through digestion using a particular enzyme (Neff et al., 1998; Kaundun and Matsumoto, 2003). Alwala et al. (2008) performed linkage mapping and genome analysis on the offspring of an interspecific cross in sugarcane using targeted region amplified polymorphism (TRAP), sequence-related amplified polymorphism (SRAP), and AFLP markers. For *S. officinarum*, the framework map had 146 connected markers in 49 LGs, and for *S. spontaneum*, it included 121 linked markers in 45 LGs.

6.3.8 RANDOM AMPLIFIED POLYMORPHIC DNA

Using brief, arbitrary primers, random PCR amplification of DNA segments is done for RAPDs. Although these markers may be produced quickly and easily, they may not be repeatable and need to be optimized (Atienzar and Jha, 2006). Sugarcane germplasm has been extensively characterized using molecular markers, which are a common application in agriculture. In these studies, phylogeny, interspecies relationships, and genetic diversity among the *Saccharum* species, related genera, and their hybrids have been analyzed through a variety of markers, including DNA-based markers, that is, RFLP, RAPD, AFLP, and microsatellite or SSR (Sharma et al., 2014; Ahmed and Gardezi, 2017).

6.3.9 INTER-SIMPLE SEQUENCE REPEAT

Using anchored primers, microsatellite sequences are amplified PCR-wise in inter-simple sequence repeats (ISSRs). They have been extensively utilized for genetic diversity evaluations and combine the benefits of SSRs and RAPDs (Guasmi et al., 2012). Due to their many benefits, including being a comparative, quick, easy, and affordable method that requires only a small amount of sugarcane DNA and no primary sequence data value for primer design, RAPD and ISSR analysis have been widely used to identify sugarcane polymorphism genomic DNA and many other plant systems (Carothers et al., 2012; Sugiyama et al., 2017).

6.3.10 SNP ARRAYS

High-density microarrays with thousands to millions of SNP markers are called SNP arrays. They make large population genotyping efficient, and they make GWAS and genomic selection possible (Chen et al., 2014). Recently, the transcriptome sequencing results of 59 F1 individuals (*S. robustum* and *S. officinarum* LA Purple) produced 83 and 105 LGs and 11,157 and 8998 SNPs, respectively (Zhang et al., 2019).

You et al. (2019) have created an Affymetrix Axiom 100K SNP array and recently reported on the target enrichment sequencing of 300 sugarcane accessions chosen from the global collection. Using single-dose SNPs for sugarcane, germplasm identification is made easier with the help of this array, which includes 31,449 single-dose SNPs and 68,648 low-dosage SNPs.

6.3.11 GENOTYPING BY SEQUENCING

Subsets of a genome are sequenced using the NGS technique known as genotyping by sequencing. It works well for both well-characterized and nonmodel species and enables the simultaneous genotyping of multiple individuals at a lower cost (Fu et al., 2015)

6.3.11.1 SUGARCANE GENOMIC SELECTION

It has been observed that there is little genetic association between cane yield (CY) and sucrose content, which can be used to increase the genetic gain for commercial sugar output (Gadallah and Mehareb, 2020; Mehareb et al., 2023). Numerous genetic areas or QTL controlled both sucrose yield and CY (Hoarau et al., 2002; Ming et al., 2002; Aitken et al., 2008). Other crops have shown the value of using genome-wide markers, like SNPs, for genotype and breeding value prediction through genomic selection techniques (Unterseer et al., 2014; Aitken et al., 2016; Roorkiwal et al., 2018; Li et al., 2019). Affymetrix Axion 100K SNP, Axiom 345K SNP, and Diversity Array Technology arrays were used for sugarcane genome prediction research (Deomano et al., 2020; Voss-Fels et al., 2021; Yadav et al., 2021). These arrays were made available recently (You et al., 2018). Numerous factors, including strong statistical models, genome-wide markers, and characteristics of training and testing populations, including the structure of populations, trait construction, trait reproducibility, genetic variation, and genetic inheritance, affect how accurate genomic prediction or selection studies are (Zhang et al., 2017; Berro et al., 2019).

6.3.12 GENE-SPECIFIC MARKERS

In order to help with tracking and selecting particular features, markers can be made to target particular genes of interest (Gupta and Rustgi, 2004). A complex mechanism comprising numerous genes and regulatory sequences that regulate biochemical activities in source-sink tissues is sugarcane's synthesis and buildup of sucrose. Among these, the invertases that affect the cell wall integrity (CWI), soluble acid invertase (SAI), sucrose phosphate synthase (SPS), and SuSy are significant (Chandra et al., 2015). The synthesis and accumulation of sucrose in sugarcane are dependent on source–sink communication, in which the source reacts to the sink's need for supply of photoassimilates. By controlling the photosynthetic efficiency of leaves, sucrose in the stalk (sink) functions as a signal and sends feedback to limit more sucrose synthesis (source). Therefore, a large number of genes and regulatory sequences, such as those encoding three invertases (SAI, CWI, and neutral invertase), SuSy, and SPS, regulate the synthesis and accumulation of sucrose. Invertase and SPS are essential for improving sink strength, which, in turn, encourages increased sucrose buildup in the sink tissues (Verma et al., 2019).

6.3.13 QUANTITATIVE TRAIL LOCUS

Low-sugar varieties and clones are typically eliminated in the initial phages of breeding programs, which suggests that the narrow range of Pol values between the two parent clones resulted from applying cross between industrial parents as the mapping population. This potential QTL needs to be verified in more crosses by looking into the same association with sugar content in order to be helpful for MAS. Since interspecific hybridization produces sugarcane, if this link is verified, the source of the QTL should be looked into, that is, it should be determined if *S. officinarum* or *S. spontaneum* is the source of the SuSy allele that causes the reduced sucrose content. There is a great deal of variation among the 11 genotypes examined in the southern examination of *S. spontaneum* genotypes using the SuSy marker (Silva and Bressiani, 2005). The efficiency of introgression programs would rise with the mapping of this segment on a commercial progeny, allowing the introduction of additional positive alleles from wild germplasm and lowering the typically associated genetic drag. The scheduling of crosses might be another useful use of these results in commercial sugarcane breeding. Because identifying high sugar genotypes will not involve environmental influences, the use of a single dosage marker linked to a significant QTL in a polymorphic cross for the selection of marker-assisted in sugarcane will permit for more precise selection.

Thirty-six sugar QTLs were found to correspond to eight nonoverlapping areas of the sorghum genome when compared with the sorghum linkage map. Both high-sugar and low-sugar QTLs were found in 75% of these locations, indicating the significance of these loci in sugar content breeding. The correct study that identifies QTLs with remarkable effects—that is, QTLs in high-sugar parents that decrease sugar content and those from low-sugar parents that raise sugar content—will determine how best to use these QTLs (Swapna and Srivastava, 2012). Aitken et al. (2006) examined QTL influencing sucrose concentration in the sugarcane cultivar Q 165.

Sugar-related mapping studies were conducted by Ming and his group using two interspecific segregating populations: (1) a set of 264 plants from *S. officinarum* Green German (GG) 9 *S. spontaneum* IND-81-146 (IND) and (2) a set of 239 plants from *S. spontaneum* PIN 84-1 (PIN) 9 *S. officinarum* Muntok Java (MJ) (Swapna and Srivastava, 2012). Precisely, 14 from GG 9 IND (8 from GG, 6 from IND) and 22 from PIN 9 MJ (18 from MJ, 4 from PIN) yielded a total of 36 marker trait correlations. Six IND QTLs and eight GG QTLs together accounted for 36% and 38.6%, respectively, of the sugar phenotypic variance. Fourteen QTLs explained a total of 65% of the variation. A part of the observed transgressive variance was accounted by three QTLs. With four PIN QTLs accounting for 33.4% of the variation and 18 MJ QTLs alone providing 45.7%, 22 QTLs from PIN 9 MJ accounted for 68.3% of the total variation.

In order to create a genetic linkage map of the sugarcane cultivar M134/75, 227 segregating clones were assessed for resistance to yellow spot in 2007. A biparental cross between M134/75 (resistant) and R570 (susceptible) produced these. A single significant QTL was found to be connected to an AFLP marker by single marker analysis. According to Aljanabi et al. (2007), this QTL may account for 23.8% of the phenotypic variance in sugarcane yellow spot resistance. In a different study, 46 QTL were found applying the interval mapping strategy, with eight QTL for sucrose concentration, 11 for fiber percentage, 13 for CY, and 14 for sugar yield. The study focused on sugarcane biomass and characteristics linked to sucrose content (Pastina et al., 2012).

One of the most effective methods for identifying causal variations for complex phenotypes in human, animal, and plant systems is the GWAS. It is possible to look into and identify several genes and alleles contributing to the same features. Since NGS technology is becoming more and more affordable, GWAS has been used extensively in the genomic era to analyze complicated traits. With up to 12 pairs of chromosomes, polyploid sugarcane may not be as amenable to GWAS as diploid varieties. The use of GWAS in sugarcane may be hampered by a number of major issues, for instance, because of the large genome sizes (approximately 10 Gbp), (1) relatively high sequencing depth is required to call sequence variants, (2) there is difficulty in determining marker dosages, and (3) the fact that majority of GWAS tools and models are made for diploids meant that dosages and gene activities cannot be included in analysis. Because of this, GWAS typically treats molecular markers in polyploids as diploids, which may not accurately reflect the genetic effects (Yang et al., 2019).

6.3.14 TARGETED REGION AMPLIFIED POLYMORPHISM

TRAP is a PCR-based marker method in which primers are designed using an arbitrary sequence and the EST sequence. In addition to being helpful in identifying species and genus-specific bands, the TRAP marker system has been shown to be a good tool for genome characterization and diversity studies (Alwala et al., 2006). This has made the system applicable to gene introgression and fingerprinting research. This has revealed the variety in the clones' sugar sequences by using fixed primers for genes involved in sucrose metabolism. Numerous bands that have been found in relation to genus, species, and/or cultivars present prospects for their application in gene introgression research, fingerprinting, and varietal identification in commercial breeding programs. It can be inferred that the presence of introns between exons increases the likelihood of polymorphism discovery in the genomic area. Furthermore, among the amplified primers, the genomic SSRs targeted more dinucleotide repeat motifs than trinucleotide repeat motifs, while the EST-SSRs targeted repeats of tri and tetra nucleotide motifs. In addition to being used for genetic evaluation, EST-SSRs can be used to target crucial quantitative qualities such as sugarcane's fiber content and sucrose accumulation, which have been shown to have functional collinearity with the selection of suitable parentage (Kalwade and Devarumath, 2014).

6.4 FUTURE OF MARKER TECHNOLOGY IN SUGARCANE

One cycle in traditional sugarcane variety improvement programs lasts 12 years on average, spanning from hybridization to cultivar release. Since methods to improve early and indirect selection in sugarcane would have a significant influence on breeding for high sugar content and might improve selection accuracy for early sugar content, molecular markers are a useful tool for breeders. Developing sugarcane varieties with high-performance agronomic features is the best way to supply growing markets for biofuels and sugar. The time and resources needed to create new sugarcane varieties may be significantly decreased by molecular markers linked to pertinent agronomic features.

Sugarcane presents a barrier for QTL mapping. Sugarcane coexists with numerous alleles (up to 12 alleles) for each gene because of its highly polyploid genome and heterozygosity. An allele is only observable when its effect surpasses the mean of all other segregated alleles in the genetic background (Raboin et al., 2006). The utilization of QTL mapping has demonstrated efficacy in identifying the loci responsible for significant characteristics in sugarcane. An EST sequence is used to build a fixed primer, and a similar-length arbitrary primer with an adenine–thymine or guanine–cytosine-rich motif (to anneal with an exon or intron, respectively) is designed. Additionally, sugarcane has been subjected to mapping experiments using SRAP (Alwala et al., 2009). Another marker technique based on conserved sequences that has also been applied to sugarcane is called conserved intron scanning primers (Khan et al., 2011).

KEYWORDS

- **sugarcane molecular marker**
- **restriction fragment length polymorphism**
- **single-nucleotide polymorphism**
- **genotype by sequencing**
- **genome**

REFERENCES

Ahmed MS, Gardezi SDA. (2017). Molecular characterization of locally adopted sugarcane (*Saccharum* Officinarum L.) varieties using microsatellite markers. *J. Anim. Plant Sci.* 27(1), 164–174.

Aitken KS. (2022). History and development of molecular markers for sugarcane breeding. *Sugar Tech.* 24, 341–353. https://doi.org/10.1007/s12355-021-01000-7.

Aitken KS, Li JC, Jackson PA, Piperidis G, McIntyre CL. (2006). AFLP analysis of genetic diversity within *Saccharum officinarum* and comparison with sugarcane varieties. *Aust. J. Agric. Res.* 57, 1167–1184.

Aljanabi SM, Parmessur Y, Kross H, Dhayan S, Saumtally S, Ramdoyal K, et al. (2007). 'Identification of a major quantitative trait locus (QTL) for yellow spot (*Mycovellosiella koepkei*) disease resistance in sugarcane. *Mol. Breed.* 19(1), 1–14.

Alwala S, Suman A, Arro JA, Veremis JC, Kimberg CA. (2006). Target region amplification polymorphism (TRAP) for assessing genetic diversity in sugarcane germplasm collections. *Crop Sci.* 46, 448–455.

Anas M, Verma KK, Haq IU, Naeem M, Li Q, Liao F, Xiang YX, Li YR. (2021a). Characterization of exotic, native and wild-type genotypes of sugarcane (*Saccharum* spp. Hybrids) for internal nitrogen use efficiency under different nitrogen levels. *Sugar Tech*. 23, 1258–1267. https://doi.org/10.1007/s12355-021-01014-1.

Anas M, Verma KK, Riaz M, Qiang L, Liao F, Liu Y, Li YR. (2021b). Physio-morphological and biochemical mechanism of nitrogen use efficiency in sugarcane (*Saccharum* spp.) genotypes under different growth stages and nitrogen levels. *J. Plant Interact.* 16, 332–343.

Assamica and sinensis varieties. (2003). *Theor. Appl. Genet.* 106, 375–383. https://doi.org/10.1007/ s00122-002-0999-9.

Atienzar FA, Jha AN. (2006). The random amplified polymorphic DNA (RAPD) assay and related techniques applied to genotoxicity and carcinogenesis studies: A critical review. *Mutat. Res*. 613, 76–102. https://doi.org/10.1016/j.mrrev.2006.06.001.

Batra J, Srinivasan S, Clements J. (2014). Single nucleotide polymorphisms (SNPs). In: *Molecular Testing in Cancer*. Springer: New York, NY, pp. 55–80.

Bhatt R, Verma KK, Jamwal NS. (2023a). Can paddy-compost eliminate N fertilization for climate-smart sugarcane? *Sugar Tech*. 25, 1036–1046. https://doi.org/10.1007/s12355-023-01270-3.

Bhatt R, Verma KK, Kumar R, Sanghera GS. (2023b). Foliar application of potassium salt of active phosphorus (PSAP) mitigates insect pests and improves yield along with sugarcane quality in response to agroclimatic conditions of Punjab. *Sugar Tech*. 25(3), 660–669.

Chandra A, Verma PK, Islam MN, Grisham MP, Jain R, Sharma A, Roopendra K, Singh K, Singh P, Verma I, Solomon S. (2015). Expression analysis of genes associated with sucrose accumulation in sugarcane (*Saccharum* spp. hybrids) varieties differing in content and time of peak sucrose storage. *Plant Biol*. 17, 608–617.

Chaudhary P, Xu M, Ahamad L, Chaudhary A, Kumar G, Adeleke BS, Verma KK, Hu D-M, Siric I, Kumar P, Popescu SM, Fayssal SA. (2023). Application of synthetic consortia for improvement of soil fertility, pollution remediation, and agricultural productivity: A review. *Agronomy*. 13, 643. https://doi.org/10.3390/agronomy13030643.

Chen GL, Zheng FJ, Lin B, Lao SB, He J, Huang Z, Zeng Y, Sun J, Verma KK. (2020). Phenolic and volatile compounds in the production of sugarcane vinegar. *ACS Omega*. 5, 30587–30595.

Chen G-L, Zheng F-J, Lin B, Yang Y-X, Fang X-C, Verma KK, Yang L-F. (2023). Vinegar: A potential source of healthy and functional food with special reference to sugarcane vinegar. *Front. Nutr.* 10, 1145862. https://doi.org/10.3389/fnut.2023.1145862.

Chen H, Xie W, He H, Yu H, Chen W, Li J, et al. (2014). A high-density SNP genotyping array for rice biology and molecular breeding. *Mol. Plant*. 7, 541–553. https://doi.org/10.1093/mp/sst135.

Cordeiro GM, Taylor GO, Henry RJ. (2000). Characterization of microsatellite markers from sugarcane (*Saccharum* spp.) a highly polyploidy species. *Plant Sci*. 155, 161–168.

Silva JA, Sorrells ME, Burnquist WL, Tanksley SD. (1993). RFLP linkage map and genome analysis of *Saccharum spontaneum*. *Genome*. 36, 782–791.

Da Silva JA, Bressiani JA. (2005). Sucrose synthase molecular marker associated with sugar content in elite sugarcane progeny. *Genet. Mol. Biol*. 28(2), 294–298.

Daniels J, Roach BT. (1987). Taxonomy and evolution. Chapter 2. In: DJ Heinz, (Eds.), *Sugarcane Improvement through Breeding*, vol. 11. Elsevier Amsterdam: Netherlands. pp. 7–84.

Deng ZN, Huang HR, Huang CM, Verma KK, Luo HB, Jiang SL, Li YJ, Cao HQ, Wu KC, Xu L. (2023). Short-term limited water irrigation influences the photosynthetic pigments, enzymatic and non-enzymatic activities in *Saccharum spontaneum* L. at vegetative stage. *Sugar Tech.* 25, 110–118. https://doi.org/10.1007/s12355-022-01175-7.

Fahmy AM, Grad WE, Mehareb EM. (2021). Ratooning ability and its relationship among yield, quality and lesser sugarcane borer (*Chilo agamemnon* bels.) in of sugarcane germplasm. *SVU Int. J. Agric. Sci.* 3(3), 40–58.

Fu Y-B, Dong Y. (2015). PaSNPg: A GBS-based pipeline for protein associated SNP discovery and genotyping in non-model species. *J. Proteomics Bioinform.* 8, 190–194. https://doi.org/10.4172/jpb.1000368.

Gadallah AFI, Mehareb EM. (2020). Yield and quality of some sugarcane varieties as affected by irrigation number. *SVU Int. J. Agric. Sci.* 2(2), 144–165.

Guo D-J, Li D-P, Chen Z-D, Yang B, Verma KK, Singh RK, Singh P, Khan Q, Sharma A, Qin Y, Zhang B-Q, Song X-P, Li Y-R. (2023). Effect of endophytic diazotroph *Enterobacter roggenkampii* ED5 on nitrogen-metabolism-related microecology in the sugarcane rhizosphere at different nitrogen levels. *Front. Microbiol.* 14, 1132016. https://doi.org/10.3389/fmicb.2023.1132016.

Guo D-J, Li D-P, Singh RK, Singh P, Sharma A, Verma KK, Qin Y, Khan Q, Lu Z, Malviya MK, Song X-P, Xing Y-X, Li Y-R (2021). Differential protein expression analysis of two sugarcane varieties in response to diazotrophic plant growth-promoting endophyte *Enterobacter roggenkampii* ED5. *Front. Plant Sci.* 12, 727741. https://doi.org/10.3389/fpls.2021.727741.

Guo DJ, Li DP, Singh RK, Singh P, Verma KK, Sharma A, Qin Y, Khan Q, Song XP, Malviya MK, Xing YX, Li YR. (2022). Comparative transcriptome analysis of two sugarcane varieties in response to diazitrophic plant growth promoting endophyte *Enterobacter roggenkampii* ED5. *J. Plant Interact.* 17(1), 75–84.

Gupta PK, Rustgi S. (2004). Molecular markers from the transcribed/expressed region of the genome in higher plants. *Funct. Integr. Genomics.* 4(3), 139–162. https://doi.org/10.1007/s10142-004-0107-0.

Hoarau JY, Offmann B, D'Hont A, Risterucci AM, Roques D, Glaszmann JC, Grivet L. (2001). Genetic dissection of a modern sugarcane varieties (*Saccharum* spp.). 1. Genome mapping with AFLP markers. *Theor. Appl. Genet.* 103, 84–97.

Hodkinson TR, Chase, MW, Lledó, MD, Salamin N, Renvoize SA. (2002). Phylogenetics of Miscanthus, *Saccharum* and related genera (Saccharinae, Andropogoneae, Poaceae) based on DNA sequences from ITS nuclear ribosomal DNA and plastid trnL intron and trnL-F intergenic spacers. *J. Plant Res.* 115, 381–392.

Huang CM, Wu KC, Verma KK, Luo HB, Deng ZN, Shi GY, Cao HQ, Wu XJ, Ye LP, Wei YW, Xu L. (2024). Isolation and expression analysis of cellulose synthase 3 (Ces3) genes from sugarcane (*Saccharum officinarum* L.). *Chem. Biol. Technol. Agric.* 11, 71. https://doi.org/10.1186/s40538-024-00597-7.

Huang HR, Gao YJ, Malviya MK, Verma KK, Solanki MK, Huang YX, Li X., Deng Y, Yan J, Tang SY, Wang LW, Xu L. (2022). Genetic diversity analysis of sugarcane (*Saccharum* spp. hybrids) among high-sucrose clones of GT series and commonly used parents by using microsatellite markers in Guangxi, China. *Sugar Tech.* 24, 397–407. https://doi.org/10.1007/s12355-021-01033-y.

Jagadesh M, Dash M, Kumari A, Singh SK, Verma KK, Kumar P, Bhatt R, Sharma SK. (2024). Revealing the hidden world of soil microbes: Metagenomic insights into plant, bacteria, and fungi interactions for sustainable agriculture and ecosystem restoration. *Microbiol. Res.* 285, 127764. https://doi.org/10.1016/j.micres.2024.127764.

Joshi A, Verma KK, Rajput VD, Arora J. (2023). Metabolic engineering of microorganisms in advancing biofuel production. In: *Advances in Lignocellulosic Biofuel Production Systems*. pp. 181–202. https://doi.org/10.1016/B978-0-323-91192-4.00016-X.

Joshi A, Verma KK, Rajput VD, Minkina T, Arora J. (2022). Recent advances in metabolic engineering of microorganisms for advancing lignocellulose-derived biofuels, *Bioengineered.* 13(4), 8135–8163. https://doi.org/10.1080/21655979.2022.2051856.

Kalia RK, Rai MK, Kalia S, Singh R, Dhawan AK. (2011). Microsatellite markers: An overview of the recent progress in plants. *Euphytica.* 177, 309–334.doi: 10.1007/s10681-010-0286-9.

Kalwade SB, Devarumath RM. (2014). Single strand conformation polymorphism of genomic and EST-SSRs marker and its utility in genetic evaluation of sugarcane. *Physiol. Mol. Biol. Plants*. 20(3), 313–321. https://doi.org/10.1007/s12298-014-0231-9.

Kaundun S, Matsumoto S. (2003). Development of CAPS markers based on three key genes of the phenylpropanoid pathway in tea, Camellia sinensis (L.) O. Kuntze, and differentiation between Assamica and sinensis varieties. *Theor. Appl. Genetics*. 106, 375–383. https://doi.org/10.1007/s00122-002-0999-9.

Khan MS, Yadav S, Srivastava S, Swapna M, Chandra A, Singh RK. (2011). Development and utilization of conserved intron-scanning primers in sugarcane. *Aust. J. Bot.* 59(1), 38–45

Laird PW. (2010). Principles and challenges of genome-wide DNA methylation analysis. *Nat. Rev. Genet.* 11, 191–203. https://doi.org/10.1038/nrg2732.

Li A-M, Liao F, Wang M, Chen Z-L, Qin C-X, Huang R-Q, Verma KK, Li Y-R, Que Y-X, Pan Y-Q, et al. (2023). Transcriptomic and proteomic landscape of sugarcane response to biotic and abiotic stressors. *Int. J. Mol. Sci.* 24, 8913. https://doi.org/10.3390/ijms24108913.

Liu P, Chandra A, Que Y, Chen PH, Grisham MP, White WH, Dalley CD, Tew TL, Pan Y-B. (2016). Identification of quantitative trait loci controlling sucrose content based on an enriched genetic linkage map of sugarcane (*Saccharum* spp. hybrids) cultivar 'LCP 85-384'. *Euphytica*. 207, 527–549.

Malviya MK, Li C-N, Lakshmanan P, Solanki MK, Wang Z, Solanki AC, Nong Q, Verma KK, Singh RK, Singh P, Sharma A, Guo D-J, Dessoky ES, Song X-P, Li Y-R. (2022). High-throughput sequencing-based analysis of rhizosphere and diazotrophic bacterial diversity among wild progenitor and closely related species of sugarcane (*Saccharum* spp. inter-specific hybrids). *Front. Plant Sci.* 13, 829337. https://doi.org/10.3389/fpls.2022.829337.

Malviya MK, Solanki MK, Li CN, Wang Z, Zeng Y, Verma KK, Singh RK, Singh P, Huang HR, Yang LT, Song XP, Li YR. (2021). Sugarcane-legume intercropping can enrich the soil microbiome and plant growth. *Front. Sustain. Food Syst.* 5, 606595. https://doi.org/10.3389/fsufs.2021.606595.

Marth GT, Korf I, Yandell MD, Yeh RT, Gu Z, Zakeri H, et al. (1999). A general approach to single-nucleotide polymorphism discovery. *Nat. Genet.* 23, 452–456. https://doi.org/10.1038/70570.

Meena VK, Sharma RK, Chand S, Kumar M, Kumar N, Jain N, et al. (2022). Elucidating molecular diversity in spring wheat (*Triticum aestivum* L. em. Thell.) under terminal heat stress environment using morpho-physiological traits and SSR markers. *Indian J. Genet. Plant Breed.* 82, 47–55. https://doi.org/10.31742/IJGPB.82.1.7.

Mehareb, Eid. M., Sakina, R. Abazied, and Fouz, F.M. Abo Elenen. (2023). Assessment of harvest age of some sugarcane genotypes by using principal component analysis. *Direct Res. J. Agric. Food Sci.* 11(8), 244–253.

Neff MM, Neff JD, Chory J, Pepper AE. (1998). DCAPS, a simple technique for the genetic analysis of single nucleotide polymorphisms: Experimental applications in *Arabidopsis thaliana* genetics. *Plant J.* 14, 387–392. https://doi.org/10.1046/j.1365-313X.1998.00124.x.

Oliveira MA, Verma KK, Bhatt R, Oliveira TBA. (2022). Impact of green and organic fertilizers on soil fertility and sugarcane productivity. In: Verma KK, et al. (Eds.), *Agro-Industrial Perspectives on Sugarcane Production under Environmental Stress*. https://doi.org/10.1007/978-981-19-3955-6_11.

Papini-Terzi FS, Rocha FR, Vêncio RZ, Felix JM, et al. (2009). Sugarcane genes associated with sucrose content. *BMC Genomics*. 10, 120. doi:10.1186/1471-2164-10-120.

Raboin L, Oliveira KM, Lecunff L, Telismart H, Roques D, Butterfield M, et al. (2006). Genetic mapping in sugarcane, a high polyploid, using bi-parental progeny: Identification of a gene controlling stalk colour and a new rust resistance gene. *Theor. Appl. Genet.* 112(7), 1382–1391.

Sharma A, Singh RK, Singh P, Vaishnav A, Guo DJ, Verma KK, Li DP, Song XP, Malviya MK, Khan N, Lakshmanan P, Li YR. (2021). Insights into the bacterial and nitric oxide-induced salt tolerance in sugarcane and their growth-promoting abilities. *Microorganisms*. 9, 2203. https://doi.org/10.3390/microorganisms9112203.

Sharma A, Song X-P , Singh RK, Vaishnav A, Gupta S, Singh P, Guo D-J, Verma KK, Li Y-R. (2022). Impact of carbendazim on cellular growth, defence system and plant growth promoting traits of Priestia megaterium ANCB-12 isolated from sugarcane rhizosphere. *Front. Microbiol.* 13, 1005942. doi:10.3389/fmicb.2022.1005942.

Sharma MS, Dobhal U, Singh P, Kumar S, Gaur AK, Singh SP, Jeena AS, Koshy EP, Kumar S. (2014). Assessment of genetic diversity among sugarcane cultivars using novel microsatellite markers. *Afr. J. Biotechnol.* 13(13), 1444–1451. https://doi.org/10.5897/AJB2013.13376.

Singh P, Singh RK, Li HB, Guo DJ, Sharma A, Verma KK, Solanki MK, Upadhyay SK, Lakshmanan P, Yang LT, Li YR. (2023). Nitrogen fixation and phytohormone stimulation of sugarcane plant through plant growth promoting diazotrophic Pseudomonas. *Biotechnol. Genetic Eng. Reviews.* 40(1), 15–35. https://doi.org/10.1080/02648725.2023.2177814.

Singh P, Singh RK, Song QQ, Li HB, Guo DJ, Malviya MK, Verma KK, Song XP, Lakshmanan P, Yang LT, Li YR. (2021). Comparative analysis of protein and differential responses of defense-related gene and enzyme activity reveals the long-term molecular responses of sugarcane inoculated with *Sporisorium scitamineum*. *J. Plant Interact.* 16(1), 12–29.

Singh RK, Singh P, Li HB, Song QQ, Guo DJ, Solanki MK, Verma KK, Malviya MK, Song XP, Lakshmanan P, Yang LT, Li YR. (2020). Diversity of nitrogen-fixing rhizobacteria associated with sugarcane: a comprehensive study of plant-microbe interactions for growth enhancement in *Saccharum* spp. *BMC Plant Biol.* 20, 220. https://doi.org/10.1186/s12870-020-02400-9.

Solanki MK, Verma KK, Dastogeer KMG, Mora-Poblete F, Mundra S. (2023). Microbial resilience in plant nutrient management towards sustainable farming. *Front. Microbiol.* 14, 1280811. https://doi.org/10.3389/fmicb.2023.1280811.

Song X-P, Mo F, Verma KK, Wei J, Zhang X, Yang L, Li YR. (2019). Effect of sugarcane smut (*Ustilago scitaminea* syd.) on ultrastructure and biochemical indices of sugarcane. *Biomed. J. Sci. Tech. Res.* 17(1), 12546–12551. https://doi.org/10.26717/BJSTR.2019.17.002950.

Song XP, Mo F, Yan M, Zhang X, Zhang B, Huang X, Huang D, Pan Y, Verma KK, Li YR. (2022). Effect of smut infection on the photosynthetic physiological characteristics and related defense enzymes of sugarcane. *Life.* 12, 1201. https://doi.org/10.3390/life12081201.

Song XP, Verma KK, Tian DD, Zhang XQ, Liang YJ, Huang X, Li CN, Li YR. (2021). Exploration of silicon functions to integrate with biotic stress tolerance and crop improvement. *Biol. Res.* 54, 19. https://doi.org/10.1186/s40659-021-00344-4.

Sreenivasan TV, Ahloowalia BS, Heinz DJ. (1987). Cytogenetics. Chapter 5. In: Heinz DJ, (Eds.), *Sugarcane Improvement through Breeding.* Elsevier: Amsterdam. pp. 211–253.

Swapna M, Srivastava S. (2012). *Molecular Marker Applications for Improving Sugar Content in Sugarcane.* Heidelberg Dordrecht London Library of Congress: Springer New York, Control Number: 2012931948. https://doi.org/10.1007/978-1-4614-2257-0.

Tian D, Qin L, Verma KK, Wei L, Li J, Li B, Zhou W, He Z, Wei D, Huang S, Long S, Huang Q, Li C, Wei S. (2023). Transcriptomic and metabolomic differences between banana varieties which are resistant or susceptible to Fusarium wilt. *PeerJ.* 11, e16549. https://doi.org/10.7717/peerj.16549.

Verma KK, Anas M, Chen Z, Rajput VD, Malviya MK, Verma CL, Singh RK, Singh P, Song XP, Li YR. (2020f). Silicon supply improves leaf gas exchange, antioxidant defense system and growth in sugarcane responsive to water limitation. *Plants.* 9, 1032, doi: 10.3390/plants9081032.

Verma KK, Joshi A, Song X-P, Liang Q, Xu L, Huang H-R, Wu K-C, Seth CS, Arora J and Li Y-R (2024a). Regulatory mechanisms of plant rhizobacteria on plants to the adaptation of adverse agroclimatic variables. *Front. Plant Sci.* 15, 1377793. https://doi.org/10.3389/fpls.2024.1377793.

Verma KK, Joshi A, Song X-P, Singh S, Kumari A, Arora J, Singh SK, Solanki MK, Seth CS, Li Y-R. (2024b). Synergistic interactions of nanoparticles and plant growth promoting rhizobacteria enhancing soil-plant systems: A multigenerational perspective. *Front. Plant Sci.* 15, 1376214. https://doi.org/10.3389/fpls.2024.1376214.

Verma KK, Li DM, Singh M, Rajput VD, Malviya MK, Minkina T, Singh RK, Singh P, Song XP, Li YR. (2020a). Interactive role of silicon and plant-rhizobacteria mitigating abiotic stresses: A new approach for sustainable agriculture and climate change. *Plants.* 9, 1055. https://doi.org/10.3390/plants9091055.

Verma KK, Rajput VD, Singh M, Verma CL, Minkina T. (2022a). Influence of waterlogging on plant physiology and molecular biology. In: *Understanding Abiotic Stresses Publisher*. Nova Science Publishers, Inc.: New York.

Verma KK, Siddiqui N, Singh M. (2022b). Understanding the plant physiological and molecular adaptation strategies to soil salinity. In: *Understanding Abiotic Stresses Publisher*: Nova Science Publishers, Inc.: New York. https://doi.org/10.52305/UBUJ4024.

Verma KK, Singh P, Song X-P, Malviya MK, Singh RK, Chen G-L, Solomon S, Li YR. (2020b). Mitigating climate change for sugarcane improvement: Role of silicon in alleviating abiotic stresses. *Sugar Tech.* 22(5), 741–749. https://doi.org/10.1007/s12355-020-00831-0.

Verma KK, Singh RK, Song QQ, Singh P, Zhang B-Q, Song X-P, Chen G-L, Li YR. (2019a). Silicon alleviates drought stress of sugarcane plants by improving antioxidant responses. *Biomed J. Sci. Tech. Res.* 17(1), 12580–12586. https://doi.org/10.26717/BJSTR.2019.17.002957.

Verma KK, Song X, Degu HD, Guo D, Joshi A, Huang H, Xu L, Singh M, Huang D, Rajput VD, Li YR. (2023a). Recent advances in nitrogen and nano-nitrogen fertilizers for sustainable crop production: A mini-review. *Chem. Biol. Technolog. Agricult.* 10, 1–14. https://doi.org/10.1186/s40538-023-00488-3.

Verma KK, Song X, Verma CL, Huang H, Singh M, Xu L, Wu J, Khan ZH, Zhang B, Li Y. (2023b). Mathematical modeling of climate and fluoride effects on sugarcane photosynthesis with silicon nanoparticles. *Plant Physiol. Biochem.* 204, 108089. https://doi.org/10.1016/j.plaphy.2023.108089.

Verma KK, Song XP, Budeguer F, Nikpay A, Enrique R, Singh M, Zhang B, Wu, JM, Li YR. (2022c). Genetic engineering: An efficient approach to mitigating biotic and abiotic stresses in sugarcane cultivation. *Plant Signal. Behav.* 17(1), e2108253.

Verma KK, Song XP, Joshi A, Rajput VD, Singh M, Sharma A, Singh RK, Li DM, Arora J, Minkina T, Li YR. (2022d). Nanofertilizer possibilities for healthy soil, water and food in future: An overview. *Front. Plant Sci.* 13, 865048. https://doi.org/10.3389/fpls.2022.865048.

Verma KK, Song XP, Joshi A, Tian DD, Rajput VD, Singh M, Arora J, Minkina TM, Li YR. (2022e). Recent trends in nano-fertilizer for advancing sustainable agriculture under the era of climate change to ensure food security for future generations. *Nanomaterials*. 12, 173. https://doi.org/10.3390/nano12010173.

Verma KK, Song XP, Kumari A, Jagadesh M, Singh SK, Bhatt R, Singh M, Seth CS, Li YR. (2024c). Climate change adaptation: Challenges for agricultural sustainability. *Plant, Cell Environ.* https://doi.org/10.1111/pce.15078.

Verma KK, Song XP, Li DM, Singh M, Wu JM, Singh RK, Sharma A, Zhang BQ, LI YR. (2022f). Silicon and soil microorganisms improves rhizospheric soil health with bacterial community, plant growth, performance and yield. *Plant Signal. Behav.* 17(1), e2104004.

Verma KK, Song XP, Lin B, Guo DJ, Singh M, Rajput VD, Singh RK, Singh P, Sharma A, Malviya MK, Chen GL, Li YR. (2022g). Silicon induced drought tolerance in crop plants: physiological adaptation strategies. *Silicon.* 14(6), 2473–2487. https://doi.org/10.1007/s12633-021-01071-x.

Verma KK, Song XP, Rajput VD, Boldyreva V, Zhang BQ, Minkina T, Li YR. (2022h). Morpho-physiological, biochemical and ultrastructural modifications on sugarcane to prolonged water deficit. In: Verma KK, et al. (Eds.), *Agro-Industrial Perspectives on Sugarcane Production under Environmental Stress*. https://doi.org/10.1007/978-981-19-3955-6_11.

Verma KK, Song XP, Rajput VD, Solomon S, Li YR, Rao GP. (2022i). *Agro-Industrial Perspectives on Sugarcane Production under Environmental Stress*. Springer Nature: Singapore, https://doi.org/10.1007/978-981-19-3955-6.

Verma KK, Song XP, Singh M, Tian DD, Rajput VD, Minkina T, Li YR. (2023c). Association of silicon and soil microorganisms induces stress mitigation, increasing plant productivity. In: de Mello Prado R. (Eds.), *Benefits of Silicon in the Nutrition of Plants*. pp. 299–328. https://doi.org/10.1007/978-3-031-26673-7_17.

Verma KK, Song XP, Tian DD, Guo DJ, Chen ZL, Zhong CS, Nikpay A, Singh M, Rajput VD, Singh RK, Minkina T, Li YR. (2021a). Influence of silicon on biocontrol strategies to manage biotic stress for crop protection, performance and improvement. *Plants*. 10, 2163. https://doi.org/10.3390/plants10102163.

Verma KK, Song XP, Tian DD, Singh M, Verma CL, Rajput VD, Singh RK, Sharma A, Singh P, Malviya MK, Li YR. (2021b). Investigation of defensive role of silicon during drought stress induced by irrigation capacity in sugarcane: physiological and biochemical characteristics. *ACS Omega*. 6, 19811–19821.

Verma KK, Song XP, Verma CL, Chen ZL, Rajput VD, Wu KC, Liao F, Chen GL, Li YR. (2021c). Functional relationship between photosynthetic leaf gas exchange in response to silicon application and water stress mitigation in sugarcane. *Biol. Res*. 54, 15. https://doi.org/10.1186/s40659-021-00338-2.

Verma KK, Song XP, Verma CL, Malviya MK, Guo DJ, Rajput VD, Sharma A, Wei KJ, Chen GL, Solomon S, Li YR. (2021d). Predication of photosynthetic leaf gas exchange of sugarcane (*Saccharum* spp.) leaves in response to leaf positions to foliar spray of potassium salt of active phosphorus under limited water irrigation. *ACS Omega*. 6, 2396–2409.

Verma KK, Song X-P, Xu L, Huang H-R, Liang Q, Seth CS, Li Y-R. (2023d). Nano-microplastic and agro-ecosystems: A mini-review. *Front. Plant Sci*. 14, 1283852. https://doi.org/10.3389/fpls.2023.1283852.

Verma KK, Song XP, Yadav G, Degu HD, Parvaiz A, Singh M, Huang HR, Mustafa G, Xu L, Li YR. (2022j). Impact of agroclimatic variables on proteo-genomics in sugarcane (*Saccharum* spp.) plant productivity. *ACS Omega*. 7, 22997–23008.

Verma KK, Song XP, Zeng Y, Guo DJ, Sing M, Rajput VD, Malviya MK, Wei KJ, Sharma A, Li DP, Chen GL, Li YR. (2021e). Foliar application of silicon boosts growth, photosynthetic leaf gas exchange, antioxidative response and resistance to limited water irrigation in sugarcane (*Saccharum officinarum* L.). *Plant Physiol. Biochem*. 166, 582–592.

Verma KK, Song XP, Zeng Y, Li DM, Guo DJ, Rajput VD, Chen GL, Barakhov A, Minkina TM, Li YR. (2020c). Characteristics and correlation of leaf stomata and its relationship with photosynthesis on *Saccharum* spp. under different irrigation and silicon application. *ACS Omega*. 5, 24145–24153.

Verma KK, Vatsal S, Singh M, Rajput VD, Minkina T. (2022k). Coping with water stress: effects and adaptive mechanisms to plants. In: *Understanding Abiotic Stresses Publisher*. Nova Science Publishers, Inc.: New York.

Verma KK, Wu KC, Verma CL, Li DM, Malviya MK, Singh RK, Singh P, Chen GL, Song XP, Li YR. (2020d). Developing mathematical model for diurnal dynamics on photosynthetic responses in sugarcane responsive to different irrigation and silicon application. *PeerJ*. 8, e10154. https://doi.org/10.7717/peerj.10154.

Verma KK, Liu X-H, Wu K-C, Singh RK, Song QQ, Malviya MK, Song X-P, Singh P, Verma CL, Li YR. (2020e). The impact of silicon on photosynthetic and biochemical responses of sugarcane under different soil moisture levels. *Silicon*. 12, 1355–1367.

Verma KK., Wu K-C., Singh P., Malviya MK., Singh RK., Song X-P., Li YR. (2019b). The protective role of silicon in sugarcane under water stress: Photosynthesis and antioxidant enzymes. *Biomed J. Sci. Tech. Res.* 15(2), 11301–11307. https://doi.org/0.26717/BJSTR.2019.15.002685.

Verma IK, Roopendra A, Sharma A, Chandra A, Kamal. (2019). Expression analysis of genes associated with sucrose accumulation and its effect on source–sink relationship in high sucrose accumulating early maturing sugarcane variety. *Physiol. Mol. Biol. Plants.* 25(1), 207–220. https://doi.org/10.1007/s12298-018-0627-z.

Vinayak V, Dhawan AK, Gupta VK. (2010). PCR Primers for identification of high sucrose *Saccharum* genotypes. *Physiol. Mol. Biol. Plants.* 16, 107–111. https://doi.org/10.1007/s12298-010-0002-1.

Williams JGK, Kubelik AR, Livak KJ, Tingey JA, Rafalski SV (1990). DNA polymorphism amplified by arbitrary primers are useful as genetic markers. *Nucleic Acids Res.* 18, 6531–6535.

Wu K-C, Huang C-M, Verma KK, Deng Z-N, Huang H-R, Pang T, Cao H-Q, Luo H-B, Jiang S-L, Xu L. (2022). Transcriptomic responses of *Saccharum spontaneum* roots in response to polyethylene glycol-6000 stimulated drought stress. *Front. Plant Sci.* 13, 992755. https://doi.org/10.3389/fpls.2022.992755.

Xu L, Deng Z-N, Wu K-C, Malviya MK, Solanki MK, Verma KK, Pang T, Li Y-J, Liu X-Y, Kashyap BK, Dessoky ES, Wang W-Z and Huang H-R. (2022). Transcriptome analysis reveals a gene expression pattern that contributes to sugarcane bud propagation induced by indole-3-butyric acid. *Front. Plant Sci*. 13, 852886. https://doi.org/10.3389/fpls.2022.852886.

Yang X, Ziliang L, James T, Sushma S, Jianping W. (2019). Genome-wide association study of multiple yield traits in a diversity panel of polyploid sugarcane (*Saccharum* spp.). *Plant Genome*. 13(1), e20006.

You Q, Yang X, Peng Z, Islam Md S, Sood S, et al. (2019). Development of an Axiom Sugarcane100K SNP array for genetic map construction and QTL identification. *Theor. Appl. Genet*. 132, 2829–2845. https://doi.org/10.1007/s00122-019-03391-4.

Zhang J, Zhang Q, Li L, Tang H, Zhang Q, Chen Y, Arrow J, Zhang X, Wang A, Miao C. (2019). Recent polyploidization events in three *Saccharum* founding species. *Plant Biotechnol.* J, 17, 264–274.

Zheng F-J, Lin B, Yang Y-X, Fang X-C, Verma KK, Chen G-L. (2023). Efficacy and functionality of sugarcane original vinegar on mice. *Front. Microbiol.* 14, 1224666. https://doi.org/10.3389/fmicb.2023.1224666.

CHAPTER 7

Optimizing Sugarcane Ratoon Yield: An Updated Overview for Nutrient Management Strategies

PRASANN KUMAR,[1*] ARADHNA KUMARI,[2] and KRISHAN K. VERMA[3]

[1]*Department of Agronomy, School of Agriculture, Lovely Professional University, Phagwara, Punjab, India*

[2]*College of Agriculture, Jawaharlal Nehru Krishi Vishwa Vidyalaya, Ganj Basoda, Vidisha, Madhya Pradesh, India*

[3]*Sugarcane Research Institute, Guangxi Academy of Agricultural Sciences/Key Laboratory of Sugarcane Biotechnology and Genetic Improvement (Guangxi), Ministry of Agriculture and Rural Affairs/ Guangxi Key Laboratory of Sugarcane Genetic Improvement, Nanning, Guangxi, China*

*Corresponding author

ABSTRACT

This study examines the enhancement of sugarcane ratoon yield by conducting a thorough comparative analysis of conventional and innovative nutrient management approaches. It seeks to improve our comprehension of the influence of nutrient strategies on sugarcane productivity throughout ratoon cycles. The chapter aims to provide valuable insights into sustainable and efficient sugarcane cultivation practices by assessing traditional and innovative methods. The chapter provides a concise overview of significant discoveries, methodological strategies, and ramifications for the agricultural sector.

Sugarcane Transformation: Biofuels, Biotechnology, and Beyond.
Krishan K. Verma, Xiu-Peng Song, Hai-Rong Huang, Aradhna Kumari, and Yang-Rui Li (Eds.)

DOI: 10.1201/9781779641571-7

7.1 INTRODUCTION

Sugarcane, scientifically known as *Saccharum officinarum*, is an essential crop globally, playing a vital role as the primary source of sugar production (Chen et al., 2020; Guo et al., 2021, 2022; Verma et al., 2023a, 2023b, 2023c). The sugarcane ratoon system is crucial for maintaining and maximizing crop productivity. Sugarcane can generate multiple harvests, referred to as ratoons, from a single planting, which set it apart from most other crops harvested once a year (Song et al., 2019, 2022; Anas et al., 2021a, 2021b; Bhatt et al., 2023a, 2023b). This phenomenon has attracted growing attention among the agricultural community because of its capacity to improve productivity and economic sustainability for sugarcane farmers (Kaab et al. 2019a; Verma et al., 2019a, 2019b; Huang et al., 2022, 2024; Yang et al. 2022). Sugarcane cultivation traditionally entails the process of replanting after every harvest. Nevertheless, the ratoon system presents a financially efficient and environmentally viable option by enabling the resurgence of shoots from the residual subterranean stalks, referred to as stubble. This practice mitigates the necessity for frequent replanting, mitigates soil erosion, and preserves water resources (Singh et al., 2020, 2021, 2023; Jagadesh et al., 2024). Gaining insight into the dynamics of the sugarcane ratoon system is crucial for optimizing agricultural practices and guaranteeing the long-term sustainability of the sugarcane industry (Verma et al., 2020a, 2020b, 2020a, 2021b, 2023d; Zheng et al., 2023). There is a lack of knowledge regarding the most efficient nutrient management strategies to optimize ratoon yields despite the potential advantages of the sugarcane ratoon system. The heterogeneity in soil conditions, climate, and agronomic practices presents a difficulty for sugarcane farmers aiming to maximize productivity in consecutive ratoon cycles (Malviya et al., 2021, 2022; Sharme et al., 2021, 2022; Song et al., 2021). The absence of comprehensive comparative studies between traditional and advanced nutrient management approaches impedes formulating evidence-based recommendations for farmers (Verma et al., 2020c, 2020d; Chaudhary et al., 2023). This study aims to fill this void by conducting a thorough investigation into the effects of nutrient management strategies on the yields of sugarcane ratoon. Conventional methods of managing nutrients in sugarcane cultivation result in inconsistent yields of ratoon crops due to imbalances in the availability of nutrients (Chen et al., 2023; Deng et al., 2023). Optimal nutrient levels and the promotion of plant health are achieved by implementing advanced nutrient management strategies, resulting in increased sugarcane productivity (Wu et al., 2022; Xu et al., 2022; Misra et al. 2023; Pongpat et al. 2023; Hoffman et al., 2024).

There is a substantial disparity in the number of crops harvested from ratoon fields that use traditional nutrient management methods compared to those that utilize advanced nutrient strategies. The interplay of soil characteristics, climate, and nutrient management substantially impacts sugarcane yield during ratoon cycles (Verma et al., 2020e, 2020f, 2022a, 2022b; Tian et al., 2023). First, it caters to the practical requirements of sugarcane farmers by offering evidence-based insights on the most efficient nutrient management strategies to maximize ratoon yields. These findings can potentially enhance profitability, decrease production expenses, and enhance the sustainability of sugarcane cultivation. Furthermore, this study improves the scientific comprehension of the complex interactions between nutrient management and crop productivity, specifically in the sugarcane ratoon system (Verma et al., 2022c, 2022d; Guo et al., 2023). Moreover, the research seeks to guide the implementation of cutting-edge agricultural practices that align with contemporary sustainability objectives by comparing conventional and advanced nutrient methodologies. This research pertains to sugarcane cultivation, particularly on the ratoon system. The study aims to examine traditional management of nutrient practices and evaluate them compared to more sophisticated methods, considering soil characteristics, climate conditions, and agronomic practices (Verma et al., 2021c, 2021d; Joshi et al., 2022, 2023). Nevertheless, it is crucial to recognize specific constraints. Regional variations may influence the research's findings, and the applicability of the results may depend on the context (Verma et al., 2021e; Li et al., 2023). External factors such as pest infestations and unforeseen weather events could also affect the study's outcomes. Notwithstanding these constraints, the research offers valuable perspectives and a basis for subsequent inquiries into enhancing sugarcane ratoon yields (Table 7.1).

7.2 SUGARCANE CULTIVATION AND RATOON SYSTEM

Sugarcane, scientifically known as *S. officinarum*, is a significant crop grown in tropical and subtropical regions. It is primarily cultivated for its abundant sucrose content, mainly used in sugar production, distinguishing it from numerous other crops (Oliveira et al., 2022; Solanki et al., 2023).

TABLE 7.1 Traditional and Advanced Nutrient Management in Sugarcane

Parameters	Traditional Nutrient Management	Advanced Nutrient Management
Nutrient application	Conventional fertilization methods include broadcast application of nitrogen, phosphorus, and potassium	Precision nutrient application based on soil testing and crop requirements, including split application and fertigation
Nutrient sources	Mostly synthetic fertilizers	Incorporation of organic amendments, use of slow-release fertilizers, and biofertilizers to enhance nutrient availability and soil health
Soil testing and analysis	Minimal soil testing, typically relying on historical data or visual assessment	Regular soil testing for nutrient status, pH, and soil health parameters. Adoption of modern soil analysis techniques like GIS-based mapping and digital soil mapping
Nutrient balance	An imbalance in nutrient application leads to potential leaching, runoff, and environmental pollution	Targeted nutrient application to match crop demand reduces the risk of nutrient loss and environmental impact
Crop monitoring	Limited monitoring of crop growth and nutrient uptake	Continuous monitoring of crop growth stages, nutrient uptake, and response to applied nutrients using remote sensing, IoT devices, and drones
Water management	Conventional irrigation methods with limited efficiency	Adoption of precision irrigation techniques like drip and sprinkler irrigation for efficient water and nutrient delivery
Crop residue management	Burning or improper disposal of crop residues	Adoption of conservation tillage and mulching practices to improve soil structure, water retention, and nutrient cycling
Integrated pest management	Reliance on chemical pesticides for pest control	Integrating biological control methods, cultural practices, and eco-friendly pesticides minimizes pest damage and promotes beneficial organisms
Yield and quality	Variable yield and quality are influenced by weather and nutrient availability	Consistent and higher yields with improved sucrose content and overall crop quality due to optimized nutrient management practices
Economic viability	High input costs with uncertain returns	Improved economic returns through efficient resource utilization, reduced input costs, and increased yields
Environmental sustainability	Potential negative impacts on soil, water, and air quality due to improper nutrient management.	It has enhanced environmental sustainability with reduced nutrient runoff, soil erosion, and greenhouse gas emissions.

Source: Adapted with permission from Das et al. (2023), Kaab et al. (2019b), and Modi et al. (2023).

Sugarcane can generate multiple harvests from a single planting by regrowing shoots from underground stalks or stubble, which is different from the usual method of replanting after each harvest. The ratoon system, a sustainable practice, has become essential in the sugarcane industry due to its economic and environmental advantages, as it minimizes the necessity for frequent replanting (Verma et al., 2022e, 2022f; Krishna et al., 2022). Throughout history, the management of nutrients in sugarcane cultivation has developed in parallel with advancements in agricultural practices. Historically, sugarcane cultivation in its early stages heavily depended on traditional farming practices and indigenous expertise. Knowledge of nutrient requirements was restricted, and fertilization practices frequently relied on experimentation (Verma et al., 2022g, 2022h; Kumari et al., 2022, 2023a, 2023b). With the advancement of scientific knowledge, the management of nutrients in sugarcane has become more methodical (Figure 7.1).

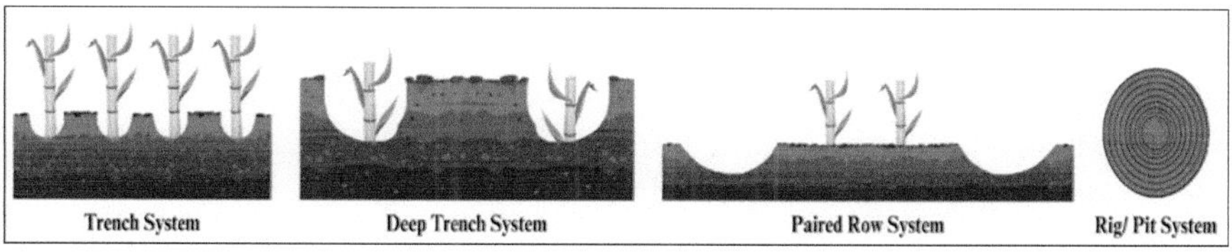

FIGURE 7.1 Different steps of sugarcane sowing.

The advent of synthetic fertilizers in the mid-20th century represented a notable breakthrough, enabling the precise administration of nutrients to optimize crop productivity. Over time, there has been a shift in focus toward enhancing the efficiency of nutrient utilization, fostering sustainability, and investigating novel methods to meet the growing demands of the sugarcane industry. Conventional nutrient management practices in sugarcane cultivation generally entail using organic and inorganic fertilizers chosen based on local knowledge and historical practices. Widely utilized nutrients consist of nitrogen, phosphorus, potassium, and micronutrients. The amount of these nutrients applied is affected by soil type, climate, and crop age (Verma et al., 2022i, 2022j, 2022k). Historically, organic sources such as animal and green manure have enhanced soil fertility and structure. Nevertheless, conventional methods may suffer from a lack of accuracy, resulting in imbalances of nutrients, deterioration of soil quality, and less-than-optimal crop yields (Akbarian et al. 2022, 2023; Singh et al., 2007; Viaud et al. 2023; Wu et al. 2023).

A comprehensive understanding of these conventional practices is essential for assessing their effectiveness and pinpointing areas that can be enhanced. Over the past few decades, advancements in agricultural science and technology have introduced sophisticated methods of managing nutrients to improve the accuracy, effectiveness, and long-term viability of sugarcane farming (Verma et al., 2024a, 2024b). Implementing precision agriculture (PA) methods, such as soil analysis, remote sensing, and precise nutrient application, has facilitated customized nutrient management strategies based on each field's unique conditions. Incorporating technological advancements, such as drip irrigation systems and controlled-release fertilizers, has enhanced the precision and efficiency of nutrient distribution, thereby reducing both resource wastage and environmental harm (Verma et al., 2024c). Biofertilizers, nanotechnology, and other innovative solutions have arisen as potential means to enhance nutrient absorption and support sustainable sugarcane cultivation. Multiple studies have investigated the correlation between nutrient management strategies and the yield of sugarcane rats. The investigations have covered different geographical regions, soil types, and climatic conditions, offering valuable insights into the factors that affect productivity in consecutive ratoon cycles.

Several studies have concentrated on conventional methods, assessing the influence of various fertilization strategies on the availability of nutrients and the performance of crops. Others have explored the intricacies of advanced nutrient management, evaluating the efficacy of PA and novel fertilization methods. Studies have examined the impact of nitrogen, phosphorus, potassium, and micronutrients on the yield of sugarcane ratoon. Nitrogen, an essential element for the growth of sugarcane, has been the main focus of numerous studies investigating the most effective rates and timing for its application. Phosphorus, which is necessary for the growth of roots and the transfer of energy, has also been closely examined for its influence on the productivity of ratoon crops. The impact of potassium on sugarcane ratoon yield has been investigated due to its essential role in promoting plant vitality and enhancing stress resistance. Although needed in smaller amounts, micronutrients are essential for various physiological processes and have been of interest in the ratoon system. Although there is a substantial body of research on the cultivation of sugarcane and the ratoon system, there are significant deficiencies in the current literature. Thorough comparative analyses are required to compare traditional and advanced nutrient management strategies for sugarcane ratoon yields directly. Numerous studies examine traditional or innovative methods, yet only a few compare their results using the same experimental framework (Dias et al., 2020; Modi et al., 2023; de Moraes Barbosa et al., 2021; Sanches et al., 2023; Shendryk et al., 2021).

Comprehending each strategy's comparative benefits and constraints is crucial in directing farmers toward the most efficient and environmentally friendly practices. Moreover, the current body of literature frequently lacks a cohesive methodology for evaluating the influence of nutrient management on sugarcane ratoon yield in various agroecosystems. The heterogeneity of soil conditions, climate, and agronomic practices hinders the generalization of findings. Future research should establish a more standardized methodology to improve the comparability and practicality of the findings. Furthermore, although research has examined the impact of individual nutrients on the productivity of sugarcane ratoon, comprehensive investigations that consider the combined effects of multiple nutrients and their interactions are necessary. A comprehensive approach is necessary to comprehend the movement and transformation of nutrients in the intricate ratoon system, considering the interdependent connections between different elements (Li et al., 2023). By addressing these deficiencies in the current body of research, we can substantially contribute to advancing knowledge in sugarcane cultivation, specifically about the ratoon system. The tool will guide sustainable nutrient management practices, offer insights for agricultural policies, and improve the overall efficiency and resilience of the sugarcane industry in response to changing environmental and economic challenges (Figure 7.2).

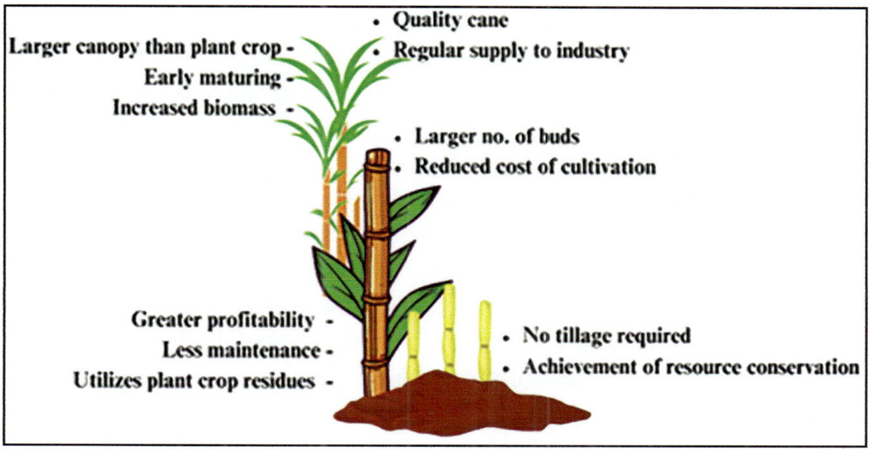

FIGURE 7.2 Sugarcane ratoon systems.

7.3 TRADITIONAL NUTRIENT MANAGEMENT IN SUGARCANE RATOON SYSTEM

The nutrient management practices in the sugarcane ratoon system are deeply rooted in historical agricultural methods. These practices have been inherited over successive generations, influenced by local knowledge, observation, and experiential learning. The fundamental principle of conventional nutrient management involves the application of both organic and inorganic fertilizers to fulfil the nutrient needs of sugarcane crops (Huang et al., 2024). Widely used nutrients in sugarcane cultivation are nitrogen (N), phosphorus (P), potassium (K), and micronutrients. Each nutrient has a distinct function to enhance nutrient levels, particularly during crucial rapid growth phases; farmers frequently depend on organic sources, such as animal manure, compost, and green manure to augment soil fertility (Boroun et al., 2023; de Oliveira Filho et al., 2021; Otto et al., 2020; Sales et al., 2022; Soulé et al., 2024).

These organic materials provide vital nutrients and enhance soil composition, water-holding capacity, and microbial function. Inorganic fertilizers, such as urea, superphosphate, and muriate of potash, are frequently employed to enhance nutrient levels, particularly during crucial growth phases. The timing and manner of nutrient application are generally determined by local expertise, prevailing weather conditions, and the developmental phase of the sugarcane crop. As a crop with high nutrient requirements, sugarcane necessitates a consistent and well-proportioned provision of vital elements throughout its growth cycle. Nitrogen is essential for promoting vegetative growth and overall plant development, whereas phosphorus facilitates root development, energy transfer, and early shoot formation. Potassium is crucial in enhancing stress tolerance, facilitating water absorption, and promoting resistance against diseases.

Micronutrients such as zinc, iron, and manganese are necessary in small quantities for various physiological processes. Traditionally, nutrients are commonly administered through the methods of broadcasting or banding. Broadcasting evenly distributes fertilizers across the entire field, ensuring a consistent spread of nutrients. Banding, however, involves the concentration of fertilizers in specific rows or trenches, enabling precise delivery of nutrients to the root zone. The selection of these methods is contingent upon soil composition, nutrient accessibility, and financial factors. Regarding timing, nutrient application is commonly categorized into three periods: preplanting, postharvest, and midgrowing season applications. Preplanting applications are intended to establish an initial nutrient level in the soil before planting sugarcane. Postharvest applications replenish nutrient levels after the main harvest, preparing the field for the subsequent ratoon crop. Applications made during the middle of the growing season offer extra nutrients to promote robust growth throughout the ratoon cycle. Although traditional nutrient management practices have successfully supported sugarcane growth for many years, they are accompanied by inherent difficulties and restrictions.

An important obstacle is the possibility of nutrient imbalances. Excessive dependence on specific fertilizers without considering the comprehensive nutrient composition of the soil can result in imbalances, which can adversely affect the well-being and productivity of plants. Moreover, the timing of nutrient applications may not consistently coincide with the precise requirements of the sugarcane crop at various stages of growth, leading to less-than-optimal utilization of nutrients. Another constraint is the dependence on organic sources, leading to nutrient content and availability variations. The gradual release of nutrients from organic materials may not always correspond to the immediate requirements of the sugarcane crop, especially during periods of rapid growth. Furthermore, transporting and utilizing organic materials can require significant manual labor, potentially leading to higher production expenses.

Soil degradation is a prevalent issue linked to conventional nutrient management. Uninterrupted farming without implementing adequate soil conservation measures can result in the loss of nutrients, erosion, and a gradual decrease in soil fertility. The deterioration of soil quality can hurt the ability to sustain sugarcane cultivation in the long run, emphasizing the importance of implementing comprehensive soil management techniques. Farmers historically utilized organic manure in a tropical region with loamy soil before planting and then used a combination of urea and muriate of potash during the growing season. The practice of placing fertilizers in trenches was favored to optimize the absorption of nutrients by the roots of sugarcane. Although the approach is simple, farmers consistently reported high yields and encountered few nutrient-related problems across multiple ratoon cycles (Colmanetti et al., 2024; Farias-Ramírez et al., 2024; Hajeb et al., 2023; Ma et al., 2024; Morel et al., 2014; Oliveira et al., 2023).

In a subtropical region characterized by sandy soil, the conventional approach to managing nutrients involves the integration of green manure cover crops into the soil before planting sugarcane. After the harvest, a well-proportioned mixture of organic and inorganic fertilizers was evenly spread throughout the field. Although this method enhanced the arrangement of soil particles and the ability to retain water, it faced difficulties with the loss of nutrients in sandy soils, requiring extra applications of nutrients during the middle of the growing season to sustain ideal nutrient levels. The case study conducted in a region characterized by clayey soil emphasized composted organic material as the predominant source of nutrients. During the preplanting and postharvest phases, farmers utilized broadcasting techniques. Although organic nutrients are released slowly, the clayey soil maintains fertility, leading to consistent ratoon yields.

Nevertheless, apprehensions arose regarding the feasibility of producing and transporting compost on a large scale. These case studies demonstrate how traditional nutrient management practices can be adjusted to suit various agroecosystems. Although there were instances of successful results, each study also identified distinct difficulties associated with soil composition, weather conditions, and nutrient accessibility, underscoring the importance of employing tailored strategies that take into account specific circumstances. The conventional approach to nutrient management in the sugarcane ratoon system has played a crucial role in maintaining sustainable agriculture for many generations. Although these practices have demonstrated resilience and adaptability, it is necessary to critically evaluate the challenges associated with nutrient imbalances, soil degradation, and variations in organic sources. To improve the sustainability and productivity of sugarcane cultivation in the ratoon system, it is essential to combine the knowledge from traditional practices with innovative approaches as the agricultural landscape changes. Further investigation is needed to examine methods of combining traditional knowledge with contemporary technology to overcome current constraints and guarantee the ongoing prosperity of sugarcane cultivation in various settings (Figure 7.3).

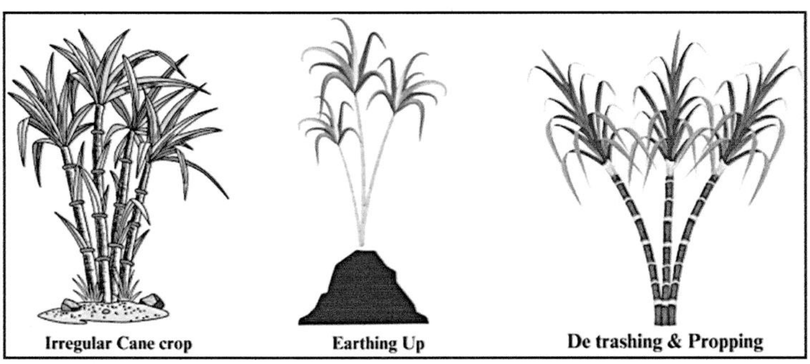

FIGURE 7.3 Intercultural operation of sugarcane.

7.4 ADVANCED NUTRIENT MANAGEMENT STRATEGIES

Sugarcane cultivation has undergone a paradigm shift with the adoption of advanced nutrient management strategies, departing from traditional methods. They utilize state-of-the-art technologies and pioneering methods to improve accuracy, productivity, and environmental friendliness. These approaches strive to enhance the efficiency of nutrient utilization, reduce the environmental footprint, and enhance crop productivity. The incorporation of PA, biofertilizers, nanotechnology, and other advanced methodologies signifies a shift from traditional practices, providing a comprehensive and technology-oriented approach to nutrient management in the sugarcane ratoon system. PA has become a revolutionary method in sugarcane cultivation. It utilizes technology to customize nutrient management practices according to the specific requirements of each field. Data-driven technologies, such as global positioning systems (GPS), remote sensing, and geographic information systems (GIS), form the foundation of PA. These technologies empower farmers to accurately delineate variations in soil characteristics, monitor the well-being of crops, and precisely administer nutrients at the appropriate time and location. State-of-the-art sensors analyze soil characteristics such as nutrient composition, moisture levels, and pH, delivering up-to-the-minute data to aid decision-making (Asrol et al., 2024; Bispo et al., 2022; Christina et al., 2021; Dengia et al., 2023; Sefeedpari et al., 2014; Shukla et al., 2022).

This data aims to generate comprehensive soil maps, which serve as a guide for farmers to optimize the application of nutrients. Satellite and drone imagery capture crop health indicators, enabling farmers to promptly detect signs of stress, disease, or nutrient deficiencies during the early stages of the growing season. This data facilitates precise interventions, guaranteeing prompt and efficient nutrient management. Variable rate technology (VRT) allows for the accurate application of fertilizers by considering spatial variations in soil conditions. VRT optimizes nutrient utilization and reduces waste by customizing application rates based on the specific attributes of each field. Tractors with GPS-guided systems enable precise planting, harvesting, and nutrient distribution. This automation process minimizes human mistakes and improves the overall effectiveness of agricultural activities. Biofertilizers and nanotechnology are innovative methods for managing nutrients in sugarcane cultivation. They utilize biological and nanoscale processes to enhance the availability and absorption of nutrients. Microbial inoculants, such as bacteria that fix nitrogen and fungi that form mycorrhizal associations, serve as the foundation for biofertilizers.

These advantageous microorganisms form symbiotic associations with the roots of sugarcane, enhancing the uptake of nutrients. Biofertilizers bolster soil health, diminish dependence on synthetic fertilizers, and augment nutrient cycling. Nanoparticles, designed with precision at the nanoscale, show potential in enhancing nutrient transportation and plant reactions. Using nanosized nutrient carriers enables precise delivery to specific plant tissues, improving nutrient uptake efficiency. In addition, nanoscale materials have the potential to impact plant physiological processes, enhancing stress tolerance and overall crop performance. Advanced nutrient management includes innovative methods like applying nutrients through the leaves, using organic nanocomposites to amend the soil, and using substances that promote plant growth. These approaches aim to improve the uptake of nutrients, reduce wastage, and encourage the sustainable utilization of nutrients in sugarcane farming. PA employs targeted nutrient applications customized to the specific requirements of each field section, thereby maximizing resource utilization and minimizing wastage (Dlamini and Zhou 2022; Fattori Junior et al., 2022; Leite et al. 2023; Park et al., 2010; Wilson et al., 2022; Zhu et al., 2023).

Advanced methodologies, such as VRT and biofertilizers, optimize nutrient utilization by adjusting the application rates to match the crop's specific needs. Biofertilizers and other innovative techniques enhance sustainable agriculture by diminishing dependence on synthetic fertilizers, mitigating environmental repercussions, and enhancing soil vitality. The utilization of nanotechnology and other sophisticated methodologies holds the promise of enhancing the absorption of nutrients, bolstering the ability of crops to withstand stress, and fostering the general well-being of sugarcane plants, thereby resulting in amplified yields. PA data facilitates informed decision-making, enabling farmers to adapt to dynamic conditions and optimize nutrient management strategies promptly. Developing sophisticated nutrient management strategies requires substantial technological, equipment, and training investments. Small-scale farmers may encounter difficulties in both accessing and implementing these technologies, although PA yields extensive data, effectively interpreting and implementing this information can pose challenges. Agricultural practitioners require the expertise and resources to convert data into practical and effective understandings.

Implementing and overseeing sophisticated nutrient management systems can be intricate, necessitating proficiency in agronomy, technology, and data analysis. Farmers may face difficulties in navigating this intricacy. Relying exclusively on sophisticated methodologies may carry potential hazards, as unanticipated variables like equipment failures or disruptions in the biofertilizer supply chain could influence the effectiveness of nutrient management approaches. Nanotechnology elicits apprehensions regarding the possible ecological ramifications, such as the enduring presence of nanoparticles in the soil and their impact on ecosystems. Addressing these concerns necessitates the implementation of responsible and sustainable practices (Alvarez-Campos et al., 2018; de Aquino et al., 2017; Han et al., 2022; Nejad et al., 2023; Oliveira et al., 2022; Mafunga et al., 2023; Poudyal et al., 2023; Reyes-Trujillo et al., 2021; Wickramasinghe et al., 2024). A tropical sugarcane farm of significant size implemented PA technologies to enhance nutrient management efficiency. Using soil and remote sensing techniques enabled the evaluation of soil conditions and the surveillance of crop vitality.

VRT facilitated accurate fertilizer application by adjusting rates according to spatial variations identified through data analysis. Implementing this strategy led to a 15% decrease in the amount of fertilizer used, 10% increase in crop yields, and enhanced overall efficiency in resource utilization. Within a subtropical area that has previously relied on synthetic fertilizers, a sugarcane farm switched to biofertilizers to enhance the soil quality. Nitrogen-fixing bacteria and mycorrhizal fungi were introduced to augment nutrient cycling and facilitate innate soil processes. Throughout three successive ratoon cycles, the farm witnessed a progressive decline in the necessity for artificial fertilizers, enhanced soil composition, and consistent crop yields, showcasing the enduring advantages of adopting biofertilizers. A research farm investigated the application of nanotechnology in the cultivation of sugarcane, with a specific emphasis on the precise delivery of nutrients. Researchers created nanoscale nutrient carriers to encapsulate vital elements, enabling accurate targeting of sugarcane roots during delivery. This method exhibited enhanced nutrient absorption efficiency, leading to improved growth, enhanced disease resistance, and a 12% boost in ratoon yields. The study emphasized the capacity of nanotechnology to transform nutrient management in sugarcane. These case studies highlight the effective execution of advanced nutrient management techniques in various agricultural ecosystems.

Although the difficulties are recognized, the advantages of PA, biofertilizers, nanotechnology, and other inventive methods demonstrate the capacity to revolutionize sugarcane farming, guaranteeing sustainability, heightened productivity, and adaptability in response to changing agricultural environments. Advanced nutrient management strategies signify a crucial transition toward a more sustainable, efficient, and technology-driven method of cultivating sugarcane in the ratoon system. Farmers can access various tools, such as PA, biofertilizers, nanotechnology, and other innovative techniques. These tools can help optimize nutrient utilization, minimize environmental harm, and improve crop performance. Combining advanced strategies with responsible and context-specific practices is crucial to fully harness the benefits of the evolving agricultural industry and ensure the ongoing success of sugarcane farming (Figure 7.4).

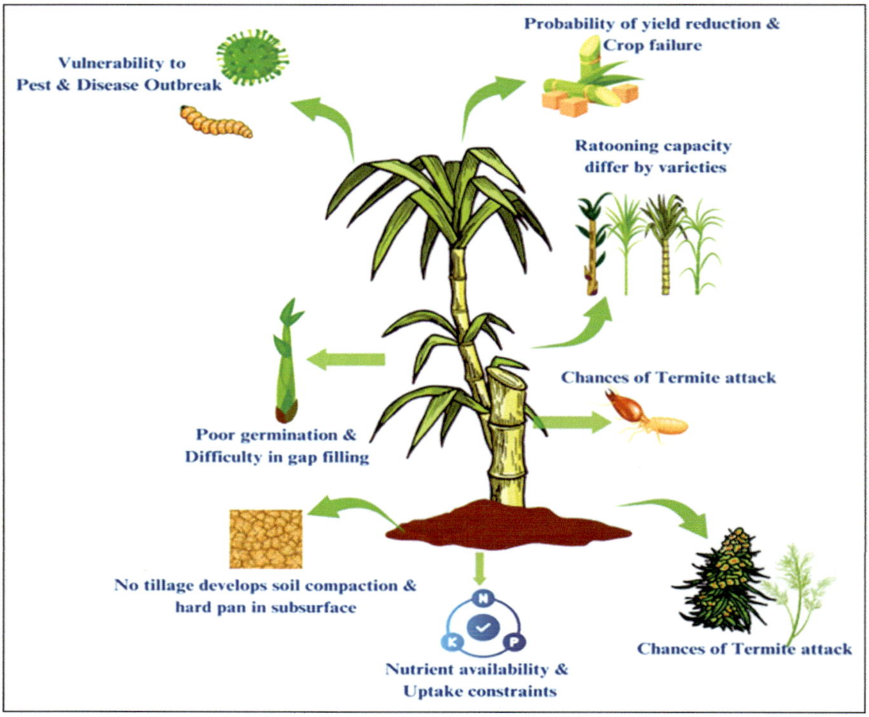

FIGURE 7.4 Poor germination and difficulties in gap-filling.

7.5 COMPARATIVE ANALYSIS

An essential task is to conduct a comparative analysis of traditional and advanced nutrient management strategies in the sugarcane ratoon system to comprehend the effectiveness, advantages, and difficulties associated with each approach. The research design employs a systematic comparative framework to assess critical parameters, specifically the differences in nutrient application between traditional broadcasting and advanced PA methods. Assessing nutrient strategies on soil health encompasses evaluating nutrient content, microbial activity, and overall fertility. Achievement of crops: An analysis of sugarcane ratoon yields, growth patterns, and overall crop health is conducted to compare the effects of traditional and advanced nutrient management techniques. Efficiency in resource utilization: Evaluating the effectiveness of nutrient and resource utilization, such as water and fertilizers, in both methods. Experimental tests are conducted in real-world conditions to evaluate the performance and effectiveness of a product or technology. We work field trials in typical sugarcane-growing areas, utilizing a randomized block design to reduce external variations. Data collection refers to gathering and recording information or data from various sources (Christina et al., 2021; Dengia et al., 2023; Ma et al., 2024; Oliveira et al., 2023; Otto et al., 2020; Sales et al., 2022; Sefeedpari et al., 2014; Shukla et al., 2022). We are collecting data on soil properties, nutrient concentrations, and crop productivity at multiple stages of the sugarcane ratoon cycle.

Using established techniques, PA technologies are applied to monitor and document nutrient application in advanced and traditional farming methods. Quantitative data analysis using statistical methods and techniques like analysis of variance to compare averages and detect noteworthy disparities between conventional and advanced nutrient management results. The data analysis phase thoroughly evaluates the gathered information to derive significant insights regarding the comparative effectiveness of traditional and advanced nutrient management strategies. Examining soil samples to determine the levels of nutrients, pH, and microbial activity to comprehend the effects of nutrient management strategies on soil well-being. To assess the effectiveness of each nutrient approach, we will measure crop growth and yield metrics, such as plant height, stalk diameter, number of tillers, and, ultimately, sugarcane yield. Efficiency in resource utilization: Evaluating the effectiveness of resource allocation, such as water and fertilizers, to identify the most optimal approach for utilizing inputs. Comparative statistical analysis uses statistical methods to compare the means and determine the statistical significance of differences between traditional and advanced nutrient management outcomes.

Time-based analysis analyzing data from multiple ratoon cycles to evaluate each nutrient management strategy's sustainability and long-term effects. The comparative analysis primarily examines the difference in crop yield between sugarcane crops managed using traditional and advanced nutrient strategies. This entails evaluating the numerical and subjective aspects of the yield from each method. Aggregate sugarcane production: Quantifying the total mass of harvested sugarcane from conventional and technologically enhanced nutrient-managed areas. The ratoon cycle examines crop yields across multiple cycles to identify fluctuations and patterns and evaluate the financial benefits of traditional and innovative nutrient management methods while considering expenses, labor, and overall profitability. Comprehending the factors that impact differences in productivity between conventional and advanced nutrient management is essential for a nuanced interpretation of the comparative analysis.

The researchers evaluate the disparity in nutrient availability between the two approaches and its influence on the growth and productivity of sugarcane. They examine soil texture, structure, and composition variations to ascertain the interaction between different soil and nutrient management strategies. Their analysis encompasses the impact of climate variables, such as temperature, precipitation, and other ecological factors, on the availability of nutrients and the productivity of sugarcane. The assessment of technological adoption and farmer expertise in implementing advanced nutrient management strategies can substantially impact outcomes (Singh et al., 2007; Akbarian et al., 2022; Das et al., 2023; Misra et al., 2023; Modi et al., 2023; de Moraes Barbosa et al., 2021; Sanches et al., 2023; Shendryk et al., 2021).

The researchers examine the effects of nutrient strategies on pest and disease pressures and how they influence the health and productivity of crops. The comparative analysis provides practical implications for sugarcane farmers, influencing decision-making and guiding sustainable practices. Farmers acquire valuable knowledge regarding the nutrient management strategy that best suits their particular circumstances, enhancing nutrient application efficiency. Comprehending the resource utilization efficiency of each strategy enables farmers to make informed decisions, minimize inefficiency, and mitigate environmental consequences. The economic analysis provides farmers with a lucid assessment of the cost–benefit ratio, facilitating financial planning and investment choices. Understanding the lasting effects of each approach aids in adopting sustainable methods that protect soil well-being and guarantee ongoing productivity.

Given their specific conditions, farmers can evaluate the practicality and advantages of implementing PA technologies and advanced nutrient management practices. The comparative analysis of traditional and advanced nutrient management in the sugarcane ratoon system thoroughly comprehends their effects on soil health, crop performance, and resource utilization efficiency. This knowledge provides sugarcane farmers with valuable information to make well-informed decisions, advocate for sustainable practices, and improve overall productivity in the ever-changing field of sugarcane cultivation (Figure 7.5).

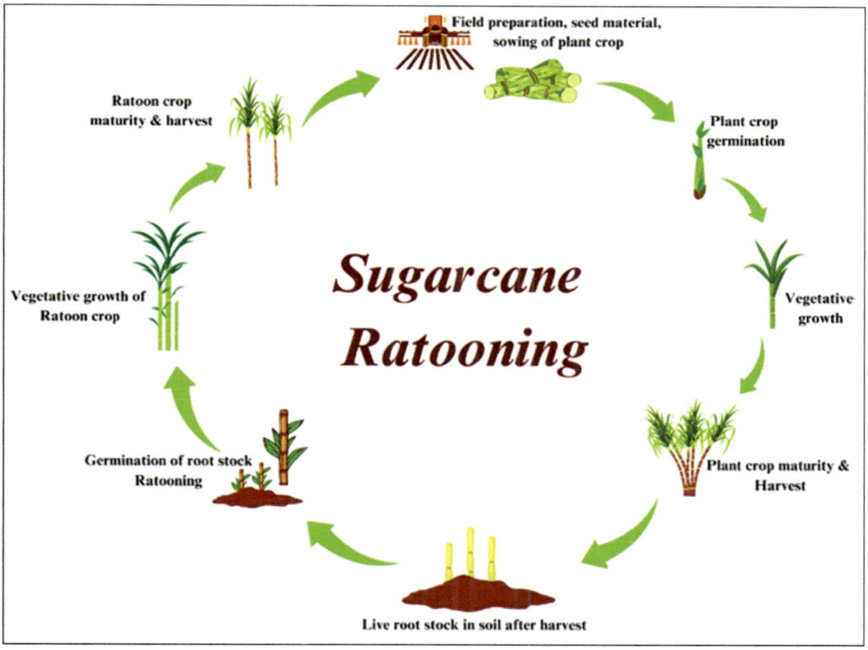

FIGURE 7.5 Sugarcane ratooning.

7.6 RECOMMENDATIONS AND BEST PRACTICES

The comparative analysis of traditional and advanced nutrient management strategies in the sugarcane ratoon system has provided valuable insights into the factors that affect soil health, crop performance, and overall productivity. Summarizing the main discoveries establishes a basis for deriving practical suggestions and optimal methods for sugarcane farmers (Figure 7.6).

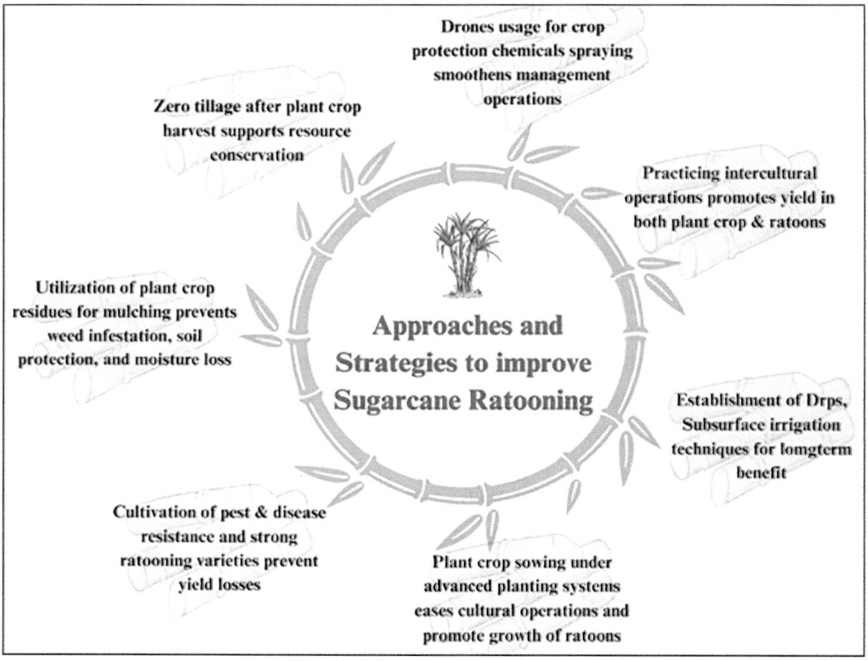

FIGURE 7.6 Approaches and strategies to improve sugarcane ratooning.

7.6.1 NUTRIENT AVAILABILITY AND SOIL HEALTH

The availability of nutrients and soil health are vital determinants of agricultural productivity and sustainability. Conventional farming practices frequently experience variations in nutrient availability, resulting in imbalances and deficiencies in the nutrient content of the soil. Nevertheless, the utilization of advanced nutrient management techniques, specifically PA, has led to a significant change in focus toward maximizing nutrient availability and improving soil health by employing precise and targeted applications. This essay examines the difficulties associated with traditional approaches, the advantages of sophisticated nutrient management, and the contribution of PA to enhancing nutrient accessibility and soil well-being.

Conventional agricultural practices commonly entail indiscriminate application of fertilizers, disregarding variations in soil nutrient levels and crop needs. This approach frequently results in excessive or insufficient utilization of specific nutrients, leading to imbalances in soil fertility. Furthermore, persistent dependence on artificial fertilizers can deteriorate soil composition, diminish microbial functioning, and contribute to environmental contamination using nutrient runoff and leaching.

Moreover, the unselective utilization of agrochemicals can disturb soil ecosystems and jeopardize the soil's long-term health. Variations in nutrient availability worsen the difficulties linked to traditional methods. The fluctuating availability of nutrients in agricultural soils is influenced by weather variability, soil erosion, and inefficient nutrient uptake by crops (Asrol et al., 2024; Boroun et al., 2023; Christina et al., 2021; Colmanetti et al., 2024; Farias-Ramírez et al., 2024; Hajeb et al., 2023; Morel et al., 2014; Oliveira et al., 2023; Sefeedpari et al., 2014; Wilson et al., 2022). Consequently, farmers face difficulties maintaining the ideal nutrient levels, reducing crop yields and quality. Advanced nutrient management approaches differ from conventional methods by focusing on enhancing nutrient accessibility and soil health through PA techniques.

These methods utilize data-driven analysis, advanced technologies, and innovative strategies to maximize nutrient use and reduce environmental negative effects. An important advantage of advanced nutrient management is the capacity to customize nutrient applications based on specific site conditions and crop needs. Soil testing and analysis are essential for evaluating nutrient, pH, and soil health indicators. Farmers can tailor fertilizer formulations and application rates to address specific deficiencies and maximize nutrient availability by integrating soil test results into nutrient management decisions. Moreover, advanced nutrient management prioritizes using alternative nutrient sources and application techniques to optimize soil fertility and reduce nutrient depletion. Organic amendments, cover crops, and precision application technologies, such as fertigation and foliar spraying, provide sustainable alternatives to synthetic fertilizers. These practices enhance the utilization of nutrients and facilitate the accumulation of organic matter in the soil, promoting the diversity of microorganisms and overall soil well-being.

PA is an innovative method of managing nutrients that utilizes advanced technologies to optimize the use of resources and increase crop productivity. Site-specific management is a key aspect of PA, where interventions are strategically implemented to address the spatial and temporal variations in soil and crop conditions. Remote sensing technologies, such as satellite imagery and drones, offer valuable information on crop health, nutrient absorption, and soil moisture levels. By analyzing these data layers, farmers can detect regions of nutrient insufficiency or surplus within their fields and implement accurate nutrient applications. VRT allows for adjusting fertilizer application rates according to spatial variations in soil properties, thereby minimizing excessive application and mitigating environmental consequences. PA focuses on spatial accuracy and highlights the significance of temporal accuracy in nutrient management. Applying nutrients at the right time during important growth stages guarantees that crops receive the required nutrients when needed most, thus maximizing their potential for yield and quality.

Continuous evaluation of crop nutrient levels is made possible through real-time monitoring systems and sensor-based technologies. This enables the implementation of adaptive management strategies and the ability to make responsive decisions. Furthermore, PA promotes the amalgamation and harmonization of farming techniques, such as efficient water distribution, strategic crop rotation, and effective pest management. Farmers can enhance agricultural sustainability by adopting a comprehensive approach to nutrient management in PA, optimizing resource allocation and improving soil health. The availability of nutrients and the health of the soil are crucial factors that determine the productivity and sustainability of agriculture. Traditional approaches frequently prove inadequate in addressing the ever-changing nutrient availability in soil, resulting in imbalances and deficiencies in soil fertility (Akbarian et al., 2022, 2023; Dias et al., 2020; Kaab et al., 2019a, 2019b; Modi et al., 2023; de Moraes Barbosa et al., 2021; Singh et al., 2007; Viaud et al., 2023; Wu et al., 2023).

Nevertheless, farmers can surmount these obstacles by employing sophisticated nutrient management techniques, specifically PA, and attain optimal utilization of nutrients and soil health. PA provides a transformative approach to nutrient management by utilizing data-driven decision-making, innovative technologies, and holistic management practices. By implementing precise nutrient applications customized to the specific conditions of each site and the needs of the crops, farmers can improve the availability of nutrients, maximize the potential for high yields, and support the long-term health and sustainability of the soil. Amidst the challenges of contemporary agriculture, PA emerges as a promising solution, leading us toward a more productive, adaptable, and ecologically conscious future.

7.6.2 CROP PERFORMANCE AND YIELD

Scientifically known as *S. officinarum*, sugarcane is an essential crop globally. It plays a significant role in the production of sugar and bioenergy. Maximizing sugarcane growth and improving yields are primary goals for both farmers and researchers. Advancements in agricultural practices have recently introduced innovative nutrient strategies, including PA and biofertilizers. These strategies show promise in improving crop performance and increasing yields. Although traditional methods have demonstrated resilience, incorporating these innovative approaches could further optimize sugarcane production efficiency. This essay examines the potential of advanced nutrient strategies to enhance the growth and yield of sugarcane. It also acknowledges the significance of optimizing traditional methods to adapt to variations and maximize the yield capacity. PA utilizes contemporary technologies, such as GIS, GPS, and remote sensing, to enhance the efficiency of agricultural input management, encompassing nutrients, water, and pesticides.

PA in sugarcane cultivation provides an accurate mapping of soil nutrient levels, allowing for customized fertilizations strategies tailored to the specific requirements of individual fields. Farmers can optimize the application of fertilizer's by precisely identifying nutrient deficiencies or surpluses, minimizing wastage and environmental repercussions while maximizing the potential crop yield. Furthermore, PA enables the immediate monitoring of crop health and growth parameters, enabling prompt interventions to address problems such as pest infestations, diseases, or water stress. This proactive approach minimizes yield losses, and sugarcane's optimal growth is guaranteed during the growing season. Moreover, incorporating data analytics and machine learning algorithms allows for creating predictive models to forecast crop performance accurately. This, in turn, improves decision-making and the allocation of resources on the farm.

Biofertilizers comprise microbial inoculants that contain advantageous microorganisms, including nitrogen-fixing bacteria, phosphate solubilizers, and plant growth-promoting rhizobacteria. These microorganisms improve the availability and absorption of nutrients by plants. Integrating biofertilizers in sugarcane cultivation can enhance soil fertility, stimulate root growth, and bolster plant resistance to environmental pressures. In addition, biofertilizers promote sustainable agriculture by decreasing the dependence on artificial fertilizers and minimizing the loss of nutrients through leaching and runoff, thereby reducing environmental pollution (Akbarian et al., 2022, 2023; Boroun et al., 2023; Christina et al., 2021; Colmanetti et al., 2024; Dengia et al., 2023; Dias et al., 2020; Farias-Ramírez et al., 2024; Hajeb et al., 2023; Kaab et al., 2019b, 2019a; Ma et al., 2024; Modi et al., 2023; de Moraes Barbosa et al., 2021; Morel et al., 2014; Oliveira et al., 2023; Shukla et al., 2022; Singh et al., 2007; Viaud et al., 2023; Wu et al., 2023).

Studies have shown that biofertilizers effectively improve sugarcane's growth and productivity. For example, introducing nitrogen-fixing bacteria through vaccination has been proven to enhance nitrogen fixation and optimize nitrogen utilization efficiency in sugarcane. This results in increased crop yields and decreased reliance on fertilizers. Phosphate-solubilizing bacteria can increase the amount of phosphorus in the soil, which, in turn, helps sugarcane plants grow better by improving root growth and nutrient absorption. Biofertilizers provide a cost-effective and environmentally friendly method for managing nutrients in sugarcane production by utilizing the advantageous interactions between plants and microbes. Although advanced nutrient strategies show potential for improving sugarcane yield, optimizing traditional methods to ensure sustainable crop production is still crucial. Conventional techniques, such as applying appropriate amounts of fertilizer, alternating crops, and preserving soil, have always played a crucial role in prosperous sugarcane farming. Nonetheless, it is imperative to modify these techniques to account for the diversity in soil conditions, climate patterns, and pest pressures.

This is essential for optimizing crop output and guaranteeing sustained productivity in the long run. Soil testing and nutrient analysis are essential practices in conventional nutrient management, exemplifying their importance. Farmers can make well-informed choices regarding the quantity and timing of fertilizer application by consistently evaluating soil fertility levels and crop nutrient demands. This approach helps to avoid both nutrient deficiencies and excessive amounts, which can hinder the potential yield of crops. Incorporating organic amendments, such as compost or manure, into soil management practices can enhance soil structure, water retention, and nutrient cycling. This, in turn, improves overall crop performance.

Moreover, adopting integrated pest management (IPM) strategies can reduce pest and disease burdens while reducing dependence on chemical pesticides. IPM integrates diverse control strategies, such as cultural practices, biological control agents, and chemical interventions, to efficiently manage pest populations while minimizing environmental harm and protecting beneficial organisms. By incorporating IPM principles into conventional sugarcane production systems, farmers can effectively preserve pest resistance and minimize crop losses, optimizing yield and profitability. Adopting a comprehensive approach that combines advanced nutrient strategies with conventional agricultural practices is essential to achieve optimal crop performance and yield in sugarcane production.

One can optimize nutrient management, enhance soil health, and bolster resistance against environmental pressures by employing PA technologies and biofertilizers (Alvarez-Campos et al., 2018; Asrol et al., 2024; Bispo et al., 2022; Dlamini and Zhou 2022; Fattori Junior et al., 2022; Leite et al., 2023; Reyes-Trujillo et al., 2021; Sefeedpari et al., 2014; Wilson et al., 2022; Zhu et al., 2023). Nevertheless, optimizing conventional techniques like balanced fertilizations, crop rotation, and IPM is crucial to adapt to variability and effectively enhance the maximum yield capacity. By integrating novel techniques with established practices, farmers can effectively increase sugarcane yield in an environmentally sustainable manner. This not only aids in ensuring global food security but also promotes economic growth and prosperity.

7.6.3 RESOURCE USE EFFICIENCY

Maximizing resource efficiency is essential for sustainable agriculture, allowing for optimal productivity while minimizing waste and environmental harm. In recent years, PA has become a promising method for improving resource utilization efficiency using advanced technologies and data-driven management techniques. PA seeks to optimize crop performance and minimize resource waste by precisely targeting inputs such as water, fertilizers, and pesticides. Nevertheless, conventional farming practices can enhance resource efficiency, particularly with precision application methods. This essay examines the role of PA and conventional methods in optimizing resource utilization in agriculture, specifically focusing on their advantages and potential collaborative effects. An avenue toward optimizing resource utilization. PA involves utilizing different technologies and methods to maximize resource utilization efficiency in agricultural production.

The core of PA lies in using data-driven tools for decision-making, including GIS, GPS, and remote sensing technologies, to oversee and control variations within fields. Through collecting and analyzing spatial data about soil properties, crop health, and environmental conditions, farmers can customize their management practices to suit the specific conditions of their fields, thereby optimizing resource utilization and crop productivity. PA offers a crucial benefit by effectively maximizing nutrient utilization by implementing focused fertilizations strategies. Through the process of mapping soil nutrient levels and employing variable rate application techniques, farmers can enhance the accuracy of fertilizer application. This allows for a more precise alignment of nutrient inputs with crop requirements and the availability of soil nutrients. This practice decreases the likelihood of excessive fertilizations, thereby reducing the amount of nutrients that are washed away and leached out while maximizing the absorption of nutrients by crops (Akbarian et al., 2022, 2023; Das et al., 2023; Dias et al., 2020; Kaab et al., 2019b; Misra et al., 2023; Modi et al., 2023; de Moraes Barbosa et al., 2021; Shendryk et al., 2021; Singh et al., 2007; Viaud et al., 2023; Wu et al., 2023).

PA can enhance nutrient use efficiency, improving crop productivity, and reducing environmental impact. Moreover, PA enables the effective management of water resources by continuously monitoring the soil's moisture levels and crops' water needs in real time. Automated irrigation systems equipped with soil moisture sensors and weather data can distribute water precisely, ensuring it is delivered to the appropriate location and at the optimal time. This technology effectively reduces water waste and maximizes crop development. Precision pesticide and herbicide applications enable farmers to target pest and weed infestations precisely, reducing chemical usage without compromising pest control efficacy. PA showcases exceptional resource efficiency through waste reduction and optimal utilization of nutrients, water, and agrochemicals. Through the utilization of data and technology, farmers can enhance crop production efficiency while reducing environmental repercussions, thereby promoting sustainable agriculture and ensuring food security.

Although PA provides substantial advantages in terms of resource efficiency, traditional agricultural practices also have a vital role in optimizing efficiency, particularly when used in conjunction with precision application techniques. Conventional methods like crop rotation, intercropping, and organic farming have been widely acknowledged for enhancing soil quality, preserving water, and decreasing dependence on external resources. Soil fertility management is an area where traditional methods can improve efficiency. Techniques such as cover cropping and organic amendments can enhance soil structure, augment organic matter content, and optimize nutrient cycling, diminishing reliance on synthetic fertilizers. By incorporating these methods alongside accurate application techniques, farmers can enhance the availability and utilization of nutrients, thus fostering sustainable soil fertility while reducing nutrient losses (Bispo et al., 2022; Boroun et al., 2023; Colmanetti et al., 2024; Dengia et al., 2023; Farias-Ramírez et al., 2024; Hajeb et al., 2023; Ma et al., 2024; Morel et al., 2014; Oliveira et al., 2023; Sales et al., 2022; Shukla et al., 2022).

Moreover, conventional approaches can enhance the control of pests and diseases by promoting biodiversity and bolstering the resilience of ecosystems. Agroecological approaches, such as IPM, prioritize utilizing biological control agents, crop diversification, and habitat manipulation to reduce pest populations while minimizing the need for chemical pesticides. By integrating precision application technologies, IPM strategies can more efficiently address pest outbreaks, resulting in decreased chemical usage and the preservation of beneficial organisms. In addition, conventional techniques provide chances to improve water utilization efficiency through the implementation of rainwater harvesting, mulching, and conservation tillage. Farmers can preserve water resources and enhance their ability to withstand drought by implementing rainwater capture and storage systems, reducing evaporation, and minimizing soil disturbance. Combining these methods with accurate irrigation technologies can optimize water management, resulting in effective water utilization and sustained crop productivity.

To summarize, achieving optimal resource utilization in agriculture necessitates a comprehensive strategy that combines PA techniques with conventional methods. Although PA provides sophisticated technologies to maximize resource utilization, conventional methods such as crop rotation, organic farming, and IPM also improve efficiency. By integrating the advantages of both methodologies and implementing accurate application methods, farmers can enhance resource efficiency, optimize crop productivity, and foster sustainable agriculture for future generations.

7.6.4 ECONOMIC CONSIDERATIONS

Economic factors are crucial in decision-making processes in contemporary agriculture. Farmers and stakeholders should thoroughly assess the expenses and advantages of various nutrient management strategies to optimize profitability and maintain sustainable production practices. Advanced techniques in managing nutrients, such as PA and biofertilizers, can save costs and increase profitability by optimizing the use of resources. Nevertheless, traditional methods also necessitate thorough economic evaluation, considering variables such as labor expenses and fertilizations expenditures. This chapter analyzes the financial implications of advanced and traditional nutrient management strategies in agriculture, focusing on their impact on farm profitability and sustainability. This is an avenue toward reducing expenses and enhancing profitability. Modern agricultural systems have increasingly adopted advanced nutrient management strategies such as PA and biofertilizers. These strategies have gained prominence because they have the potential to improve efficiency and productivity while also reducing input costs. According to economic analysis, these innovative approaches provide opportunities for reducing costs and increasing profitability compared to traditional methods.

PA allows farmers to maximize resource utilization by accurately directing inputs such as fertilizers, pesticides, and water to meet the specific needs of crops. Farmers can customize their management practices to suit the specific conditions of their fields by utilizing technologies such as GPS, GIS, and remote sensing, thereby reducing inefficiencies and optimizing crop production. The economic advantages of PA are apparent in the decreased expenses for inputs, diminished labor demands, and augmented crop yields, resulting in enhanced profitability for farmers. Similarly, biofertilizers provide a cost-efficient substitute for synthetic fertilizers by introducing beneficial microorganisms, thereby improving soil fertility and stimulating plant growth. Biofertilizers enhance nutrient accessibility and absorption by leveraging the mutualistic interactions between plants and microorganisms, diminishing reliance on chemical fertilizers and their accompanying expenses.

Economic research has shown that using biofertilizers can lead to substantial financial savings while preserving or even augmenting crop productivity, thereby enhancing the profitability of agricultural operations. Furthermore, implementing sophisticated nutrient management techniques is crucial in promoting sustainable agriculture by reducing environmental harm and conserving resources. PA minimizes nutrient runoff and leaching, safeguards water quality and mitigates pollution. Simultaneously, biofertilizers foster soil health and biodiversity, decreasing dependence on synthetic inputs and bolstering ecosystem resilience (Akbarian et al., 2022, 2023; Das et al., 2023; Dias et al., 2020; Kaab et al., 2019b; Modi et al., 2023; Modi et al., 2023; de Moraes Barbosa et al., 2021; Pongpat et al., 2023; Shendryk et al., 2021; Singh et al., 2007; Viaud et al., 2023; Wu et al., 2023).

The environmental advantages result in farmers experiencing long-term economic benefits due to decreased expenses related to regulatory compliance and enhanced opportunities to sell sustainably grown crops.

7.6.4.1 THE SIGNIFICANCE OF THOROUGH ECONOMIC EVALUATION

Although advanced nutrient management strategies show potential for economic advantages, conventional approaches are still widely used in agricultural systems. Therefore, a thorough economic evaluation is necessary to ensure profitability and sustainability. Conventional practices like applying fertilizers uniformly and irrigating uniformly may result in increased costs and labor requirements. As a result, farmers need to assess their economic viability in different production scenarios carefully. Farmers frequently utilize fertilizers according to standardized guidelines or historical methods in traditional nutrient management without considering the specific variations in soil fertility or crop needs within individual fields. This method can result in excessive use of fertilizers, leading to the runoff of nutrients, degradation of soil, and higher production expenses. To reduce these risks, farmers must perform comprehensive economic evaluations, considering fertilizer costs, labor expenses, crop prices, and potential yield, to ascertain the most effective nutrient management strategy for their operations. Labor costs are a major part of production expenses in traditional farming, especially for tasks that require manual work like planting, harvesting, and maintaining the fields.

Economic analysis is crucial for assessing the cost-effectiveness of labor-saving technologies and mechanization choices, which have the potential to decrease labor demands and enhance operational efficiency. Farmers can enhance cost-effectiveness and profitability by implementing labor-saving techniques while sustaining or augmenting crop yields. Furthermore, the expenses associated with fertilization are a substantial financial burden for farmers who practice conventional nutrient management. Economic evaluations aid farmers in maximizing the efficiency of fertilizer usage by considering factors such as nutrient accessibility, soil productivity, crop absorption rates, and market values. Farmers can minimize input expenses while maximizing returns on investment, thereby enhancing long-term profitability and sustainability through precise estimation of nutrient requirements and subsequent adjustment of fertilizer rates. The primary focus in agricultural nutrient management is on economic considerations, which are crucial in guiding farmers' decisions to optimize profitability and sustainability. Promising opportunities for cost savings and improved profitability can be achieved by implementing advanced nutrient management strategies, such as PA and biofertilizers (de Aquino et al., 2017; Asrol et al., 2024; Bispo et al., 2022; Leite et al., 2023; Reyes-Trujillo et al., 2021; Sefeedpari et al., 2014; Shukla et al., 2022; Zhu et al., 2023).

These strategies optimize resource utilization and minimize environmental impact. Nevertheless, traditional methods also necessitate a thorough economic evaluation to guarantee profitability in different production circumstances. Farmers can utilize comprehensive financial analyses and implement cost-effective nutrient management practices to improve agricultural systems' profitability, sustainability, and resilience.

7.6.5 ENVIRONMENTAL IMPACT

Sugarcane cultivation plays a vital role in worldwide agricultural production, substantially contributing to the production of sugar and bioenergy. Nevertheless, the implementation of intensive farming techniques can result in detrimental environmental consequences, such as the degradation of soil, contamination of water sources, and the decline of biodiversity. There is an increasing acknowledgment of the necessity to implement sustainable nutrient management practices to reduce environmental harm and maintain long-term productivity. Sugarcane farming can benefit from PA and the judicious application of biofertilizers to reduce its environmental footprint effectively. This essay examines the ecological advantages of PA and biofertilizers. The document offers practical recommendations for sugarcane farmers to implement efficient and sustainable nutrient management practices within the ratoon system, minimizing the discharge and infiltration of nutrients. PA employs sophisticated technologies such as GIS, GPS, and remote sensing to maximize resource efficiency and reduce environmental harm.

PA plays a crucial role in mitigating nutrient runoff and leaching, safeguarding water quality, and maintaining the health of ecosystems. PA allows farmers to optimize the application of fertilizers and irrigation water by accurately mapping soil properties and crop variability. This ensures that nutrient inputs are tailored to match crop requirements and soil nutrient levels, resulting in more efficient use of resources. This focused strategy reduces the use of excessive nutrients, thereby mitigating the risk of nutrient runoff into water bodies, which can result in eutrophication and the proliferation of harmful algal blooms. Moreover, PA enables variable rate application methods, which involve modifying fertilizer rates based on specific conditions within each field. This practice effectively minimizes nutrient losses and mitigates environmental harm. Moreover, PA enables the immediate monitoring of crop health and growth parameters, allowing prompt interventions to tackle pest infestations, diseases, and water stress. Farmers can minimize the necessity for chemical inputs, such as pesticides and herbicides, thereby decreasing environmental pollution and fostering the resilience of ecosystems through the optimization of agricultural practices.

7.6.5.1 PROMOTING ECOLOGICAL RESPONSIBILITY

Biofertilizers provide a viable and environmentally friendly substitute for synthetic fertilizers. They exploit the advantageous symbiotic relationships between plants and microbes to optimize nutrient accessibility and absorption. In contrast to chemical fertilizers, biofertilizers do not leach into water bodies and pollute groundwater. They enhance soil health and fertility by decomposing organic matter and cycling nutrients. An important ecological advantage of biofertilizers is their capacity to enhance soil composition and moisture retention, thereby mitigating soil erosion and runoff. Biofertilizers improve soil aggregation and porosity, increasing water infiltration rates and decreasing surface runoff. This helps to prevent soil erosion and maintain soil fertility. In addition, biofertilizers aid in the process of carbon sequestration in the soil, thereby helping to mitigate climate change by decreasing the release of greenhouse gases. Furthermore, using biofertilizers fosters environmental stewardship by decreasing the dependence on artificial inputs and mitigating chemical contamination. Contrary to chemical fertilizers, which can build up in the soil and water systems over time, biofertilizers can be broken down by natural processes, posing a negligible threat to ecosystems and human well-being.

Farmers can actively advance sustainable agriculture and protect the environment for future generations by integrating biofertilizers into nutrient management practices. Farmers can perform periodic soil analysis to evaluate the fertility status and identify nutrient deficiencies in sugarcane plantations. They can utilize the findings to modify fertilizer application rates and customize nutrient management practices according to specific field conditions (Akbarian et al., 2022, 2023; Das et al., 2023; Dias et al., 2020; Kaab et al., 2019b; Modi et al., 2023; de Moraes Barbosa et al., 2021; Pongpat et al., 2023; Shendryk et al., 2021; Singh et al., 2007; Viaud et al., 2023; Wu et al., 2023). Farmers can utilize PA methods, such as variable rate application and soil moisture monitoring, to optimize fertilizer utilization and reduce nutrient losses. They can administer fertilizers at the appropriate timing and in the correct dosage to fulfil crop needs and minimize ecological repercussions. They can incorporate organic amendments, such as compost or manure, and chemical fertilizers to enhance soil health and fertility. Farmers can utilize biofertilizers, containing advantageous microorganisms, to augment nutrient accessibility and foster sustainable soil management practices. To enhance soil structure, mitigate erosion, and optimize nutrient cycling, it is advisable to integrate cover crops and crop rotation into the ratoon system. One effective approach is introducing nitrogen-fixing cover crops, replenishing soil nitrogen levels and minimizing reliance on synthetic fertilizers.

To minimize soil disruption and erosion, it is advisable to implement conservation tillage methods, such as minimum tillage or no-till farming. Ensure the soil is protected and enhance water infiltration by covering it with crop residues or cover crops. Implementing effective and environmentally friendly methods of managing nutrients is crucial for reducing the negative environmental effects and ensuring the long-term viability of sugarcane farming. PA methods, such as variable rate application and soil moisture monitoring, can effectively mitigate nutrient runoff and leaching, safeguard water quality, and promote ecosystem health.

Cautiously utilizing biofertilizers provides a sustainable substitute for synthetic fertilizers, enhancing soil health and fertility while reducing chemical contamination. By adopting feasible recommendations for sustainable nutrient management in the ratoon system, sugarcane farmers can improve their environmental stewardship and contribute to a more sustainable agricultural future.

7.6.6 SOIL TESTING AND ANALYSIS

The successful cultivation of sugarcane is highly dependent on the quality and richness of the soil, as it directly affects the plant's growth and productivity. Soil testing and analysis are crucial for farmers as they offer vital information about nutrient levels, pH balance, and soil health. Farmers can customize their nutrient management strategies by implementing regular soil tests to cater to sugarcane crops' precise requirements, guaranteeing optimal growth conditions and maximizing crop yields. This essay explores the significance of soil testing and analysis in sugarcane farming, emphasizing its role in tailored nutrient management and promoting sustainable agricultural practices. Soil testing is essential for optimizing nutrient management in sugarcane farming. Soil tests assess important soil parameters, including nutrient levels, pH, and organic matter content, to provide valuable information about the soil's fertility and ability to support crop growth. Comprehending these factors enables farmers to make well-informed decisions regarding fertilizer application rates, timing, and amendments, ensuring that sugarcane crops receive the essential nutrients for optimal development. Regular soil testing is crucial because soil fertility is constantly changing.

Soil characteristics exhibit substantial variation over time and in different field locations, influenced by weather conditions, cropping history, and management practices. Periodic soil testing allows farmers to track variations in soil nutrient levels and adapt their nutrient management strategies accordingly, thus averting nutrient deficiencies or surpluses that may hinder crop growth and yield. Soil testing is crucial in tailoring nutrient management to specific needs, which is one of its main advantages. Through the examination of soil test results, farmers can create customized fertilizations strategies that address the precise nutrient needs of sugarcane crops during various stages of growth. By employing a focused strategy, nutrients are administered with optimal efficiency, thereby reducing waste and environmental consequences while simultaneously maximizing the absorption of nutrients by plants.

Soil testing offers valuable information about the presence of vital nutrients like nitrogen (N), phosphorus (P), potassium (K), and micronutrients. This enables farmers to make precise adjustments to fertilizer compositions and application rates. For instance, when soil test results reveal a lack of nitrogen, farmers can enhance nitrogen levels by administering nitrogen-rich fertilizers or organic amendments. Likewise, focused fertilizations techniques can effectively tackle phosphorus and potassium insufficiencies, guaranteeing a well-balanced nutrient supply for the most favorable crop development. In addition, soil testing assists farmers in regulating soil pH, which is crucial for nutrient accessibility and plant absorption (Alvarez-Campos et al., 2018; Asrol et al., 2024; Bispo et al., 2022; Dlamini and Zhou 2022; Fattori Junior et al., 2022; Han et al., 2022; Oliveira et al., 2022; Park et al., 2010; Sefeedpari et al., 2014; Wilson et al., 2022; Zhu et al., 2023).

The pH of soil impacts the ability of essential nutrients to dissolve and the functioning of soil microorganisms, which, in turn, affects the process of nutrient cycling and the uptake of nutrients by plants. Farmers can regulate soil acidity or alkalinity by applying lime or other amendments to monitor pH levels based on regular soil testing. This practice ensures the creation of ideal pH conditions for sugarcane growth. Soil testing improves nutrient management and encourages sustainable agricultural practices by reducing environmental impact and enhancing resource efficiency. By adhering to soil test recommendations and applying fertilizers accordingly, farmers can prevent excessive nutrient application, thus minimizing the likelihood of nutrient runoff and leaching into water bodies. This contributes to the preservation of water quality and the health of ecosystems by reducing environmental pollution and encouraging the practice of sustainable land management. Furthermore, farmers can improve soil health and productivity in the long run by carefully managing soil fertility through precise nutrient management. Optimal nutrient levels in healthy soils promote vigorous root growth, enhance the ability to absorb and retain water, and improve the cycling of nutrients and the activity of microorganisms.

The soil health benefits enhance the sustainability of crop production by decreasing reliance on external inputs and enhancing resistance to environmental pressures like drought or disease. Soil testing and analysis are crucial methods for tailoring nutrient management in sugarcane farming, offering vital information for enhancing soil fertility and encouraging sustainable agricultural practices. Farmers can create customized fertilizations strategies that effectively address sugarcane crops' precise nutrient needs through regular soil tests and careful analysis of the findings, promoting ideal growth and maximizing yields. In addition, soil testing plays a role in environmental stewardship by reducing nutrient losses and environmental pollution and enhancing long-term soil health and productivity. Sugarcane farmers can improve their agricultural systems' sustainability, profitability, and resilience by including soil testing in their nutrient management practices.

7.6.7 PA ADOPTION

Given the increasing difficulties in modern agriculture, it has become crucial for farmers to embrace PA technologies. These technologies are essential for farmers who want to efficiently utilize resources, reduce environmental harm, and maximize crop yields. GPS-guided systems and VRT are notable technologies that can potentially transform farming practices significantly. This essay promotes the exploration and adoption of PA technologies, specifically highlighting the advantages of GPS-guided systems and VRT in improving the accuracy of nutrient application, minimizing waste, and enhancing overall efficiency in resource utilization. Utilizing PA technologies, such as GPS-guided systems and VRT, is essential for contemporary farmers seeking to enhance efficiency and sustainability in their operations. These technologies provide unparalleled accuracy and authority in farming practices, allowing farmers to customize their management strategies to the unique requirements of each field and crop. Farmers can discover fresh avenues for enhancing resource utilization, minimizing input expenses, and boosting profitability by exploring and adopting PA technologies. The rationale for embracing PA technologies lies in their capacity to maximize nutrient application precision, minimize wastage, and enhance overall efficiency in resource utilization.

GPS-guided systems employ satellite technology to accurately monitor the position and motion of agricultural machinery in the field. By utilizing this technology, farmers can uphold consistent spacing and alignment when carrying out planting, spraying, and harvesting activities.

This helps reduce unnecessary overlaps or gaps, ensuring that inputs such as fertilizers, pesticides, and water are evenly distributed. VRT enhances PA by enabling farmers to modify input application rates according to field-specific conditions and crop needs. VRT combines data on soil fertility, nutrient levels, and crop health with GPS information to allow farmers to generate personalized prescription maps. These maps guide the application of inputs at different rates across the field with precision. By employing a focused strategy, resource utilization is optimized, input expenses are reduced, and the ecological footprint is minimized, leading to the adoption of more sustainable and efficient agricultural methods (Akbarian et al., 2022, 2023; Dias et al., 2020; Kaab et al., 2019b; Misra et al., 2023; Modi et al., 2023; de Moraes Barbosa et al., 2021; Otto et al., 2020; Sanches et al., 2023; Shendryk et al., 2021; Singh et al., 2007; Viaud et al., 2023; Wu et al., 2023).

Furthermore, PA technologies enhance data-based decision-making and immediate crop performance monitoring, allowing farmers to promptly detect and resolve problems like nutrient deficiencies, pest infestations, and water stress. By collecting and analyzing soil and crop conditions data, farmers can enhance their management practices, make well-informed decisions regarding inputs and resources, and maximize yields while minimizing risks and uncertainties. Multiple case studies and success stories demonstrate PA technologies' significant and positive effects on farm productivity, profitability, and sustainability. For instance, a study carried out by researchers at a vast sugarcane plantation in Brazil discovered that the implementation of GPS-guided systems and VRT resulted in substantial enhancements in crop yield and financial gains. By implementing the precise application of fertilizers and irrigation water tailored to the specific conditions of each field, the farm has the potential to decrease input expenses, enhance crop yields, and optimize resource utilization, thereby yielding significant economic and environmental advantages.

Similarly, a study on a wheat farm in the United States showcased the efficacy of PA technologies in enhancing nitrogen management and mitigating environmental harm. The farm effectively reduced nitrogen losses through leaching and denitrification by adopting a variable rate nitrogen application strategy that relies on soil test results and yield potential maps. This approach maintained or enhanced yields, improved farm profitability, decreased nitrogen runoff, and reduced greenhouse gas emissions. As a result, the farm made significant contributions to environmental sustainability. Adopting PA technologies, such as GPS-guided systems and VRT, can greatly transform modern farming practices and advance sustainability in agriculture. PA technologies allow farmers to achieve higher yields, lower input costs, and greater profitability by improving nutrient application accuracy, reducing waste, and enhancing overall resource utilization efficiency. Additionally, these technologies help to minimize environmental impact and promote long-term sustainability. Farmers can discover fresh prospects for creativity and achievement in response to changing agricultural difficulties by examining and applying PA technologies.

7.6.8 BALANCED FERTILIZATION PRACTICES

Optimal growth and productivity in sustainable agriculture rely heavily on implementing balanced fertilizations practices, guaranteeing that crops receive the necessary nutrients in appropriate quantities and proportions. By implementing a judicious application of nutrients tailored to the specific requirements of the crop while refraining from excessive utilization of artificial fertilizers, farmers can enhance soil well-being, foster sustainable productivity, and mitigate environmental harm. This chapter promotes adopting balanced fertilizations practices, highlighting their significance in attaining enduring sustainability in agriculture. The rationale behind advocating for balanced fertilization practices lies in the understanding that nutrient imbalances have the potential to negatively impact soil health, crop productivity, and environmental sustainability. By strategically applying nutrients in a balanced manner based on the specific requirements of the crop, farmers can enhance the absorption and utilization of nutrients, leading to healthier soils, increased crop yields, and minimize negative effects on the environment. Furthermore, by reducing the overuse of artificial fertilizers, we can effectively mitigate the negative impacts of nutrient leaching, runoff, and pollution, thereby protecting the quality of water and the health of ecosystems. The justification for implementing balanced fertilizations practices is their capacity to enhance soil health, ensure sustainable productivity, and minimize environmental repercussions.

Optimal fertilizations entails providing crops with a precise combination of vital nutrients, such as nitrogen (N), phosphorus (P), potassium (K), and micronutrients, to fulfil their nutritional needs throughout the entire growth period. This guarantees crops can obtain the necessary nutrients for optimal growth and development, resulting in increased crop yields, improved quality, and enhanced ability to withstand drought, disease, and pests. Ensuring a proper balance of nutrients is crucial for sustaining soil fertility and productivity in the long run. Overuse of specific nutrients, like nitrogen or phosphorus, can cause imbalances in nutrients and degradation of soil, leading to lower crop yields, heightened vulnerability to pests and diseases, and reduced ability to retain water and nutrients (Alvarez-Campos et al., 2018; de Aquino et al., 2017; Asrol et al., 2024; Bispo et al., 2022; Dlamini and Zhou 2022; Fattori Junior et al., 2022; Park et al., 2010; Poudyal et al., 2023; Reyes-Trujillo et al., 2021; Sefeedpari et al., 2014; Wickramasinghe et al., 2024; Wilson et al., 2022).

Through the strategic application of nutrients, farmers can prevent nutrient imbalances and enhance soil health, organic matter accumulation, and beneficial microbial activity. This, in turn, improves soil structure, fertility, and productivity. Balanced fertilization practices can lead to reduced environmental impact and enhanced sustainability in agriculture. Excessive application of synthetic fertilizers can lead to leaching of these chemicals into groundwater or runoff into surface water bodies. This can lead to nutrient pollution and eutrophication, harming aquatic ecosystems and human health. Farmers can decrease nutrient losses, alleviate environmental pollution, and protect water quality by reducing the use of synthetic fertilizers and implementing balanced fertilizations methods. This will help conserve natural resources and maintain the integrity of ecosystems. Furthermore, employing balanced fertilization techniques promotes the sustainable management of nutrients and the cycling of nutrients in agricultural ecosystems. Through the optimization of nutrient use efficiency and the reduction of nutrient losses, balanced fertilizations contribute to the conservation of limited resources, the mitigation of greenhouse gas emissions, and the promotion of sustainable agricultural intensification. By adopting this approach, farmers can increase their crop production and financial gains while reducing their ecological footprint and improving their ability to withstand the effects of climate change and other difficulties.

Multiple case studies and success stories provide evidence of the advantages of implementing balanced fertilization practices for improving soil health, crop productivity, and environmental sustainability. For instance, research on maize farms in Sub-Saharan Africa revealed that applying a well-balanced combination of nitrogen, phosphorus, and potassium fertilizers resulted in a significant increase in maize yields, up to

30%, compared to plots that were not fertilized. In addition, implementing a balanced fertilizations strategy has been shown to enhance soil fertility, optimize nutrient utilization, and minimize nutrient losses. This, in turn, contributes to improved farm profitability and sustainability (Akbarian et al., 2022; Boroun et al., 2023; Colmanetti et al., 2024; Dias et al., 2020; Modi et al., 2023; de Moraes Barbosa et al., 2021; Oliveira et al., 2023; de Oliveira Filho et al., 2021; Otto et al., 2020; Sales et al., 2022; Sanches et al., 2023; Shendryk et al., 2021; Soulé et al., 2024; Wu et al., 2023).

Similarly, research conducted on rice farms in Asia demonstrated that implementing balanced fertilization practices, which involve applying nitrogen, phosphorus, and potassium based on the specific needs of the crops, resulted in enhanced rice yields and improved efficiency in nutrient utilization, all while minimizing the negative effects on the environment. Through optimizing nutrient management and minimizing nutrient losses via balanced fertilizations, farmers successfully attained increased yields, decreased input costs, and reduced greenhouse gas emissions, making significant contributions to the sustainable production of rice and environmental stewardship. Adopting balanced fertilizations practices is crucial for enhancing soil health, ensuring sustainable productivity, and reducing the environmental footprint in agriculture. By employing a judicious application of nutrients by the specific needs of crops while refraining from excessive reliance on artificial fertilizers, farmers can enhance the absorption and utilization of nutrients, enhance crop yields, and mitigate environmental contamination.

Furthermore, balanced fertilization promotes the sustainable management of nutrients, the cycling of nutrients, and the integrity of ecosystems, thereby enhancing agricultural systems' long-term sustainability and resilience. By implementing balanced fertilization practices, farmers can attain increased crop yields, reduced expenses on inputs, and enhanced profitability, all while ensuring the preservation of soil health, water quality, and environmental sustainability for future generations.

7.6.9 INTEGRATION OF BIOFERTILIZERS

Incorporating biofertilizers into nutrient management practices signifies a substantial change in contemporary agriculture, providing sustainable alternatives to synthetic fertilizers while enhancing soil health, crop productivity, and environmental sustainability. Nitrogen-fixing bacteria and mycorrhizal fungi are essential components of biofertilizers that are crucial in increasing nutrient availability, enhancing soil composition, and promoting overall health in agricultural ecosystems. This essay promotes incorporating nitrogen-fixing bacteria and mycorrhizal fungi into nutrient management practices, emphasizing their revolutionary capacity for sustainable agriculture. Incorporating biofertilizers into nutrient management stems from the requirement to shift toward more sustainable and ecologically conscious agricultural systems. Although synthetic fertilizers successfully increase crop productivity, they frequently result in nutrient imbalances, soil deterioration, and environmental contamination. In contrast, biofertilizers provide environmentally friendly and natural methods to enhance soil fertility, improve nutrient cycling, and ensure long-term sustainability in agriculture.

Nitrogen-fixing bacteria, such as Rhizobia and Azotobacter, are essential for biological nitrogen fixation. This process involves the conversion of atmospheric nitrogen into a usable form for plant growth and development. The bacteria establish symbiotic relationships with leguminous plants, such as soybeans, peas, and lucerne, by inhabiting specialized structures known as nodules on the roots of the plants. The nodules contain nitrogen-fixing bacteria that transform atmospheric nitrogen into ammonium, which the plant subsequently absorbs and utilizes as a nitrogen source. The incorporation of nitrogen-fixing bacteria into nutrient management practices provides multiple benefits for the promotion of sustainable agriculture. First and foremost, it decreases the dependence on synthetic nitrogen fertilizers, which require a significant amount of energy to manufacture and contribute to releasing greenhouse gases.

Additionally, it enhances soil fertility by restoring nitrogen levels and facilitating nutrient cycling within the agroecosystem. Furthermore, it improves the yield and ability of crops to withstand environmental challenges, such as drought or nutrient deficiencies, by maintaining a consistent nitrogen supply throughout the entire growth period. Mycorrhizal fungi establish symbiotic relationships with the roots of most plant species, aiding in the absorption of nutrients and promoting plant growth and vitality. The fungi facilitate the expansion of plant roots into the soil, thereby enhancing the surface area for nutrient uptake and augmenting the plant's capacity to acquire water, phosphorus, and other vital nutrients. The plant reciprocates by supplying carbohydrates to the fungi, nourishing their growth and metabolic processes. Incorporating mycorrhizal fungi into nutrient management practices provides numerous advantages for soil health, plant growth, and ecosystem sustainability. First, it improves the arrangement and bonding of soil particles, thereby enhancing the ability of water to penetrate and be retained in the soil while reducing soil erosion and compression.

Additionally, it enhances the process of nutrient cycling and improves the efficiency of nutrient utilization by aiding in the absorption of phosphorus, potassium, and micronutrients by plants. Furthermore, it improves the productivity of crops and ensures a consistent yield by increasing the availability of nutrients and enhancing the ability of plants to withstand environmental pressures. Incorporating nitrogen-fixing bacteria and mycorrhizal fungi into nutrient management practices can revolutionize agricultural systems, fostering sustainability, resilience, and the overall health of ecosystems. By utilizing the potential of these advantageous microorganisms, farmers can diminish their reliance on artificial fertilizers, enhance soil fertility and composition, and augment crop productivity and profitability. In addition, biofertilizers play a role in preserving the environment by reducing the amount of nutrients that are washed away and causing pollution. They also support a variety of living organisms and help counteract climate change by storing carbon in the soil. To summarize, incorporating nitrogen-fixing bacteria and mycorrhizal fungi into nutrient management practices presents a significant opportunity for advancing sustainable agriculture.

By utilizing the innate capabilities of these advantageous microorganisms, farmers can enhance soil fertility, optimize nutrient cycling, and foster long-term sustainability in agricultural ecosystems. Furthermore, the incorporation of biofertilizers aids in the preservation of the environment by diminishing the reliance on artificial fertilizers, mitigating nutrient contamination, and improving soil quality and biodiversity. Farmers can develop robust, efficient, and ecologically sound agricultural systems to support current and future generations by adopting biofertilizers as essential elements of nutrient management strategies (Dias et al., 2020; Kaab et al., 2019b; Misra et al., 2023; Modi et al., 2023; Otto et al., 2020; Sanches et al., 2023; Shendryk et al., 2021; Soulé et al., 2024; Viaud et al., 2023).

7.6.10 TAILORED APPLICATION TIMING

Sugarcane, a crucial sugar and bioenergy production crop, necessitates meticulous nutrient management at every stage of its growth to attain maximum yields and quality. Comprehending the specific nutritional needs during different stages of growth is essential for adjusting the application timing. To optimize nutrient absorption and enhance crop growth, farmers can strategically time nutrient applications to coincide with crucial stages of plant development. This chapter examines the significance of customized timing for applying nutrients to enhance sugarcane growth and the strategies for efficiently managing nutrient applications. Sugarcane progresses through various growth stages, each of which requires specific nutrients. The stages encompassed in this process are germination, tillering, vigorous growth, maturity, and ripening. During germination, the young plant depends on the stored resources within the seed until it develops roots that can absorb nutrients. Tillering is the stage at which the plant begins to form stalks, and it necessitates sufficient nitrogen, phosphorus, and potassium (NPK) for strong shoot growth. Rapid vegetative growth is a defining feature of grand growth, which requires an increased nitrogen supply to support the development of leaves and stalks. As the plant matures, its nutrient requirements change to prioritize reproductive development and the accumulation of sugars. Customizing nutrient applications to these specific growth stages promotes optimal sugarcane growth.

By customizing the timing of application, nutrients are made accessible to the plant during its most crucial stages, thereby optimizing the efficiency of nutrient absorption and minimizing any potential nutrient wastage. Optimizing the timing of nutrient application improves the ability of crops to withstand stress, increases their potential for producing higher yields, and enhances the sugar content. Furthermore, by synchronizing nutrient applications with growth phases, the negative effects on the environment, such as nutrient leaching and runoff, can be minimized, thus fostering sustainable practices in sugarcane production. To achieve economic and environmental sustainability, farmers can enhance resource utilization by comprehending the nutrient requirements of sugarcane at every stage of its growth cycle. Various approaches can be utilized to enhance nutrient management practices and synchronize the timing of nutrient application with the different stages of sugarcane growth (de Aquino et al., 2017; Asrol et al., 2024; Bispo et al., 2022; Fattori Junior et al., 2022; Han et al., 2022; Leite et al., 2023; Sefeedpari et al., 2014; Shukla et al., 2022; Wilson et al., 2022).

Soil testing and plant tissue analysis offer valuable information about nutrient availability and plant absorption, which helps make fertilizer recommendations. By employing slow-release fertilizers or controlled-release formulations, a consistent and prolonged provision of nutrients is achieved, thereby minimizing the potential for nutrient leaching and improving nutrient utilization efficiency. Foliar feeding is a method of providing additional nutrients to plants through their leaves. This can be done during important growth periods to supply nutrients and correct deficiencies quickly. Furthermore, implementing PA technologies, such as variable rate application, allows for the precise delivery of nutrients based on the spatial variability present within the field. This approach optimizes the allocation of resources and enhances crop productivity. Nitrogen is a crucial nutrient that is necessary for the growth of sugarcane and the accumulation of sucrose. Optimizing nitrogen application timing is crucial to prevent excessive vegetative growth that may reduce sugar yield. Dividing nitrogen applications into multiple doses throughout the growing season ensures that nutrients are available when plants need them, encouraging balanced vegetative and reproductive growth.

Applying nitrogen during the tillering and grand growth stages promotes shoot growth and the canopy's expansion. On the other hand, applying nitrogen later in the season helps accumulate sucrose during the maturity stage. Farmers can maximize the potential for high crop yield and sugar quality while reducing negative environmental effects by coordinating the timing of nitrogen applications with the different stages of sugarcane growth. Sugarcane relies on phosphorus and potassium for essential functions such as root growth, stalk durability, and overall plant well-being. Phosphorus enhances the initial growth of roots and the formation of tillers, thereby supporting robust plant development. To achieve optimal root development and improve nutrient uptake efficiency, it is recommended to apply phosphorus before planting or during the early growth stages. Potassium plays a vital role in the growth of stalks, the absorption of water, and the ability to resist diseases. Strategically timing the application of potassium during the growth stages of plants enhances the development of strong stalks and increases the ability of crops to withstand environmental pressures (Akbarian et al., 2022; Dias et al., 2020; Hoffman et al., 2024; Kaab et al., 2019b; Modi et al., 2023; de Moraes Barbosa et al., 2021; Sanches et al., 2023; Shendryk et al., 2021; Singh et al., 2007; Viaud et al., 2023).

To optimize sugarcane growth and yield, farmers can improve the vigor of the crop and maximize its potential by effectively managing phosphorus and potassium levels during key growth stages. The sugarcane plant relies on iron, zinc, manganese, and copper for crucial physiological processes. Micronutrient deficiencies, even in small amounts, can have a substantial impact on the growth and productivity of crops. Applying micronutrients directly to the leaves at crucial growth stages can rectify deficiencies and guarantee ideal plant nourishment. Additionally, including micronutrients in soil amendments or fertigation programs can effectively prevent nutrient imbalances and facilitate a harmonious uptake of nutrients. To enhance the health and productivity of sugarcane, farmers can optimize it by considering both the macronutrients and micronutrients needed. Optimizing nutrient uptake and promoting robust sugarcane growth throughout the crop cycle requires precise application timing. By synchronizing nutrient applications with crucial growth stages, farmers can optimize crop yield, improve sugar quality, and reduce negative effects on the environment. By incorporating soil testing, PA technologies, and balanced fertilization practices, it becomes possible to effectively manage nutrients in a way specifically tailored to meet the unique needs of sugarcane. To foster a prosperous sugarcane industry, farmers can effectively boost sugarcane productivity by sustainably implementing these strategies.

7.6.11 ECONOMIC EVALUATION

Effectively controlling nutrients is of utmost significance in sustainable agriculture. Implementing nutrient management practices is crucial for achieving maximum crop growth and has substantial implications for both economic profitability and environmental sustainability. This thorough cost–benefit analysis examines the economic assessment of agricultural nutrient management practices. By carefully examining the expenses and advantages of different nutrient management methods, our objective is to offer valuable knowledge for making informed economic choices essential for sustainable farming practices and financial security. Nitrogen, phosphorus, and potassium are indispensable for the growth and productivity of crops. Nevertheless, inadequate handling of these nutrients can result in economic and environmental difficulties. Overapplication of fertilizer can cause nutrients to wash away, causing water pollution and harmful impacts on ecosystems. Furthermore, using ineffective strategies for managing nutrients can result in avoidable expenses for farmers while reducing agricultural productivity over time.

Therefore, it is crucial to implement efficient nutrient management strategies to maximize resource utilization and guarantee economic viability in agriculture. Activities related to nutrient management practices entail diverse tasks, each accompanied by its specific expenses. The main costs associated with nutrient management involve the purchase of fertilizers, which make up a substantial portion of the expenses for farmers. The prices of fertilizers fluctuate based on the specific type and nutrient composition. In addition, the costs associated with transportation and application also contribute to the total expenditure. Soil testing is crucial for accurately determining the nutrient needs of crops. Although initial expenses are associated with soil testing, farmers can customize their nutrient management strategies based on the unique characteristics of their soil. This allows them to maximize the efficiency of fertilizer usage and minimize unnecessary waste. Adopting sophisticated nutrient management techniques like PA or controlled-release fertilizers may necessitate technological, equipment, and infrastructure investments. While the initial investments may be significant, they frequently result in long-term cost reductions and productivity improvements. The expenses related to nutrient management activities, such as soil sampling, fertilizer application, and equipment maintenance, substantially impact the overall costs due to labor expenses. Inadequate management of nutrients, such as soil degradation, water pollution, and greenhouse gas emissions, can harm the environment.

Although farmers may not bear these costs directly, they can lead to regulatory measures, fines, or damage to their reputation, which can indirectly affect the profitability of their farms. Implementing efficient nutrient management strategies provides numerous advantages that improve farms' financial viability and ecological sustainability (Alvarez-Campos et al., 2018; de Aquino et al., 2017; Asrol et al., 2024; Bispo et al., 2022; Dlamini and Zhou 2022; Fattori Junior et al., 2022; Han et al., 2022; Sefeedpari et al., 2014; Wilson et al., 2022; Zhu et al., 2023). Effective nutrient management guarantees crops receive the essential nutrients in optimal amounts and at the correct timings, thereby maximizing their yield potential. Farmers can enhance the overall effectiveness of fertilizer usage by aligning nutrient inputs with crop needs, which reduces nutrient losses caused by leaching, volatilization, or runoff. Implementing precision nutrient management techniques enables farmers to decrease input expenses by preventing excessive application of fertilizers and minimizing inefficiency. Implementing sustainable nutrient management practices helps to maintain soil health, preserve water quality, and protect biodiversity, ensuring the long-term availability of natural resources for future generations.

Following proper nutrient management practices guarantees adherence to environmental regulations and standards, thereby minimizing the likelihood of penalties or sanctions. A thorough cost–benefit analysis is crucial for assessing the economic feasibility of nutrient management practices. The study aims to measure the monetary value of the expenses and advantages of various nutrient management methods and then compare them to determine their relative profitability. Essential stages in performing a cost–benefit analysis include listing and categorizing all pertinent expenditures and advantages linked to each nutrient management technique, encompassing explicit costs, workforce involvement, ecological repercussions, and enhancements in productivity.

Allocate monetary values to the identified costs and benefits whenever feasible. This may entail approximating input expenses by considering prevailing market prices, evaluating enhancements in yield by utilizing crop price projections, and assessing environmental costs through economic valuation techniques. To incorporate the concept of the time value of money, one must calculate the present value of future costs and benefits by applying a discounting method. This facilitates an equitable comparison of cash flows that transpire at distinct time intervals and permits a pragmatic evaluation of long-term profitability. Perform a sensitivity analysis to evaluate how uncertainties or variations in important parameters affect the cost–benefit analysis results. This facilitates the identification of crucial factors that impact the economic viability of nutrient management practices and enables the formulation of strategies to minimize risks. Utilize the cost–benefit analysis findings to make well-informed choices regarding implementing or continuing particular nutrient management practices (Akbarian et al., 2022, 2023; Dias et al., 2020; Kaab et al., 2019b; Modi et al., 2023; de Moraes Barbosa et al., 2021; Otto et al., 2020; Pongpat et al., 2023; Sanches et al., 2023; Shendryk et al., 2021; Soulé et al., 2024; Viaud et al., 2023; Wu et al., 2023).

When making decisions, it is important to consider factors such as profitability, environmental sustainability, risk tolerance, and regulatory compliance. An economic assessment of nutrient management practices is crucial for advancing sustainable agriculture and guaranteeing the financial sustainability of farming operations. Farmers can make well-informed choices that improve productivity, profitability, and environmental responsibility by meticulously evaluating the expenses and advantages of various nutrient management methods. Furthermore, policymakers and stakeholders can utilize the findings from cost–benefit analyses to develop incentives, policies, and extension programs that promote adoption of optimal nutrient management practices throughout agricultural areas. Incorporating economic factors into nutrient management decisions is crucial for successfully attaining farm productivity and environmental sustainability objectives, particularly in light of the changing challenges of climate change, limited resources, and regulatory demands.

7.6.12 ADAPTATION OF PA

The utilization of advanced technologies and data-driven methods in PA has significantly transformed contemporary farming practices. Nevertheless, it is imperative to fully comprehend the potential of adapting PA techniques to align with specific local conditions and farmers' knowledge. This integration strategy improves the functionality and efficiency of PA, guaranteeing its broad acceptance and implementation among farmers. This discussion examines the justification for implementing PA in local settings and the methods for customizing techniques to align with specific conditions and knowledge. Agricultural landscapes exhibit substantial variations across different regions due to disparities in soil composition, climatic patterns, topographical features, and cultivation methods. What is effective in one domain may not be appropriate in another. Therefore, it is crucial to customize PA methods to suit specific local circumstances to accommodate these differences and maximize their efficiency. Farmers possess invaluable traditional knowledge and expertise acquired over generations of farming in specific localities (Alvarez-Campos et al., 2018; de Aquino et al., 2017; Bispo et al., 2022; Dlamini and Zhou 2022; Fattori Junior et al., 2022; Han et al., 2022; Leite et al., 2023; Nejad et al., 2023; Reyes-Trujillo et al., 2021; Wickramasinghe et al., 2024; Zhu et al., 2023).

By incorporating modern technology, integrating PA with indigenous knowledge acknowledges and respects farmers' expertise and enhances its relevance and acceptance among them. Numerous agricultural communities encounter limitations in terms of resources, including restricted technological advancements, financial resources, and access to infrastructure. Customizing PA to suit specific local conditions necessitates considering these limitations and formulating pragmatic solutions that are viable and economical for farmers to adopt. Sustainable agricultural practices are designed to suit each region's distinct environmental conditions, aiming to minimize detrimental effects on ecosystems and natural resources. Applying PA methods to specific local conditions enables the efficient utilization of resources, minimizes environmental impact, and fosters long-term sustainability. For farmers to widely adopt PA investments, they must be economically feasible. Customizing methods to suit specific circumstances guarantees that the advantages surpass the expenses, thereby optimizing the gains from investments and improving the overall profitability of agricultural activities. To enhance data accuracy, incorporate localized data using on-farm sensors, remote sensing technologies, and participatory approaches that engage farmers.

Localized data offers precise information regarding soil, climate, and crop conditions, facilitating focused interventions. Create customized decision support systems (DSS) to cater to local farmers' unique requirements and preferences. These DSS should consider crop preferences, farming practices, resource availability, and risk tolerance levels to offer personalized recommendations for optimizing inputs and maximizing yields. Employ adaptive management strategies to enable immediate modifications in response to evolving local circumstances. This may entail employing VRT to modify inputs such as irrigation, fertilizers, and pesticides based on the spatial variability of the field. Deliver training and extension services to farmers to augment their comprehension of PA techniques and their implementation in specific local circumstances. This encompasses hands-on demonstrations, interactive workshops, and information-sharing platforms that empower farmers to make well-informed decisions (Akbarian et al., 2023; Dias et al., 2020; Kaab et al., 2019b; Modi et al., 2023; de Moraes Barbosa et al., 2021; Sanches et al., 2023; Shendryk et al., 2021; Singh et al., 2007; Wu et al., 2023).

Promote cooperation among farmers, researchers, extension agents, agribusinesses, and policymakers to develop solutions that tackle specific local issues and utilize their combined knowledge and skills. Collaborative partnerships guarantee PA initiatives are tailored to the specific context, widely accepted by society, and economically viable in the long term. In areas with restricted availability of resources and infrastructure, programs such as the "Digital Farmer Field School" in Kenya employ mobile technology and participatory methods to provide customized agronomic guidance to small-scale farmers. These initiatives enhance productivity, resilience, and food security by incorporating PA into conventional farming methods.

Precision irrigation technologies, such as drip irrigation and soil moisture sensors, are designed to enhance water use efficiency and reduce water wastage in arid and semiarid regions that experience water scarcity. These technologies are combined with weather forecasting models and crop water requirement algorithms to offer customized irrigation scheduling recommendations to farmers. PA techniques are customized to cater to the specific needs of high-value crops, including fruits, vegetables, and ornamentals, in horticultural and specialty crop production. This encompasses implementing precise planting techniques, customized nutrient management practices, and vigilant pest monitoring strategies specifically designed to meet the distinct requirements of every crop species and cultivar.

Customizing PA methods to suit specific local conditions and knowledge is crucial for optimizing their practicality, efficiency, and adoption by farmers. Combining PA with indigenous knowledge, localized data collection, customized decision support systems, adaptive management strategies, and collaborative partnerships makes it feasible to customize PA initiatives to tackle various agricultural landscapes' varied requirements and difficulties. In conclusion, this integration strategy guarantees that PA plays a role in sustainable intensification, environmental stewardship, and inclusive rural development, thereby promoting a more resilient and equitable agricultural sector in the future.

7.7 CONCLUSION

The comparative analysis of traditional and advanced nutrient management strategies in the sugarcane ratoon system has yielded valuable insights into optimizing yield, promoting sustainable farming practices, and integrating technological advancements with conventional wisdom. As we wrap up this investigation, several essential considerations arise, influencing the sugarcane cultivation and nutrient management approach. The distinction between traditional and advanced nutrient management is not a binary decision of correctness but rather a pursuit of equilibrium. The resilience and adaptability of conventional practices, deeply ingrained in agricultural heritage, are evident. Conversely, advanced strategies, characterized by accuracy and effectiveness, demonstrate the capacity for significant and impactful alterations. The crucial factor lies in achieving equilibrium, recognizing the merits of each approach, and seamlessly incorporating them to achieve the best possible results.

PA is being adopted as a catalyst for optimizing the yield of sugarcane ratoon. The utilization of GPS-guided systems, VRT, and data-driven decision-making has showcased unparalleled accuracy in the application of nutrients. This reduces the amount of wasted resources and improves overall efficiency, giving farmers a solid tool to navigate the complexities of modern agriculture. By facilitating mutualistic associations between crops and advantageous microorganisms, biofertilizers offer a sustainable substitute for synthetic fertilizers. Biofertilizers are crucial in sustainable sugarcane cultivation by promoting soil health, increasing nutrient availability, and potentially reducing environmental impact. The comparative analysis has also revealed inherent challenges in conventional and innovative nutrient management. Traditional practices encounter challenges associated with imbalances in nutrients, degradation of soil, and inconsistencies in organic sources. On the other hand, more complex strategies present difficulties related to the adoption of technology and interpretation of data, as well as the potential danger of relying too heavily on them. Acknowledging these difficulties creates opportunities for conducting research, developing new ideas, and implementing flexible strategies to overcome barriers and take advantage of favorable circumstances.

The practical recommendations for sugarcane farmers derived from this comparative analysis involve adopting a comprehensive and all-encompassing approach. Instructions include conducting soil testing, adopting PA techniques, implementing balanced fertilizations practices, and integrating biofertilizers. Moreover, performing an economic assessment of farming practices and investing in training programs provide farmers with the knowledge and tools to make well-informed decisions that align with their agricultural and financial goals. The consistent pursuit of comprehensive solutions involves incorporating conventional and innovative methods for managing nutrients. An integrated approach is built upon preserving local wisdom, adapting PA to local contexts, and promoting collaboration between traditional farmers and researchers. This synergy represents more than just a mere coexistence but a seamless integration that optimizes the advantages of both paradigms. Enhancing sugarcane ratoon yield does not end with this comparative analysis; it is a starting point for future research efforts.

Further exploration is necessary in essential areas such as longitudinal studies, in-depth soil microbial analyses, climate-responsive nutrient management, and investigations into farmers' adoption behavior. Integrating agronomy, economics, and social sciences will lead to precise and tailored solutions. The comparative analysis has revealed various potential options for sugarcane farmers aiming to maximize the yield of ratoon crops. This is not a process of completely disregarding traditions or unquestioningly adopting technology; instead, it is a detailed examination of how tradition and innovation can combine to establish a sustainable, resilient, and productive future for sugarcane farming. As farmers navigate the ever-changing agricultural environment, equipped with knowledge, flexibility, and a dedication to responsible land management, maximizing the sugarcane ratoon yield becomes an objective and a collaborative effort toward a prosperous future in agriculture.

KEYWORDS

- **traditional**
- **nutrient**
- **sustainable**
- **productivity**
- **conventional**
- **management**
- **no poverty**
- **zero hunger**

REFERENCES

Akbarian S, Xu C, Wang W, Stephen G, Samsung L. (2022). Sugarcane yields prediction at the row level using a novel cross-validation approach to multi-year multispectral images. *Comput. Electron. Agric*. 198, 107024. https://doi.org/10.1016/j.compag.2022.107024.

Akbarian S, Jamnani MR, Xu C, Wang W, Lim S. (2023). Plot level sugarcane yield estimation by machine learning on multispectral images: A case study of Bundaberg, Australia. *Inf. Process. Agric*. 11(4), 476–487. https://doi.org/10.1016/j.inpa.2023.06.004.

Alvarez-Campos O, Lang TA., Bhadha JH, Mabry McCray J, Glaz B, Daroub, SH. (2018). Biochar and mill ash improve yields of sugarcane on a sand soil in Florida. *Agr. Ecosyst. Environ.* 253, 122–130. https://doi.org/10.1016/j.agee.2017.11.006.

Anas M, Verma KK, Haq IU, Naeem M, Li Q, Liao F, Xiang YX, Li YR. (2021a). Characterization of exotic, native and wild-type genotypes of sugarcane (*Saccharum* spp. hybrids) for internal nitrogen use efficiency under different nitrogen levels. *Sugar Tech.* 23, 1258–1267. https://doi.org/10.1007/s12355-021-01014-1.

Anas M, Verma KK, Riaz M, Qiang L, Liao F, Liu Y, Li YR. (2021b). Physio-morphological and biochemical mechanism of nitrogen use efficiency in sugarcane (*Saccharum* spp.) genotypes under different growth stages and nitrogen levels. *J. Plant Interact.* 16, 332–343.

Asrol M, Marimin, Machfud, Yani M, Rohayati. (2024). A multi-criteria model of supply chain sustainability assessment and improvement for sugarcane agroindustry. *Heliyon*. 10(7), e28259. https://doi.org/10.1016/j.heliyon.2024.e28259.

Bhatt R, Verma KK, Jamwal NS. (2023a). Can paddy-compost eliminate N fertilization for climate-smart sugarcane? *Sugar Tech.* 25, 1036–1046. https://doi.org/10.1007/s12355-023-01270-3.

Bhatt R, Verma KK, Kumar R, Sanghera GS. (2023b). Foliar application of potassium salt of active phosphorus (PSAP) mitigates insect pests and improves yield along with sugarcane quality in response to agroclimatic conditions of Punjab. *Sugar Tech.* 25(3), 660–669.

Bispo RC, Hernandez FBT, Gonçalves IZ, Neale CMU, Teixeira AHC. (2022). Remote sensing based evapotranspiration modeling for sugarcane in Brazil using a hybrid approach. *Agric. Water Manag.* 271, 107763. https://doi.org/10.1016/j.agwat.2022.107763.

Boroun M, Mohammad G, Abd Ali N, Babak B. (2023). Use of imperialist competitive algorithm for optimization of energy productivity and damage assessment in sugar industry: A case study. *Environ. Sustain. Indic*. 19, 100263. https://doi.org/10.1016/j.indic.2023.100263.

Chaudhary P, Xu M, Ahamad L, Chaudhary A, Kumar G, Adeleke BS, Verma KK, Hu D-M, Siric I, Kumar P, Popescu SM, Fayssal SA. (2023). Application of synthetic consortia for improvement of soil fertility, pollution remediation, and agricultural productivity: A review. *Agronomy*. 13, 643. https://doi.org/10.3390/agronomy13030643.

Chen GL, Zheng FJ, Lin B, Lao SB, He J, Huang Z, Zeng Y, Sun J, Verma KK. (2020). Phenolic and volatile compounds in the production of sugarcane vinegar. *ACS Omega*. 5, 30587–30595.

Chen G-L, Zheng F-J, Lin B, Yang Y-X, Fang X-C, Verma KK, Yang L-F. (2023). Vinegar: A potential source of healthy and functional food with special reference to sugarcane vinegar. *Front. Nutr*. 10, 1145862. https://doi.org/10.3389/fnut.2023.1145862.

Christina M, Jones MR, Versini A, Mézino M, Le Mézo L, Auzoux S, Soulié JC, Poser C, Gérardeaux E. (2021). Impact of climate variability and extreme rainfall events on sugarcane yield gap in a tropical Island. *Field Crops Res.* 274, 108326. https://doi.org/10.1016/j.fcr.2021.108326.

Colmanetti MAA, Cuadra SV, Lamparelli RAC, Cabral OMR, de Castro Victoria D, de Almeida Monteiro JEB, de Freitas HC, Galdos MV, Marafon AC, de Andrade Junior AS, dos Anjos e Silva SD, Buffon VB, Hernandes TAD, le Maire G. (2024). Modeling sugarcane development and growth within ECOSMOS biophysical model. *Eur. J. Agron.* 154, 127061. https://doi.org/10.1016/j.eja.2023.127061.

Das A, Kumar M, Kushwaha A, Dave R, Dakhore KK, Chaudhari K, Bhattacharya BK. (2023). Machine learning model ensemble for predicting sugarcane yield through synergy of optical and SAR remote sensing. *Remote Sens. Appl. Soc. Environ.* 30, 100962. https://doi.org/10.1016/j.rsase.2023.100962.

de Aquino GS, de Conti Medina C, da Costa DC, Shahab M, Santiago AD. (2017). Sugarcane straw management and its impact on production and development of ratoons. *Ind. Crops Prod.* 102, 58–64. https://doi.org/10.1016/j.indcrop.2017.03.018.

de Moraes Barbosa A, Zilliani RR, Tiritan CS, Souza GM, de Almeida Silva M. (2021). Energy conversion efficiency in sugarcane cultivars as a function of production environments in Brazil. *Renew. Sustain. Energy Rev.* 150, 111500. https://doi.org/10.1016/j.rser.2021.111500.

de Oliveira Filho, ASB, de Mello Prado R, Teixeira GCM, Rocha AMS, de Souza Junior JP, de Cássia Piccolo M, Rocha JR. (2021). Silicon attenuates the effects of water deficit in sugarcane by modifying physiological aspects and C:N:P stoichiometry and its use efficiency. *Agric. Water Manag.* 255, 107006. https://doi.org/10.1016/j.agwat.2021.107006.

de Oliveira IN, de Souza ZM, Bolonhezi D, Totti MCV, de Moraes MT, Lovera LH, de Souza Lima E, Esteban DAA, Oliveira CF. (2022). Tillage systems impact on soil physical attributes, sugarcane yield and root system propagated by pre-sprouted seedlings. *Soil Tillage Res.* 223, 105460. https://doi.org/10.1016/j.still.2022.105460.

Deng ZN, Huang HR, Huang CM, Verma KK, Luo HB, Jiang SL, Li YJ, Cao HQ, Wu KC, Xu L. (2023). Short-term limited water irrigation influences the photosynthetic pigments, enzymatic and non-enzymatic activities in *Saccharum spontaneum* L. at vegetative stage. *Sugar Tech.* 25, 110–118. https://doi.org/10.1007/s12355-022-01175-7.

Dengia, A, Dechassa N, Wogi L, Amsalu B. (2023). A simplified approach to satellite-based monitoring system of sugarcane plantation to manage yield decline at Wonji-Shoa Sugar Estate, Central Ethiopia. *Heliyon.* 9(8), e18982. https://doi.org/10.1016/j.heliyon.2023.e18982.

Dias, HB, Inman-Bamber G, Everingham Y, Sentelhas PC, Bermejo R, Christodoulou D. (2020). Traits for canopy development and light interception by twenty-seven Brazilian sugarcane varieties. *Field Crops Res.* 249, 107716. https://doi.org/10.1016/j.fcr.2020.107716.

Dlamini, NE, Zhou M. (2022). Soils and seasons effect on sugarcane ratoon yield. *Field Crops Res.* 284, 108588. https://doi.org/10.1016/j.fcr.2022.108588.

Farias-Ramírez, AJ, Duarte SN, Moreno-Pizani MA, de Oliveira Costa J, da Silva Barros TH, Coelho RD. (2024). Combined effect of silicon and nitrogen doses applied to planting furrows on sugar, biomass and energy water productivity of sugarcane (*Saccharum* spp.). *Agric. Water Manag.* 296, 108796. https://doi.org/10.1016/j.agwat.2024.108796.

Junior F, Martins I, dos Santos Vianna M, Marin FR. (2022). Assimilating leaf area index data into a sugarcane process-based crop model for improving yield estimation. *Eur. J. Agron.* 136, 126501. https://doi.org/10.1016/j.eja.2022.126501.

Guo D-J, Li D-P, Chen Z-D, Yang B, Verma KK, Singh RK, Singh P, Khan Q, Sharma A, Qin Y, Zhang B-Q, Song X-P, Li Y-R. (2023). Effect of endophytic diazotroph *Enterobacter roggenkampii* ED5 on nitrogen-metabolism-related microecology in the sugarcane rhizosphere at different nitrogen levels. *Front. Microbiol.* 14, 1132016. https://doi.org/10.3389/fmicb.2023.1132016.

Guo D-J, Li D-P, Singh RK, Singh P, Sharma A, Verma KK, Qin Y, Khan Q, Lu Z, Malviya MK, Song X-P, Xing Y-X, Li Y-R (2021). Differential protein expression analysis of two sugarcane varieties in response to diazotrophic plant growth-promoting endophyte *Enterobacter roggenkampii* ED5. *Front. Plant Sci.* 12, 727741. https://doi.org/10.3389/fpls.2021.727741.

Guo DJ, Li DP, Singh RK, Singh P, Verma KK, Sharma A, Qin Y, Khan Q, Song XP, Malviya MK, Xing YX, Li YR. (2022). Comparative transcriptome analysis of two sugarcane varieties in response to diazitrophic plant growth promoting endophyte *Enterobacter roggenkampii* ED5. *J. Plant Interact.* 17(1), 75–84.

Hajeb, M, Hamzeh S, Alavipanah SK, Neissi L, Verrelst J. (2023). Simultaneous retrieval of sugarcane variables from sentinel-2 data using Bayesian regularized neural network. *Int. J. Appl. Earth Obs. Geoinf.* 116, 103168. https://doi.org/10.1016/j.jag.2022.103168.

Han SY, Bishop TFA, Filippi P. (2022). Data-driven, early-season forecasts of block sugarcane yield for precision agriculture. *Field Crops Res.* 276, 108360. https://doi.org/10.1016/j.fcr.2021.108360.

Hoffman, N, Singels A, Joshi S. (2024). Aerial phenotyping for sugarcane yield and drought tolerance. *Field Crops Res.* 308, 109275. https://doi.org/10.1016/j.fcr.2024.109275.

Huang CM, Wu KC, Verma KK, Luo HB, Deng ZN, Shi GY, Cao HQ, Wu XJ, Ye LP, Wei YW, Xu L. (2024). Isolation and expression analysis of cellulose synthase 3 (Ces3) genes from sugarcane (*Saccharum officinarum* L.). *Chem. Biol. Technol. Agric.* 11, 71. https://doi.org/10.1186/s40538-024-00597-7.

Huang HR, Gao YJ, Malviya MK, Verma KK, Solanki MK, Huang YX, Li X., Deng Y, Yan J, Tang SY, Wang LW, Xu L. (2022). Genetic diversity analysis of sugarcane (*Saccharum* spp. hybrids) among high-sucrose clones of GT series and commonly used parents by using microsatellite markers in Guangxi, *China. Sugar Tech.* 24, 397–407. https://doi.org/10.1007/s12355-021-01033-y.

Jagadesh M, Dash M, Kumari A, Singh SK, Verma KK, Kumar P, Bhatt R, Sharma SK. (2024). Revealing the hidden world of soil microbes: Metagenomic insights into plant, bacteria, and fungi interactions for sustainable agriculture and ecosystem restoration. *Microbiol. Res.* 285, 127764. https://doi.org/10.1016/j.micres.2024.127764.

Joshi A, Verma KK, Rajput VD, Arora J. (2023). Metabolic engineering of microorganisms in advancing biofuel production. In: *Advances in Lignocellulosic Biofuel Production Systems*. pp. 181–202. https://doi.org/10.1016/B978-0-323-91192-4.00016-X.

Joshi A, Verma KK, Rajput VD, Minkina T, Arora J. (2022). Recent advances in metabolic engineering of microorganisms for advancing lignocellulose-derived biofuels, *Bioengineered.* 13(4), 8135–8163. https://doi.org/10.1080/21655979.2022.2051856.

Kaab A, Sharifi M, Mobli H, Nabavi-Pelesaraei A, Chau K-W. (2019a). Combined life cycle assessment and artificial intelligence for prediction of output energy and environmental impacts of sugarcane production. *Sci. Total Environ.* 664, 1005–1019. https://doi.org/10.1016/j.scitotenv.2019.02.004.

Kaab A, Sharifi M, Mobli H, Nabavi-Pelesaraei A, Chau K-W. (2019b). Use of optimization techniques for energy use efficiency and environmental life cycle assessment modification in sugarcane production. *Energy*. 181, 1298–1320. https://doi.org/10.1016/j.energy.2019.06.002.

Krishna D, Sachan HK, Verma KK. (2022). Abiotic stresses in plants: An overview. In: *Understanding Abiotic Stresses Publisher*. Nova Science Publishers, Inc.: New York.

Kumari A, Dash M, Singh SK, Jagadesh M, Mathpal B, Mishra PK, Pandey SK, Verma KK. (2023a). Soil microbes: A natural solution for mitigating the impact of climate change. *Environ. Monit. Assess.* 195, 1436. https://doi.org/10.1007/s10661-023-11988-y.

Kumari A, Lakshmi GA, Krishna GK, Patni B, Prakash S, Bhattacharyya M, Singh SK, Verma KK. (2022). Climate change and its impact on crops: A comprehensive investigation for sustainable agriculture. *Agronomy*. 12, 3008. https://doi.org/10.3390/agronomy12123008.

Kumari A, Singh SK, Mathpal B, Verma KK, Garg VK, Bhattacharyya M, Bhatt R. (2023b). The biosynthesis, mechanism of action, and physiological functions of melatonin in horticultural plants: A review. *Horticulturae*. 9, 913. https://doi.org/10.3390/horticulturae9080913.

Leite LG, Chacon-Orozco JG, Shapiro-Ilan DI, Baldo FB, Cardoso JFM. (2023). Effects of temperature for optimizing production and storage of *Steinernema rarum* in a novel biphasic process, and efficacy of the nematode against sphenophorus levis. *Biol. Control.* 187, 105381. https://doi.org/10.1016/j.biocontrol.2023.105381.

Li A-M, Liao F, Wang M, Chen Z-L, Qin C-X, Huang R-Q, Verma KK, Li Y-R, Que Y-X, Pan Y-Q, et al. (2023). Transcriptomic and proteomic landscape of sugarcane response to biotic and abiotic stressors. *Int. J. Mol. Sci.* 24, 8913. https://doi.org/10.3390/ijms24108913.

Ma F, Wang M, Yan N, Adnan M, Jiang F, Hu Q, He G, Shen Y, Wan Y, Yang Y, Liang S, Chen X, Zhang M, Huang J. (2024). A fast and efficient phenotyping method to estimate sugarcane stalk bending properties using near-infrared spectroscopy. *Eur. J. Agron.* 154, 127107. https://doi.org/10.1016/j.eja.2024.127107.

Mafunga, WP, Ferrer SRD, Stark A. (2023). Towards a bioeconomy: How sugarcane fibre price structure stimulates farm-level decisions and sugarcane biorefinery feedstock supply. *Renew. Sustain. Energy Rev.* 183, 113432. https://doi.org/10.1016/j.rser.2023.113432.

Malviya MK, Li C-N, Lakshmanan P, Solanki MK, Wang Z, Solanki AC, Nong Q, Verma KK, Singh RK, Singh P, Sharma A, Guo D-J, Dessoky ES, Song X-P, Li Y-R. (2022). High-throughput sequencing-based analysis of rhizosphere and diazotrophic bacterial diversity among wild progenitor and closely related species of sugarcane (*Saccharum* spp. inter-specific hybrids). *Front. Plant Sci.* 13, 829337. https://doi.org/10.3389/fpls.2022.829337.

Malviya MK, Solanki MK, Li CN, Wang Z, Zeng Y, Verma KK, Singh RK, Singh P, Huang HR, Yang LT, Song XP, Li YR. (2021). Sugarcane-legume intercropping can enrich the soil microbiome and plant growth. *Front. Sustain. Food Syst.* 5, 606595. https://doi.org/10.3389/fsufs.2021.606595.

Misra V, Mall AK, Ansari SA, Raheem A, Tripathi MK, Ansari MI. (2023). Silicon as a beneficial nutrient for productivity augmentation and abiotic/biotic stress tolerance in sugarcane. *Biocatal. Agric. Biotechnol. 54,* 102944. https://doi.org/10.1016/j.bcab.2023.102944.

Modi RU, Chandel AK, Chandel NS, Dubey K, Subeesh, A, Singh AK, Jat D, Kancheti M. (2023a). State-of-the-art computer vision techniques for automated sugarcane lodging classification. *Field Crops Res.* 291, 108797. https://doi.org/10.1016/j.fcr.2022.108797.

Modi, RU, Kancheti M, Subeesh A, Raj C, Singh AK, Chandel NS, Dhimate AS, Singh MK, Shweta Singh. (2023b). An automated weed identification framework for sugarcane crop: A deep learning approach. *Crop Protect.* 173, 106360. https://doi.org/10.1016/j.cropro.2023.106360.

Morel, J, Bégué A, Todoroff P, Martiné J-F, Lebourgeois V, Petit M. (2014). Coupling a sugarcane crop model with the remotely sensed time series of FIPAR to optimise the yield estimation. *Eur. J. Agron.* 61, 60–68. https://doi.org/10.1016/j.eja.2014.08.004.

Nejad MS, Almassi M, Ghahderijani M. (2023). Life cycle energy and environmental impacts in sugarcane production: A case study of amirkabir sugarcane agro-industrial company in Khuzestan province. *Results Eng.* 20, 101545. https://doi.org/10.1016/j.rineng.2023.101545.

Oliveira MA, Verma KK, Bhatt R, Oliveira TBA. (2022). Impact of green and organic fertilizers on soil fertility and sugarcane productivity. In: Verma KK, et al. (Eds.), *Agro-Industrial Perspectives on Sugarcane Production under Environmental Stress.* https://doi.org/10.1007/978-981-19-3955-6_11.

Oliveira BG, Lourenço KS, Carvalho JLN, Gonzaga LC, Teixeira MC, Tamara AF, Soares JR, Cantarella H. (2023). New trends in sugarcane fertilization: Implications for NH3 volatilization, N_2O emissions and crop yields. *J. Environ. Manag.* 342, 118233. https://doi.org/10.1016/j.jenvman.2023.118233.

Otto R, Pereira GL, Tenelli S, Carvalho JLN, Lavres J, de Castro SAQ, Lisboa IP, Sermarini RA. (2020). Planting legume cover crop as a strategy to replace synthetic n fertilizer applied for sugarcane production. *Ind. Crops Prod.* 156, 112853. https://doi.org/10.1016/j.indcrop.2020.112853.

Park SE, Webster TJ, Horan HL, James AT, Thorburn PJ. 2010. A legume rotation crop lessens the need for nitrogen fertiliser throughout the sugarcane cropping cycle. *Field Crops Res.* 119(2), 331–141. https://doi.org/10.1016/j.fcr.2010.08.001.

Pongpat P, Mahmood A, Ghani HU, Silalertruksa T, Gheewala SH. (2023). Optimization of food-fuel-fibre in biorefinery based on environmental and economic assessment: The case of sugarcane utilization in Thailand. *Sustain. Prod. Consum.* 37, 398–411. https://doi.org/10.1016/j.spc.2023.03.013.

Poudyal, C, Sandhu H, Ampatzidis Y, Odero DC, Arbelo OC, Cherry RH, Costa LF. (2023). Prediction of morpho-physiological traits in sugarcane using aerial imagery and machine learning. *Smart Agric. Technol.* 3, 100104. https://doi.org/10.1016/j.atech.2022.100104.

Reyes-Trujillo, A, Daza-Torres MC, Galindez-Jamioy CA, Rosero-García EE, Muñoz-Arboleda F, Solarte-Rodriguez E. (2021). Estimating canopy nitrogen concentration of sugarcane crop using in situ spectroscopy. *Heliyon.* 7(3), e06566. https://doi.org/10.1016/j.heliyon.2021.e06566.

Sales LR, Silva AO, Sales FR, Rodrigues TL, Barbosa MV, dos Santos JV, Kemmelmeier K, Siqueira JO, Carneiro MAC. (2022). On farm inoculation of native arbuscular mycorrhizal fungi improves efficiency in increasing sugarcane productivity in the field. *Rhizosphere*. 22, 100539. https://doi.org/10.1016/j.rhisph.2022.100539.

Sanches GM, de Oliveira Bordonal R, Magalhães PSG, Otto R, Chagas MF, de Fátima Cardoso T, dos Santos Luciano AC. (2023). Towards greater sustainability of sugarcane production by precision agriculture to meet ethanol demands in South-Central Brazil based on a life cycle assessment. *Biosyst. Eng.* 229, 57–68. https://doi.org/10.1016/j.biosystemseng.2023.03.013.

Sefeedpari P, Shokoohi Z, Behzadifar Y. (2014). Energy use and carbon dioxide emission analysis in sugarcane farms: A survey on haft-tappeh sugarcane agro-industrial company in Iran. *J. Clean. Prod.* 83, 212–219. https://doi.org/10.1016/j.jclepro.2014.07.048.

Sharma A, Singh RK, Singh P, Vaishnav A, Guo DJ, Verma KK, Li DP, Song XP, Malviya MK, Khan N, Lakshmanan P, Li YR. (2021). Insights into the bacterial and nitric oxide-induced salt tolerance in sugarcane and their growth-promoting abilities. *Microorganisms*. 9, 2203. https://doi.org/10.3390/microorganisms9112203.

Sharma A, Song X-P, Singh RK, Vaishnav A, Gupta S, Singh P, Guo D-J, Verma KK, Li Y-R. (2022). Impact of carbendazim on cellular growth, defence system and plant growth promoting traits of Priestia megaterium ANCB-12 isolated from sugarcane rhizosphere. *Front. Microbiol.* 13, 1005942. doi:10.3389/fmicb.2022.1005942.

Shendryk Y, Robert D, Peter T. (2021). Integrating satellite imagery and environmental data to predict field-level cane and sugar yields in Australia using machine learning. *Field Crops Res*. 260, 107984. https://doi.org/10.1016/j.fcr.2020.107984.

Shukla SK, Jaiswal VP, Sharma L, Tiwari R, Pathak AD, Gaur A, Awasthi SK, Srivastava A. (2022). Trash management and *Trichoderma harzianum* influencing photosynthesis, soil carbon sequestration, and growth and yield of sugarcane ratoon in subtropical India. *Eur. J. Agron.* 141, 126631. https://doi.org/10.1016/j.eja.2022.126631.

Singh P, Singh RK, Li HB, Guo DJ, Sharma A, Verma KK, Solanki MK, Upadhyay SK, Lakshmanan P, Yang LT, Li YR. (2023). Nitrogen fixation and phytohormone stimulation of sugarcane plant through plant growth promoting diazotrophic Pseudomonas. *Biotechnol. Genet. Eng. Rev.* 40(1), 15–35. https://doi.org/10.1080/02648725.2023.2177814.

Singh P, Singh RK, Song QQ, Li HB, Guo DJ, Malviya MK, Verma KK, Song XP, Lakshmanan P, Yang LT, Li YR. (2021). Comparative analysis of protein and differential responses of defense-related gene and enzyme activity reveals the long-term molecular responses of sugarcane inoculated with *Sporisorium scitamineum. J. Plant Interact.* 16(1), 12–29.

Singh RK, Singh P, Li HB, Song QQ, Guo DJ, Solanki MK, Verma KK, Malviya MK, Song XP, Lakshmanan P, Yang LT, Li YR. (2020). Diversity of nitrogen-fixing rhizobacteria associated with sugarcane: A comprehensive study of plant-microbe interactions for growth enhancement in *Saccharum* spp. *BMC Plant Biol.* 20, 220. https://doi.org/10.1186/s12870-020-02400-9.

Singh PN, Shukla SK, Bhatnagar VK. (2007). Optimizing soil moisture regime to increase water use efficiency of sugarcane (*Saccharum* spp. hybrid complex) in subtropical India. *Agric. Water Manag.* 90(1):95–100. https://doi.org/10.1016/j.agwat.2007.02.008.

Solanki MK, Verma KK, Dastogeer KMG, Mora-Poblete F, Mundra S. (2023). Microbial resilience in plant nutrient management towards sustainable farming. *Front. Microbiol.* 14, 1280811. https://doi.org/10.3389/fmicb.2023.1280811.

Song X-P, Mo F, Verma KK, Wei J, Zhang X, Yang L, Li YR. (2019). Effect of sugarcane smut (*Ustilago scitaminea* Syd.) on ultrastructure and biochemical indices of sugarcane. *Biomed. J. Sci. Tech. Res.* 17(1), 12546–12551. https://doi.org/10.26717/BJSTR.2019.17.002950.

Song XP, Mo F, Yan M, Zhang X, Zhang B, Huang X, Huang D, Pan Y, Verma KK, Li YR. (2022). Effect of smut infection on the photosynthetic physiological characteristics and related defense enzymes of sugarcane. *Life*. 12, 1201. https://doi.org/10.3390/life12081201.

Song XP, Verma KK, Tian DD, Zhang XQ, Liang YJ, Huang X, Li CN, Li YR. (2021). Exploration of silicon functions to integrate with biotic stress tolerance and crop improvement. *Biol. Res*. 54, 19. https://doi.org/10.1186/s40659-021-00344-4.

Soulé, M., A. Mansuy, J. Chetty, S. Auzoux, P. Viaud, M. Schwartz, A. Ripoche, B. Heuclin, and M. Christina. (2024). Effect of crop management and climatic factors on weed control in sugarcane intercropping systems. *Field Crops Res.* 306, 109234. https://doi.org/10.1016/j.fcr.2023.109234.

Tian D, Qin L, Verma KK, Wei L, Li J, Li B, Zhou W, He Z, Wei D, Huang S, Long S, Huang Q, Li C, Wei S. (2023). Transcriptomic and metabolomic differences between banana varieties which are resistant or susceptible to Fusarium wilt. *PeerJ*. 11, e16549. https://doi.org/10.7717/peerj.16549.

Verma KK, Anas M, Chen Z, Rajput VD, Malviya MK, Verma CL, Singh RK, Singh P, Song XP, Li YR. (2020f). Silicon supply improves leaf gas exchange, antioxidant defense system and growth in sugarcane responsive to water limitation. *Plants*. 9, 1032, https://doi.org/10.3390/plants9081032.

Verma KK, Joshi A, Song X-P, Liang Q, Xu L, Huang H-r, Wu K-C, Seth CS, Arora J and Li Y-R (2024a). Regulatory mechanisms of plant rhizobacteria on plants to the adaptation of adverse agroclimatic variables. *Front. Plant Sci.* 15, 1377793. https://doi.org/10.3389/fpls.2024.1377793.

Verma KK, Joshi A, Song X-P, Singh S, Kumari A, Arora J, Singh SK, Solanki MK, Seth CS, Li Y-R. (2024b). Synergistic interactions of nanoparticles and plant growth promoting rhizobacteria enhancing soil-plant systems: A multigenerational perspective. *Front. Plant Sci.* 15, 1376214. https://doi.org/10.3389/fpls.2024.1376214.

Verma KK, Li DM, Singh M, Rajput VD, Malviya MK, Minkina T, Singh RK, Singh P, Song XP, Li YR. (2020a). Interactive role of silicon and plant-rhizobacteria mitigating abiotic stresses: A new approach for sustainable agriculture and climate change. *Plants*. 9, 1055. https://doi.org/10.3390/plants9091055.

Verma KK, Rajput VD, Singh M, Verma CL, Minkina T. (2022a). Influence of waterlogging on plant physiology and molecular biology. In: *Understanding Abiotic Stresses Publisher*. Nova Science Publishers, Inc.: New York.

Verma KK, Siddiqui N, Singh M. (2022b). Understanding the plant physiological and molecular adaptation strategies to soil salinity. In: *Understanding Abiotic Stresses Publisher*. Nova Science Publishers, Inc.: New York. https://doi.org/10.52305/UBUJ4024

Verma KK, Singh P, Song X-P, Malviya MK, Singh RK, Chen G-L, Solomon S, Li YR. (2020b). Mitigating climate change for sugarcane improvement: Role of silicon in alleviating abiotic stresses. *Sugar Tech.* 22(5), 741–749. https://doi.org/10.1007/s12355-020-00831-0.

Verma KK, Singh RK, Song QQ, Singh P, Zhang B-Q, Song X-P, Chen G-L, Li YR. (2019a). Silicon alleviates drought stress of sugarcane plants by improving antioxidant responses. *Biomed J. Sci. Tech. Res.* 17(1), 12580–12586. https://doi.org/10.26717/BJSTR.2019.17.002957.

Verma KK, Song X, Degu HD, Guo D, Joshi A, Huang H, Xu L, Singh M, Huang D, Rajput VD, Li YR. (2023a). Recent advances in nitrogen and nano-nitrogen fertilizers for sustainable crop production: A mini-review. *Chem. Biol. Technolog. Agricult.* 10, 1–14. https://doi.org/10.1186/s40538-023-00488-3.

Verma KK, Song X, Verma CL, Huang H, Singh M, Xu L, Wu J, Khan ZH, Zhang B, Li Y. (2023b). Mathematical modeling of climate and fluoride effects on sugarcane photosynthesis with silicon nanoparticles. *Plant Physiol. Biochem.* 204, 108089. https://doi.org/10.1016/j.plaphy.2023.108089.

Verma KK, Song XP, Budeguer F, Nikpay A, Enrique R, Singh M, Zhang B, Wu, JM, Li YR. (2022c). Genetic engineering: An efficient approach to mitigating biotic and abiotic stresses in sugarcane cultivation. *Plant Signal. Behav.* 17(1), e2108253.

Verma KK, Song XP, Joshi A, Rajput VD, Singh M, Sharma A, Singh RK, Li DM, Arora J, Minkina T, Li YR. (2022d). Nanofertilizer possibilities for healthy soil, water and food in future: An overview. *Front. Plant Sci.* 13, 865048. https://doi.org/10.3389/fpls.2022.865048.

Verma KK, Song XP, Joshi A, Tian DD, Rajput VD, Singh M, Arora J, Minkina TM, Li YR. (2022e). Recent trends in nano-fertilizer for advancing sustainable agriculture under the era of climate change to ensure food security for future generations. *Nanomaterials.* 12, 173. https://doi.org/10.3390/nano12010173.

Verma KK, Song XP, Kumari A, Jagadesh M, Singh SK, Bhatt R, Singh M, Seth CS, Li YR. (2024c). Climate change adaptation: Challenges for agricultural sustainability. *Plant, Cell Environ.* 48(4), 2522–2533. https://doi.org/10.1111/pce.15078

Verma KK, Song XP, Li DM, Singh M, Wu JM, Singh RK, Sharma A, Zhang BQ, LI YR. (2022f). Silicon and soil microorganisms improves rhizospheric soil health with bacterial community, plant growth, performance and yield. *Plant Signal. Behav.* 17(1), e2104004.

Verma KK, Song XP, Lin B, Guo DJ, Singh M, Rajput VD, Singh RK, Singh P, Sharma A, Malviya MK, Chen GL, Li YR. (2022g). Silicon induced drought tolerance in crop plants: Physiological adaptation strategies. *Silicon.* 14(6), 2473–2487. https://doi.org/10.1007/s12633-021-01071-x.

Verma KK, Song XP, Rajput VD, Boldyreva V, Zhang BQ, Minkina T, Li YR. (2022h). Morpho-physiological, biochemical and ultrastructural modifications on sugarcane to prolonged water deficit. In: Verma KK, et al. (Eds.), *Agro-Industrial Perspectives on Sugarcane Production under Environmental Stress*. https://doi.org/10.1007/978-981-19-3955-6_11.

Verma KK, Song XP, Rajput VD, Solomon S, Li YR, Rao GP. (2022i). *Agro-Industrial Perspectives on Sugarcane Production under Environmental Stress*. Springer Nature: Singapore, https://doi.org/10.1007/978-981-19-3955-6.

Verma KK, Song XP, Singh M, Tian DD, Rajput VD, Minkina T, Li YR. (2023c). Association of silicon and soil microorganisms induces stress mitigation, increasing plant productivity. In: de Mello Prado, R. (Eds.), *Benefits of Silicon in the Nutrition of Plants*. pp. 299–328. https://doi.org/10.1007/978-3-031-26673-7_17.

Verma KK, Song XP, Tian DD, Guo DJ, Chen ZL, Zhong CS, Nikpay A, Singh M, Rajput VD, Singh RK, Minkina T, Li YR. (2021a). Influence of silicon on biocontrol strategies to manage biotic stress for crop protection, performance and improvement. *Plants*. 10, 2163. https://doi.org/10.3390/plants10102163.

Verma KK, Song XP, Tian DD, Singh M, Verma CL, Rajput VD, Singh RK, Sharma A, Singh P, Malviya MK, Li YR. (2021b). Investigation of defensive role of silicon during drought stress induced by irrigation capacity in sugarcane: Physiological and biochemical characteristics. *ACS Omega*. 6, 19811–19821.

Verma KK, Song XP, Verma CL, Chen ZL, Rajput VD, Wu KC, Liao F, Chen GL, Li YR. (2021c). Functional relationship between photosynthetic leaf gas exchange in response to silicon application and water stress mitigation in sugarcane. *Biol. Res*. 54, 15. https://doi.org/10.1186/s40659-021-00338-2.

Verma KK, Song XP, Verma CL, Malviya MK, Guo DJ, Rajput VD, Sharma A, Wei KJ, Chen GL, Solomon S, Li YR. (2021d). Predication of photosynthetic leaf gas exchange of sugarcane (*Saccharum* spp.) leaves in response to leaf positions to foliar spray of potassium salt of active phosphorus under limited water irrigation. *ACS Omega*. 6, 2396–2409.

Verma KK, Song X-P, Xu L, Huang H-R, Liang Q, Seth CS, Li Y-R. (2023d). Nano-microplastic and agro-ecosystems: A mini-review. *Front. Plant Sci.* 14, 1283852. https://doi.org/10.3389/fpls.2023.1283852.

Verma KK, Song XP, Yadav G, Degu HD, Parvaiz A, Singh M, Huang HR, Mustafa G, Xu L, Li YR. (2022j). Impact of agroclimatic variables on proteo-genomics in sugarcane (*Saccharum* spp.) plant productivity. *ACS Omega*. 7, 22997–23008.

Verma KK, Song XP, Zeng Y, Guo DJ, Sing M, Rajput VD, Malviya MK, Wei KJ, Sharma A, Li DP, Chen GL, Li YR. (2021e). Foliar application of silicon boosts growth, photosynthetic leaf gas exchange, antioxidative response and resistance to limited water irrigation in sugarcane (*Saccharum officinarum* L.). *Plant Physiol. Biochem*. 166, 582–592.

Verma KK, Song XP,Zeng Y, Li DM, Guo DJ, Rajput VD, Chen GL, Barakhov A, Minkina TM, Li YR. (2020c). Characteristics and correlation of leaf stomata and its relationship with photosynthesis on *Saccharum* spp. under different irrigation and silicon application. *ACS Omega*. 5, 24145–24153.

Verma KK, Vatsal S, Singh M, Rajput VD, Minkina T. (2022k). Coping with water stress: effects and adaptive mechanisms to plants. In: *Understanding Abiotic Stresses Publisher*. Nova Science Publishers, Inc.: New York.

Verma KK, Wu KC, Verma CL, Li DM, Malviya MK, Singh RK, Singh P, Chen GL, Song XP, Li YR. (2020d). Developing mathematical model for diurnal dynamics on photosynthetic responses in sugarcane responsive to different irrigation and silicon application. *PeerJ*. 8, e10154. https://doi.org/10.7717/peerj.10154.

Verma KK, Liu X-H, Wu K-C, Singh RK, Song QQ, Malviya MK, Song X-P, Singh P, Verma CL, Li YR. (2020e). The impact of silicon on photosynthetic and biochemical responses of sugarcane under different soil moisture levels. *Silicon*. 12, 1355–1367.

Verma KK., Wu K-C., Singh P., Malviya MK., Singh RK., Song X-P., Li YR. (2019b). The protective role of silicon in sugarcane under water stress: Photosynthesis and antioxidant enzymes. *Biomed J. Sci. Tech. Res.* 15(2), 1–7. https://doi.org/0.26717/BJSTR.2019.15.002685.

Viaud P, Benjamin H, Philippe L, Mathias C, Antoine V, Alizé M, Julien C, Krishna N. (2023). Sugarcane yield response to legume intercropped: A meta-analysis. *Field Crops Res.* 295, 108882. https://doi.org/10.1016/j.fcr.2023.108882.

Wickramasinghe KP, Kong C-Y, Lin X-Q, Zhao P-F, Mehdi F, Li X-J, Liu X-L, Mao J, Lu X. (2024). Photoperiodic and lighting treatments for flowering control and its genetic regulation in sugarcane breeding. *Heliyon.* 10(7), e28531. https://doi.org/10.1016/j.heliyon.2024.e28531.

Wilson BE, Salgado LD, Villegas JM. (2022). Optimizing chemical control for *Diatraea saccharalis* (Lepidoptera: Crambidae) in sugarcane. *Crop Prot.* 152, 105843. https://doi.org/10.1016/j.cropro.2021.105843.

Wu K-C, Huang C-M, Verma KK, Deng Z-N, Huang H-R, Pang T, Cao H-Q, Luo H-B, Jiang S-L, Xu L. (2022). Transcriptomic responses of *Saccharum spontaneum* roots in response to polyethylene glycol-6000 stimulated drought stress. *Front. Plant Sci.* 13, 992755. https://doi.org/10.3389/fpls.2022.992755.

Wu T, Zhang Q, Wu J, Liu Q, Su J, Li H. (2023). An improved YOLOv5s model for effectively predict sugarcane seed replenishment positions verified by a field re-seeding robot. *Comput. Electron. Agric.* 214, 108280. https://doi.org/10.1016/j.compag.2023.108280.

Xu L, Deng Z-N, Wu K-C, Malviya MK, Solanki MK, Verma KK, Pang T, Li Y-J, Liu X-Y, Kashyap BK, Dessoky ES, Wang W-Z, Huang H-R. (2022). Transcriptome analysis reveals a gene expression pattern that contributes to sugarcane bud propagation induced by indole-3-butyric acid. *Front. Plant Sci.* 13, 852886. https://doi.org/10.3389/fpls.2022.852886.

Yang L, Zhou Y, Meng B, Li H, Zhan J, Xiong H, Zhao H, Cong W, Wang X, Zhang W, Lakshmanan P, Deng Y, Shi X, Chen X, Zhang F. (2022). Reconciling productivity, profitability and sustainability of small-holder sugarcane farms: A combined life cycle and data envelopment analysis. *Agric. Syst.* 199, 103392. https://doi.org/10.1016/j.agsy.2022.103392.

Zheng F-J, Lin B, Yang Y-X, Fang X-C, Verma KK, Chen G-L. (2023). Efficacy and functionality of sugarcane original vinegar on mice. *Front. Microbiol.* 14, 1224666. https://doi.org/10.3389/fmicb.2023.1224666.

Zhu L, Liu X, Wang Z, Tian L. (2023). High-precision sugarcane yield prediction by integrating 10-m Sentinel-1 VOD and Sentinel-2 GRVI indexes. *Eur. J. Agron.* 149, 126889. https://doi.org/10.1016/j.eja.2023.126889.

CHAPTER 8

Assessing the Genetic Integrity of Sugarcane Germplasm

MAYUR DARVHANKAR,[1] ARADHNA KUMARI,[2]
CHANDRA MOHAN MEHTA,[1] and BHUPENDRA MATHPAL[1*]

[1]Department of Agronomy, School of Agriculture, Lovely Professional University, Phagwara, Punjab, India

[2]College of Agriculture, Jawaharlal Nehru Krishi Vishwavidyalaya, Ganj Basoda, Vidisha, Madhya Pradesh, India

*Corresponding author

ABSTRACT

Sugarcane plays a significant role in the world's economy due to its use for sugar production, bioethanol, and energy. It is an efficient C_4 crop that contributes nearly 80% of table sugar and around 60% of ethanol globally. It is a major cash crop of tropical and subtropical regions. As the increasing human population has put extreme pressure on fossil fuel reserves, significant improvement is required in sugarcane production. Biodiversity in germplasm and identification of superior traits are important to develop the genetically improved cultivars of sugarcane that can grow under changing environmental conditions. Multiple sets of chromosomes present in it and responsible for genetic diversity and complexity, which make it difficult to manipulate. *Saccharum spontaneum*, a perennial grass and a wild relative of sugarcane, gives a notable genetic diversity to the crop. This perennial wild grass possesses important traits that can be incorporated into cultivated species of sugarcane to make them resistant against biotic and abiotic stresses. The breeding program should focus on

Sugarcane Transformation: Biofuels, Biotechnology, and Beyond.
Krishan K. Verma, Xiu-Peng Song, Hai-Rong Huang, Aradhna Kumari, and Yang-Rui Li (Eds.)

DOI: 10.1201/9781779641571-8

the utilization of available biodiversity and important traits for the further improvement of the sugarcane species in changing climatic regimes. This chapter highlights the available germplasm biodiversity and its utilization of sugarcane.

8.1 INTRODUCTION

Saccharum spp. hybrid is a bioenergy crop, which is being used for its food, fodder, and the production of energy products. Initially, it was identified as a wild grass and, unlike other cereal crops, it was selected for its stem, which contains a high concentration of sucrose (Babu et al., 2022; Verma et al., 2023a, 2023b). The crop is primarily suited to be cultivated under tropical and subtropical climatic conditions of more than 100 countries. Total area under sugarcane cultivation is estimated to be around 27.1 million ha with a total production of 1.9 billion metric tons per year (FAO, 2013; Zang et al., 2020; Bhatt et al., 2023a; Huang et al., 2024). Sugarcane is a C_4 crop with stronger and vigorous growth habit than other crops of tropical climatic conditions. It has a great potential to attain large biomass because of its high photosynthetic efficiency, which makes it the best crop to contribute around 70% of the world's sugar and to produce ethanol-based biofuels (Verma et al., 2019a, 2019b; Chen et al., 2020, 2023; Bhatt et al., 2023b). Because of its high economic importance, sugarcane plays an important role in boosting the economy of many countries by fulfilling the demand of around 80% global sugar and acts as a source to supply 40% of the world's ethanol (Lam et al., 2009; Singh et al., 2020, 2021, 2023).

Sugarcane productivity was increased after the green revolution because of the cultivation of improved varieties and the advancement of agronomic practices (Anas et al., 2021a, 2021b; Hale et al., 2022). But, increasing demand for sugar and biofuel due to increasing population posed a challenge for ongoing breeding programs to develop new varieties with improved resistance against biotic and abiotic stresses, enhanced ratooning ability, and improved yield potential (Song et al., 2019, 2021, 2022; Park et al., 2024). Genetic diversity plays a crucial role in developing such improved varieties having high potential to act as a feedstock for the production of sugar and biofuel (Hoang et al., 2015; Verma et al., 2020a, 2020b; Sharma et al., 2021; Deng et al., 2023).

There are six species of sugarcane that are considered important. Out of six, two are wild (*Saccharum spontaneum* L. and *Saccharum robustum* Brandes & Jesw. Ex Gressl) while four are domesticated (*Saccharum officinarum* L., *Saccharum sinense* Roxburgh, *Saccharum barberi* Jeswiet, and *Saccharum edule* Hassk (Moore et al., 2013). The current cultivars of sugarcane were developed in the early period of the 1900s and are a result of crosses of less than 20 of *S. spontaneum* and *S. officinarum* clones as parents (Raboin et al., 2006). It resulted in dominance of a few sugarcane varieties over a large growing area, which finally led to monoculture and made these varieties susceptible to environmental stresses. Similarly, varieties developed in the late 20th century were slightly genetically different but also had a narrow parental base (Santos et al., 2012). So, to improve the flexibility of sugarcane varieties, there is a need to enhance the germplasm diversity by the identification of various landraces and wild relatives with unique genetic traits (Crystian et al., 2018).

Occurrence of less genetic diversity among sugarcane varieties is a major concern for breeding program in changing environmental conditions (Sindhu et al., 2011). There is a need to focus on the introgression of genes from related wild races, such as *S. spontaneum*, *S. robustum*, *Erianthus* spp., *Miscanthus* spp., *Narenga* spp., and *Sclerostachya* spp., to improve the germplasm diversity and performance of sugarcane cultivars (Govindaraj et al., 2012). Among all the wild races, *S. spontaneum*, *S. robustum*, and *Erianthus arundinaceus* have shown enormous capabilities to improve the genetic base and can act as a source for increased biomass production, improved rationing ability, tolerance against biotic and abiotic stresses (Ram et al., 2001; Nair, 2012; Govindaraj, 2017). Various sugarcane species and allied genera have been collected by the World Collection of Sugarcane and allied Grasses, which is kept up to date by the USDA-ARS Subtropical Horticulture Research Station's National Germplasm Repository in Miami, Florida, USA (Fickett et al., 2020). Out of the total 1164 accessions collected at WCSRG, the majority are of *Saccharum* species, including 307 of *S. spontaneum* accessions, 158 of *S. officinarum* accessions, 127 of S. hybrid accessions, 81 of *S. robustum* accessions, 48 of *S. sinense* accessions, and 33 of *S. barberi*. All these accessions belong to various agroclimatic regions of the world and are believed to contain genes contribute to resistance against different climatic and biotic stresses (Nayak et al., 2014; Xiong et al., 2022).

8.2 GENETIC DIVERSITY IN SUGARCANE

8.2.1 CENTRE OF ORIGIN AND DOMESTICATION

The genus *Saccharum* traditionally comprises different species (D'Hont et al., 1998). According to Irvine (1999), the genus should be whittled down to just two species, with *S. spontaneum* remaining a distinct species and *S. robustum*, *S. edule*, *S. barberi*, *S. sinense*, and *S. officinarum* grouped together as the species *S. officinarum*. Selective breeding of species within the *Saccharum* genus, including crossings between *S. officinarum* and *S. spontaneum*, produced commercial sugarcane hybrid cultivars (Cox, Hogarth and Smith, 2000; Lakshmann et al., 2005). Given that *S. officinarum* is a species that is exclusively cultivated, it has a connection to human activity. Indonesia/New Guinea area is considered as the center of origin for *S. officinarum* since 8000 B.C. (Daniels and Roach 1987; Fauconnier, 1993). The selection of sweet varieties of *S. robustum* led to the evolution of *S. officinarum*. Originally, canes were utilized for archery, fencing, and house construction (Daniels and Roach, 1987). The distribution of *S. officinarum* occurred due to human migration to Southeast Asia, India, Pacific, and the Mediterranean nearly 500 BC (Fauconnier, 1993). Amplified fragment length polymorphism marker studies (Aitken et al., 2006) support the hypothesis that Papua New Guinea is the hub of diversification for *S. officinarum* (Daniels and Roach, 1987).

8.2.2 CENTER OF DIVERSITY IN SUGARCANE

There are opportunities to improve plant traits through genetic variety among species of plants (Hemaprabha et al., 2022). In order to effectively utilize genetic resources in breeding programs and address the risks posed by environmental variations, it is imperative that genetic diversity assessed (Zhang et al., 2019). Crop plant improvement is significantly impacted by knowledge of genetic diversity among appropriate cultivars or elite breeding materials (Singh et al., 2017). Genetic diversity can be obtained through pedigree analysis, morphological traits, and molecular breeding (Tolera et al., 2023). Six main species were considered to comprise the Saccharum genus: four cultivated species (*S. officinarum*, *Barberi*, *Sinense*, and *Edulle*), and two wild species (*S. spontaneum* and *S. robustum*). Utilized two duplicate collections of "*Saccharum* Complex" germplasm, together referred to as the World Collection of Sugarcane and Related Grasses (WCSRG). There are two WCSRGs: one in Miami, Florida, USA, and the other in Coimbatore, India. The WCSRG is maintained in the United States by the National Germplasm Repository, which is located at the USDA-ARS Subtropical Horticulture Research Station in Miami, Florida (Comstock et al., 1995). According to Erdding and Roach (1987), this WCSRG may have a high level of genetic diversity and several beneficial alleles for a wide range of morphological features, biomass yield components, responses to biotic and abiotic challenges, and many other quality traits (Figure 8.1).

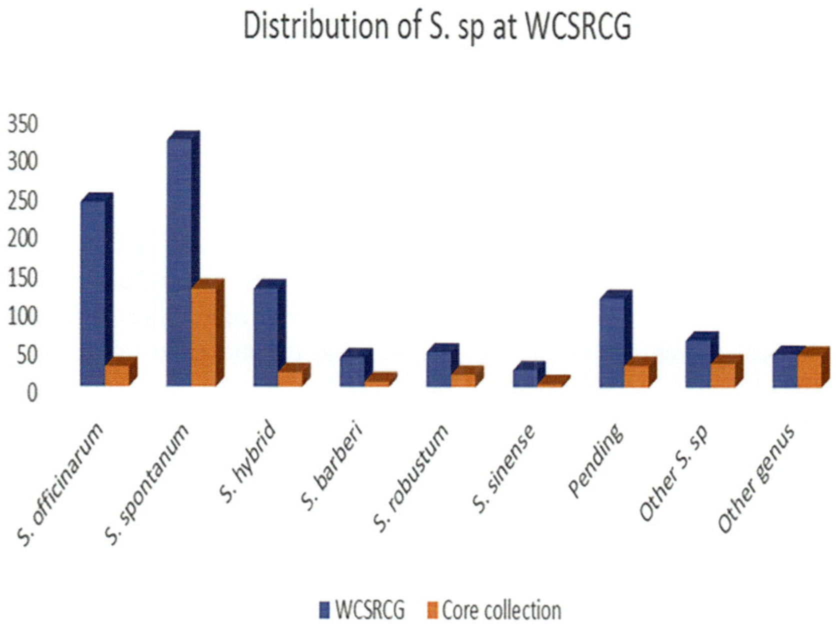

FIGURE 8.1 Distribution of *Saccharum* species at World Collection of Sugarcane and Related Grasses (WCSRG).

Complete characterization and application of the WCSRG in breeding programs is a challenging task due to the high number and genetically complicated accessions. Its application requires a core collection, which is a condensed assembly of the full collection with optimum genetic diversity and minimized repetition (Brown, 1989). For *Saccharum* spp., a core collection of this type would yield a selection of typical accessions and enable comprehensive phenotypic, physiological, and genetic analysis. Currently used sugarcane varieties (*Saccharum* spp. hybrids) are widely complex interspecific hybrids developed from wild and nonwild cultivars (Wang et al., 2022). Within the *Saccharum* species, intra-specific hybridization and selection are employed to improve species-germplasm. The interspecific hybridization process involves the advantage of the improved species-hybrids. Additionally, attempts are being made to utilize *S. barberi, S. sinense, S. spontaneum,* and *Erianthus arundinaceus* as female hybridization parents in an effort to increase the hybrids' cytoplasmic base. Superior hybrids might be introduced into the breeding pools. Interspecific and intergeneric hybrids are constantly generated and assessed. A total of nearly 300 hybrids have been assessed in subtropical Karnal, Haryana, India, and some of the better hybrids have been included in the breeding programs.

8.2.3 GERMPLASM CONSERVATION

Collection, conservation, and utilization of germplasm are essential components for crop improvement programs (Frison and Putter 1993). Conservation of germplasm is of prime importance for the balance of biological system diversification, food security, and safety (Chandran, 2010). Since the majority of plant species are becoming endangered over time, protecting plant resources is crucial. According to Mukherjee (1957), genetic resources have the potential to be a sustainable supply of agricultural produce that may be used to reduce poverty and maintain the economic conditions of the populace through the effective production of food crops. In order to protect gene resources for systemic preservation and utilization at the national and international levels, germplasm conservation is necessary (Taylor and Dukic, 1993).

The most apparent illustration of preservation is seen in on-farm conservation, which supports the course of evolution that gives rise to genetic heterogeneity (Bajaj, 1995). Gene variations can be found both *in situ*, or in a natural setting, and *ex situ*, or in gene banks derived from lab cultures (Putra and Damayanti, 2012). A vast array of the most valuable agricultural fields has been set aside for the gene bank and located in state-of-the-art aseptic facilities. Collections from both national and international programs—such as NAGRAB, OSTROM, and IITTA—are among the varied offerings. A scientific study has shown that even though there are roughly three lakh species of higher plants in the globe, just 1% of them are now in use. Only 8–10 crops, including varieties of millet, rye, rice, and wheat, produce around 80% of the world's food. The majority of agricultural growth in the modern world depends on a diverse array of genetic resources with two distinct types of value. A number of characteristics are taken into account by plant genes and genotypes, including drought tolerance, insect and pest resistance, and plant structure, function, and color. These traits serve as an immediate source of plant genotype conservation for desired qualities (Huang et al., 2022; Li et al., 2023; Tian et al., 2023). Second, genetic variety or variation in genes guarantees future requirements. Consequently, they support the agricultural sector on a local and national basis.

8.3 UTILIZATION OF SUGARCANE GERMPLASM FOR VARIOUS TRAITS

8.3.1 DROUGHT

The climate significantly influences the growth rate of the sugarcane crop (Binbol et al., 2006; Verma et al., 2020c, 2020d, 2023c, 2023d). The temperature range in which sugarcane is cultivated is 18–40°C, and the temperature requirements change depending on the stage of growth (Vandana et al., 2020; Guo et al., 2023). The rate of mineralization in the soil and the rate at which plants absorb nutrients are both accelerated by an increase in soil temperature (Malviya et al., 2021, 2022; Wu et al., 2022). Conversely, greater temperatures speed up the rate at which leaves transpire. Unfavorable climate conditions have an impact on sugarcane production and are becoming more frequent and intense. Since dry summer periods reduce output even during the rainy season, water deficit is reported as more severe stress on sugarcane farming (Verma et al., 2020e, 2020f; Xu et al., 2022; Contiliani et al., 2022). For example, increasing losses in sugarcane yield were caused by water restriction. Because sugarcane crops are negatively impacted by climate change, the sugar and bioenergy sectors work to develop sustainable cultivars (Verma et al., 2021a, 2021b, 2024a, 2024b, 2024c).

Stomatal closure and drastically decreased cell division and photosynthesis are the most frequent drought-induced plant responses (Ferreira et al., 2017; Verma et al., 2021c, 2022a, 2022b). Different approaches have been applied to assess the stress tolerance, including *in vitro* demonstrations, artificial climate boxes, greenhouse studies, and the hypertonic solution (PEG) method (Guo et al., 2021, 2022; Oliveira et al., 2022; Xu et al., 2023). Morphological, physiological, and biochemical indices are utilized to select resistance evaluation features (Silva et al., 2007; Rhein et al., 2011; Abbas et al., 2014). Membership function analysis (Chen et al., 2012), the drought resistance index (Wang et al., 2015), the drought resistance coefficient (Wang et al., 2015), and cluster analysis (Liu et al., 2015; Sharma et al., 2022) are the most crucial components for an in-depth evaluation of resistance. The number of drought-tolerant germplasm are available in the world and India with specific characteristics (Table 8.1).

TABLE 8.1 List of Sugarcane Germplasm for Drought Tolerance

Name of Genotypes	Country	Characteristic	References
Co 99004	India	Less transpiration and high photosynthetic rate	Gomathi et al. (2011)
Co 99008	India	Less transpiration and high photosynthetic rate	Gomathi et al. (2011)
IACSP94-2094	World	Oxidoreductase activity, regulation of transcription, amino acid metabolism, and translation	Contiliani et al. (2022)
CP 80-1842	China	Positive correlation of the membership function value of drought resistance (MFVD) with drought drought-tolerant coefficient	Xu et al. (2023)
Mintang 69-421	China	Positive correlation of the membership function value of drought resistance (MFVD) with drought-tolerant coefficient	Xu et al. (2023)
Yuetang 57-423	China	Positive correlation of membership function value of drought resistance (MFVD) with drought-tolerant coefficient	Xu et al. (2023)
L116	China	Positive correlation of the membership function value of drought resistance (MFVD) with drought-tolerant coefficient	Xu et al. (2023)
Yuenong 79-177	China	Positive correlation of the membership function value of drought resistance (MFVD) with drought-tolerant coefficient	Xu et al. (2023)
FR 93-1054	China	Positive correlation of the membership function value of drought resistance (MFVD), with drought-tolerant coefficient	Xu et al. (2023)

TABLE 8.1 *(Continued)*

Name of Genotypes	Country	Characteristic	References
Yuetang 75-25	China	Positive correlation of the membership function value of drought resistance (MFVD) with drought-tolerant coefficient	Xu et al. (2023)
Q200	China	Positive correlation of the membership function value of drought resistance (MFVD) with drought-tolerant coefficient	Xu et al. (2023)
Neijiang 57-416	China	Positive correlation of the membership function value of drought resistance (MFVD) with drought-tolerant coefficient	Xu et al. (2023)
Yunzhe 06-407	China	Positive correlation of the membership function value of drought resistance (MFVD) with drought-tolerant coefficient	Xu et al. (2023)
Co 86032	India	High cane yield and sucrose content	Mahadevaiah et al. (2021)
CoM 0265	India	High cane yield and sucrose content	Mahadevaiah et al. (2021)

8.3.2 SALINITY

The negative effects of saline soil have been reported in several plant species (Cruz et al., 2018; Jagadesh et al., 2024). Accumulation of salts interferes with a plant's ability to absorb water, thereby resulting in increased osmotic potential and reduced absorption of water by plant roots (Machado and Serralheiro 2017; Chaudhary et al., 2023). According to Vasantha and Gomathi (2012), sodium carbonate is the primary source of alkalinity, whereas the chlorides and sulfates of sodium, calcium, magnesium, and potassium are the salts that mainly contribute to salinity. The sugarcane genotypes that are tolerant exhibit improved osmotic adjustment, reduced photosynthesis, and biomass yield. A sugarcane genotype must exhibit the following characteristics in order to function as a tolerant type: biomass production, long-term maintenance of the water status, osmotic adjustment, and maintenance of a high photosynthetic rate (Zheng et al., 2023; Solanki et al., 2023). The germplasm showing tolerances to saline conditions is listed in Table 8.2.

TABLE 8.2 List of Sugarcane Germplasm for Salinity Tolerance

Name of Genotypes	Country	Characteristic	References
Co 0238	India	High proline conc. Under salinity stress	Preet et al. (2024)
Co 13035	India	High proline conc. Under salinity stress	Preet et al. (2024)
SP 80-1816	Brazil	Highest biomass accumulation	Silva et al. (2022)
Co 94012	India	Good quality jaggery	Vasantha and Gomathi (2012)
Co 99004	India	Good quality jaggery	Vasantha and Gomathi (2012)

8.3.3 HIGH TEMPERATURE

Sugarcane is one of the world's major C_4 crops that mainly grow in the tropics and subtropics regions (Kohila and Gomathi 2018). Weather and climate-related events are the key factors for sugarcane yield globally, especially in terms of different developing nations (Zhao and Li, 2015). According to Kaushal et al. (2016), an adverse temperature may have a major impact on leaf photosynthetic responses, nutrient uptake and accumulation, and cell membrane integrity on sugarcane plant leaves. Excess air temperature downregulated photosynthetic pigments, stability of chlorophyll, values of soil plant analysis development, optimum chlorophyll fluorescence efficiency of PSII, photosynthetic CO_2 assimilation rate, relative leaf water content, and nitrate reductase, sucrose-metabolizing enzymatic activities (Kohila and Gomathi, 2018). Higher temperature (above 32°C) results in shorter internodes and higher moisture content (Bonnett et al., 2006). All the genotypes along with their characteristic for tolerance to high temperature are listed in Table 8.3.

TABLE 8.3 List of Sugarcane Germplasm for Tolerance to High Temperature

Genotypes	Country	Characteristic	References
Co 99004	India	Lower levels of lipid peroxidation and membrane injury index are accompanied by higher levels of chlorophyll, CSI, RWC, total phenols, NR, and sucrose-metabolizing enzyme activities, soluble sugar content, and leaf gas exchange	Kohila and Gomathi (2018)
Lho 83-153	Pakistan	Higher membrane stability, lower lipid peroxidation of membranes	Abbas et al. (2013)
HSF-242	Pakistan	Higher membrane stability, lower lipid peroxidation of membranes	Abbas et al. (2013)
S-2002-US-133	Pakistan	Higher membrane stability, lower lipid peroxidation of membranes	Abbas et al. (2013)

8.3.4 LOW TEMPERATURE/FREEZING TOLERANCE

Temperature plays a crucial role in the synthesis and accumulation of sucrose in most of the C_4 plants, such as sugarcane (Pathak et al., 2018). In C_4 plants, the phosphoenol pyruvate carboxykinase synthesis is temperature-sensitive (Hatch and Slack, 1998). Juice acidity increases in cold temperatures because of stored sucrose conversion in the stalk and the production of organic acids (Panje et al., 1968). In subtropical areas during winter, when the temperature goes below 10°C, the activities of sucrose-synthesizing and hydrolyzing enzymes are affected which resulting drop of sucrose recovery (Solomon et al., 1997). Although the possibility of time with calendar months can be somewhat significant, the quality of sugarcane is primarily determined by the weather sequences that the crop experiences at harvest (Prasada, 1997; Ram et al., 1973). Cultivar HoCP04-838 contained the highest trans-aconitic acid in juice before the freeze, and the cultivar rankings persisted after the freeze. The commercial cultivar Co 0238 and the wild genome introgressed hybrids (AS04-635, AS04-1687, GU07-2276, GU07-3849, AS04-245, AS04-2097, and GU07-3774) showed excellent/good potential for cold tolerance (Kumar et al., 2022).

8.3.5 DISEASES

Sugarcane crop requires near about 12–14 months to complete its life cycle, and it suffers from adverse agroclimatic conditions during this long duration. Specifically, pathogens and insect pests are the major factors to downregulate its yield (20%). Fungal pathogens pose the most significant challenge during pathogenic diseases (Singh 2017; Joshi et al., 2022, 2023). It has been observed that over 100 fungi can cause diseases in sugarcane (Subhani et al., 2008). The most important sugarcane diseases are wilt, red rot, whip smut, leaf blast, sugarcane mosaic virus, pineapple disease, ratoon stunting disease, leaf scald, mottled stripe, and pokkah boeng. Numbers of disease resistance germplasm are listed in Table 8.4.

8.3.6 INSECT PEST

Insect pests are a major problem for sugarcane producers throughout the world (Dessoky et al., 2021). Lepidoptera stem borers are one of the

TABLE 8.4 List of Sugarcane Germplasm for Resistance to Different Diseases

Genotypes	Country	Characteristic	References
Baragua	India	Stem rot resistance	Vishwanathan et al. (2017)
Koelz 11131	India	Stem rot resistance	Vishwanathan et al. (2017)
Koelz 11132	India	Stem rot resistance	Vishwanathan et al. (2017)
RB867515	Brazil	Resistance to smut	Cursi et al. (2022)
Cop 11436	India	Resistance to smut	Minnatullah et al. (2016)
Cop 11437	India	Resistance to smut	Minnatullah et al. (2016)
Cop 11438	India	Resistance to smut and wilt	Minnatullah et al. (2016)
Cop 11439	India	Resistance to smut	Minnatullah et al. (2016)
Cop 11440	India	Resistance to smut	Minnatullah et al. (2016)
CoSe 11451	India	Resistance to smut	Minnatullah et al. (2016)
BO 153	India	Resistance to smut	Minnatullah et al. (2016)
CoSe 11455	India	Resistance to smut	Minnatullah et al. (2016)
BO 130BO 91	India	Resistance to smut	Minnatullah et al. (2016)
CoP 9301	India	Resistance to smut and wilt	Minnatullah et al. (2016)
CoP 08436	India	Resistance to smut	Minnatullah et al. (2016)
CoSe 09452	India	Resistance to smut	Minnatullah et al. (2016)
BO 155	India	Resistance to smut and wilt	Minnatullah et al. (2016)
CoSe 10451	India	Resistance to smut	Minnatullah et al. (2016)
CoSe 10452	India	Resistance to smut and wilt	Minnatullah et al. (2016)
CoSe 10453	India	Resistance to smut	Minnatullah et al. (2016)
BO 154	India	Resistance to smut and wilt	Minnatullah et al. (2016)
CoP 09437	India	Resistance to smut	Minnatullah et al. (2016)
YZ 01-1413	China	Resistance to smut	Su et al. (2016)
LC 05-136	China	Resistance to smut	Su et al. (2016)
YT 96-86	China	Resistance to smut	Su et al. (2016)

major pests of sugarcane (Weng et al., 2011). The stem borer (*Diatraea saccharalis*) is found in South America, Central America, the Caribbean, and the southern United States (Rossato et al., 2011). *Emmalocera depressalis* (root borer) is found in India and Pakistan. *Chiloterrenellus* (Sugarcane top borer) is found in Bangladesh, Thailand, and Australia (Goebel and Way, 2003). *Sesamia inferens* (pink borer) is found in Asia, Cambodia, China, Hong Kong, and India. *Eoreuma loftini* (Mexican rice borer) and *Sesamia cretica* (pink stem borer) are found in the Mediterranean basin and extend via the Middle East and Arabia to Pakistan, northern

India, and northern Africa, with the latter a pair reaching as far south as northern Kenya and northern Cameroon. About 25–30% of the yield is lost as a result of these borers (Kalunke et al., 2009). Large numbers of germplasms are developed for insect pest resistance (Table 8.5).

TABLE 8.5 List of Sugarcane Germplasm for Resistance to Different Insect Pest

Name of Genotypes	Country	Characteristic	References
CoVC 99463	India	Less susceptible to early shoot borer	Umashankar et al. (2017)
Co 86032	India	Less susceptible to early shoot borer	Umashankar et al. (2017)
IM76-227	Brazil	Resistance to spittlebugs	Valverde et al. (2018)
H. Kawandang	Brazil	Resistance to spittlebugs	Valverde et al. (2018)
IN8473	Brazil	Resistance to spittlebugs	Valverde et al. (2018)
Co 8014	India	Less susceptible to early shoot borer	

8.3.7 BIOMASS AND RENEWABLE ENERGY

Enhancing sugar yield and stress tolerance were the main objectives for sugarcane genetic advancement (Verma et al., 2021d, 2021e, 2022c; Zhao et al., 2022). Sugarcane and related grasses are also used for the production of biofuel and biomass. Sugarcane can produce a nearly 100 t/ha yield, which is double or more than other crops yields. Related grasses with similar capacities, such as *S. spontaneum* and *Erianthus,* can also generate large amounts of biomass (Jackson 2005). *Erianthus* produces almost 100 t/ha of biomass, is tolerant to drought and flooding, and can be cultivated in less than ideal circumstances. Compared to sugarcane, which yields 32 t/ha of wet bagasse, *Erianthus* produces 60 t/ha and contains 25–30% fiber (Joaquim et al., 2011). *Erianthus* is a perfect energy crop because of its multiratoon potential, resistance to biotic and abiotic stress, and high fiber content. While switch grass is widely used in the USA for energy generation, *Miscanthus* is already cultivated as an energy crop in Europe. Given its greater biomass yield and improved adaptability, *Erianthus*—basically a tropical grass—has a far higher potential as an energy crop (Meena et al., 2024).

Superior clones of the *Erianthus* germplasm at SBI have been identified after it was evaluated for biomass potential. A few of them were imported for extensive farming to serve as raw materials for the paper industry. Additionally, efforts were made to increase the amount of fiber in *Erianthus* by selfing the high-fiber clones and identifying progenies with 30% fiber (Vilarinho et al., 2018). *S. spontaneum* and *Erianthus* have also been incorporated into commercial cultivars to create energy canes with significant biomass potential. Energy canes of two types, type I containing more than 20% fiber and more than 15% sucrose, and type II containing more than 25% fiber and less than 15% sucrose, have been produced and are undergoing experiments in factories to see if they can be cultivated as energy plantations. Erianthus's high biomass potential makes it a prime option for cellulosic ethanol production as well (Botha, 2009).

8.4 MOLECULAR BREEDING IN SUGARCANE

Traditional breeding enhancement is very expensive, time-consuming, and labor-intensive due to the low fertility and genetic complexity of sugarcane cultivated naturally (Budeguer et al., 2021; Kumari et al., 2022, 2023a, 2023b). Sugarcane is a promising option for crop enhancement by genetic engineering because of its vegetative propagation, which facilitates steady transfer and replication of transgenes (Verma et al., 2022d, 2022e, 2022f, 2022g, 2022h; Krishna et al., 2022). Furthermore, the lack of resistance in the crop germplasm restricts the ability of conventional breeding to generate insect resistance in sugarcane (Verma et al., 2022i, 2022j, 2022k). Insect-resistant gene introduction, such as the introduction of Bt genes, is one of the cost-effective methods for increasing the resistance of various plants to insects. List of different genotypes, which developed through molecular breeding or transgenic breeding for biotic and abiotic stress, are listed in Tables 8.6–8.8.

8.5 CONCLUSION

Adopting appropriate variety and management practices plays a key role in improving the yield and productivity of the sugarcane crop. A number of germplasms with genetic variability are available today for cultivation. Presence of this genetic variability provides a base for breeders to improve the quality, yield, and performance of sugarcane under various biotic and abiotic stresses. The collection of sugarcane germplasm is well characterized, documented, and maintained properly according to various

TABLE 8.6 List of Transgenic Sugarcane Engineered for Diseases Resistance

Genotypes/ Varieties	Disease Targeted	Transgene	Method Applied	References
Q95, Q153, and Q155	SCMV	CP	Particle bombardment	Joyce et al. (1998)
CP65-357 and CP72-1210	SrMV	CP	Particle bombardment	Ingelbrecht et al. (1999)
Q63 and Q87	Leaf scald	Albicidin detoxifying	Particle bombardment	Zhang et al. (1999)
Ja60-5 and B4362	Brown rust	Glucanase and a chitinase	Agrobacterium	Enriquez et al. (2000)
Q124	FDV	Segment 9 of ORF 1	Particle bombardment	McQualter et al. (2004)
CP 84-1198 and CP 80-1827	SCMV	CP	Particle bombardment	Gilbert et al. (2005)
CP 92-1666	SCYLV	CP	Particle bombardment	Gilbert et al. (2009)
H62-4671	SCYLV	CP	Particle bombardment	Zhu et al. (2011)
ROC22	SrMV	CP	Agrobacterium	Guo et al. (2015)
CoJ 83	Red rot	β-1,3-glucanase	Agrobacterium	Nayyar et al. (2017)
Bululawang	SCMV	CP	Agrobacterium	Apriasti et al. (2018)
SPF-234 and NSG-311	SCMV	CP	Particle bombardment	Aslam et al. (2018)
S2006SP-93	Red rot	Chitinase class-II	Particle bombardment	Tariq et al. (2018)

TABLE 8.7 List of Transgenic Sugarcane Engineered for Insect Pest Resistance

Genotypes/ Cultivars	Targeted Pest	Transgene	Method Used	References
Ja 60-5	*D. saccharalis*	*cry1Ab*	Electroporation	Arencibia et al. (1997)
SP 80-1842	*D. saccharalis*	*cry1Ab*	Particle bombardment	Braga et al. (2001, 2003)
CP 65-357	*Eoreuma loftini, D. saccharalis*	Snowdrop lectin	Paint-sprayer delivery	Sétamou et al. (2002)
SP 80-1842 and SP 80-3280	*D. saccharalis*	Soybean Kunitz trypsin inhibitor (skti) Soybean	Particle bombardmen	Falco and Silva-Filho (2003)
YT 79-177 and ROC16	*P. venosatus*	Synthetic-cry1Ac	Particle bombardment	Weng et al. (2006)
ROC25	*D. saccharalis*	Fusion Amaranthus viridis agglutinin and skti genes	Agrobacterium	Deng et al. (2008)

TABLE 8.7 *(Continued)*

Genotypes/ Cultivars	Targeted Pest	Transgene	Method Used	References
SP80-185	*Sphenophorus levis*	HIS Cane CPI-1	Plasmid transformation	Ribeiro et al. (2008)
Co C671	*C. infuscatellus, C. sacchariphagus*	*cry1Aa3*	Agrobacterium	Kalunke et al. (2009)
FN15 and ROC22	*D. saccharalis*	*cry1Ac*	Particle bombardment	Zhou et al. (2018)
ROC22	*C. sacchariphagus*	*cry2A*	Particle bombardment	Gao et al. (2018)

TABLE 8.8 List of Transgenic Sugarcane Engineered for Abiotic Stress

Genotypes/ Cultivars	Abiotic Stress	Transgene	Method Applied	References
ROC10	Drought	*Tsase*	Agrobacterium	Zhang et al. (2006)
RB855536	Cold	*ipt*	Biolistic	Belintani et al. (2012)
RB855156	Salinity	*P5CS*	Biolistic	Guerzoni et al. (2014)
Co 86032	Drought/ Salinity	*pUBI*	Agrobacterium/ biobalistic	Augustine et al. (2015b)
ROC22	Cold	*SoTUA*	Agrobacterium	Chen et al. (2021)

agroclimatic zones of India as well as their resistance to biotic and abiotic stresses. As sugar and energy requirements are increasing day by day due to an increased population, more efforts are required to identify the germplasm with improved traits. The large genetic variability present across the sugarcane species and in their related wild genera offers enormous possibilities of further improvement, which will be helpful to meet the future need of human beings.

KEYWORDS

- **adverse environmental variables**
- **biomass**
- **genetic variations**
- ***Saccharum* spp**
- **pathogenic diseases**

REFERENCES

Abbas SR, Ahmad SD, Sabir SM, Shah AH. (2014). Detection of drought tolerant sugarcane genotypes (*Saccharum officinarum*) using lipid peroxidation, antioxidant activity, glycine-betaine and proline contents. *J. Soil. Sci. Plant. Nut.* 14, 233–243.

Abbas SR, Shah AH, Rao A. (2013). Classification of sugarcane genotypes based on heat stress and morphological parameters. *J. Agric. Stud.* 1 (2), 114–122.

Aitken K, Jackson PA, McIntyre CL. (2006). Quantitative trait loci identified for sugar related traits in a sugarcane (*Saccharum* spp.) cultivar x *Saccharum officinarum* population. *Theor. Appl. Genet.* 112 (7), 1 306–1 317.

Anas M, Verma KK, Haq IU, Naeem M, Li Q, Liao F, Xiang YX, Li YR. (2021a). Characterization of exotic, native and wild-type genotypes of sugarcane (*Saccharum* spp. Hybrids) for internal nitrogen use efficiency under different nitrogen levels. *Sugar Tech.* 23, 1258–1267. https://doi.org/10.1007/s12355-021-01014-1

Anas M, Verma KK, Riaz M, Qiang L, Liao F, Liu Y, Li YR. (2021b). Physio-morphological and biochemical mechanism of nitrogen use efficiency in sugarcane (*Saccharum* spp.) genotypes under different growth stages and nitrogen levels. *J. Plant. Interact.* 16, 332–343.

Apriasti R, Widyaningrum S, Hidayati WN, Sawitri WD, Darsono N, Hase T, Sugiharto B. (2018). Full sequence of the coat protein gene is required for the induction of pathogen-derived resistance against sugarcane mosaic virus in transgenic sugarcane. *Mol. Biol Rep.* 45, 2749–2758. https://doi.org/10.1007/s11033-018-4326-1.

Arencibia A, Vazquez RI, Prieto D, Tellez P, Carmona ER, Coego A, Hernandez L, De la Riva GA Selman-Housein G. (1997). Transgenic sugarcane plants resistant to stem borer attack. *Mol. Breed.* 3, 247–255. https://doi.org/10.1023/A:1009616318854.

Aslam U, Tabassum B, Nasir IA, Khan A., Husnain T. (2018). A virus derived short hairpin RNA confers resistance against sugarcane mosaic virus in transgenic sugarcane. *Transgenic Res.* 27, 203–210. https://doi.org/10.1007/s11248-018-0066-1.

Augustine SM, Ashwin Narayan J, Syamaladevi DP, Appunu C, Chakravarthi M, Ravichandran V, Tuteja N. Subramonian N. (2015b). Overexpression of EaDREB2 and pyramiding of EaDREB2 with the pea DNA helicase gene (PDH45) enhance drought and salinity tolerance in sugarcane (Saccharum spp. hybrid). *Plant. Cell. Rep.* 34, 247–263. https://doi.org/10.1007/s00299-014-1704-6.

Babu KSD, Janakiraman V, Palaniswamy H, Kasirajan L, Gomathi R, Ramkumar TR. (2022). A short review on sugarcane: Its domestication, molecular manipulations and future perspectives. *Genet. Resour. Crop. Evol.* 69, 2623–2643.

Bajaj Y. (1995). Cryopreservation of plant cell, tissue, and organ culture for the conservation of germplasm and biodiversity. In: *Cryopreservation of Plan Germplasm I.* Springer; pp. 3–28.

Belintani NG, Guerzoni JTS, Moreira RMP, Vieira, LGE. (2012). Improving low-temperature tolerance in sugarcane by expressing the IPT gene under a cold inducible promoter. *Biol. Plant.* 56, 71–77. https://doi.org/10.1007/s10535-012-0018-1.

Berding N. Roach BT. (1987). Germplasm collection, maintenance, and use. In: Heinz DJ. (Ed.), *Sugarcane Improvement Through Breeding*. Amsterdam: Elsevier; pp. 143–210.

Bhatt R, Verma KK, Jamwal NS. (2023a). Can paddy-compost eliminate N fertilization for climate-smart sugarcane? *Sugar. Tech.* 25, 1036–1046.https://doi.org/10.1007/s12355-023-01270-3

Bhatt R, Verma KK, Kumar R, Sanghera GS. (2023b). Foliar application of potassium salt of active phosphorus (PSAP) mitigates insect pests and improves yield along with sugarcane quality in response to agroclimatic conditions of Punjab. *Sugar Tech.* 25(3), 660–669.

Binbol NL, Adebayo AA, Kwon-Ndung EH. (2006). Influence of climatic factors on the growth and yield of sugarcane at Numan, *Nigeria. Clim. Res.* 32 (3), 247–252.

Bonnett GD, Hewitt ML, Glassop D. (2006). Effects of high temperature on the growth and composition of sugarcane internodes. *Aust. J. Agric. Res.* 57, 1087–1095.

Botha FC. (2009). Energy yield and cost in a sugarcane biomass system. *Proc. Austr. Soc. Sugar Cane Technol.* 31, 1–10.

Braga DPV, Arrigoni EDB, Burnquist WL, Silva-filho MC, Ulian EC. (2001). A new approach for control of Diatraeasaccharalis (Lepidoptea: Crambiae) through the expression of an insecticidal cryIa(b) protein in transgenic sugarcane," In: *International Society of Sugar Cane Technologists. Proceedings of the XXIV Congress, Brisbane*, Vol. 2 (Mackay, QLD: Australian Society of Sugar Cane Technologists. pp. 331–336.

Braga DPV, Arrigoni EDB, Silva-Filho MC, Ulian EC. (2003). Expression of the Cry1Ab protein in genetically modified sugarcane for the control of Diatraeasaccharalis (Lepidoptera: Crambidae). *J. New Seeds.* 5, 209–221. https://doi.org/10.1300/J153v05n02_07.

Brown A. (1989). Core collections: A practical approach to genetic resources management. *Genome*. 31, 818–824.

Budeguer F, Enrique R, Perera MF, Racedo J, Castagnaro AP, Noguera AS, Welin B. (2021). Genetic transformation of sugarcane, current status and future prospects. *Front. Plant Sci.* 12, 768609. https://doi.org/10.3389/fpls.2021.768609.

Chandran K. (2010). In vitro multiplication and conservation of Saccharum germplasm. *Indian. J. Plant. Genet.* Res. 23, 65–68.

Chaudhary P, Xu M, Ahamad L, Chaudhary A, Kumar G, Adeleke BS, Verma KK, Hu D-M, Siric I, Kumar P, Popescu SM, Fayssal SA. (2023). Application of synthetic consortia for improvement of soil fertility, pollution remediation, and agricultural productivity: A review. *Agronomy*. 13, 643. https://doi.org/10.3390/agronomy13030643.

Chen GL, Zheng FJ, Lin B, Lao SB, He J, Huang Z, Zeng Y, Sun J, Verma KK. (2020). Phenolic and volatile compounds in the production of sugarcane vinegar. *ACS Omega*. 5, 30587–30595.

Chen G-L, Zheng F-J, Lin B, Yang Y-X, Fang X-C, Verma KK, Yang L-F. (2023). Vinegar: A potential source of healthy and functional food with special reference to sugarcane vinegar. *Front. Nutr*. 10, 1145862. https://doi.org/10.3389/fnut.2023.1145862

Chen J, Khan Q, Sun B, Tang L, Yang L, Zhang B, Xiu X, Dong D. Li, Y. (2021). Overexpression of sugarcane SoTUA gene enhances cold tolerance in transgenic sugarcane. *Agron. J.* 20, 2–21. https://doi.org/10.1002/agj2.20618.

Chen XJ, Min DH, Yasira TA, Hu YG. (2012). Evaluation of 14 morphological, yield-related and physiological traits as indicators of drought tolerance in Chinese winter bread wheat revealed by analysis of the membership function value of drought tolerance (MFVD). *Field. Crop. Res.* 137, 195–201.

Comstock J, Schnellt R, Miller J. (1995). Current status of the world sugarcane germplasm collection in Florida. In: Croft BJ, Piggin CM, Wallis ES, Hogarth DM (Eds.), *Sugarcane Germplasm Conservation and Exchange*. : ACIAR Proceedings: Canberra; pp. 17–18.

Contiliani DF, Nebo JFCO, Ribeiro RV, Andrade LM, Junior RFP, Lembke CG, Machado RS, Silva DN, Belloti M, de Souza GM, Perecin D, Pereira TC, Pires RCM, Fontoura PR, Landell MGA, Figueira A, Creste S. (2022). Leaf transcriptome profiling of contrasting sugarcane genotypes for drought tolerance under field conditions. *Sci. Rep.* 12. 9153. https://doi.org/10.1038/s41598-022-13158-5.

Contiliani DF, Nebo JFCO, Ribeiro RV, Andrade LM, Junior RFP, Lembke CG, Machado RS, Silva DN, Belloti M, Souza GM, Perecin D, Pereira TC, Pires RCM, Fontoura PR, Landell MGA, Figueira A, Creste S. (2022). Leaf transcriptome profiling of contrasting sugarcane genotypes for drought tolerance under field conditions. *Sci. Rep.* 12(1), 9153. https://doi.org/10.1038/s41598-022-13158-5.

Cox M, Hogarth M, Smith G. (2000). Cane breeding and improvement, Chapter 5, In: Hogarth M, Allsopp P. (eds.), *Manual of Cane Growing, Bureau of Sugar Experimental Stations*, Indooroopilly, Australia; pp. 91–108.

Cruz FJR, Ferreira JDC, Santos DMM. (2018). Low salt stress affects physiological parameters and sugarcane plant growth. *Aust. J. Crop Sci.* 12(8), 1272–1279.

Crystian D, Santos JM, Barbosa GVS, Almeida C. (2018). Genetic diversity trends in sugarcane germplasm: Analysis in the germplasm bank of the RB varieties. *Crop Breed Appl. Biotechnol.* 18, 426–431. http://dx.doi.org/10.1590/1984-70332018v18n4n62.

Cursi DE, Hoffmann HP, Barbosa GVS, Bressiani JA, Gazaffi R, Chapola RG, Fernandes Junior AR, Balsalobre TWA, Diniz CA, Santos J.M, Carneiro MS. (2022). History and current status of sugarcane breeding, germplasm development and molecular genetics in Brazil. *Sugar Tech.* 24, 112–133. https://doi.org/10.1007/s12355-021-00951-1

D'Hont A, Ison D, Alix K, Roux C, Glaszmann JC. (1998). Determination of basic chromosome numbers in the genus Saccharum by physical mapping of ribosomal RNA genes. *Genome*. 41(2), 221–225.

Daniels J, Roach BT. (1987). Taxonomy and evolution", Chapter 2, In: Heinz DJ. (Ed.), *Sugarcane Improvement through Breeding*. Elsevier: Amsterdam, The Netherlands; vol. 11, pp. 7–84.

Deng ZN, Huang HR, Huang CM, Verma KK, Luo HB, Jiang SL, Li YJ, Cao HQ, Wu KC, Xu L. (2023). Short-term limited water irrigation influences the photosynthetic pigments, enzymatic and non-enzymatic activities in *Saccharum spontaneum* L. at vegetative stage. *Sugar. Tech.* 25, 110–118. https://doi.org/10.1007/s12355-022-01175-7

Deng ZN, Wei YW, Lu WL, Li YR. (2008). Fusion insect-resistant gene mediated by matrix attachment region (MAR) sequence in transgenic sugarcane. *Sugar. Tech.* 10, 87–90. https://doi.org/10.1007/s12355-008-0015-z.

Dessoky ES, Ismail RM, Elarabi NI, Abdelhadi AA, Abdallah NA. (2021). Improvement of sugarcane for borer resistance using Agrobacterium mediated transformation of cry1Ac gene. *GM. Crops. & Food.* 1, 12(1), 47–56.

Enriquez GA, Trujillo LE, Menendez C, Vazquez RI, Tiel K, Dafhnis F, Arrieta J, Selman G, Hernandez L. (2000). Sugarcane (Saccharun hybrid) genetic transformation mediated by Agrobacterium tumefaciens: Production of transgenic plants expressing proteins with agronomic and industrial value, In Arencibia AD. (Ed.), *Plant Genetic Engineering: Towards the Third Millenium*, Amsterdam: Elsevier; pp. 76–81.

Falco MC, Silva-Filho MC. (2003). Expression of soybean proteinase inhibitors in transgenic sugarcane plants: Effects on natural defense against Diatraeasaccharalis. *Plant. Physiol. Bioch.* 41, 761–766. https://doi.org/10.1016/S0981- 9428(03)00100-1.

FAO. FAOSTAT. (2015). FAO Statistical Databases 2013. Available at: http://faostat3.fao.org/.

Fauconnier R. (1993). *Sugar Cane*, Macmillan Press Ltd.: London; pp. 1–140.

Ferreira TH, Tsunada MS, Bassi D, Araújo P, Righetto GL, Lakshmanan P, Menossi M. (2017). Sugarcane water stress tolerance mechanisms and its implications on developing biotechnology solutions. *Front. Plant. Sci.* 8, 1077. https://doi.org/10.3389/fpls.2017.01077.

Fickett ND, Ebrahimi L, Parco AP, Gutierrez AV, Hale AL, Pontif MJ, Todd J, Kimbeng CA, Hoy JW, Ayala-Silva T, Gravois KA, Baisakh N. (2020). An enriched sugarcane diversity panel for utilization in genetic improvement of sugarcane. *Sci. Rep.* 10, 13390. https://doi.org/10.1038/s41598-020-70292-8.

Frison EA, Putter CAJ. (1993). FAO/IBPGR Technical guidelines for the safe movement of sugarcane germplasm. Food and Agriculture Organization of the United Nations, Rome/International Board for Plant Genetic Resources, Rome.

Gao S, Yang Y, Xu, L, Guo J, Su, Y, Wu, Q, Wang C, Que Y. (2018). Particle bombardment of the cry2A gene cassette induces stem borer resistance in sugarcane. *Int. J. Mol. Sci.* 19, 1692. https://doi.org/10.3390/ijms19061692.

Gilbert RA, Gallo-Meagher M, Comstock JC, Miller J.D, Jain M. Abouzid A. (2005). Agronomic evaluation of sugarcane lines transformed for resistance to sugarcane mosaic virus strain E. *Crop. Sci.* 45, 2060–2067.

Gilbert RA, Glynn NC, Comstock JC, Davis MJ. (2009). Agronomic performance and genetic characterization of sugarcane transformed for resistance to sugarcane yellow leaf virus. *Field. Crops. Res.* 111, 39–46.

Goebel FR and Way M. (2003). Investigation of the impact of Eldanasaccharina (Lepidoptera: pyralidae) on sugarcane yield in field trials in Zululand. *Proc. S. Afr. Sugar Technol. Assoc.* 7, 256–65.

Gomathi R, Vasantha S, Hemaprabha G, Alarmelu S, Shanthi RM. (2011). Evaluation of elite sugarcane clones for drought tolerance. *J. Sugarcane. Res.* 1(1), 55–62.

Govindaraj P. (2017). Energy canes as a feed stock for biofuel industries. *Indian Farm.* 67 (2), 61–63.

Govindaraj P, Balamurugan A, Natarajan US. (2012). Identification of intergeneric hybrids between *Erianthus arundinaceus* and *Saccharum spontaneum* through STMS markers. *Int. Sugar J.* 114, 350–358.

Guerzoni JTS, Belintani NG, Moreira RMP, Hoshino AA, Domingues DS, Filho JCB, Vieira LGE. (2014). Stress-induced Δ1-pyrroline-5-carboxylate synthetase (P5CS) gene confers tolerance to salt stress in transgenic sugarcane. *Acta. Physiolog. Plant.* 36, 2309–2319. https://doi.org/10.1007/s11738-014-1579-8.

Guo D-J, Li D-P, Chen Z-D, Yang B, Verma KK, Singh RK, Singh P, Khan Q, Sharma A, Qin Y, Zhang B-Q, Song X-P, Li Y-R. (2023). Effect of endophytic diazotroph *Enterobacter roggenkampii* ED5 on nitrogen-metabolism-related microecology in the sugarcane rhizosphere at different nitrogen levels. *Front. Microbiol.* 14, 1132016. https://doi.org/10.3389/fmicb.2023.1132016

Guo D-J, Li D-P, Singh RK, Singh P, Sharma A, Verma KK, Qin Y, Khan Q, Lu Z, Malviya MK, Song X-P, Xing Y-X and Li Y-R (2021). Differential protein expression analysis of two sugarcane varieties in response to diazotrophic plant growth-promoting endophyte *Enterobacter roggenkampii* ED5. *Front. Plant Sci.* 12, 727741. https://doi.org/10.3389/fpls.2021.727741

Guo DJ, Li DP, Singh RK, Singh P, Verma KK, Sharma A, Qin Y, Khan Q, Song XP, Malviya MK, Xing YX, Li YR. (2022). Comparative transcriptome analysis of two sugarcane varieties in response to diazitrophic plant growth promoting endophyte *Enterobacter roggenkampii* ED5. *J. Plant Interact.* 17(1), 75–84.

Guo J, Gao S, Lin Q, Wang H, Que Y, Xu L. (2015). Transgenic sugarcane resistant to Sorghum mosaic virus based on coat protein gene silencing by RNA interference. *Bio. Med. Res. Int.* 015, 861907. https://doi.org/10.1155/2015/861907.

Hale AL, Todd JR, Gravois KA, Mollov D, Malapi-Wight M, Momotaz A, Laborde C, Goenaga R, Kimbeng C, Solis A, Waguespack H. (2022). Sugarcane breeding programs in the USA. *Sugar Tech.* 24, 97–111. https://doi.org/10.1007/s12355-021-01018-x.

Hatch MD, Slack CR. (1998). C4 *Photosynthesis Discovery, Resolution, Recognition and Significance in Discoveries in Plant Biology*, 1st edn.; Kung S-D, Yang S-F. (Eds.), World Scientific Publishing Co Inc.: River Edge, NJ, USA. ISBN 9810213131.

Hemaprabha G, Mohanraj K, Jackson PA, Lakshmanan P, Li, A, Huang DL, Ram B. (2022). Sugarcane genetic diversity and major germplasm collections. *Sugar Tech.* 24, 279–297.

Hoang NV, Furtado A, Botha FC, Simmons BA, Henry RJ. (2015). Potential for genetic improvement of sugarcane as a source of biomass for biofuels. *Front. Bioeng. Biotechnol.* 3. 182. https://doi.org/10.3389/fbioe.2015.00182.

Huang CM, Wu KC, Verma KK, Luo HB, Deng ZN, Shi GY, Cao HQ, Wu XJ, Ye LP, Wei YW, Xu L. (2024). Isolation and expression analysis of cellulose synthase 3 (*Ces3*) genes from sugarcane (*Saccharum officinarum* L.). *Chem. Biol. Technol. Agric.* 11, 71. https://doi.org/10.1186/s40538-024-00597-7

Huang HR, Gao YJ, Malviya MK, Verma KK, Solanki MK, Huang YX, Li X, Deng Y, Yan J, Tang SY, Wang LW, Xu L. (2022). Genetic diversity analysis of sugarcane (*Saccharum* spp. Hybrids) among high-sucrose clones of GT series and commonly used parents by using microsatellite markers in Guangxi, China. *Sugar Tech.* 24, 397–407. https://doi.org/10.1007/s12355-021-01033-y

Ingelbrecht IL, Irvine JE, Mirkov TE. (1999). Posttranscriptional gene silencing in transgenic sugarcane. Dissection of homology-dependent virus resistance in a monocot that has a complex polyploid genome. *Plant. Physiol.* 119, 1187–1197. https://doi.org/10.1104/pp.119.4.1187.

Irvine JE. (1999). Saccharum species as horticultural classes. *Theor. Appl. Genet.* 98(2), 186–194.

Jackson P. (2005). Breeding for improved sugar content in sugarcane. *Field. Crops. Res.* 2, 277–290.

Jagadesh M, Dash M, Kumari A, Singh SK, Verma KK, Kumar P, Bhatt R, Sharma SK. (2024). Revealing the hidden world of soil microbes: Metagenomic insights into plant, bacteria, and fungi interactions for sustainable agriculture and ecosystem restoration. *Microbiol. Res.* 127764. https://doi.org/10.1016/j.micres.2024.127764

Joaquim E, Isaias C, Helena L, Carlos E, Celso A. (2011). Life cycle assessment of Brazilian sugarcane products: GHG emissions and energy use. *Biof. Bioprod. Bioref.* 5, 519–532.

Joshi A, Verma KK, Rajput VD, Arora J. (2023). Metabolic engineering of microorganisms in advancing biofuel production. In: *Advances in Lignocellulosic Biofuel Production Systems*. pp. 181–202. https://doi.org/10.1016/B978-0-323-91192-4.00016-X

Joshi A, Verma KK, Rajput VD, Minkina T, Arora J. (2022). Recent advances in metabolic engineering of microorganisms for advancing lignocellulose-derived biofuels, *Bioengineered*. 13(4), 8135–8163. https://doi.org/10.1080/21655979.2022.2051856.

Joyce PA, McQualter RB, Bernard MJ, Smith GR. (1998). Engineering for resistance to SCMV in sugarcane. *Acta. Horticulturae*. 461, 385–392. https://doi.org/10.17660/ActaHortic.1998.461.44.

Kalunke MR, Archana MK, Babu KH, Prasad DT. (2009). Agrobacterium-mediated transformation of sugarcane for borer resistance using Cry1Aa3 gene and one-step regeneration of transgenic plants. *Sugar Tech.* 11(4), 355–59. https://doi.org/10.1007/s12355-009-0061-1.

Kalunke RM, Kolge AM, Babu KH, Prasad DT. (2009). Agrobacterium mediated transformation of sugarcane for borer resistance using Cry 1Aa3 gene and one-step regeneration of transgenic plants. *Sugar Tech.* 11, 355–359. https://doi.org/10.1007/s12355-009-0061-1.

Kaushal N, Bhandari K, Siddique KHM, Nayyar H. (2016). Food crops face rising temperatures: An overview of responses, adaptive mechanisms, and approaches to improve heat tolerance. *Cogent. Food Agric.* 2, 1–42.

Krishna D, Sachan HK, Verma KK. (2022). Abiotic stresses in plants: An overview. In book: In: *Understanding Abiotic Stresses Publisher*: Nova Science Publishers, Inc.: New York.

Kumar R, Meena MR, Dhansu P, Karuppaiyan R, Appunu C, Kulshreshtha N, Kaushik P, Ram B. (2022). Winter tolerance potential of genetically diverse sugarcane clones under subtropical climate of northern India. *Sustainability*. 14(18), 11757. https://doi.org/10.3390/su141811757.

Kumari A, Dash M, Singh SK, Jagadesh M, Mathpal B, Mishra PK, Pandey SK, Verma KK. (2023a). Soil microbes: A natural solution for mitigating the impact of climate change. *Environ. Monit. Assess.* 195, 1436. https://doi.org/10.1007/s10661-023-11988-y

Kumari A, Lakshmi GA, Krishna GK, Patni B, Prakash S, Bhattacharyya M, Singh SK, Verma KK. (2022). Climate change and its impact on crops: A comprehensive investigation for sustainable agriculture. *Agronomy*. 12, 3008. https://doi.org/10.3390/agronomy12123008

Kumari A, Singh SK, Mathpal B, Verma KK, Garg VK, Bhattacharyya M, Bhatt R. (2023b). The biosynthesis, mechanism of action, and physiological functions of melatonin in horticultural plants: A review. *Horticulturae.* 9, 913. https://doi.org/10.3390/horticulturae9080913

Lakshmanan P, Geijskes RJ, Aitken KS, Grof CLP, Bonnett GD, Smith GR. (2005). Sugarcane biotechnology: The challenges and opportunities. *Vitro Cellul. Develop. Biol.* 41 (4), 345–363.

Lam E, Shine J. Jr, Da Silva J, Lawton M, Bonos S, Calvino M, Carrer H, Silva-Filho MC, Glynn N, Helsel Z, Ma, J, Richard E, Souza GM, Ray M. (2009). Improving sugarcane for biofuel: engineering for an even better feedstock. *GCB Bioenergy*. 1, 251–255. https://doi.org/10.1111/j.1757-1707.2009.01016.x.

Li A-M, Liao F, Wang M, Chen Z-L, Qin C-X, Huang R-Q, Verma KK, Li Y-R, Que Y-X, Pan Y-Q, et al. (2023). Transcriptomic and proteomic landscape of sugarcane response to biotic and abiotic stressors. *Int. J. Mol. Sci.* 24, 8913. https://doi.org/10.3390/ijms24108913

Liu JY, Basnayake JPA, Chen XK, Zhao J, Zhao PF, Yang LH, Bai YD, Xia HM, Zan FG, Qin W, Yang K, Yao L, Zhao LP, Zhu JR, Lakshmanan P, Zhao XD, Fan YH. (2016). Growth and yield of sugarcane genotypes are strongly correlated across irrigated and rainfed environments. *Field. Crop. Res.* 196, 418–425.

Machado RMA, Serralheiro RP. (2017). Soil salinity: Effect on vegetable crop growth. Management practices to prevent and mitigate soil salinization. *Horticulturae.* 3(30), https://doi.org/10.3390/horticulturae3020030.

Mahadevaiah C, Hapase P, Sreenivasa V, Hapase R, Swamy HKM, Anilkumar C, Mohanraj K, Hemaprabha G, Ram B. (2021). Delineation of genotype × environment interaction for identification of stable genotypes for tillering phase drought stress tolerance in sugarcane. *Scient. Rep.* 11,18649.

Malviya MK, Li C-N, Lakshmanan P, Solanki MK, Wang Z, Solanki AC, Nong Q, Verma KK, Singh RK, Singh P, Sharma A, Guo D-J, Dessoky ES, Song X-P, Li Y-R. (2022). High-throughput sequencing-based analysis of rhizosphere and diazotrophic bacterial diversity among wild progenitor and closely related species of sugarcane (*Saccharum* spp. inter-specific hybrids). *Front. Plant Sci.* 13, 829337. https://doi.org/*10.3389/fpls.2022.829337.*

Malviya MK, Solanki MK, Li CN, Wang Z, Zeng Y, Verma KK, Singh RK, Singh P, Huang HR, Yang LT, Song XP, Li YR. (2021). Sugarcane-legume intercropping can enrich the soil microbiome and plant growth. *Front. Sustain. Food syst.* 5, 606595. https://doi.org/10.3389/fsufs.2021.606595.

McQualter RB, Dale JL, Hardin RH, McMahon JA, Smith GR. (2004). Production and evaluation of transgenic sugarcane containing a Fiji disease virus (FDV) genome segment S9-derived synthetic resistance gene. *Aust. J. Agric. Res.* 55, 139–145.

Meena MR, Govindaraj P, Kumar RA, Elayaraja K, Appunu C, Kumar R, Chhabra ML, Kulshreshtha N, Hemaprabha G. (2024). Biomass and energy potential of *Erianthus arundinaceus* and *Saccharum spontaneum*-derived novel sugarcane hybrids in rainfed environments. *BMC Plant Biology*. 24, 198. https://doi.org/10.1186/s12870-024-04885-0.

Moore PH, Paterson AH, Tew T. (2013). Sugarcane: The crop, the plant, and domestication. In: Moore PH, Botha FC. (Eds.), *Sugarcane: Physiology, Biochemistry and Functional Biology.* Wiley: New York; pp. 1–15. https://doi.org/10.1002/9781118771280.ch1

Mukherjee SK. (1957). Origin and distribution of Saccharum. *Botanical. Gazette.* 119, 55–61.

Nair NV. (2012). Sugarcane genetic resources-status, potential and role in sugarcane improvement. *J. Sugarc. Res.* 2(2), 1–8.

Nathanael D, Ebrahimi FL, Parco AP, Gutierrez AV, Hale AL, Pontif MJ, Todd J, Kimbeng CA, Hoy JW, Ayala-Silva T, Gravois KA, Baisakh N. (2020). An enriched sugarcane diversity panel for utilization in genetic improvement of sugarcane. *Sci. Rep.* 10, 13390. https://doi.org/10.1038/s41598-020-70292-8.

Nayak SN, Song J, Villa A, Pathak B, Ayala-Silva T, Yang X, Todd J, Glynn NC, Kuhn DN, Glaz B, Gilbert RA, Comstock JC, Wang J. (2014). Promoting utilization of Saccharum spp. genetic resources through genetic diversity analysis and core collection construction. *PLoS One.* 9, e110856. https://doi.org/10.1371/journal.pone.0110856.

Nayyar S, Sharma B. K, Kaur A, Kalia A, Sanghera GS, Thind KS, Yadav IS, Sandhu JS. (2017). Red rot resistant transgenic sugarcane developed through expression of β-1,3-glucanase gene. *PLoS One.* 12, e0179723. https,//doi.org/10.1371/journal.pone.0179723.

Oliveira MA, Verma KK, Bhatt R, Oliveira TBA. (2022). Impact of green and organic fertilizers on soil fertility and sugarcane productivity. In: Verma KK, et al. (eds.), *Agro-Industrial Perspectives on Sugarcane Production Under Environmental Stress*. https://doi.org/10.1007/978-981-19-3955-6_11.

Panje RR, Mathur PS, Motiwala MP. (1968). Studies in the germination of sugarcane: On the effect of cutting cane into two portions on the germination of its bud. In *Proceedings of the 13th Congress ISSCT,* Taipei, Taiwan, p. 635.

Park S, Zhang D, Ali GS. (2024). Assessing the genetic integrity of sugarcane germplasm in the USDA-ARS national plant germplasm system collection using single-dose SNP markers. *Front. Plant. Sci.* 14, 1337736. https://doi.org/10.3389/fpls.2023.1337736.

Pathak S, Singh P, Singh M, Sharma B. (2018). Impact of temperature and humidity on sugar recovery in Uttar Pradesh. *Sugar Tech.* 21, 176–181.

Preet K, Dhansu P, Sehrawat N, Kumar R, Chinnaswamy A, Krishnapriya V, Arunkumar R, Rana R, Kumar S, Joon V. (2024). Morpho-physiological analysis of salinity tolerance in sugarcane genotypes. *Plant Physiol. Rep*. https://doi.org/10.1007/s40502-024-00782-8.

Putra LK, Damayanti TA. (2012). Major diseases affecting sugar production in Indonesia. *Func. Plant Sci. Biotech.* 6(2), 124–129.

Raboin LM, Oliveira KM, Lecunff L, Telismart H, Roques D, Butterfield M, Hoarau JY, D'Hont A. (2006). Genetic mapping in sugarcane, a high polyploid, using bi-parental progeny: Identification of a gene controlling stalk colour and a new rust resistance gene. *TAG Theor. Appl. Genet.* 112(7), 1382–1391. https://doi.org/10.1007/s00122-006-0240-3.

Ram B, Sreenivasan TV, Sahi BK, Singh N. (2001). Introgression of low temperature tolerance and red rot resistance from Erianthus in sugarcane. *Euphytica*. 122(1), 145–153.

Ram P, Singh A, Singh PP. (1973). Pattern of sugar accumulation and yield in sugarcane varieties planted in autumn and spring seasons. *Indian J. Agron.* 18, 189–192.

Rao PKK. (1997). 75 *Years of Agricultural Research at Anaakapalle*. The Platinum Jubilee Committee: Anakapalle, India; pp. 1913–1988

Rhein AFDL, Santos DMMD, Carlin SD. (2011). Nitrate reductase enzyme activity and free proline contents in sugarcane roots underwater and acid stress in soil. *Semina: Ciencia. Agrarias.* 32, 1345–1359.

Ribeiro CW, Soares CA, Falco MC, Chabregas SM, Ulian EC, Cotrin SS, Carmona AK, Santana LA, Oliva MLV, Henrique-Silva F. (2008). Production of a His-tagged cane cystatin in transgenic sugarcane and subsequent purification. *Biotechnol. Progress.* 24, 1060–1066. https://doi.org/10.1007/s12355-008-0048-3.

Rossato JAS, Fernandes OA, Mutton MJR, Higley LG, Madaleno LL. (2011). Sugarcane response to two biotic stressors: *Diatraea saccharalis* and *Mahanarva fimbriolata*. *Int. Sugar. J.* 113, 453–55.

Santos JM, Duarte Filho LSC, Soriano ML, Silva PP, Nascimento VX, Barbosa GVS, Todaro AR, RamalhoNeto CE, Almeida C. (2012). Genetic diversity of the main progenitors of sugarcane from the RIDESA germplasm bank using SSR markers. *Ind. Crops. Prod.* 40, 145–150.

Setamou M, Bernal JS, Legaspi JC, Mirkov TE, Legaspi BCJr. (2002). Evaluation of lectin-expressing transgenic sugarcane against stalkborers (Lepidoptera: pyralidae): effects on life history parameters. *J. Econom. Entomol.* 2, 469–477. https://doi.org/10.1603/0022-0493-95.2.469.

Sharma A, Singh RK, Singh P, Vaishnav A, Guo DJ, Verma KK, Li DP, Song XP, Malviya MK, Khan N, Lakshmanan P, Li YR. (2021). Insights into the bacterial and nitric oxide-induced salt tolerance in sugarcane and their growth-promoting abilities. *Microorganisms*. 9, 2203. https://doi.org/10.3390/microorganisms9112203.

Sharma A, Song X-P , Singh RK, Vaishnav A, Gupta S, Singh P, Guo D-J, Verma KK, Li Y-R. (2022). Impact of carbendazim on cellular growth, defence system and plant growth promoting traits of *Priestia megaterium* ANCB-12 isolated from sugarcane rhizosphere. *Front. Microbiol.* 13, 1005942. https://doi.org/10.3389/fmicb.2022.1005942

Silva AA, Rubio ZCC, Linhares PCA, Silva KR, Pimental GV, marchiori PER (2022). Genotypic variation of sugarcane for salinity tolerance: Morphological and physiological responses. *Ciência e Agrotecnol.* 46, e000122. https://doi.org/10.1590/1413-7054202246000122.

Silva MDA, Jifon JL, Silva JAGD, Sharma V. (2007). Use of physiological parameters as fast tools to screen for drought tolerance in sugarcane. *Braz. J. Plant Physiol.* 19, 193–201.

Sindhu R, Govindaraj P, Balamurugan A, Appunu C. (2011). Assessment of genetic diversity among sugarcane hybrids (Saccharum spp. complex) commercially grown in tropical India using STMS markers. *J. Plant Biochem. Biotechnol.* 20 (1), 118–124.

Singh P, Singh RK, Li HB, Guo DJ, Sharma A, Verma KK, Solanki MK, Upadhyay SK, Lakshmanan P, Yang LT, Li YR. (2023). Nitrogen fixation and phytohormone stimulation of sugarcane plant through plant growth promoting diazotrophic *Pseudomonas*. *Biotechnol. Genetic Eng. Reviews*. 1–21. https://doi.org/10.1080/02648725.2023.2177814

Singh P, Singh RK, Song QQ, Li HB, Guo DJ, Malviya MK, Verma KK, Song XP, Lakshmanan P, Yang LT, Li YR. (2021). Comparative analysis of protein and differential responses of defense-related gene and enzyme activity reveals the long-term molecular responses of sugarcane inoculated with *Sporisorium scitamineum*. *J. Plant Interact.* 16(1), 12–29.

Singh RK, Singh P, Li HB, Song QQ, Guo DJ, Solanki MK, Verma KK, Malviya MK, Song XP, Lakshmanan P, Yang LT, Li YR. (2020). Diversity of nitrogen-fixing rhizobacteria associated with sugarcane: A comprehensive study of plant-microbe interactions for growth enhancement in *Saccharum* spp. *BMC Plant Biol.* 20, 220. https://doi.org/10.1186/s12870-020-02400-9.

Singh P, Singh SP, Tiwari AK, Sharma BL. (2017). Genetic diversity of sugarcane hybrid cultivars by RAPD markers. *3 Biotech.* 7(3), 222. https://doi.org/10.1007/s13205-017-0855-x.

Solanki MK, Verma KK, Dastogeer KMG, Mora-Poblete F, Mundra S. (2023). Microbial resilience in plant nutrient management towards sustainable farming. *Front. Microbiol.* 14, 1280811. https://doi.org/10.3389/fmicb.2023.1280811

Solomon S, Srivastava AK, Srivastava BL, Madan VK. (1997). Pre milling sugar losses and their management in sugarcane. In *Technical Bulletin*, ICAR-IISR: Lucknow, India,

Song X-P, Mo F, Verma KK, Wei J, Zhang X, Yang L, Li YR. (2019). Effect of sugarcane smut (*Ustilago scitaminea* Syd.) on ultrastructure and biochemical indices of sugarcane. *Biomed J. Sci. Tech. Res.* 002950. https://doi.org/10.26717/BJSTR.2019.17.002950.

Song XP, Mo F, Yan M, Zhang X, Zhang B, Huang X, Huang D, Pan Y, Verma KK, Li YR. (2022). Effect of smut infection on the photosynthetic physiological characteristics and related defense enzymes of sugarcane. *Life* 12, 1201. https://doi.org/10.3390/life12081201

Song XP, Verma KK, Tian DD, Zhang XQ, Liang YJ, Huang X, Li CN, Li YR. (2021). Exploration of silicon functions to integrate with biotic stress tolerance and crop improvement. *Biol. Res.* 54, 19. https://doi.org/10.1186/s40659-021-00344-4

Su, Y, Wang Z, Xu, L, Peng Q, Liu F, Li, Z, Que Y. (2016). Early selection for smut resistance in sugarcane using pathogen proliferation and changes in physiological and biochemical indices. *Front. Plant. Sci.* 28 (7), 1133. https://doi.org/10.3389/fpls.2016.01133.

Tariq M, Khan A, Tabassum B, Toufiq N, Bhatti MU, Riaz S, Nasir IA, Husnain T. (2018). Antifungal activity of chitinase II against Colletotrichum falcatum Went. causing red rot disease in transgenic sugarcane. *Turkish J. Biol.* 42, 45.

Taylor PWJ, Dukic S. (1993). Development of an in vitro culture technique for conservation of Saccharum sp. hybrid germplasm. *Plant. Cell. Tissue. Organ Culture.* 34, 217–222.

Tian D, Qin L, Verma KK, Wei L, Li J, Li B, Zhou W, He Z, Wei D, Huang S, Long S, Huang Q, Li C, Wei S. (2023). Transcriptomic and metabolomic differences between banana varieties which are resistant or susceptible to *Fusarium wilt. PeerJ.* 11, e16549. https://doi.org/10.7717/peerj.16549

Tolera B, Gedebo A, Tena E. (2023). Genetic diversity of sugarcane (Saccharum spp.) genotypes based on agro-morphological and biochemical traits. *Cogent Food Agric.* 9, 1. https://doi.org/10.1080/23311932.2023.2254141.

Umashankar HG, Patel VN, Nagaraja T, Vijaykumar L, Sugeetha S. (2017). Screening of sugarcane genotypes for resistance against sugarcane early shoot borer, *Chilo infuscatellus* Snellen. *Int. J. Environ. Agricul. Biotechnol.* 2(6), 3224–3228.

Valverde AHP, Soares BO, Tomaz AC, Pimentel GV, Peternelli LA, Barbosa MHP. (2018). A new methodology for large-scale screening sugarcane resistance to *Mahanarva fimbriolata* (Hemiptera: Cercopidae). *Bragantia.* 77(4): 599–608.

Vasantha S, Gomathi R. (2012). Growth and development of sugarcane under salinity. *J. Sugarcane. Res.* 2 (1), 1–10.

Vandana P, Singh D, Srivastava S, Ram GD. (2020). Effect of climate change on sugarcane crop: A review. *J. Pharmacog. Phytoch.* SP6, 255–261.

Verma KK, Anas M, Chen Z, Rajput VD, Malviya MK, Verma CL, Singh RK, Singh P, Song XP, Li YR. (2020f). Silicon supply improves leaf gas exchange, antioxidant defense system and growth in sugarcane responsive to water limitation. *Plants*. 9, 1032, https://doi.org/10.3390/plants9081032.

Verma KK, Joshi A, Song X-P, Liang Q, Xu L, Huang H-R, Wu K-C, Seth CS, Arora J and Li Y-R (2024a). Regulatory mechanisms of plant rhizobacteria on plants to the adaptation of adverse agroclimatic variables. *Front. Plant. Sci.* 15, 1377793. https://doi.org/10.3389/fpls.2024.1377793

Verma KK, Joshi A, Song X-P, Singh S, Kumari A, Arora J, Singh SK, Solanki MK, Seth CS, Li Y-R. (2024b). Synergistic interactions of nanoparticles and plant growth promoting rhizobacteria enhancing soil-plant systems: a multigenerational perspective. *Front. Plant. Sci.* 15, 1376214. https://doi.org/10.3389/fpls.2024.1376214

Verma KK, Li DM, Singh M, Rajput VD, Malviya MK, Minkina T, Singh RK, Singh P, Song XP, Li YR. (2020a). Interactive role of silicon and plant-rhizobacteria mitigating abiotic stresses: A new approach for sustainable agriculture and climate change. *Plants.* 9, 1055. https://doi.org/10.3390/plants9091055.

Verma KK, Rajput VD, Singh M, Verma CL, Minkina T. (2022a). Influence of waterlogging on plant physiology and molecular biology. In: *Understanding Abiotic Stresses*, Nova Science Publishers, Inc.: New York.

Verma KK, Siddiqui N, Singh M. (2022b). Understanding the plant physiological and molecular adaptation strategies to soil salinity. In: *Understanding Abiotic Stresses*, Nova Science Publishers, Inc., New York. https://doi.org/10.52305/UBUJ4024

Verma KK, Singh P, Song X-P, Malviya MK, Singh RK, Chen G-L, Solomon S, Li YR. (2020b). Mitigating climate change for sugarcane improvement: Role of silicon in alleviating abiotic stresses. *Sugar. Tech.* 22(5), 741–749. https://doi.org/10.1007/s12355-020-00831-0.

Verma KK, Singh RK, Song QQ, Singh P, Zhang B-Q, Song X-P, Chen G-L, Li YR. (2019a). Silicon alleviates drought stress of sugarcane plants by improving antioxidant responses. *Biomed. J. Sci. Tech. Res.* 002957. https://doi.org/10.26717/BJSTR.2019.17.002957

Verma KK, Song X, Degu HD, Guo D, Joshi A, Huang H, Xu L, Singh M, Huang D, Rajput VD, Li YR. (2023a). Recent advances in nitrogen and nano-nitrogen fertilizers for sustainable crop production: A mini-review. *Chem. Biol. Technol. Agricult.* 10, 1–14. https://doi.org/10.1186/s40538-023-00488-3

Verma KK, Song X, Verma CL, Huang H, Singh M, Xu L, Wu J, Khan ZH, Zhang B, Li Y. (2023b). Mathematical modeling of climate and fluoride effects on sugarcane photosynthesis with silicon nanoparticles. *Plant Physiol. Biochem.* 204, 108089. https://doi.org/10.1016/j.plaphy.2023.108089

Verma KK, Song XP, Budeguer F, Nikpay A, Enrique R, Singh M, Zhang B, Wu, JM, Li YR. (2022c). Genetic engineering: An efficient approach to mitigating biotic and abiotic stresses in sugarcane cultivation. *Plant. Signal. Behav.* 17(1), e2108253.

Verma KK, Song XP, Joshi A, Rajput VD, Singh M, Sharma A, Singh RK, Li DM, Arora J, Minkina T, Li YR. (2022d). Nanofertilizer possibilities for healthy soil, water and food in future: An overview. *Front. Plant. Sci.* 13, 865048. https://doi.org/10.3389/fpls.2022.865048

Verma KK, Song XP, Joshi A, Tian DD, Rajput VD, Singh M, Arora J, Minkina TM, Li YR. (2022e). Recent trends in nano-fertilizer for advancing sustainable agriculture under the era of climate change to ensure food security for future generations. *Nanomaterials* 12, 173. https://doi.org/10.3390/nano12010173.

Verma KK, Song XP, Kumari A, Jagadesh M, Singh SK, Bhatt R, Singh M, Seth CS, Li YR. (2024c). Climate change adaptation: Challenges for agricultural sustainability. *Plant. Cell. Environ.* https://doi.org/10.1111/pce.15078

Verma KK, Song XP, Li DM, Singh M, Wu JM, Singh RK, Sharma A, Zhang BQ, LI YR. (2022f). Silicon and soil microorganisms improves rhizospheric soil health with bacterial community, plant growth, performance and yield. *Plant. Signal. Behav.* 17(1), e2104004.

Verma KK, Song XP, Lin B, Guo DJ, Singh M, Rajput VD, Singh RK, Singh P, Sharma A, Malviya MK, Chen GL, Li YR. (2022g). Silicon induced drought tolerance in crop plants: Physiological adaptation strategies. *Silicon.* 14(6), 2473–2487. https://doi.org/10.1007/s12633-021-01071-x

Verma KK, Song XP, Rajput VD, Boldyreva V, Zhang BQ, Minkina T, Li YR. (2022h). Morpho-physiological, biochemical and ultrastructural modifications on sugarcane to prolonged water deficit. In: Verma KK, et al. (Eds.), *Agro-Industrial Perspectives on Sugarcane Production Under Environmental Stress*. https://doi.org/10.1007/978-981-19-3955-6_11.

Verma KK, Song XP, Rajput VD, Solomon S, Li YR, Rao GP. (2022i). Agro-industrial perspectives on sugarcane production under environmental stress. Springer Nature Singapore, https://doi.org/10.1007/978-981-19-3955-6

Verma KK, Song XP, Singh M, Tian DD, Rajput VD, Minkina T, Li YR. (2023c). Association of silicon and soil microorganisms induces stress mitigation, increasing plant productivity. In: R. de Mello Prado (Ed.), *Benefits of Silicon in the Nutrition of Plants*. pp. 299–328. https://doi.org/10.1007/978-3-031-26673-7_17

Verma KK, Song XP, Tian DD, Guo DJ, Chen ZL, Zhong CS, Nikpay A, Singh M, Rajput VD, Singh RK, Minkina T, Li YR. (2021a). Influence of silicon on biocontrol strategies to manage biotic stress for crop protection, performance and improvement. *Plants*. 10, 2163. https://doi.org/10.3390/plants10102163

Verma KK, Song XP, Tian DD, Singh M, Verma CL, Rajput VD, Singh RK, Sharma A, Singh P, Malviya MK, Li YR. (2021b). Investigation of defensive role of silicon during drought stress induced by irrigation capacity in sugarcane: Physiological and biochemical characteristics. *ACS Omega.* 6, 19811–19821.

Verma KK, Song XP, Verma CL, Chen ZL, Rajput VD, Wu KC, Liao F, Chen GL, Li YR. (2021c). Functional relationship between photosynthetic leaf gas exchange in response to silicon application and water stress mitigation in sugarcane. *Biol. Res.* 54, 15. https://doi.org/10.1186/s40659-021-00338-2

Verma KK, Song XP, Verma CL, Malviya MK, Guo DJ, Rajput VD, Sharma A, Wei KJ, Chen GL, Solomon S, Li YR. (2021d). Predication of photosynthetic leaf gas exchange of sugarcane (*Saccharum* spp.) leaves in response to leaf positions to foliar spray of potassium salt of active phosphorus under limited water irrigation. *ACS Omega.* 6, 2396–2409.

Verma KK, Song X-P, Xu L, Huang H-R, Liang Q, Seth CS, Li Y-R. (2023d). Nano-microplastic and agro-ecosystems: A mini-review. *Front. Plant Sci.* 14, 1283852. https://doi.org/10.3389/fpls.2023.1283852

Verma KK, Song XP, Yadav G, Degu HD, Parvaiz A, Singh M, Huang HR, Mustafa G, Xu L, Li YR. (2022j). Impact of agroclimatic variables on proteo-genomics in sugarcane (*Saccharum* spp.) plant productivity. *ACS Omega*, 7, 22997–23008.

Verma KK, Song XP, Zeng Y, Guo DJ, Sing M, Rajput VD, Malviya MK, Wei KJ, Sharma A, Li DP, Chen GL, Li YR. (2021e). Foliar application of silicon boosts growth, photosynthetic leaf gas exchange, antioxidative response and resistance to limited water irrigation in sugarcane (*Saccharum officinarum* L.). *Plant. Physiol. Biochem.* 166, 582–592.

Verma KK, Song XP, Zeng Y, Li DM, Guo DJ, Rajput VD, Chen GL, Barakhov A, Minkina TM, Li YR. (2020c). Characteristics and correlation of leaf stomata and its relationship with photosynthesis on *Saccharum* spp. under different irrigation and silicon application. . 5, 24145–24153.

Verma KK, Vatsal S, Singh M, Rajput VD, Minkina T. (2022k). Coping with water stress: Effects and adaptive mechanisms to plants. In: *Understanding Abiotic Stresses*, Nova Science Publishers, Inc.: New York.

Verma KK, Wu KC, Verma CL, Li DM, Malviya MK, Singh RK, Singh P, Chen GL, Song XP, Li YR. (2020d). Developing mathematical model for diurnal dynamics on photosynthetic responses in sugarcane responsive to different irrigation and silicon application. *PeerJ*. 8, e10154. https://doi.org/10.7717/peerj.10154.

Verma KK, Liu X-H, Wu K-C, Singh RK, Song QQ, Malviya MK, Song X-P, Singh P, Verma CL, Li YR. (2020e). The impact of silicon on photosynthetic and biochemical responses of sugarcane under different soil moisture levels. *Silicon*. 12, 1355–1367.

Verma KK, Wu K-C, Singh P, Malviya MK, Singh RK, Song X-P, Li YR. (2019b). The protective role of silicon in sugarcane under water stress: Photosynthesis and antioxidant enzymes. *Biomed J. Sci. Tech. Res.* 002685. https://doi.org/0.26717/BJSTR.2019.15.002685

Vilarinho AC, Baudet L, Gomes G. (2018). Biomass yield and energy potential of energy cane grown in Southern Europe. *Ind. Crops. Products*. 118, 45–51.

Viswanathan R, Malathi P, Chandran K, Gopi R. (2017). Screening of sugarcane germplasm for red rot resistance. *J. Sugar. Res.* 7(2), 100–111.

Wang LF, Wu, J, Jing RL, Cheng XZ, Wang SM. (2015). Identification of mung bean germplasm resources resistant to drought at adult stage. *Acta. Agronom. Sinica*. 41, 1287–1294.

Weng LX, Deng H, Xu JL, Li, Q, Wang LH, Jiang Z, Zhang HB, Li, Q, Zhang LH. (2006). Regeneration of sugarcane elite breeding lines and engineering of stem borer resistance. *Pest Manag. Sci.* 62, 178–187. https://doi.org/10.1002/ps.1144.

Weng LX, Deng HH, Xu, JL, Li, Q, Zhang YQ, Jiang ZD, Zhang LH. (2011). Transgenic sugarcane plants expressing high levels of modified cry1Ac provide effective control against stem borers in field trials. *Transg. Res.* 20(4), 759–72. https://doi.org/10.1007/s11248-010-9456-8.

Wu K-C, Huang C-M, Verma KK, Deng Z-N, Huang H-R, Pang T, Cao H-Q, Luo H-B, Jiang S-L, Xu L. (2022). Transcriptomic responses of *Saccharum spontaneum* roots in response to polyethylene glycol–6000 stimulated drought stress. *Front. Plant Sci.* 13, 992755. https://doi.org/10.3389/fpls.2022.992755

Xiong H, Chen Y, Gao SJ, Pan YB, Shi A. (2022). Population structure and genetic diversity analysis in sugarcane (Saccharum spp. hybrids) and six related saccharum species. *Agronomy*. 12, 412. https://doi.org/10.3390/agronomy12020412.

Xu L, Deng Z-N, Wu K-C, Malviya MK, Solanki MK, Verma KK, Pang T, Li Y-J, Liu X-Y, Kashyap BK, Dessoky ES, Wang W-Z and Huang H-R. (2022). Transcriptome analysis reveals a gene expression pattern that contributes to sugarcane bud propagation induced by indole-3-butyric acid. *Front. Plant Sci.* 13, 852886. https://doi.org/10.3389/fpls.2022.852886

Xu, C, Mao J, Li, X, Burner DM, Li, C, Hussin SHH, Lin X, Liu H, Zhao P, Lu, X, Liu X. (2023). Evaluation of drought tolerance in sugarcane genotypes using the membership function value of drought tolerance (MFVD). *Euphytica.* 219,€37. https://doi.org/10.1007/s10681-023-03167-4

Zan F, Zhang Y, Wu, Z, Zhao J, Wu, C, Zhao Y, Chen X, Zhao L, Qin W, Yao L, Xia H, Zhao P, Yang K., Liu J, Yang X. (2020). Genetic analysis of agronomic traits in elite sugarcane (Saccharum spp.) germplasm. *PLoS One.* 15(6): e0233752. https:// doi.org/10.1371/journal.pone.0233752.

Zhang J, Zhang Q, Li L, Tang H, Ming R. (2019). Recent polyploidization events in three Saccharum founding species. *Plant Biotechnol. J.* 17, 264–274.

Zhang L, Xu J, Birch RG. (1999). Engineered detoxification confers resistance against a pathogenic bacterium. *Nat. Biotechnol.* 17,1021.

Zhang SZ, Yang BP, Feng CL, Chen RK, Luo JP, Cai WW, Liu FH. (2006). Expression of the *Grifola frondosa* Trehalose synthase gene and improvement of drought-tolerance in sugarcane (*Saccharum officinarum* L.). *J. Integr. Plant. Biol.* 48(4), 453–459. https://doi.org/10.1111/j.1744-7909.2006.00246.x

Zhao D, Li YR. (2015). Climate change and sugarcane production: Potential impact and mitigation strategies. *Int. J. Agron.* ID: 547386. http://dx.doi.org/10.1155/2015/547386.

Zhao Y, Liu J, Huang H, Zan F, Zhao P, Zhao J, Deng J, Wu, C. (2022). Genetic improvement of sugarcane (*Saccharum* spp.*)* contributed to high sucrose content in China based on an analysis of newly developed varieties. *Agriculture*. 12(11): 1789. https://doi.org/10.3390/agriculture12111789Zheng F-J, Lin B, Yang Y-X, Fang X-C, Verma KK, Chen G-L. (2023). Efficacy and functionality of sugarcane original vinegar on mice. *Front. Microbiol.* 14, 1224666. https://doi.org/10.3389/fmicb.2023.1224666

Zhou D, Liu X, Gao S, Guo J, Su Y, Ling H, Wang C, Li Z, Xu L, Que Y. (2018). Foreign cry1Ac gene integration and endogenous borer stress-related genes synergistically improve insect resistance in sugarcane. *BMC Plant Biology.* 18, 342. https://doi.org/10.1186/s12870-018-1536-6.

Zhu YJ, McCafferty H, Osterman G, Lim S, Agbayani R, Lehrer A, Schenck S, Komor E. (2011). Genetic transformation with untranslatable coat protein gene of sugarcane yellow leaf virus reduces virus titers in sugarcane. *Transgenic Res.* 20, 503–512. https://doi.org/10.1007/s11248-010-9432-3.

CHAPTER 9

Bioethanol Production: Integrated Technological and Economic Perspectives

DHARMENDRA KUMAR and CHANDRA SHEKHAR SETH*

Department of Botany, University of Delhi, New Delhi, Delhi, India

*Corresponding author

ABSTRACT

Bioethanol, a renewable fuel derived from biomass feedstocks, has emerged as a viable substitute for traditional fossil fuels. This chapter provides a comprehensive assessment of the technical and economic aspects of bioethanol production. It delves into the intricate processes involved, including feedstock pretreatment, enzymatic hydrolysis, fermentation, and downstream processing. The chapter explores the latest advancements in enzyme engineering, metabolic engineering of microorganisms, and process integration strategies aimed at improving efficiency and cost-effectiveness. Additionally, it examines the economic considerations shaping the industry, such as feedstock costs, capital investments, operating expenses, coproduct valorization opportunities, and the role of government policies. Furthermore, the chapter highlights future perspectives, including the exploration of sustainable feedstock sources, mitigation of environmental and social impacts, and the importance of ensuring economic viability through technological innovations and supportive policy frameworks. Overall, this chapter provides a comprehensive understanding of the aspects influencing the progress of the bioethanol industry as a sustainable and economically viable renewable energy source.

Sugarcane Transformation: Biofuels, Biotechnology, and Beyond.
Krishan K. Verma, Xiu-Peng Song, Hai-Rong Huang, Aradhna Kumari, and Yang-Rui Li (Eds.)

DOI: 10.1201/9781779641571-9

9.1 INTRODUCTION

The world's heavy reliance on fossil fuels for energy has given rise to a multitude of pressing environmental, economic, and geopolitical concerns (Demichelis et al., 2020; Jarunglumlert et al., 2021; Verma et al., 2024a, 2024b, 2024c; Jagadesh et al., 2024). Fossil fuel combustion increases atmospheric concentrations of greenhouse gases, which exacerbate climate change worldwide (Chen et al., 2020; Huang et al., 2024). Additionally, fossil fuel reserves are finite and unevenly distributed across the globe, leading to issues of energy security and potential conflicts over dwindling resources (Jarunglumlert et al., 2021; Zheng et al., 2023). These challenges have driven global efforts to develop and adopt renewable and sustainable energy sources that can reduce our survival on fossil fuels while mitigating environmental impression (Kumari et al., 2022, 2023a, 2023b; Deng et al., 2023). Among the various renewable energy options, biofuels obtained from biomass such as organic matter from plants and other organisms have emerged as a promising avenue (Chong et al., 2020; Jarunglumlert et al., 2021; Jain et al., 2024; Singh et al., 2017; Guo et al., 2023).

Bioethanol, an alcohol-based fuel produced through the fermentation of sugars from biomass feedstocks, has acquired noteworthy attention as a leading candidate for transportation fuel due to its compatibility with existing infrastructure and the relative technological maturity of its production processes (Verma et al., 2019a, 2019b, 2023a; Joshi et al., 2022). Bioethanol can be mass-produced from a variety of feedstocks, including sugarcane, corn, wheat, and lignocellulosic materials such as agricultural byproducts, woody biomass, and devoted energy crops (Jarunglumlert et al., 2021; Regis et al., 2023; Vasilakou et al., 2023; Verma et al., 2023b, 2023c, 2023d). It offers several advantages over conventional gasoline, including reduced greenhouse gas emissions, potential for domestic production to enhance energy security, support for the agricultural sector, and compatibility with existing gasoline-powered vehicles (Anas et al., 2021a, 2021b; Sharma et al., 2021; Guerrero et al., 2023; Chaudhary et al., 2024).

However, the widespread adoption and commercial viability of bioethanol production hinge on addressing several technical and economic challenges. From a technical standpoint, the efficient conversion of biomass feedstocks into fermentable sugars and subsequently bioethanol requires optimized pretreatment, hydrolysis, fermentation, and distillation processes. Advancements in areas such as enzyme engineering, metabolic engineering of fermentation microorganisms, and process integration are crucial for improving yields, efficiency, and cost-effectiveness (Figure 9.1). Moreover, the choice of biomass feedstocks and their availability, logistics, and pretreatment requirements have a significant impact on the technical feasibility and economics of bioethanol production (Chaturvedi et al., 2006; Singh et al., 2020, 2021; Guerrero et al., 2023; Santoyo-Castelazo et al., 2023). While first-generation feedstocks like sugarcane and corn have well-established supply chains, their use raises concerns over food security and land-use competition. Lignocellulosic materials offer a promising alternative as second-generation feedstocks but require more complex and costly pretreatment processes (Jarunglumlert et al., 2021; Joshi et al., 2023; Guerrero et al., 2023; Santoyo-Castelazo et al., 2023).

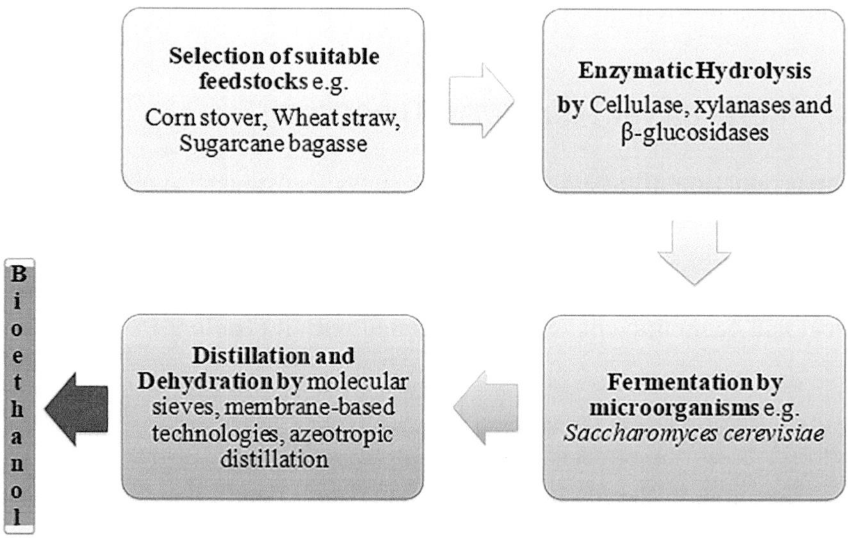

FIGURE 9.1 Schematic process of bioethanol production.

From an economic perspective, the viability of bioethanol production is influenced by factors such as feedstock costs, capital investments for production facilities, operating expenses, and the potential for coproduct valorization. Government policies, including mandates, subsidies, and tax incentives, also play an important role in creating the economic landscape for bioethanol production (Song et al., 2019, 2021; Joseph et al., 2023; Regis et al., 2023). This chapter aims to provide a comprehensive overview of the technical aspects of bioethanol production, delving into the intricate processes involved, the latest advancements, and the challenges that must be addressed to ensure efficient and cost-effective operations. Additionally, it will explore the economic considerations that shape the development of the bioethanol industry, including feedstock costs, capital investments, operating expenses, coproduct valorization, and the role of government policies. By examining the technical and economic dimensions of bioethanol production, this chapter seeks to equip readers with a holistic understanding of the factors influencing the growth and commercial viability of this renewable fuel source. Furthermore, it will highlight future perspectives and potential solutions to overcome the challenges faced by the industry, showcasing the ongoing efforts to harness the full potential of bioethanol as a sustainable alternative to fossil fuels.

9.2 BIOETHANOL PRODUCTION PROCESSES

The production of bioethanol from biomass feedstocks involves a series of complex processes, each playing a crucial role in converting the raw materials into a viable fuel source. Understanding these processes is essential for optimizing bioethanol production and addressing the technical and economic challenges associated with it (Table 9.1).

9.2.1 FEEDSTOCK HANDLING AND PRETREATMENT

The first step in bioethanol production is the selection and preparation of suitable biomass feedstocks. Common feedstocks include sugarcane, corn, wheat, and lignocellulosic materials such as agricultural residues (e.g., corn stover, wheat straw, sugarcane bagasse), forestry residues, and dedicated energy crops (e.g., switchgrass, miscanthus) (Malviya et al., 2021, 2022; Jayakumar et al., 2023; Singh et al., 2023). Pretreatment is a crucial step, particularly for lignocellulosic materials, as it aims to break down the complex structures of biomass and increase the accessibility of fermentable sugars (Li et al., 2023; Singh et al.,

TABLE 9.1 A Comprehensive Overview of Various Bioethanol Production Processes, Highlighting The Technical and Economic Aspects Crucial for Assessing their Development and Viability

Process	Feedstock	Conversion Technology	Key Steps	Technical Challenges	Economic Considerations
First-generation bioethanol	Sugarcane, corn	Fermentation	Milling, hydrolysis, fermentation, distillation	Feedstock competition with food, enzyme costs	High capital cost, competitive with food crops
Second-generation bioethanol	Lignocellulosic biomass	Enzymatic hydrolysis and fermentation	Pretreatment, hydrolysis, fermentation, distillation	Pretreatment efficiency, high enzyme costs	Lower feedstock cost, high processing cost
Thermochemical conversion	Various biomass types	Gasification and fermentation	Gasification, syngas cleanup, fermentation	High temperature process, tar formation	High capital and operating costs
Algal bioethanol	Algae	Fermentation	Cultivation, harvesting, hydrolysis, fermentation	Cultivation technology, water usage	High production cost, limited commercial scale
Hybrid processes	Mixed feedstocks	Combined biological and thermochemical	Pretreatment, hydrolysis, gasification, fermentation	Process integration, technological complexity	Potential cost reductions, complex

2023; Shukla et al., 2023; 2023). Various pretreatment techniques are employed, including:

- Physical pretreatment: Techniques such as milling, grinding, and steam explosion are used to reduce the particle size and disrupt the biomass structure.
- Chemical pretreatment: Processes like acid hydrolysis, alkaline pretreatment, and oxidative delignification employ chemical agents to solubilize and remove lignin and hemicellulose fractions, exposing the cellulose for subsequent enzymatic hydrolysis.
- Biological pretreatment: This approach utilizes microorganisms or enzymes to degrade specific components of the biomass, facilitating the release of fermentable sugars.

The choice of pretreatment method depends on factors such as the type of feedstock, desired sugar yields, and economic considerations. Optimizing pretreatment processes is crucial for maximizing sugar recovery, while minimizing the formation of inhibitory compounds that can hinder subsequent fermentation steps.

9.2.2 ENZYMATIC HYDROLYSIS

After pretreatment, the feedstock undergoes enzymatic hydrolysis, where enzymes catalyze the breakdown of polysaccharides into fermentable sugars. This step is particularly crucial for converting cellulosic and hemicellulosic materials into fermentable substrates (Favaretto et al., 2023). The hydrolysis process typically involves the use of cellulase enzymes, which are produced by various microorganisms, such as fungi and bacteria. These enzymes work synergistically to degrade cellulose into glucose monomers. Additional enzymes, like xylanases and β-glucosidases, may also be employed to break down hemicellulose into xylose and other fermentable sugars (Favaretto et al., 2023; Samantaray et al., 2024). Advancements in enzyme engineering and production have played a significant role in improving the efficiency and cost-effectiveness of enzymatic hydrolysis. Strategies such as genetic engineering, directed evolution, and metagenomics have led to the development of more robust and efficient enzyme cocktails, capable of operating under a wider range of conditions and processing diverse feedstocks (Favaretto et al., 2023; Joseph et al., 2023; Samantaray et al., 2024).

9.2.3 FERMENTATION

The sugars obtained from enzymatic hydrolysis are then subjected to fermentation, where microorganisms, typically yeast strains like *Saccharomyces cerevisiae*, convert the sugars into bioethanol. The choice of microorganism and fermentation conditions (e.g., temperature, pH, aeration) significantly impacts the efficiency and yields of the fermentation process (Shenbagamuthuraman and Nanthagopal, 2023; Tenkolu et al., 2024). Recent developments in metabolic engineering and synthetic biology have enabled the creation of genetically modified microorganisms with enhanced fermentation capabilities. These include strains capable of fermenting a broader range of sugars (e.g., xylose, arabinose), exhibiting higher tolerance to inhibitors, and producing higher ethanol yields (Hawaz et al., 2024; Samantaray et al., 2024). Additionally, advanced fermentation strategies, such as simultaneous saccharification and fermentation (SSF) and consolidated bioprocessing (CBP), have been explored to improve process efficiency and reduce costs. SSF combines enzymatic hydrolysis and fermentation in a single step, while CBP utilizes microorganisms capable of producing the necessary enzymes for hydrolysis and fermentation, eliminating the need for separate enzyme production and addition (Shenbagamuthuraman and Nanthagopal, 2023; Samantaray et al., 2024).

9.2.4 DISTILLATION AND DEHYDRATION

The fermentation broth containing bioethanol and other byproducts is subjected to distillation to separate the ethanol from the remaining components. The distillation process concentrates the ethanol to a certain level, typically around 95% purity. After distillation, dehydration techniques are employed to achieve fuel-grade bioethanol with minimal water content (Tsai et al., 2023). Common dehydration methods include:

- Molecular sieves: These porous materials selectively adsorb water molecules, allowing the production of anhydrous ethanol with water content below 0.5%.
- Membrane-based technologies: Techniques like pervaporation and vapor permeation utilize semipermeable membranes to separate water from the ethanol–water mixture, producing highly concentrated ethanol.
- Azeotropic distillation: This method involves the addition of a third component, such as benzene or cyclohexane, to the ethanol–water mixture, forming a new azeotropic mixture with a different boiling point, enabling more effective separation of ethanol and water.

The choice of dehydration technique depends on factors such as energy consumption, capital costs, and the desired purity levels of the final bioethanol product. Throughout the bioethanol production process, various opportunities for process integration and optimization exist. For instance, the integration of enzymatic hydrolysis and fermentation (SSF or CBP) can improve process efficiency and reduce costs. Additionally, the utilization of byproducts and waste streams from the process, such as lignin, can enhance the overall economic viability of bioethanol production through coproduct valorization.

Ongoing research and development efforts focus on improving each step of the bioethanol production process, from optimizing pretreatment techniques to developing more efficient enzymes, engineering superior fermentation microorganisms, and exploring novel distillation and dehydration methods. These advancements aim to increase yields, reduce energy consumption, and improve the overall cost-effectiveness of bioethanol production, making it a more viable and sustainable alternative to fossil fuels.

9.3 TECHNICAL CONSIDERATIONS

While the general processes involved in bioethanol production are well-established, several technical considerations must be addressed to ensure efficient and cost-effective operations. These considerations encompass various aspects, including feedstock availability and logistics, pretreatment and hydrolysis efficiency, fermentation performance, and downstream processing.

9.3.1 FEEDSTOCK AVAILABILITY AND LOGISTICS

The availability and logistics of biomass feedstocks play a crucial role in the technical feasibility of bioethanol production. Factors such as geographic distribution, seasonal variations, and transportation costs significantly impact the overall efficiency and cost-effectiveness of the process. For instance, the use of sugarcane as a feedstock is often concentrated in tropical and subtropical regions, where climatic conditions are favorable for its cultivation. This can lead to logistical challenges in transporting the feedstock to distant bioethanol production facilities, increasing transportation costs and potential losses during transit (Bhatt et al., 2023a, 2023b; Jayakumar et al., 2023; Chen et al., 2023; Singh et al., 2023).

Similarly, the use of agricultural residues as feedstocks presents challenges related to their dispersed nature and seasonal availability (Guo et al., 2021, 2022; Huang et al., 2022). Efficient collection, storage, and transportation systems are necessary to ensure a consistent supply of feedstock throughout the year. Densification techniques, such as pelletization or briquetting, can be employed to improve the bulk density and transportability of these residues. Dedicated energy crops, such as switchgrass and miscanthus, offer potential solutions to address feedstock availability and logistics. These crops can be cultivated specifically for bioethanol production, allowing for better control over supply and distribution networks. However, their large-scale adoption may require significant investments in infrastructure and land resources (Krishna et al., 2022; Sharma et al., 2022). To mitigate these challenges, strategic planning and optimization of biomass supply chains are crucial (Song et al., 2022; Wu et al., 2022; Tian et al., 2023). This may involve the establishment of decentralized pretreatment facilities or mobile preprocessing units closer to the feedstock sources, reducing transportation costs, and losses associated with hauling bulky biomass over long distances (Shukla et al., 2023; Singh et al., 2023).

9.3.2 PRETREATMENT AND HYDROLYSIS EFFICIENCY

Optimizing pretreatment and hydrolysis processes is essential for maximizing sugar yields and minimizing the formation of inhibitory compounds, which can negatively impact subsequent fermentation steps. Advancements in pretreatment technologies, such as the development of more efficient and cost-effective chemical or biological pretreatment methods, can significantly improve the overall efficiency and economics of bioethanol production. For example, the use of ionic liquids, deep eutectic solvents, or genetically engineered microorganisms for pretreatment has shown promising results in enhancing the accessibility of cellulose and hemicellulose for enzymatic hydrolysis. Furthermore, the development of tailored pretreatment strategies for specific feedstocks can help address the diverse compositional and structural characteristics of different biomass sources. This ensures that pretreatment conditions are optimized for each feedstock, maximizing sugar yields while minimizing the formation of inhibitory compounds (Favaretto et al., 2023; Samantaray et al., 2024).

In the realm of enzymatic hydrolysis, continuous efforts are being made to improve enzyme performance and reduce costs. Strategies such as enzyme engineering, directed evolution, and metagenomics have led to the development of more robust and efficient cellulase and hemicellulase enzyme cocktails. These enzymes exhibit improved thermal stability, pH tolerance, and catalytic activity, enabling more efficient hydrolysis of pretreated biomass. Additionally, the integration of pretreatment and enzymatic hydrolysis processes, through techniques like simultaneous saccharification and pretreatment, can enhance overall process efficiency and reduce costs associated with separate unit operations (Samantaray et al., 2024).

9.3.3 FERMENTATION PERFORMANCE

The selection of robust and efficient fermentation microorganisms, along with optimized fermentation conditions, is critical for achieving high bioethanol yields and productivity. Traditional fermentation strains, such as *S. cerevisiae*, have been extensively studied and engineered for improved performance. However, the utilization of lignocellulosic feedstocks presents additional challenges due to the presence of inhibitory compounds generated during pretreatment processes. These inhibitors can impair the growth and fermentation performance of traditional microorganisms, necessitating the development of more tolerant strains (Shenbagamuthuraman and Nanthagopal, 2023; Tenkolu et al., 2024).

Metabolic engineering and synthetic biology approaches have enabled the creation of genetically modified microorganisms with enhanced fermentation capabilities (Chaudhary et al., 2023; Solanki et al., 2023). These include strains capable of fermenting a broader range of sugars (e.g., xylose, arabinose), exhibiting higher tolerance to inhibitors, and producing higher ethanol yields. Additionally, the exploration of nonconventional fermentation microorganisms, such as bacteria and filamentous fungi, has opened up new avenues for improving fermentation efficiency and addressing specific challenges associated with different feedstocks (Xu et al., 2022; Samantaray et al., 2024). Furthermore, the optimization of fermentation conditions, including temperature, pH, aeration, and nutrient supplementation, plays a crucial role in maximizing bioethanol productivity and yield. Advanced fermentation strategies, such as fed-batch or continuous fermentation, can also be employed to improve process performance and reduce downtime (Tenkolu et al., 2024).

9.3.4 DOWNSTREAM PROCESSING

Efficient downstream processing techniques, such as distillation and dehydration, are necessary to obtain fuel-grade bioethanol while minimizing energy consumption and associated costs. Advancements in distillation technologies, including the development of energy-efficient distillation columns and the integration of heat recovery systems, can significantly reduce the energy requirements and operational costs of this process step (Popescu et al., 2021). In the realm of dehydration, the development of novel membrane-based technologies, such as pervaporation and vapor permeation, has shown promise in achieving high ethanol purity with reduced energy consumption compared to traditional methods like molecular sieves or azeotropic distillation. Additionally, the integration of distillation and dehydration processes, through techniques like membrane-assisted distillation or extractive distillation, can further improve overall process efficiency and reduce energy requirements (Popescu et al., 2021; Tsai et al., 2023).

9.3.5 PROCESS INTEGRATION AND INTENSIFICATION

Beyond optimizing individual unit operations, the integration and intensification of the entire bioethanol production process offer significant opportunities for improving technical performance and cost-effectiveness. Process integration involves the systematic evaluation and optimization of material and energy flows throughout the production process, ensuring efficient utilization of resources and minimizing waste streams. This can include strategies such as heat integration, where waste heat from one process step is recovered and utilized in other stages, thereby reducing overall energy consumption (Gómez-Castro et al., 2023). Process intensification, on the other hand, focuses on the development of novel and compact process technologies that can enhance process efficiency, reduce equipment size, and improve overall process economics. Examples of process intensification strategies include reactive distillation, where reaction and separation operations are combined in a single unit, and the use of microreactors or membrane reactors for more efficient and controlled processing. By addressing these technical considerations and leveraging the latest advancements in feedstock logistics, pretreatment, hydrolysis, fermentation, downstream processing, and process integration, the bioethanol industry can enhance the overall efficiency, sustainability, and economic viability of bioethanol production (Kumakiri et al., 2021).

9.4 ECONOMIC CONSIDERATIONS

While the technical aspects of bioethanol production are crucial, the economic viability of the process is equally important for its widespread adoption and commercial success. Several economic factors play a pivotal role in shaping the development of the bioethanol industry, including feedstock costs, capital investments, operating expenses, coproduct valorization, and government policies.

9.4.1 FEEDSTOCK COSTS

The cost of biomass feedstocks is a significant factor influencing the economics of bioethanol production. These costs can vary substantially depending on the type of feedstock, geographic location, agricultural practices, and competition with other industries for the same feedstocks. For first-generation feedstocks like sugarcane and corn, the costs are primarily driven by factors such as land availability, crop yields, and the prices of agricultural inputs like fertilizers, pesticides, and irrigation water (Jayakumar et al., 2023). In regions with well-established sugarcane or corn industries, the feedstock supply chains are typically more developed, which can lead to lower costs compared to regions where these crops are not widely cultivated. Lignocellulosic feedstocks, such as agricultural residues, forestry waste, and dedicated energy crops, offer the potential for lower feedstock costs due to their abundance and the opportunity to utilize waste materials. However, the costs associated with collecting, transporting, and preprocessing these feedstocks can be significant, potentially offsetting the perceived cost advantages. To mitigate feedstock costs, strategies such as developing dedicated energy crops with higher yields and better adaptability to marginal lands, optimizing supply chains for efficient feedstock logistics, and exploring alternative low-cost feedstocks are being pursued (Jayakumar et al., 2023; Singh et al., 2023).

9.4.2 CAPITAL INVESTMENTS AND OPERATING EXPENSES

The establishment of bioethanol production facilities requires substantial capital investments in equipment and infrastructure. The specific capital costs can vary depending on factors such as the production capacity, feedstock type, and the chosen technologies for pretreatment, hydrolysis, fermentation, and downstream processing. For example, facilities utilizing lignocellulosic feedstocks generally require more capital-intensive pretreatment and hydrolysis equipment compared to those using starch or sugar-based feedstocks (Dempfle et al., 2021). Additionally, the integration of advanced technologies, such as CBP or membrane-based dehydration systems, can further increase the initial capital requirements. Careful consideration must be given to the selection of appropriate production scales and technologies to ensure an optimal balance between capital costs and operational efficiencies. Economies of scale can be leveraged by opting for larger production capacities, but this must be weighed against the increased risk and initial capital outlay. Furthermore, the development of modular and scalable bioethanol production facilities can offer flexibility and reduced upfront capital costs, allowing for gradual capacity expansion as demand increases and technologies mature (Aletheia et al., 2024).

The ongoing operating expenses associated with bioethanol production can significantly impact the overall economics of the process. These expenses include costs related to feedstock transportation, energy consumption, labor, maintenance, and other consumables such as enzymes and chemicals. Energy consumption, particularly for pretreatment, distillation, and dehydration processes, can represent a substantial portion of the operating costs. Optimizing energy efficiency through process integration, heat recovery systems, and the adoption of energy-efficient technologies is crucial for minimizing these costs. Labor costs can also be a significant factor, especially in regions with higher labor rates. Automation and process control strategies can help reduce labor requirements and associated costs, while also improving process consistency and efficiency. Additionally, the costs of enzymes and chemical inputs used in pretreatment and hydrolysis processes can impact the overall operating expenses. Continuous efforts to develop more cost-effective enzymes and explore alternative pretreatment methods can help mitigate these costs (Dempfle et al., 2021; Aletheia et al., 2024).

9.4.3 COPRODUCT VALORIZATION

The bioethanol production process often generates various byproducts and waste streams, which can be valorized to create additional revenue streams and improve the overall economics of the process. For instance, the lignin fraction separated during pretreatment can be utilized as a fuel source for generating process heat and electricity, reducing energy costs and associated greenhouse gas emissions. Additionally, the residual biomass after fermentation, known as distillers' grains, can be used as animal feed or as a feedstock for other value-added products, such as biochemicals or bioplastics. Coproduct valorization opportunities can vary depending on the feedstock type and the specific bioethanol production process employed. Identifying and developing efficient strategies for coproduct utilization can significantly enhance the economic viability of bioethanol production facilities (Zheng et al., 2024).

9.4.4 GOVERNMENT POLICIES AND INCENTIVES

Government policies, including mandates, subsidies, tax incentives, and regulatory frameworks, can play a crucial role in shaping the economic landscape for bioethanol production. Renewable fuel mandates, such as the Renewable Fuel Standard in the United States and the Renewable Energy Directive in the European Union, create demand for biofuels like bioethanol and provide market certainty for producers. Subsidies and tax incentives can help offset the higher production costs associated with bioethanol, particularly during the initial stages of technology development and commercialization. These incentives can take various forms, such as production tax credits, investment tax credits, or direct subsidies for feedstock production or conversion facilities (Godinho et al., 2022).

Additionally, regulatory frameworks governing issues such as land-use policies, sustainability criteria, and greenhouse gas emission standards can influence the economic viability of different bioethanol production pathways. Furthermore, government support for research and development initiatives can drive technological advancements and cost reductions, enhancing the long-term competitiveness of bioethanol production. However, it is important to strike a balance between government support and market forces, ensuring that the bioethanol industry remains economically sustainable without excessive reliance on subsidies or mandates. By carefully considering and addressing these economic factors, including feedstock costs, capital investments, operating expenses, coproduct valorization opportunities, and supportive government policies, the bioethanol industry can navigate the economic challenges and pave the way for widespread adoption as a viable and sustainable alternative to fossil fuels (Usmani et al., 2023).

9.5 FUTURE PERSPECTIVES AND CHALLENGES

The development of bioethanol production has made significant strides in recent years, driven by the need for sustainable and renewable energy sources. However, the industry continues to face several challenges that must be addressed to ensure its long-term viability and widespread adoption. This section explores the future perspectives and challenges associated with the technical and economic aspects of bioethanol production.

9.5.1 TECHNOLOGICAL ADVANCEMENTS

Ongoing research and development efforts hold the key to unlocking the full potential of bioethanol production. Advancements in various areas, such as feedstock processing, enzyme engineering, metabolic engineering of microorganisms, and process integration, can significantly improve the efficiency, sustainability, and cost-effectiveness of the entire production process.

- Pretreatment and hydrolysis: The development of novel pretreatment techniques and more efficient enzyme cocktails can enhance the accessibility of fermentable sugars from lignocellulosic biomass. Strategies such as ionic liquid pretreatment, deep eutectic solvents, and the use of CBP organisms have shown promising results in improving sugar yields while reducing costs and energy consumption.
- Fermentation and metabolic engineering: Advances in metabolic engineering and synthetic biology have enabled the creation of genetically modified microorganisms with enhanced fermentation capabilities. These include strains capable of fermenting a broader range of sugars (e.g., xylose, arabinose), exhibiting higher tolerance to inhibitors, and producing higher ethanol yields. Additionally, the exploration of nonconventional fermentation microorganisms, such as bacteria and filamentous fungi, presents opportunities for addressing specific challenges associated with different feedstocks and process conditions.
- Process integration and intensification: Integrating various unit operations, such as pretreatment, hydrolysis, and fermentation, can lead to more efficient and cost-effective processes. Strategies like simultaneous SSF and CBP have the potential to reduce energy consumption, minimize the formation of inhibitory compounds, and improve overall process economics.
- Downstream processing: Improvements in distillation and dehydration technologies, including the development of energy-efficient distillation columns, membrane-based dehydration systems, and the integration of heat recovery systems, can significantly reduce energy requirements and operational costs.
- Biorefinery concepts: The integration of bioethanol production with biorefineries, where multiple value-added products are obtained from biomass feedstocks, presents an opportunity to enhance the economic viability and sustainability of the entire process. By coproducing biochemicals, bioplastics, and other high-value products alongside bioethanol, biorefineries can maximize resource utilization and generate additional revenue streams.

While these technological advancements hold significant promise, their successful implementation and commercialization will require continued research efforts, pilot-scale demonstrations, and effective technology transfer mechanisms to bridge the gap between laboratory-scale innovations and industrial-scale operations.

9.5.2 SUSTAINABLE FEEDSTOCK SOURCES

The choice of biomass feedstocks for bioethanol production has far-reaching implications for the sustainability and economic viability of the entire process. While first-generation feedstocks like sugarcane and corn have well-established supply chains, their use raises concerns over food security and land-use competition with food production. To address these concerns, the development and utilization of sustainable feedstock sources, such as dedicated energy crops, agricultural and forestry residues, and municipal solid waste, are crucial. These second-generation feedstocks offer the potential to reduce competition with food crops and minimize the environmental impact associated with land-use changes. However, the large-scale adoption of lignocellulosic feedstocks presents its own set of challenges. These include the logistical complexities of collecting and transporting dispersed feedstocks, the development of efficient pretreatment and conversion technologies tailored for specific feedstock compositions, and the need for sustainable feedstock management practices to maintain soil health and biodiversity (Guimarães et al., 2023).

Ongoing research efforts focus on identifying and developing dedicated energy crops with high biomass yields, tolerance to marginal lands and adverse environmental conditions, and favorable compositional characteristics for bioethanol production. Additionally, optimizing supply chains for efficient feedstock logistics and exploring innovative feedstock preprocessing techniques are crucial for ensuring a consistent and cost-effective supply of biomass.

9.5.3 ENVIRONMENTAL AND SOCIAL IMPACTS

The environmental and social impacts of large-scale bioethanol production must be carefully evaluated and mitigated to ensure the long-term sustainability of the industry. Concerns over land-use changes, water consumption, and potential conflicts with food production are valid and require comprehensive assessments and sustainable practices (Verma et al., 2020a, 2020b, 2022a, 2022b). Land-use changes, such as the conversion of natural habitats or carbon-rich ecosystems like forests and peatlands for biomass cultivation, can result in significant greenhouse gas emissions and biodiversity loss. Appropriate land-use policies, zoning regulations, and sustainable agricultural practices must be implemented to minimize these impacts. Water consumption is another critical consideration, particularly in regions with limited water resources (Verma et al., 2021a, 2021b, 2021c, 2022c, 2022d). The cultivation of biomass feedstocks and bioethanol production processes can have significant water footprints. Efforts to improve water use efficiency, implement water recycling and treatment systems, and explore drought-tolerant feedstock varieties are essential for mitigating water-related impacts (Verma et al., 2021d, 2021e, 2022e, 2022f, 2022g; Joseph et al., 2023; Regis et al., 2023).

Additionally, the potential impact of bioethanol production on food security and agricultural commodity prices should be carefully monitored and addressed. While the use of lignocellulosic feedstocks can alleviate competition with food crops, mechanisms to ensure equitable access to food and prevent price volatility should be established (Verma et al., 2020c, 2020d, 2022h, 2022i). Engaging with local communities, conducting comprehensive environmental and social impact assessments, and developing inclusive and sustainable policies are crucial for addressing these challenges and ensuring the responsible development of the bioethanol industry (Regis et al., 2023).

9.5.4 ECONOMIC VIABILITY AND POLICY SUPPORT

Despite the technological advancements and potential sustainability benefits, the economic viability of bioethanol production remains a significant challenge (Verma et al., 2020e, 2020f, 2022j). Factors such as fluctuating feedstock costs, high capital investments, and competition with conventional fossil fuels can impact the competitiveness of bioethanol in the energy market. Continued government support through policies, incentives, and regulatory frameworks can play a crucial role in fostering the growth of the bioethanol industry, particularly during the initial stages of technology development and commercialization. Renewable fuel mandates, production subsidies, tax incentives, and investment in research and development can help create a favorable environment for bioethanol production and encourage private sector investment (Guerrero et al., 2023; Santoyo-Castelazo et al., 2023; Regis et al., 2023).

However, it is essential to strike a balance between government support and market forces, ensuring that the bioethanol industry remains economically sustainable without excessive reliance on subsidies or mandates. Policies should be designed to promote technological innovation, efficiency improvements, and cost reductions, ultimately enabling bioethanol to compete on a level playing field with conventional fuels. Additionally, the implementation of carbon pricing mechanisms, such as carbon taxes or emissions trading schemes, can provide an economic incentive for the adoption of low-carbon fuels like bioethanol, thereby improving their competitiveness in the market. By addressing these future perspectives and challenges through concerted efforts from researchers, policymakers, industry stakeholders, and society at large, the bioethanol industry can continue to progress toward a more sustainable and economically viable future, contributing to the global transition toward renewable energy sources and mitigating the impacts of climate change (Verma et al., 2022k; Guerrero et al., 2023; Santoyo-Castelazo et al., 2023).

9.6 CONCLUSION

The development of bioethanol production represents a pivotal step toward a more sustainable and renewable energy future. As the world grapples with the challenges of climate change, energy security, and the depletion of finite fossil fuel resources, the adoption of bioethanol as a transportation fuel offers a promising solution. Throughout this chapter, we have explored the intricate technical processes involved in bioethanol production, from feedstock selection and pretreatment to enzymatic hydrolysis, fermentation, and downstream processing. We have also delved into the economic considerations that shape the viability of this industry, including feedstock costs, capital investments, operating expenses, coproduct valorization opportunities, and the role of government policies. Despite the significant progress made in this field, several challenges persist, necessitating continued research, innovation, and collaborative efforts from stakeholders across various sectors.

Technological advancements in areas such as enzyme engineering, metabolic engineering of microorganisms, and process integration hold the potential to enhance efficiency, reduce costs, and improve the overall sustainability of bioethanol production. Additionally, the exploration of sustainable feedstock sources, such as dedicated energy crops, agricultural residues, and municipal solid waste, is crucial for addressing concerns over food security and land-use competition. Mitigating the environmental and social impacts of large-scale bioethanol production through sustainable practices, comprehensive assessments, and inclusive policies is equally important. Furthermore, ensuring the economic viability of bioethanol production remains a significant challenge that requires the concerted efforts of policymakers, industry stakeholders, and researchers.

Supportive government policies, incentives, and regulatory frameworks can play a pivotal role in fostering the growth of this industry, particularly during the initial stages of technology development and commercialization. As we navigate the path toward a more sustainable energy future, the development of bioethanol production represents a promising avenue. By addressing the technical and economic challenges highlighted in this chapter, and embracing a holistic approach that balances technological innovations, sustainable practices, and economic considerations, we can unlock the full potential of bioethanol as a renewable and environmentally friendly transportation fuel. The transition toward a bioethanol-based economy will require perseverance, collaboration, and a unwavering commitment to sustainable development. However, the rewards of such efforts are immense, paving the way for a cleaner, more secure, and more resilient energy landscape for generations to come.

ACKNOWLEDGMENT

Chandra Shekhar Seth is thankful to the Department of Botany, University of Delhi, and Institution of Eminence, University of Delhi, India for providing necessary support and facilities in research and academics.

KEYWORDS

- **alcohol-based fuel**
- **feedstock**
- **fermentation**
- **lignocellulosic**
- **microorganisms**
- **saccharification**

REFERENCES

Aletheia SP Ahmad S, Kelvin K, Kuntum K, Muhammad MR. (2024). Techno-economic analysis of biodiesel and bioethanol production from Chlorella sp. algae biomass. In *E3S Web Conf.* vol. 503, EDP Sciences; p. 02004.

Anas M, Verma KK, Haq IU, Naeem M, Li Q, Liao F, Xiang YX, Li YR. (2021a) Characterization of exotic, native and wild-type genotypes of sugarcane (saccharum spp. hybrids) for internal nitrogen use efficiency under different nitrogen levels. *Sugar. Tech* 23, 1258–1267. https://doi.org/10.1007/s12355-021-01014-1

Anas M, Verma KK, Riaz M, Qiang L, Liao F, Liu Y, Li YR. (2021b). Physio-morphological and biochemical mechanism of nitrogen use efficiency in sugarcane (Saccharum spp.) genotypes under different growth stages and nitrogen levels. *J. Plant. Interact.* 16, 332–343.

Bhatt R, Verma KK, Jamwal NS. (2023a). Can paddy-compost eliminate N fertilization for climate-smart sugarcane? *Sugar. Tech.* 25, 1036–1046. https://doi.org/10.1007/s12355-023-01270-3

Bhatt R, Verma KK, Kumar R, Sanghera GS. (2023b). Foliar application of potassium salt of active phosphorus (PSAP) mitigates insect pests and improves yield along with sugarcane quality in response to agroclimatic conditions of Punjab. *Sugar. Tech.* 25(3), 660–669.

Chaudhary P, Xu M, Ahamad L, Chaudhary A, Kumar G, Adeleke BS, Verma KK, Hu D-M, Siric I, Kumar P, Popescu SM, Fayssal SA. (2023). Application of synthetic consortia for improvement of soil fertility, pollution remediation, and agricultural productivity: A review. *Agronomy.* 13, 643. https://doi.org/10.3390/agronomy13030643.

Chen GL, Zheng FJ, Lin B, Lao SB, He J, Huang Z, Zeng Y, Sun J, Verma KK. (2020). Phenolic and volatile compounds in the production of sugarcane vinegar. *Amer. Chem. Soc.* 5, 30587–30595.

Chen G-L, Zheng F-J, Lin B, Yang Y-X, Fang X-C, Verma KK, Yang L-F. (2023). Vinegar: A potential source of healthy and functional food with special reference to sugarcane vinegar. *Front. Nutr.* 10, 1145862. https://doi.org/10.3389/fnut.2023.1145862

Chong TY, Siang AC, Chin TO, Lee YW, Chern RG, Inn ST, Henry CYF, Man KL, Steven L. (2020). Techno-economic evaluation of third-generation bioethanol production utilizing the macroalgae waste: A case study in Malaysia. the single letter. *Energy*. 210, 118491.

Demichelis F, Maddalena L, Marco C, Silvia F. (2020). Technical, economic and environmental assessment of bioethanol biorefinery from waste biomass. *J. Clean. Prod.* 277, 124111.

Dempfle D, Oliver K, Michael H-PS. (2021). Techno-economic assessment of bioethanol production from lignocellulose by consortium-based consolidated bioprocessing at industrial scale. *Biotechnology.* 65, 53–60.

Deng ZN, Huang HR, Huang CM, Verma KK, Luo HB, Jiang SL, Li YJ, Cao HQ, Wu KC, Xu L. (2023). Short-term limited water irrigation influences the photosynthetic pigments, enzymatic and non-enzymatic activities in Saccharum spontaneum L. at vegetative stage. *Sugar. Tech.* 25, 110–118. https://doi.org/10.1007/s12355-022-01175-7

Favaretto DPC, Alan R, Julia RL, Ana CMS, Tauane L, Luiza DB, Vandré BB, Luciane MC, Helen T. (2023). Fruit residues as biomass for bioethanol production using enzymatic hydrolysis as pretreatment. *World J. Microbiol. Biotechnol.* 39(6), 144.

Guerrero AB, Edmundo M. (2023). Techno-economic and life cycle analysis of bioethanol production. In: *Bioethanol: A Green Energy Substitute for Fossil Fuels*. Cham: Springer International Publishing; pp. 305–337.

Guimarães CEC Francisco SN, Viviane de CB, Jean GAN, Roberta BRV, Paulo GS Jr, Ana KSB et al. (2023), Sustainable bioethanol production from first-and second-generation sugar-based feedstocks: Advanced bibliometric analysis. *Bioresour. Technol. Rep.* 23, 101543.

Guo D-J, Li D-P, Chen Z-D, Yang B, Verma KK, Singh RK, Singh P, Khan Q, Sharma A, Qin Y, Zhang B-Q, Song X-P, Li Y-R. (2023). Effect of endophytic diazotroph *Enterobacter roggenkampii* ED5 on nitrogen-metabolism-related microecology in the sugarcane rhizosphere at different nitrogen levels. *Nat. Inst. Health (NIH) Catalog.* 14, 1132016. https://doi.org/10.3389/fmicb.2023.1132016

Guo D-J, Li D-P, Singh RK, Singh P, Sharma A, Verma KK, Qin Y, Khan Q, Lu Z, Malviya MK, Song X-P, Xing Y-X and Li Y-R. (2021). Differential protein expression analysis of two sugarcane varieties in response to diazotrophic plant growth-promoting endophyte *Enterobacter roggenkampii* ED5. *Front. Plant Sci.* 12, 727741. https://doi.org/10.3389/fpls.2021.727741

Guo DJ, Li DP, Singh RK, Singh P, Verma KK, Sharma A, Qin Y, Khan Q, Song XP, Malviya MK, Xing YX, Li YR. (2022). Comparative transcriptome analysis of two sugarcane varieties in response to diazitrophic plant growth promoting endophyte *Enterobacter roggenkampii* ED5. *J. Plant Interact.* 17(1), 75–84.

Hawaz E, Mesfin T, Anteneh T, Solomon K, Dereje B, Gessesse K, Teun B et al. (2024). Bioethanol production from sugarcane molasses by co-fermentation of *Saccharomyces cerevisiae* isolate TA2 and *Wickerhamomyces anomalus* isolate HCJ2F-19. *Ann. Microbiol.* 74(1), 13.

Huang CM, Wu KC, Verma KK, Luo HB, Deng ZN, Shi GY, Cao HQ, Wu XJ, Ye LP, Wei YW, Xu L. (2024). Isolation and expression analysis of cellulose synthase 3 (*Ces3*) genes from sugarcane (*Saccharum officinarum* L). *Chem. Biol. Technol. Agric.* 11, 71. https://doi.org/10.1186/s40538-024-00597-7

Huang HR, Gao YJ, Malviya MK, Verma KK, Solanki MK, Huang YX, Li X, Deng Y, Yan J, Tang SY, Wang LW, Xu L. (2022). Genetic diversity analysis of sugarcane (*Saccharum spp. Hybrids*) among high-sucrose clones of GT series and commonly used parents by using microsatellite markers in Guangxi. *China. Sugar. Tech.* 24, 397-407. https://doi.org/10.1007/s12355-021-01033-y

Jagadesh M, Dash M, Kumari A, Singh SK, Verma KK, Kumar P, Bhatt R, Sharma SK. (2024). Revealing the hidden world of soil microbes: Metagenomic insights into plant, bacteria, and fungi interactions for sustainable agriculture and ecosystem restoration. *Microbiol. Res.* 127764. https://doi.org/10.1016/j.micres.2024.127764

Jain S, Shushil K. (2024). A comprehensive review of bioethanol production from diverse feedstocks: Current advancements and economic perspectives. *Energy.* 131130.

Jarunglumlert T, Chattip P. (2021).Net energy analysis and techno-economic assessment of co-production of bioethanol and biogas from cellulosic biomass. *Ferment.* 7(4), 229.

Jayakumar M, Gadissa T, Gindaba K, Bizuneh G, Selvakumar P, Abdisa J, Gurunathan B, Beula IJ, Arivalagan P. (2023). Bioethanol production from agricultural residues as lignocellulosic biomass feedstock's waste valorization approach: A comprehensive review. *Sci. Total Environ.* 879, 163158.

Joseph AM, Yeswanth T, Devashish S, Bandaru K. (2023).Techno-economic feasibility and exergy analysis of bioethanol production from waste. *Energ. Convers. Manag-X* 18, 100358.

Joshi A, Verma KK, Rajput VD, Arora J. (2023). Metabolic engineering of microorganisms in advancing biofuel production. In: *Advances in Lignocellulosic Biofuel Production Systems*. pp. 181–202. https://doi.org/10.1016/B978-0-323-91192-4.00016-X

Joshi A, Verma KK, Rajput VD, Minkina T, Arora J. (2022). Recent advances in metabolic engineering of microorganisms for advancing lignocellulose-derived biofuels. *Bioengineered.* 13(4), 8135–8163. https://doi.org/10.1080/21655979.2022.2051856.

Krishna D, Sachan HK, Verma KK. (2022). Abiotic stresses in plants: An overview. In: *Understanding Abiotic Stresses Publisher*, Nova Science Publishers, Inc.: New York.

Kumakiri I, Morihisa Y, Ryotaro T, Yu S, Worapon K, Jun WL, Masayuki M, Mamoru Y. (2021). Process intensification in bio-ethanol production—Recent developments in membrane separation. *Processes.* 9(6), 1028.

Kumari A, Dash M, Singh SK, Jagadesh M, Mathpal B, Mishra PK, Pandey SK, Verma KK. (2023a). Soil microbes: A natural solution for mitigating the impact of climate change. *Environ. Monit. Assess*. 195, 1436. https://doi.org/10.1007/s10661-023-11988-y

Kumari A, Lakshmi GA, Krishna GK, Patni B, Prakash S, Bhattacharyya M, Singh SK, Verma KK. (2022). Climate change and its impact on crops: A comprehensive investigation for sustainable agriculture. *Agronomy.* 12, 3008. https://doi.org/10.3390/agronomy12123008

Kumari A, Singh SK, Mathpal B, Verma KK, Garg VK, Bhattacharyya M, Bhatt R. (2023b). The biosynthesis, mechanism of action, and physiological functions of melatonin in horticultural plants: A review. *Horticulturae.* 9, 913. https://doi.org/10.3390/horticulturae9080913

Li A-M, Liao F, Wang M, Chen Z-L, Qin C-X, Huang R-Q, Verma KK, Li Y-R, Que Y-X, Pan Y-Q, et al. (2023). Transcriptomic and proteomic landscape of sugarcane response to biotic and abiotic stressors. *Int. J. Mol. Sci.* 24, 8913. https://doi.org/10.3390/ijms24108913

Malviya MK, Li C-N, Lakshmanan P, Solanki MK, Wang Z, Solanki AC, Nong Q, Verma KK, Singh RK, Singh P, Sharma A, Guo D-J, Dessoky ES, Song X-P, Li Y-R. (2022). High-throughput sequencing-based analysis of rhizosphere and diazotrophic bacterial diversity among wild progenitor and closely related species of sugarcane (saccharum spp. inter-specific hybrids). *Front. Plant Sci.* 13, 829337. https://doi.org/10.3389/fpls.2022.829337.

Malviya MK, Solanki MK, Li CN, Wang Z, Zeng Y, Verma KK, Singh RK, Singh P, Huang HR, Yang LT, Song XP, Li YR. (2021). Sugarcane-legume intercropping can enrich the soil microbiome and plant growth. *Front. Sustain. Food Syst.* 5, 606595. https://doi.org/10.3389/fsufs.2021.606595.

Regis F, Alessandro H, Antonio M, Debora F. (2023). A techno-economic assessment of bioethanol production from switchgrass through biomass gasification and syngas fermentation. *Energy.* 274(2023), 127318.

Samantaray B, Sonali M, Rashmi R, Mishra B, Chandra B, Hrudayanath T. (2024). Bioethanol production from agro-wastes: A comprehensive review with a focus on pretreatment, enzymatic hydrolysis, and fermentation. *Int. J. Green Energy.* 21(6), 1398–1424.

Santoyo-Castelazo E, Santoyo E, Zurita-García L, Camacho Luengas DA, Solano-Olivares K. (2023). Life cycle assessment of bioethanol production from sugarcane bagasse using a gasification conversion process: Bibliometric analysis, systematic literature review and a case study. *Appl. Therm. Eng.* 219, 119414.

Sharma A, Singh RK, Singh P, Vaishnav A, Guo DJ, Verma KK, Li DP, Song XP, Malviya MK, Khan N, Lakshmanan P, Li YR. (2021). Insights into the bacterial and nitric oxide-induced salt tolerance in sugarcane and their growth-promoting abilities. *Microbes.* 9, 2203. https://doi.org/10.3390/microorganisms9112203.

Sharma A, Song X-P, Singh RK, Vaishnav A, Gupta S, Singh P, Guo D-J, Verma KK, Li Y-R. (2022). Impact of carbendazim on cellular growth, defence system and plant growth promoting traits of priestia megaterium ancb-12 isolated from sugarcane rhizosphere. *Front. Microbiol.* 13, 1005942. https://doi.org/10.3389/fmicb.2022.1005942

Shenbagamuthuraman V, Nanthagopal K. (2023). Microwave irradiation pretreated fermentation of bioethanol production from *Chlorella vulgaris* biomasses: Comparative analysis of response surface methodology and artificial neural network techniques. *Bioresour. Technol.* 390, 129867.

Shukla A, Kumar D, Girdhar M, Kumar A, Goyal A, Malik T, Mohan A. (2023). Strategies of pretreatment of feedstocks for optimized bioethanol production: Distinct and integrated approaches. *Biotechnol. Biofuels Bioprod.* 16(1), 44.

Singh A, Prajapati P, Vyas S, Gaur VK, Sindhu R, Binod P, Kumar V. et al. (2023). A comprehensive review of feedstocks as sustainable substrates for next-generation biofuels. *Bioenergy. Res.* 16(1), 105–122.

Singh P, Singh RK, Li HB, Guo DJ, Sharma A, Verma KK, Solanki MK, Upadhyay SK, Lakshmanan P, Yang LT, Li YR. (2023). Nitrogen fixation and phytohormone stimulation of sugarcane plant through plant growth promoting diazotrophic pseudomonas. *Biotechnol. Genet. Eng. Rev.* 1–21. https://doi.org/10.1080/02648725.2023.2177814

Singh P, Singh RK, Song QQ, Li HB, Guo DJ, Malviya MK, Verma KK, Song XP, Lakshmanan P, Yang LT, Li YR. (2021). Comparative analysis of protein and differential responses of defense-related gene and enzyme activity reveals the long-term molecular responses of sugarcane inoculated with *Sporisorium scitamineum*. *J. Plant. Int.* 16(1), 12–29.

Singh RK, Singh P, Li HB, Song QQ, Guo DJ, Solanki MK, Verma KK, Malviya MK, Song XP, Lakshmanan P, Yang LT, Li YR. (2020). Diversity of nitrogen-fixing rhizobacteria associated with sugarcane: A comprehensive study of plant-microbe interactions for growth enhancement in Saccharum spp. *BMC. Plant. Biol.* 20, 220. https://doi.org/10.1186/s12870-020-02400-9.

Solanki MK, Verma KK, Dastogeer KMG, Mora-Poblete F, Mundra S. (2023). Microbial resilience in plant nutrient management towards sustainable farming. *Front. Microbiol.* 14, 1280811. https://doi.org/10.3389/fmicb.2023.1280811

Song X-P, Mo F, Verma KK, Wei J, Zhang X, Yang L, Li YR. (2019). Effect of sugarcane smut (*Ustilago scitaminea* Syd.) on ultrastructure and biochemical indices of sugarcane. *Biomed J. Sci. Tech. Res.* 002950. https://doi.org/10.26717/BJSTR.2019.17.002950.

Song XP, Mo F, Yan M, Zhang X, Zhang B, Huang X, Huang D, Pan Y, Verma KK, Li YR. (2022). Effect of smut infection on the photosynthetic physiological characteristics and related defense enzymes of sugarcane. *Life.* 12, 1201. https://doi.org/10.3390/life12081201

Song XP, Verma KK, Tian DD, Zhang XQ, Liang YJ, Huang X, Li CN, Li YR. (2021). Exploration of silicon functions to integrate with biotic stress tolerance and crop improvement. *Biological. Research.* 54, 19. https://doi.org/10.1186/s40659-021-00344-4

Tenkolu GA, Kumsa DK, Gadissa TG. (2024). Optimization of fermentation condition in bioethanol production from waste potato and product characterization. *Biomass Convers. Biorefin.* 14(4), 5205–5223.

Tian D, Qin L, Verma KK, Wei L, Li J, Li B, Zhou W, He Z, Wei D, Huang S, Long S, Huang Q, Li C, Wei S. (2023). Transcriptomic and metabolomic differences between banana varieties which are resistant or susceptible to Fusarium wilt. *PeerJ.* 11, e16549. https://doi.org/10.7717/peerj.16549

Tsai C-Y, Tiffany A, Zong YK, Jaka S, Vincentius SKA. (2023). Toward a flexible design for the bioethanol dehydration using extractive distillation. Part 1: Steady-state design and optimization. *Ind. Eng. Chem. Res.* 62(51), 22043–22057.

Vasilakou K, Philippe N, Gwenny T, Pieter B, Steven VP. (2023). Assessing the future of second-generation bioethanol by 2030—A techno-economic assessment integrating technology learning curves. *Appl. Energy.* 344, 121263.

Vasilakou K, Philippe N, Pieter B, Steven VP. (2023). Geospatial environmental techno-economic assessment of pretreatment technologies for bioethanol production. *Renew Sustain. Energy Rev., Renew. Sust. Energ.* 187, 113743.

Verma KK, Anas M, Chen Z, Rajput VD, Malviya MK, Verma CL, Singh RK, Singh P, Song XP, Li YR. (2020f). Silicon supply improves leaf gas exchange, antioxidant defense system and growth in sugarcane responsive to water limitation. *Plants.* 9, 1032, https://doi.org/10.3390/plants9081032.

Verma KK, et al. (Ed.). *Agro-Industrial Perspectives on Sugarcane Production Under Environmental Stress*. https://doi.org/10.1007/978-981-19-3955-6_11.

Verma KK, Joshi A, Song X-P, Liang Q, Xu L, Huang H-r, Wu K-C, Seth CS, Arora J and Li Y-R. (2024a). Regulatory mechanisms of plant rhizobacteria on plants to the adaptation of adverse agroclimatic variables. *Front. Plant. Sci.* 15, 1377793. https://doi.org/10.3389/fpls.2024.1377793

Verma KK, Joshi A, Song X-P, Singh S, Kumari A, Arora J, Singh SK, Solanki MK, Seth CS, Li Y-R. (2024b). Synergistic interactions of nanoparticles and plant growth promoting rhizobacteria enhancing soil-plant systems: A multigenerational perspective. *Front. Plant Sci.* 15, 1376214. https://doi.org/10.3389/fpls.2024.1376214

Verma KK, Li DM, Singh M, Rajput VD, Malviya MK, Minkina T, Singh RK, Singh P, Song XP, Li YR. (2020a). Interactive role of silicon and plant-rhizobacteria mitigating abiotic stresses: A new approach for sustainable agriculture and climate change. *Plants* 9, 1055. https://doi.org/10.3390/plants9091055.

Verma KK, Liu X-H, Wu K-C, Singh RK, Song QQ, Malviya MK, Song X-P, Singh P, Verma CL, Li YR. (2020e). The impact of silicon on photosynthetic and biochemical responses of sugarcane under different soil moisture levels. *Silicon* 12, 1355–1367.

Verma KK, Rajput VD, Singh M, Verma CL, Minkina T. (2022a). Influence of waterlogging on plant physiology and molecular biology. In: *Understanding Abiotic Stresses Publisher*, Nova Publishers: New York.

Verma KK, Siddiqui N, Singh M. (2022b). Understanding the plant physiological and molecular adaptation strategies to soil salinity. In: *Understanding Abiotic Stresses Publisher*. Nova Publishers: New York. https://doi.org/10.52305/UBUJ4024

Verma KK, Singh P, Song X-P, Malviya MK, Singh RK, Chen G-L, Solomon S, Li YR. (2020b). Mitigating climate change for sugarcane improvement: Role of silicon in alleviating abiotic stresses. *Sugar. Tech.* 22(5), 741–749. https://doi.org/10.1007/s12355-020-00831-0.

Verma KK, Singh RK, Song QQ, Singh P, Zhang B-Q, Song X-P, Chen G-L, Li YR. (2019a). Silicon alleviates drought stress of sugarcane plants by improving antioxidant responses. *Biomed J. Sci. Tech. Res.* 002957. https://doi.org/10.26717/BJSTR.2019.17.002957

Verma KK, Song X, Degu HD, Guo D, Joshi A, Huang H, Xu L, Singh M, Huang D, Rajput VD, Li YR. (2023a). Recent advances in nitrogen and nano-nitrogen fertilizers for sustainable crop production: A mini-review. *Chem. Agric. Biotechnol.* 10, 1–14. https://doi.org/10.1186/s40538-023-00488-3

Verma KK, Song X, Verma CL, Huang H, Singh M, Xu L, Wu J, Khan ZH, Zhang B, Li Y. (2023b). Mathematical modeling of climate and fluoride effects on sugarcane photosynthesis with silicon nanoparticles. *Plant. Physiol. Biochem.* 204, 108089. https://doi.org/10.1016/j.plaphy.2023.108089

Verma KK, Song XP, Budeguer F, Nikpay A, Enrique R, Singh M, Zhang B, Wu, JM, Li YR. (2022c). Genetic engineering: An efficient approach to mitigating biotic and abiotic stresses in sugarcane cultivation. *Plant. Signal. Behav.* 17(1), e2108253.

Verma KK, Song XP, Joshi A, Rajput VD, Singh M, Sharma A, Singh RK, Li DM, Arora J, Minkina T, Li YR. (2022d). Nanofertilizer possibilities for healthy soil, water and food in future: An overview. *Front. Plant Sci.* 13, 865048. https://doi.org/10.3389/fpls.2022.865048

Verma KK, Song XP, Joshi A, Tian DD, Rajput VD, Singh M, Arora J, Minkina TM, Li YR. (2022e). Recent trends in nano-fertilizer for advancing sustainable agriculture under the era of climate change to ensure food security for future generations. *Nanomaterials.* 12, 173. https://doi.org/10.3390/nano12010173.

Verma KK, Song XP, Kumari A, Jagadesh M, Singh SK, Bhatt R, Singh M, Seth CS, Li YR. (2024c). Climate change adaptation: Challenges for agricultural sustainability. *Plant. Cell. Environ.* https://doi.org/10.1111/pce.15078

Verma KK, Song XP, Li DM, Singh M, Wu JM, Singh RK, Sharma A, Zhang BQ, LI YR. (2022f). Silicon and soil microorganisms improves rhizospheric soil health with bacterial community, plant growth, performance and yield. *Plant. Signal. Behav.* 17(1), e2104004.

Verma KK, Song XP, Lin B, Guo DJ, Singh M, Rajput VD, Singh RK, Singh P, Sharma A, Malviya MK, Chen GL, Li YR. (2022g). Silicon induced drought tolerance in crop plants: Physiological adaptation strategies. *Silicon.* 14(6), 2473–2487. https://doi.org/10.1007/s12633-021-01071-x

Verma KK, Song XP, Rajput VD, Boldyreva V, Zhang BQ, Minkina T, Li YR. (2022h). Morpho-physiological, biochemical and ultrastructural modifications on sugarcane to prolonged water deficit. In: Verma KK. et al. (eds.), *Agro-Industrial*

Perspectives on Sugarcane Production Under Environmental Stress. https://doi.org/10.1007/978-981-19-3955-6_11.

Verma KK, Song XP, Rajput VD, Solomon S, Li YR, Rao GP. (2022i). Agro-industrial perspectives on sugarcane production under environmental stress. *Springer Nature* https://doi.org/10.1007/978-981-19-3955-6

Verma KK, Song XP, Singh M, Tian DD, Rajput VD, Minkina T, Li YR. (2023c). Association of silicon and soil microorganisms induces stress mitigation, increasing plant productivity. In: de mello prado R. (Ed.), *Silicon (Si)*. pp. 299–328. https://doi.org/10.1007/978-3-031-266737_17

Verma KK, Song XP, Tian DD, Guo DJ, Chen ZL, Zhong CS, Nikpay A, Singh M, Rajput VD, Singh RK, Minkina T, Li YR. (2021a). Influence of silicon on biocontrol strategies to manage biotic stress for crop protection, performance and improvement. *Plants*. 10, 2163. https://doi.org/10.3390/plants10102163

Verma KK, Song XP, Tian DD, Singh M, Verma CL, Rajput VD, Singh RK, Sharma A, Singh P, Malviya MK, Li YR. (2021b). Investigation of defensive role of silicon during drought stress induced by irrigation capacity in sugarcane: Physiological and biochemical characteristics. *ACS Omega*. 6, 19811–19821.

Verma KK, Song XP, Verma CL, Chen ZL, Rajput VD, Wu KC, Liao F, Chen GL, Li YR. (2021c). Functional relationship between photosynthetic leaf gas exchange in response to silicon application and water stress mitigation in sugarcane. *Biological Research*. 54, 15. https://doi.org/10.1186/s40659-021-00338-2

Verma KK, Song XP, Verma CL, Malviya MK, Guo DJ, Rajput VD, Sharma A, Wei KJ, Chen GL, Solomon S, Li YR. (2021d). Predication of photosynthetic leaf gas exchange of sugarcane (Saccharum spp.) leaves in response to leaf positions to foliar spray of potassium salt of active phosphorus under limited water irrigation. *ACS Omega*. 6, 2396–2409.

Verma KK, Song X-P, Xu L, Huang H-R, Liang Q, Seth CS, Li Y-R. (2023d). Nano-microplastic and agro-ecosystems: A mini-review. *Front. Plant Sci.* 14, 1283852. https://doi.org/10.3389/fpls.2023.1283852

Verma KK, Song XP, Yadav G, Degu HD, Parvaiz A, Singh M, Huang HR, Mustafa G, Xu L, Li YR. (2022j). Impact of agroclimatic variables on proteo-genomics in sugarcane (Saccharum spp.) plant productivity. *ACS. Omega*. 7, 22997–23008.

Verma KK, Song XP, Zeng Y, Guo DJ, Sing M, Rajput VD, Malviya MK, Wei KJ, Sharma A, Li DP, Chen GL, Li YR. (2021e). Foliar application of silicon boosts growth, photosynthetic leaf gas exchange, antioxidative response and resistance to limited water irrigation in sugarcane (*Saccharum officinarum* L). *Plant. Physiol. Biochem*. 166, 582–592.

Verma KK, Song XP, Zeng Y, Li DM, Guo DJ, Rajput VD, Chen GL, Barakhov A, Minkina TM, Li YR. (2020c). Characteristics and correlation of leaf stomata and its relationship with photosynthesis on *Saccharum* spp. under different irrigation and silicon application. *ACS Omega*. 5, 24145–24153.

Verma KK, Vatsal S, Singh M, Rajput VD, Minkina T. (2022k). Coping with water stress: Effects and adaptive mechanisms to plants. In: *Understanding Abiotic Stresses Publisher*, Nova Science Publishers, Inc.: New York.

Verma KK, Wu K-C, Singh P, Malviya MK, Singh RK, Song X-P, Li YR. (2019b). The protective role of silicon in sugarcane under water stress: Photosynthesis and antioxidant enzymes. *Biomed. J. Sci. Tech. Res. 002685*.. https://doi.org/0.26717/BJSTR.2019.15.002685

Verma KK, Wu KC, Verma CL, Li DM, Malviya MK, Singh RK, Singh P, Chen GL, Song XP, Li YR. (2020d). Developing mathematical model for diurnal dynamics on photosynthetic responses in sugarcane responsive to different irrigation and silicon application. *PeerJ.* e10154, 8, https://doi.org/10.7717/peerj.10154.

Wu K-C, Huang C-M, Verma KK, Deng Z-N, Huang H-R, Pang T, Cao H-Q, Luo H-B, Jiang S-L, Xu L. (2022). Transcriptomic responses of *Saccharum spontaneum* roots in response to polyethylene glycol-6000 stimulated drought stress. *Front. Plant Sci*. 13, 992755. https://doi.org/10.3389/fpls.2022.992755

Xu L, Deng Z-N, Wu K-C, Malviya MK, Solanki MK, Verma KK, Pang T, Li Y-J, Liu X-Y, Kashyap BK, Dessoky ES, Wang W-Z, Huang H-R. (2022). Transcriptome analysis reveals a gene expression pattern that contributes to sugarcane bud propagation induced by Indole-3-Butyric Acid. *Front. Plant Sci.* 13, 852886. https://doi.org/10.3389/fpls.2022.852886

Zheng F-J, Lin B, Yang Y-X, Fang X-C, Verma KK, Chen G-L. (2023). Efficacy and functionality of sugarcane original vinegar on mice. *Front. Microbiol*. 14, 1224666. https://doi.org/10.3389/fmicb.2023.1224666

CHAPTER 10

Modifying Plants for Biofuel and Biomaterial Production

BHUMIKA SINGH LODHI

Department of Plant Breeding and Genetics, Jawaharlal Nehru Krishi Vishwa Vidyalaya, Jabalpur, Madhya Pradesh, India

ABSTRACT

This chapter provides a comprehensive analysis of approaches and discoveries in plant modification for the sustainable production of biofuels and biomaterials. It focuses on increasing plant biomass for efficient conversion, exploring how biotechnology, genomics, and genetic engineering have transformed the bioenergy sector. The text underscores lignocellulosic biomass's vital role as a feedstock and addresses key challenges, such as the high costs of chemical pretreatment and enzymatic hydrolysis, proposing biotechnological solutions as viable alternatives. The chapter demonstrates how genetic alterations can reduce cell wall recalcitrance. It details successful efforts to modify lignin content, composition, and structure, as well as hemicellulose biosynthesis. These modifications are shown to significantly improve saccharification yields and increase ethanol production without compromising plant integrity. The discussion emphasizes the necessity of developing dedicated bioenergy crops with ideal traits, such as high C6-sugar content and low levels of fermentation inhibitors. Genetic modification is presented as a powerful strategy, for instance by overexpressing enzymes involved in polysaccharide production to increase carbohydrate availability. Simultaneously, reducing recalcitrance through the overexpression of

Sugarcane Transformation: Biofuels, Biotechnology, and Beyond.
Krishan K. Verma, Xiu-Peng Song, Hai-Rong Huang, Aradhna Kumari, and Yang-Rui Li (Eds.)

DOI: 10.1201/9781779641571-10

carbohydrate-active enzymes is explored, with attention paid to maintaining plant health and yield. Ultimately, the chapter offers a thorough assessment of advancements in developing optimized feedstocks for a green bioeconomy.

10.1 INTRODUCTION

On the front lines of the "green revolution," biofuels and biomaterials offer a sustainable solution for today's ever-expanding environmental and resource problems. These changes have been made possible by advances in plant biotechnology that have magnified plants' potential as energy or materials sources (Chen et al., 2020, 2023; Zheng et al., 2023; Huang et al., 2024). In essence, developing more efficient biomass for biofuel means using genetic engineering to innovate plants, so we are seeking a mutually beneficial future in which the natural world and technology exist in harmony (Xu et al., 2022; Wu et al., 2022; Deng et al., 2023; Jagadesh et al., 2024; Verma et al., 2024a, 2024b, 2024c). Besides low levels of carbon-dioxide, these new innovations could lead to new prosperity and growth, and represent a "dense" pattern of economic development that is dependent on sustainable methods of meeting our needs for energy and material (UNEP, 2007; Singh et al., 2020; Verma et al., 2020a, 2020b; 2023a, 2023b, 2023c; Oliveira et al., 2022; Bhatt et al., 2023a, 2023b).

10.1.1 SO WHAT ARE BIOFUELS AND BIOMATERIALS?

Biofuels are a type of fuel produced from such biological raw materials as plants and organic waste. Renewable and able to replace traditional fossil fuels in vehicles, heating, or other applications, it is also biodegradable (Guo et al., 2021, 2022; Verma et al., 2020c, 2023d; Huang et al., 2022; Li et al., 2023). Ethanol and biodiesel are among the biofuels obtained via fermentation and transesterification.

Renewable fuels made from biological sources, usually plants' waste, are known as biofuels. They are primarily divided into groups according to their source and mode of production and are seen as a sustainable substitute for fossil fuels (Carere et al., 2008; Verma et al., 2019a, 2019b; Anas et al., 2021a, 2021b; Malviya et al., 2021, 2022) (Figure 10.1).

The main categories of biofuels are as follows:

FIGURE 10.1 Illustrating types of biofuels.

10.1.1.1 BIOFUELS OF THE FIRST GENERATION

- **Bioethanol:** Mostly made from fermentable sugar found in plants including maize, beets, and sugarcane. It serves as an additive or a replacement for fuel.
- **Biodiesel:** Produced via transesterification, a process that uses vegetable or animal fats. Diesel engines employ it.
- **Biogas:** Anaerobic digestion of organic materials, such as plant and animal waste, produces biogas. Carbon dioxide and methane make up the majority.

10.1.1.2 BIOFUELS OF SECOND GENERATION

- **Cellulosic ethanol:** Made from biomass that is lignocellulosic, such as grasses, wood, and inedible plant components. It does not compete with food crops for agricultural land, in contrast to first generation.
- **Biomass-to-liquid fuels:** Consists of a variety of biofuels made from different biomass sources, such as dimethyl ether and synthetic biofuels.

10.1.1.3 BIOFUELS OF THE THIRD GENERATION

- **Algae-derived biofuels:** Made from algae that may be grown in ponds or photo-bioreactors and can provide significant amounts of oil. Due to its ability to thrive in a range of settings, the production of biofuels from algae may be more sustainable than that of other biofuels.

10.1.1.4 FUELS OF THE FOURTH GENERATION

These are made by genetically altering organisms in order to maximize fuel output. Although this technology is still in its infancy, it has the potential to produce biofuels that are both sustainable and productive (Joshi et al., 2022, 2023; Sharma et al., 2021, 2022; Tian et al., 2023; Singh et al., 2021, 2023). All kinds of biofuels have unique benefits and drawbacks (Verma et al., 2020d, 2021a, 2021b). First-generation biofuels, for example, are simpler to make but could compete with food crops for land; second-generation biofuels solve this problem but need more processing work. Though they take novel techniques, third- and fourth-generation biofuels need further study and development to be economically viable. The worldwide drive to switch to renewable energy sources and lower greenhouse gas emissions needs to include biofuels (Chaudhary et al., 2023; Guo et al., 2023; Solanki et al., 2023). Biomaterials, however, are those materials composed partly or wholly of plants, animals, and microorganisms (Verma et al., 2020e, 2020f; 2021c, 2022a, 2022b, 2022c). They are used for such things as medical devices, packaging materials, and construction materials. Oftentimes, biomaterials are biodegradable and sustainable, providing an environmentally friendly alternative to synthetic products (Song et al., 2019, 2021, 2022; Krishna et al., 2022).

The main forms of biomaterials are as follows:

10.1.1.5 ORGANIC BIOMATERIALS

- **Collagen and gelatin:** The exceptional biocompatibility and biodegradability of collagen and gelatin make them widely utilized in medical applications, widely used in tissue engineering, cosmetics, and wound treatments.
- **Cellulose:** Occurring in plant cell walls, cellulose and its derivatives have several applications in the fields of medicine, textiles, and packaging.

- **Chitosan:** Made from the chitin found in crab shells, this biodegradable film is utilized in food packaging, medication delivery systems, and wound healing.

10.1.1.6 ARTIFICIAL BIOMATERIALS

- **Polymers** extracted from plants are utilized in tissue engineering scaffolds, controlled drug delivery systems, and surgical sutures (Brandon et al., 2020). Examples of polymers are polycaprolactone, polyglycolic acid, and polylactic acid.
- **Ceramics:** Due to their compatibility with bone tissue, bioceramics such as hydroxyapatite and bioactive glass are utilized in dental applications and bone grafts.

10.1.1.7 COMPOSITE BIOMATERIALS

- Combinations of various materials to improve qualities like strength, durability, and biocompatibility (e.g., polymers with metals or ceramics) are utilized in dental fillings, prosthetic devices, and bone grafting.

10.1.1.8 BIOMATERIALS FOR REGENERATION

Materials intended to encourage organ and tissue regeneration. They are employed in tissue engineering and regenerative medicine and frequently contain growth factors or cells.

10.1.1.9 MATERIALS DERIVED FROM BIOMATERIALS

Derived from biorenewable biological resources, including proteins, polysaccharides, and plant oils. They are part of an expanding field that aims to lessen its influence on the environment and are utilized in bioplastics, textiles, and sustainable packaging (Verma et al., 2021d, 2021e; 2022d, 2022e, 2022f). Biomaterials are essential to the advancement of medical technology, the reduction of environmental impact, and the promotion of sustainable industrial practices because they are sustainable, efficient, and biocompatible materials (Kumari et al., 2022, 2023a, 2023b). These attributes drive the development and implementation of biomaterials (Table 10.1).

TABLE 10.1 Comparison Between Biofuel and Biomaterials

Aspect	Biofuels	Biomaterials
Definition	Fuels made from biological materials, used as an alternative or supplement to petroleum-based fuels	Materials derived from biological sources or inspired by them, used in various industries, especially in medicine
Source	Derived from organic matter such as plants (sugarcane, corn), algae, and agricultural or urban waste	Sourced from natural substances like collagen, cellulose, chitosan, or synthetically designed to mimic natural materials
Primary use	Used as renewable energy sources for transportation, heating, and electricity generation	Used in medical implants, tissue engineering, sustainable packaging, and environmentally friendly products
Examples	Ethanol, biodiesel, biogas, and cellulosic ethanol	Collagen for medical applications, bioplastics, bio-based textiles, and biodegradable packaging
Environmental impact	Reduces greenhouse gas emissions, supports energy independence, but may compete with food production	Biodegradable and eco-friendly, reducing reliance on synthetic materials, but challenges in scalability and cost
Technological focus	Focused on improving yield and efficiency of production, reducing land use, and developing advanced biofuels	Advances in biocompatibility, durability, and integration with living tissues, focus on sustainable and green production

Biofuels and biomaterials have a complex economic environment that is shaped by global trends, technological developments, legislative frameworks, and local resources. Biofuels and biomaterials play key roles in the shift to a more robust and sustainable global economy (Loqué et al., 2015; Verma et al., 2022g, 2022h). In this regard, sugarcane is essential to the biofuel sector, especially when it comes to producing bioethanol. Sugarcane varieties with increased sugar content and biomass have been created through genetic breakthroughs, greatly increasing the efficiency of bioethanol production (Verma et al., 2022i, 2022j, 2022k; Huang et al., 2024). Breeding efforts have produced variants in Brazil, the world's largest producer of sugarcane ethanol, that not only boost the yield of ethanol per hectare but also improve resistance to pests and illnesses, minimizing the impact on the environment and promoting sustainability.

Similar to this, sugarcane in India is a crucial feedstock for the manufacturing of bioethanol, supporting the Ethanol Blended Petrol Program of the government. India hopes to increase its output of bioethanol through genetic advancements and selective breeding, promoting energy security and creating jobs in rural areas. Furthermore, the utilization of agricultural leftovers such as sugarcane bagasse and others as biomaterials in the creation of bioplastics and biocomposites highlights the crops' dual significance in the biofuel and biomaterial industries (Gírio et al., 2010).

Genetically modified crops such as corn, soybeans, and jatropha have also been enhanced to produce biofuels. In the United States, corn is mostly used for bioethanol, but breeding for increased starch content and disease resistance has resulted in notable yield gains. The main source of biodiesel, soybeans, gains from genetic modifications that raise its oil content. In order to lessen rivalry with food crops, jatropha, which is grown in a number of tropical and subtropical regions, including India, is being genetically modified to produce nonedible oil for biodiesel.

These illustrations highlight the revolutionary possibilities of genetics and plant breeding in the biofuels and biomaterials industries. These scientific activities foster a sustainable future by improving the production, resilience, and specialized features of sugarcane and other crops, while also supporting the economic viability of biofuels and biomaterials. This methodical approach to crop improvement demonstrates a dedication to tackling the world's sustainability issues and highlights the crucial role that agricultural science plays in rural development, environmental preservation, and food security.

10.2 BIOTECHNOLOGY IN SUSTAINABLE PRODUCTION

As the world looks for greener substitutes for fossil fuels and synthetic materials, biotechnology plays an increasingly important role in sustainable production, particularly in the fields of biofuel and biomaterial development. Advanced techniques and instruments provided by biotechnology can improve the productivity, sustainability, and efficiency of bio-based goods.

10.2.1 GENETIC ENGINEERING FOR IMPROVED BIOFUEL PLANTS

The genetic engineering of crops like sugarcane, maize, and algae to boost productivity and lower resource inputs is a good example of biotechnology in sustainable biofuel production. To improve the energy balance and sustainability of bioethanol production, for instance, genetically modified sugarcane with a greater sugar content or altered metabolic pathways can generate more bioethanol per unit of land. Algae have also undergone genetic modification to enhance their lipid content, which makes them better suited for the manufacture of biodiesel (Sticklen, 2008). These changes lessen the rivalry between food crops and biofuel crops, while simultaneously increasing the efficiency of biofuel production.

10.2.2 CRISPR-CAS9 IN BIOFUEL PRODUCTION

Biofuel crops' genetic makeup has been altered through the use of CRISPR–Cas9 gene editing technology, allowing for precise modifications to enhance features including biomass composition, growth rate, and tolerance to drought. For instance, altering the lignin composition and structure of crops such as switchgrass increases the biomass's receptiveness to biofuel conversion processes, hence augmenting the total production of biofuel (Himmel and Bayer, 2009).

10.2.3 MICROBIAL FACTORIES FOR BIOMATERIALS

The use of microbes as factories to create biomaterials, such as bioplastics, has been made possible by biotechnology. Genetically modified bacteria, such as Escherichia coli, can now synthesize polyhydroxyalkanoates, a biodegradable bioplastic derived from renewable resources. This methodology not only offers a sustainable route for the production of biomaterials, but it also permits the modification of the polymer characteristics for particular uses.

10.2.4 ENZYME ENGINEERING FOR THE PRODUCTION OF BIOMATERIALS

Biotechnology-based enzyme creation and optimization improve biomass processing into biomaterials. For example, enhanced activity and stability of engineered cellulases and ligninases improve the breakdown of lignocellulosic biomass, increasing the efficiency of producing bio-based textiles and composites (Chen and Peng, 2013). Because enzymatic processing uses less harsh chemicals, it adheres to sustainability and green chemistry principles.

10.3 LIGNOCELLULOSIC BIOMASS

Plant biomass that is mostly made up of cellulose, hemicellulose, and lignin—the primary structural elements of plant cell walls—is referred to as lignocellulosic biomass. This kind of biomass may be obtained from a variety of sources, including forestry waste (sawdust and wood chips), grassy and woody plants, specialized energy crops (miscanthus and switchgrass), and agricultural wastes (straw and maize stover). Because of its extensive availability and sustainability, lignocellulosic biomass is regarded as an essential renewable resource for the production of biofuels, biomaterials, and biochemicals. Because lignocellulosic biomass is carbon-neutral, its usage does not result in net greenhouse gas emissions, in contrast to fossil fuels (Abramson et al., 2010). Its application in the production of bioenergy and bioproducts is a viable route for lowering dependency on nonrenewable resources, slowing down global warming, and moving toward a circular economy (Park et al., 2016).

Plant cell walls are made of lignocellulose, a complex and durable structure that is essential for providing mechanical strength, defense, and resistance to microbial invasions. The growth, structural integrity, and overall productivity of the sugarcane plant—a major provider of lignocellulosic biomass—are all dependent on this composite material.

10.3.1 LIGNOCELLULOSE FORMATION

10.3.1.1 CELLULOSE SYNTHESIS

Long chains of cellulose are formed by the polymerization of glucose molecules in the cytoplasm. After that, these chains are sent to the cell wall, where they combine to form microfibrils. The cellulose synthase complexes found in the plasma membrane catalyze this activity. The cellulose microfibrils act as a framework for other components and provide the cell walls with their tensile strength.

10.3.1.2 INTEGRATION OF HEMICELLULOSE AND PECTIN

Hemicellulose is a heterogeneous collection of polysaccharides that is produced in the Golgi apparatus and secreted into the cell wall, where it attaches itself to the cellulose microfibril surface. Hemicelluloses in sugarcane, including xylans, mannans, and glucans, combine with cellulose and lignin to form a matrix that interconnects the cellulose fibrils. Another polysaccharide produced in the Golgi, pectin, is also released into the cell wall where it adds flexibility and hydrates the cell wall matrix by gelling and hydrating it (Figure 10.2).

10.3.1.3 LIGNIN DEPOSITION

In sugarcane, lignification, or the polymerization of monolignols (coniferyl, sinapyl, and *p*-coumaryl alcohol), occurs in the cell wall. Peroxidase and laccase enzymes catalyze oxidative coupling processes, which cause this to happen. In vascular tissues and sclerenchyma cells, lignin deposition is very noticeable. Here, it provides stiffness, impermeability, and resistance against infections and mechanical stress. In sugarcane, the lignification process is critical for providing structural support that allows the plant to grow tall and resist lodging, which is necessary to maximize biomass output and photosynthetic efficiency.

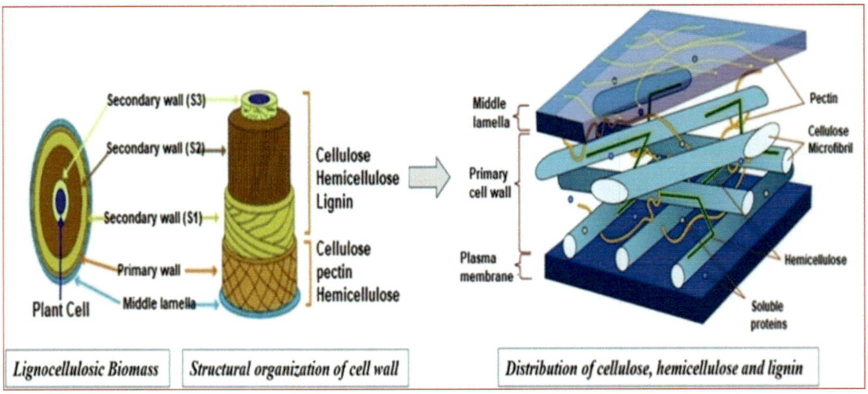

FIGURE 10.2 Structural organization and distribution of cellulose, hemicellulose, and lignin in plant cell wall.

Source: Reprinted from Ilnicka and Lukaszewicz, 2015.

Comprehending the process of lignocellulose development in sugarcane holds noteworthy consequences for the production of biofuel and genetic engineering. To improve biomass digestibility without sacrificing plant integrity, for example, lignin content and composition are being changed as part of attempts to increase sugarcane's capacity to produce bioethanol (Donohoe et al., 2017). The goal of genetic engineering techniques is to maximize sugarcane's lignocellulosic biomass for more effective conversion into biofuels. Examples of these efforts include downregulating the genes involved in lignin biosynthesis or overexpressing those involved in cellulose and hemicellulose synthesis.

10.3.2 MODIFYING THE CONTENT AND COMPOSITION OF LIGNIN

10.3.2.1 GENETIC ENGINEERING

The ability to specifically modify genes involved in the lignin biosynthesis pathway has been made possible by advances in genetic engineering.

- Deletion or overexpression of lignin biosynthesis genes—Lignin content can be decreased by suppressing certain genes involved in the production of lignin precursors. For instance, it has been demonstrated that downregulating the expression of 4-coumarate:CoA ligase or cinnamoyl–CoA reductase reduces the amount of lignin in transgenic plants (Furtado et al., 2014; Chabannes et al., 2001).
- Overproduction of enzymes that degrade lignin-enzymes that break down lignin naturally in the environment, such as peroxidases and laccases, are essential. Scientists want to replicate this breakdown process in bioenergy crops by overexpressing these enzymes, which will make the lignin less complicated and easier to break down in the processes involved in producing ethanol (Figure 10.3).

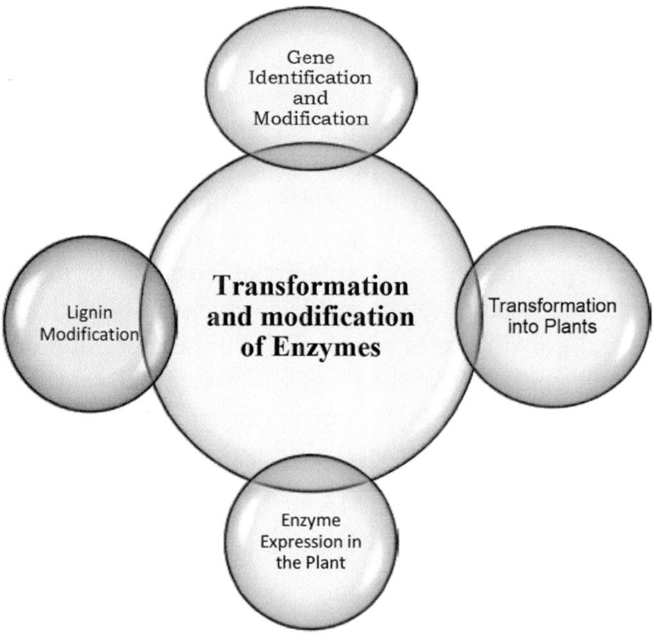

FIGURE 10.3 Process of identification and insertion of gene for over enzyme production.

10.3.2.2 BREEDING PROGRAM

In the effort to create plant varieties with optimal features for the production of biofuel and biomaterials is to use both traditional and marker-assisted breeding approaches in breeding programs. In order to increase the effectiveness of biomass conversion processes, these methods concentrate on identifying and promoting characteristics linked to decreased lignin concentration and modified lignin composition (Grabber et al., 2004).

Conventional breeding techniques: Plant phenotypic selection is the foundation of conventional breeding, which has been the mainstay of agricultural advancement for millennia.

- Phenotypic selection is the process of identifying, from observable features, plants that have desired attributes (in this case, lower lignin content). Because environmental influences might have an impact on phenotypic expression, this procedure can be difficult.
- Hybridization and crossbreeding is mating particular plant types with other cultivars that have complimentary desired features. By combining good genes from several parent plants, this technique may result in offspring with a decreased lignin content as well as other desirable traits like stress tolerance or enhanced yield.
- It is possible to stabilize the presence of desirable features in the plant population by repeating the selection process over a number of generations. This is known as successive generations of selection. With regard to the desired lignin qualities, this iterative procedure progressively improves the crop's genetic composition.
- The main problem with traditional breeding is that it takes a long time to produce noticeable changes in lignin characteristics; several growth cycles are needed. Furthermore, lignin's intricate genetic control frequently calls for a careful selection strategy since changing the lignin's composition and content might unintentionally affect other vital plant processes.

Marker-assisted selection (MAS): MAS is a more advanced and accurate method of plant breeding.

- Genetic markers identification: Locating certain DNA sequences linked to desired lignin properties using cutting-edge genomic technologies. These markers function as tell-tale signs of the existence of genes that affect the composition and concentration of lignin.

- Using these genetic markers, rapid screening may be done to find individuals in huge plant populations that have the required genetic composition (Himmel et al., 2007). Comparing this procedure to traditional ones, the identification of potential plants is accelerated dramatically.
- By using marker data to guide breeding choices, MAS makes it easier to combine desirable features with reduced lignin content, enhancing the breeding process' efficiency and effectiveness.
- By concentrating on the genetic potential of the plants rather than on visible features, which are subject to change based on environmental factors, MAS provides a strategic edge. This method works especially well for characteristics that are late in plant growth or are challenging to quantify directly.
- Combining MAS with traditional breeding methods helps maximize the benefits of each strategy. Breeders may use the genetic variety and local adaptation of plants using this combination method, which also speeds up the breeding process considerably by utilizing molecular markers.

10.3.2.3 CHEMICAL AND BIOLOGICAL PRETREATMENTS

One important component of the plant cell wall that gives it stiffness and resistance to microbial attack is lignin, whose structural integrity is intended to be changed by these pretreatment procedures. Pretreatments increase cellulose and hemicellulose accessibility by altering lignin (Mosier et al., 2005). These carbohydrate polymers are fermented into sugars for biofuel production or are turned into useful biomaterials.

10.3.2.3.1 Chemical Pretreatments

Steam explosion: This process breaks down the structure of the cell wall by putting biomass to high-pressure steam and then quickly decompressing it. The enzymatic digestibility of cellulose can be improved by partially depolymerizing lignin and hemicellulose due to the high temperature and pressure of the steam explosion. Because it uses less chemicals than other pretreatment techniques, steam explosion is one of the most economical and environmentally benign procedures.

Acid pretreatment: To improve cellulose accessibility, lignin structure is changed and hemicellulose is hydrolyzed into fermentable sugars using sulfuric acid or other diluted acids. Although acid pretreatment works well for many different kinds of biomass, it raises the cost of the process since it needs equipment that is resistant to corrosion and neutralization procedures before enzymatic hydrolysis.

Alkaline pretreatment: To increase the porosity of the biomass, alkaline materials such as calcium or sodium hydroxide are employed to remove lignin and hemicellulose. This process works especially well with high-lignin biomass, but it might cause problems for wastewater treatment because it requires a lot of washing to get rid of alkaline residues.

10.3.2.3.2 Biological Pretreatments

Microbiological: By utilizing the enzymes generated by specific fungi and bacteria, lignin and hemicellulose are broken down without the harsh conditions needed for chemical treatments. Chemical approaches can be more harshly avoided by using an enzymatic pretreatment that is extremely specific and can be adjusted to target particular links within lignin and hemicellulose.

Microbial consortia: By utilizing a number of enzymatic processes, microbial consortia—groups of microorganisms that cooperate with one another—can break down lignin and hemicellulose. Adaptable to different kinds of biomass, this method replicates the natural breakdown processes.

10.3.3 MANIPULATING HEMICELLULOSE BIOSYNTHESIS

An advanced tactic to improve biomass quality for the generation of biofuel and biomaterials is to manipulate the hemicellulose biosynthesis in plants such as sugarcane. Hemicellulose is a hetero-polysaccharide that is present in plant cell walls and plays a role in the chemical and physical barriers that restrict the ability of cellulolytic enzymes to access cellulose fibers during the biomass processing process. The ease of deconstruction of plant biomass into fermentable sugars is influenced by its composition and cross-linking with other components of the cell wall, such as lignin.

10.3.3.1 GENETIC ENGINEERING

The ability to specifically manipulate genes involved in hemicellulose production pathways has been made possible. For instance, changing the amount of a certain glycosyltransferase enzyme in sugarcane can change the structure and makeup of xylans, which are the main hemicellulose type in grasses (Cantarel et al., 2009). These alterations may result in less hemicellulose and lignin cross-linking, improving cellulose's extractability. Because xylans are particularly good at attaching cellulose fibers and interacting with lignin, research has concentrated on lowering the xylan level in sugarcane hemicellulose. Increased saccharification efficiency has been demonstrated in transgenic sugarcane lines with downregulated xylan biosynthesis genes, without appreciable effects on growth or yield. The degree to which hemicellulose is acetylated also influences how it interacts with lignin and cellulose. Scientists have created biomass that requires less aggressive chemical pretreatments for efficient saccharification by modifying the expression of hemicellulose acetyltransferases in sugarcane.

10.3.3.2 CRISPR/CAS9 GENOME EDITING

A precise method for altering the genes involved in hemicellulose production is provided by the CRISPR/Cas9 system. Researchers can produce sugarcane varieties with modified hemicellulose that is less resistant to enzymatic hydrolysis by deleting or altering genes that encode enzymes involved in the polymerization and modification of hemicellulose subunits (Arruda, 2012). The process is as follows:

- *Design of Guide RNA (gRNA)*: The first stage is to create a gRNA that complements a particular DNA sequence found in a target gene that is involved in the synthesis of hemicellulose. The Cas9 nuclease is guided by this gRNA to the exact spot in the genome where the modification is wanted to be made (Jung et al., 2012a).
- *Cas9 nuclease introduction*: At the targeted genomic region, the Cas9 protein initiates a double-strand break (DSB) under the guidance of the gRNA. The gRNA sequence controls the specificity of

this cut, enabling accurate targeting of the genes involved in the polymerization and modification of hemicellulose subunits.

- *DNA repair and gene modification*: The plant cell's built-in DNA repair mechanisms, such as homology-directed repair (HDR) or nonhomologous end joining (NHEJ), are triggered upon the introduction of the DSB. Gene function may be eliminated by NHEJ-induced insertions or deletions (indels) at the cut site. With the help of a repair template, HDR can precisely modify genes by introducing particular mutations or insertions.

10.3.3.3 BREEDING PROGRAM

Breeding efforts are essential for choosing and refining sugarcane types with improved bioconversion efficiency. These programs include both conventional and marker-assisted approaches. Through the use of desired mutations and natural genetic variation, these projects seek to create crops with the ideal hemicellulose composition for the manufacture of biofuel.

10.3.3.3.1 Traditional Breeding Techniques

Phenotypic selection: This method involves choosing plants according to observable features that may imply a desirable hemicellulose composition, such as growth rate, yield, and physical attributes. To find the phenotypes that are associated with higher bioconversion efficiency, a lot of field research and laboratory analysis are needed.

10.3.3.3.2 Crossing and Hybridization

In an effort to pass on the advantageous hemicellulose composition to their progeny in addition to other desired features, breeders cross specific types with traits like pest resistance or drought tolerance. This mechanism depends on meiosis, a natural genetic recombination process that creates new genetic combinations that can be passed down across generations.

10.3.3.3.3 Evaluation and Selection

Through biochemical testing and pilot-scale fermentation experiments, the offspring of these crossings are assessed for their hemicellulose content and bioconversion efficiency. Throughout several generations, this crossing and selection cycle is repeated in order to accumulate the desired features in the sugarcane genome.

10.3.3.3.4 MAS

Identifying molecular markers: Sophisticated genomics methods, such QTL mapping and genome-wide association studies, are employed to find markers connected to genes or genomic areas that affect the composition of hemicellulose. Plant genomes containing advantageous alleles can be predicted by these markers.

Marker screening: Following their identification, DNA samples from breeding populations are checked to see if any of the markers are present. This considerably expedites the breeding process by enabling the early selection of individuals likely to display the desired hemicellulose content.

Introgression and pyramiding: Selective alleles linked to the ideal hemicellulose composition can be introduced (introgressed) into superior sugarcane cultivars using MAS. Additionally, by stacking various alleles that are individually selected for by a corresponding marker, marker-assisted pyramiding combines several advantageous features into a single variety.

10.4 SACCHARIFICATION

An essential biochemical step in the transformation of lignocellulosic biomass into fermentable sugars that may be fermented to produce ethanol or other biofuels, or utilized to make a variety of biomaterials, is saccharification (Wang et al., 2013). By means of enzymatic hydrolysis, complex polysaccharides like cellulose and hemicellulose are broken down into simpler, fermentable sugars, mostly glucose, in this process (Bhatia et al., 2017).

10.4.1 SACCHARIFICATION MECHANISMS

10.4.1.1 HYDROLYSIS BY ENZYMES

Enzymatic hydrolysis, in which enzymes catalyze the conversion of cellulose and hemicellulose into sugars, is the fundamental process of saccharification. Two primary categories of enzymes are usually involved in this process:

- *Cellulases enzyme:* They catalyze the breakdown of cellulose into glucose units. Endoglucanases are cellulose chain-cutting enzymes that break down cellulose chains into smaller polysaccharides by cutting their internal connections. Cellobiohydrolases, on the other hand, cleave cellobiose units off the ends of cellulose chains, whereas β-glucosidases hydrolyzecellobiose into glucose.
- *Hemicellulases:* A sophisticated mixture of enzymes that disassemble hemicellulose, a heteropolymer made up of several sugars, into its monomeric constituents. Xylanases, which hydrolyzemannans, and mannanases, which break down xylan, the primary hemicellulose component in many plants, into xylose, are important hemicellulases (Chandra et al., 2007).

10.4.1.2 CHEMICAL HYDROLYSIS

Using acids or alkalis, chemical hydrolysis can also saccharify biomass, albeit it is less popular than enzymatic hydrolysis because of costs and environmental issues. For instance, sulfuric acid is used in acid hydrolysis to change cellulose into glucose and hemicellulose into various sugars that can be fermented.

10.4.2 FACTORS AFFECTING SACCHARIFICATION

Biomass pretreatment: By reducing cellulose crystallinity, removing lignin and hemicellulose barriers, and enhancing the biomass's porosity, pretreatment of the biomass attempts to improve enzyme accessibility and has a significant impact on saccharification efficiency.

Enzyme composition and loading: The kinds and amounts of enzymes utilized, as well as their cooperative effects, have a major impact on the results of saccharification.

Process parameters: Since each enzyme has a unique range of activity, it is essential to maintain ideal pH, temperature, and reaction time to maximize saccharification effectiveness.

10.4.3 RELEVANCE TO THE PRODUCTION OF BIOFUELS

An essential stage in the synthesis of cellulosic biofuels is saccharification. The overall output and financial viability of the procedures used to produce biofuels are directly impacted by its efficiency. In order to increase the viability and sustainability of biofuel production, research is now being conducted on enhancing saccharification yields through genetic modification of feedstock (e.g., reducing lignin content in plants), optimizing pretreatment techniques, and improving enzyme formulations. As sustainable bioenergy solutions are sought, it is critical to maximize the yield and conversion efficiency of biofuel crops. Enhancing the efficiency of converting biomass into biofuels is just as important as optimizing the amount of biomass a crop can produce per unit of land. This is a comprehensive examination of these elements, which are essential for increasing crop productivity for biofuel.

10.4.4 SIGNIFICANCE YIELD IN BIOFUEL CROP PRODUCTIVITY

Biofuel crop viability and sustainability as a renewable energy source are directly impacted by its output. High-yielding crops generate more biomass on the same amount of land, increasing the amount of raw materials available to make biofuels without requiring the addition of new agricultural land. This is crucial for lessening the impact on natural ecosystems and the competition between the production of biofuel and food crops. Moreover, the cost per unit of produced biofuel can be considerably reduced by high-yielding crops, increasing the competitiveness of biofuels over fossil fuels and other renewable energy sources.

10.4.5 PRODUCTION OF SPECIFIC BIOENERGY CROPS

Creating plant varieties especially intended for the generation of biofuel is the goal of the creation of specialized bioenergy crops. These crops have undergone genetic modification or breeding to develop characteristics that make them more suitable as feedstocks for biofuels, like:

- Enhanced biomass production: To ensure a greater amount of feedstock for the conversion of biomass into biofuel, characteristics that promote fast growth and high biomass output are chosen.
- Decreased input requirements: creating crops that can survive with less water, fertilizer, and pesticide use, which will lessen the impact of biofuel crops on the environment and their production costs.
- Adaptability to marginal areas: It is possible to increase the cultivable area for bioenergy crops without taking away food crops by breeding crops that can thrive on marginal areas that are unsuitable for food production.

10.4.6 INCREASING THE CARBOHYDRATE CONTENT VIA GENETIC MANIPULATION

Plant biomass is mostly composed of carbohydrates, namely cellulose and hemicellulose, which are then transformed into biofuels. The bioconversion process can be made much more efficient by genetically modifying biofuel crops to increase their carbohydrate content:

- Targeting biosynthetic routes: It is possible to genetically alter the routes that produce cellulose and hemicellulose in order to direct more energy into the synthesis of carbohydrates in plants. This will increase the amount of fermentable sugars produced per unit of biomass.
- Lowering lignin content: Lignin prevents sugars from being extracted from biomass. Saccharification yields and the overall efficiency of biofuel production can be increased by engineering crops with a lower lignin concentration, which will increase the accessibility of cellulose and hemicellulose.
- Changing the carbohydrate composition: The conversion efficiency can be increased by changing the cellulose and hemicellulose's chemical makeup to make them easier to hydrolyze. For example, biofuel yields can be increased by increasing the fraction of some hemicellulose sugars that ferment more easily.

In the realm of sustainable energy, the idea of genetically modified bioenergy crops for the synthesis of enzymes is revolutionary and provides a revolutionary substitute for conventional microbial bioreactors. Using this technique, bioenergy crops are altered to directly create cellulolytic enzymes, which will speed up the process of breaking down biomass into fermentable sugars.

10.5 FUTURE PERSPECTIVES AND CHALLENGES

The development and use of sustainable biofuel production technologies are contingent upon tackling the forthcoming views and issues in the ever-evolving sector of bioenergy. The main issues that require attention are covered in this part, which also looks at potential future research topics and technological advancements (Li et al., 2014). These issues include resolving yield penalties and environmental and ecological concerns.

10.5.1 SURMOUNTING YIELD PENALTIES

The possibility of yield penalties is one of the main obstacles to genetically altering crops for improved biofuel production. These costs may result from shifting resources away from development and reproduction and toward the synthesis of enzymes or precursors for biofuels.

- Mitigation techniques: Future studies should concentrate on creating techniques that reduce any adverse effects on agricultural productivity. This might involve the control of gene expression both spatially and temporally, such as when biofuel-related features are expressed exclusively in nonessential organs or during particular developmental stages. Furthermore, yield penalties may be lessened by the creation of metabolic pathways that are more effective and consume less resources.
- Advanced methods of breeding: Through the use of genome editing using CRISPR/Cas systems or speed breeding, genetic engineering, and sophisticated breeding techniques may be used to quickly produce and screen for plant varieties that retain high yield while having the necessary features for the generation of biofuel.

10.5.2 ECOLOGICAL AND ENVIRONMENTAL ASPECTS TO TAKE INTO ACCOUNT

Careful management is needed to prevent negative ecological and environmental effects of bioenergy crop development, such as loss of biodiversity, water shortages, and altered land use.

- *Sustainable crop management:* Bioenergy crops may be made more sustainable by using techniques like intercropping, crop rotation, and the use of perennials. These methods promote biodiversity, lessen erosion, and preserve the health of the soil.
- *Life-cycle evaluation:* It is crucial to carry out thorough life cycle assessments (LCAs) in order to assess the environmental impact of producing biofuels, from crop cultivation to fuel burning. LCAs can assist in pinpointing areas in need of development and guarantee that biofuels outperform fossil fuels in terms of net environmental benefit.
- *Genetic confinement:* To avoid the unintentional introduction of genetically engineered characteristics into wild populations, which might have unanticipated ecological repercussions, research into genetic confinement techniques is essential.

10.5.3 PROSPECTS FOR FUTURE TECHNOLOGIES AND RESEARCH APPROACHES

Current constraints in the production of biofuels can be overcome with the help of innovations in biotechnology and process engineering.

- Synthetic biology: Creating new metabolic pathways that can more effectively convert biomass into biofuels or directly create hydrocarbons in bioenergy crops through the use of synthetic biology has the potential to completely transform this sector.
- Nanotechnology: Using nanotechnology to stabilize and immobilize enzymes can improve their effectiveness and recyclable nature, which is important for biomass conversion processes.
- Combination with additional renewable technology: More sustainable and effective energy systems may result from investigating the integration of biofuel production with other renewable energy sources, such solar and wind power. The carbon footprint of producing biofuel, for example, may be decreased by processing biomass using renewable energy sources.
- Advanced pretreatment technologies: Increasing the sustainability and economic viability of biofuel production will depend on the development of kinder, more efficient pretreatment techniques that use less energy and chemical input.

Given the present issues in sustainability, efficiency, and cost-effectiveness, biofuels and biomaterials have a bright future thanks to continuous research and technical advancements. The potential of biofuels and biomaterials to lower greenhouse gas emissions, boost rural economies, and promote a circular economy is predicted to drive their increasing importance in the global energy mix as efforts are made to move away from fossil fuels. Realizing this potential, though, will need a well-rounded strategy that supports economic viability, encourages innovation, and takes the environment into account. To meet the challenges of producing biofuel and biomaterials and make sure that these bio-based solutions are created responsibly and sustainably, policymakers, researchers, and industry stakeholders need to operate together. To sum up, the development of biofuels and biomaterials as viable substitutes for fossil fuels is a fascinating and rapidly evolving area that presents a route toward a future that is both energy-secure and sustainable. Biofuels and biomaterials can contribute significantly to reaching global sustainability targets and transforming the energy environment into one that is cleaner and greener with sustained innovation and careful application.

KEYWORDS

- **biomass and biofuel**
- **bioethanol**
- **organic-artificial biomaterial**
- **sustainable production**
- **genetic improvement**

REFERENCES

Abramson M, Shoseyov O, Shani Z. (2010). Plant cell wall reconstructiontoward improved lignocellulosic production and processability. *Plant Sci.* 178, 61–72.

Anas M, Verma KK, Haq IU, Naeem M, Li Q, Liao F, Xiang YX, Li YR. (2021a). Characterization of exotic, native and wild-type genotypes of sugarcane (*Saccharum* spp. hybrids) for internal nitrogen use efficiency under different nitrogen levels. *Sugar Tech.* 23, 1258–1267. https://doi.org/10.1007/s12355-021-01014-1

Anas M, Verma KK, Riaz M, Qiang L, Liao F, Liu Y, Li YR. (2021b). Physio-morphological and biochemical mechanism of nitrogen use efficiency in sugarcane (*Saccharum* spp.) genotypes under different growth stages and nitrogen levels. *J. Plant Interact.* 16, 332–343.

Arruda P. (2012). Genetically modified sugarcane for bioenergy generation. *Curr. Opin. Biotechnol*. 23, 315–322.

Bhatia R, Gallagher JA, Gomez LD, Bosch M. (2017). Genetic engineering of grass cell wall polysaccharides for biorefining. *Plant Biotechnol. J*. 15, 1071–1092.

Bhatt R, Verma KK, Jamwal NS. (2023a). Can paddy-compost eliminate N fertilization for climate-smart sugarcane? *Sugar Tech*. 25, 1036–1046. https://doi.org/10.1007/s12355-023-01270-3

Bhatt R, Verma KK, Kumar R, Sanghera GS. (2023b). Foliar application of potassium salt of active phosphorus (PSAP) mitigates insect pests and improves yield along with sugarcane quality in response to agroclimatic conditions of Punjab. *Sugar Tech*. 25(3), 660–669.

Brandon, AG, Scheller HV. (2020). Engineering of bioenergy crops: Dominant genetic approaches to improve polysaccharide properties and composition in biomass. *Front. Plant Sci*. 11, 282.

Cantarel BL, Coutinho PM, Rancurel C, Bernard T, Lombard V, Henrissat B. (2009). The Carbohydrate-Active EnZymes database (CAZy): An expert resource for glycogenomics. *Nucleic Acids Res*. 37, D233–D238.

Carere C, Sparling R, Cicek N, Levin D. (2008). Third generation biofuels via direct cellulose fermentation. *Int. J. Mol. Sci*. 9, 1342–1360.

Chabannes M, Barakate A, Lapierre C, Marita JM, Ralph J, Pean M, Danoun S, Halpin C, Grima-Pettenati J, Boudet AM (2001). Strong decrease in lignin content without significant alteration of plant development is induced by simultaneous down-regulation of cinnamoyl CoA reductase (CCR) and cinnamyl alcohol dehydrogenase (CAD) in tobacco plants. *Plant J*. 28, 257–270.

Chandra RP, Bura R, Mabee WE, Berlin A, Pan X, Saddler JN (2007). Substrate pretreatment: The key to effective enzymatic hydrolysis of lignocellulosics? *Adv. Biochem. Eng. Biotechnol*. 108, 67–93.

Chaudhary P, Xu M, Ahamad L, Chaudhary A, Kumar G, Adeleke BS, Verma KK, Hu D-M, Siric I, Kumar P, Popescu SM, Fayssal SA. (2023). Application of synthetic consortia for improvement of soil fertility, pollution remediation, and agricultural productivity: A review. *Agronomy* 13, 643. https://doi.org/10.3390/agronomy13030643.

Chen GL, Zheng FJ, Lin B, Lao SB, He J, Huang Z, Zeng Y, Sun J, Verma KK. (2020). Phenolic and volatile compounds in the production of sugarcane vinegar. *ACS Omega* 5, 30587–30595.

Chen G-L, Zheng F-J, Lin B, Yang Y-X, Fang X-C, Verma KK, Yang L-F. (2023). Vinegar: A potential source of healthy and functional food with special reference to sugarcane vinegar. *Front. Nutr*. 10, 1145862. https://doi.org/10.3389/fnut.2023.1145862

Chen P, Peng L. (2013). CHAPTER 6. the diversity of lignocellulosic biomass resources and their evaluation for use as biofuels and chemicals. In: Sun, J, Ding, S-Y, Peterson JD. (Eds.), *Biological Conversion of Biomass for Fuels and Chemicals: Explorations from Natural Utilization Systems Energy and Environment Series*. Royal Society of Chemistry: Cambridge; pp. 83–113.

Deng ZN, Huang HR, Huang CM, Verma KK, Luo HB, Jiang SL, Li YJ, Cao HQ, Wu KC, Xu L. (2023). Short-term limited water irrigation influences the photosynthetic pigments, enzymatic and non-enzymatic activities in *Saccharum spontaneum* L. at vegetative stage. *Sugar Tech*. 25, 110–118. https://doi.org/10.1007/s12355-022-01175-7

Donohoe BS, Wei H, Mittal A, Shollenberger T, Lunin VV, Himmel ME, et al. (2017). Towards an understanding of enhanced biomass digestibility by in planta expression of a family 5 glycoside hydrolase. *Sci. Rep.* 7, 4389.

Furtado A, Lupoi JS, Hoang NV, Healey A, Singh S, Simmons BA, Henry RJ. (2014). Modifying plants for biofuel and biomaterial production. *Plant Biotechnol. J.* 12(9), 1246–1258.

Gírio FM, Fonseca C, Carvalheiro F, Duarte LC, Marques S, BogelŁukasik, R. (2010). Hemicelluloses for fuel ethanol: A review. *Bioresour. Technol.* 101, 4775–4800.

Grabber JH, Ralph J, Lapierre C, Barrière Y. (2004). Genetic and molecular basis of grass cell-wall degradability. I. Lignin-cell wall matrix interactions. *Comptes Rendus. Biol.* 327, 455–465.

Guo D-J, Li D-P, Chen Z-D, Yang B, Verma KK, Singh RK, Singh P, Khan Q, Sharma A, Qin Y, Zhang B-Q, Song X-P, Li Y-R. (2023). Effect of endophytic diazotroph *Enterobacter roggenkampii* ED5 on nitrogen-metabolism-related microecology in the sugarcane rhizosphere at different nitrogen levels. *Front. Microbiol.* 14, 1132016. https://doi.org/10.3389/fmicb.2023.1132016

Guo D-J, Li D-P, Singh RK, Singh P, Sharma A, Verma KK, Qin Y, Khan Q, Lu Z, Malviya MK, Song X-P, Xing Y-X and Li Y-R. (2021). Differential protein expression analysis of two sugarcane varieties in response to diazotrophic plant growth-promoting endophyte *Enterobacter roggenkampii* ED5. *Front. Plant Sci.* 12, 727741. https://doi.org/10.3389/fpls.2021.727741

Guo DJ, Li DP, Singh RK, Singh P, Verma KK, Sharma A, Qin Y, Khan Q, Song XP, Malviya MK, Xing YX, Li YR. (2022). Comparative transcriptome analysis of two sugarcane varieties in response to diazitrophic plant growth promoting endophyte *Enterobacter roggenkampii* ED5. *J. Plant Interact.* 17(1), 75–84.

Himmel ME, Bayer EA. (2009). Lignocellulose conversion to biofuels: Current challenges, global perspectives. *Curr. Opin. Biotechnol.* 20, 316–317.

Himmel ME, Ding SY, Johnson DK, Adney WS, Nimlos MR, Brady JW, Foust TD. (2007). Biomass recalcitrance: Engineering plants and enzymes for biofuels production. *Science* 315, 804–807.

Huang CM, Wu KC, Verma KK, Luo HB, Deng ZN, Shi GY, Cao HQ, Wu XJ, Ye LF, Wei YW, Xu L. (2024). Isolation and expression analysis of cellulose synthase 3 (*Ces3*) genes from sugarcane (*Saccharum officinarum* L.). *Chem. Biol. Technol. Agric.* 11, 71. https://doi.org/10.1186/s40538-024-00597-7

Huang HR, Gao YJ, Malviya MK, Verma KK, Solanki MK, Huang YX, Li X, Deng Y, Yan J, Tang SY, Wang LW, Xu L. (2022). Genetic diversity analysis of sugarcane (*Saccharum* spp. Hybrids) among high-sucrose clones of GT series and commonly used parents by using microsatellite markers in Guangxi, China. *Sugar Tech.* 24, 397–407. https://doi.org/10.1007/s12355-021-01033-y

Ilnicka A, Lukaszewicz JP. (2015). Discussion remarks on the role of wood and chitin constituents during carbonization. *Front. Mater.* 2. https://doi.org/10.3389/fmats.2015.00020

Jagadesh M, Dash M, Kumari A, Singh SK, Verma KK, Kumar P, Bhatt R, Sharma SK. (2024). Revealing the hidden world of soil microbes: Metagenomic insights into plant, bacteria, and fungi interactions for sustainable agriculture and ecosystem restoration. *Microbiol. Res.* 285, 127764. https://doi.org/10.1016/j.micres.2024.127764

Joshi A, Verma KK, Rajput VD, Arora J. (2023). Metabolic engineering of microorganisms in advancing biofuel production. In: *Advances in Lignocellulosic Biofuel Production Systems*; pp. 181–202. https://doi.org/10.1016/B978-0-323-91192-4.00016-X

Joshi A, Verma KK, Rajput VD, Minkina T, Arora J. (2022). Recent advances in metabolic engineering of microorganisms for advancing lignocellulose-derived biofuels. *Bioengineered* 13(4), 8135–8163. https://doi.org/10.1080/21655979.2022.2051856.

Jung JH, Fouad WM, Vermerris W, Gallo M, Altpeter F. (2012a). RNAi suppression of lignin biosynthesis in sugarcane reduces recalcitrance for biofuel production from lignocellulosic biomass. *Plant Biotechnol. J.* 10, 1067–1076.

Krishna D, Sachan HK, Verma KK. (2022). Abiotic stresses in plants: An overview. In: *Understanding Abiotic Stresses*. Nova Science Publishers, Inc.: New York.

Kumari A, Dash M, Singh SK, Jagadesh M, Mathpal B, Mishra PK, Pandey SK, Verma KK. (2023a). Soil microbes: A natural solution for mitigating the impact of climate change. *Environ. Monit. Assess.* 195, 1436. https://doi.org/10.1007/s10661-023-11988-y

Kumari A, Lakshmi GA, Krishna GK, Patni B, Prakash S, Bhattacharyya M, Singh SK, Verma KK. (2022). Climate change and its impact on crops: A comprehensive investigation for sustainable agriculture. *Agronomy* 12, 3008. https://doi.org/10.3390/agronomy12123008

Kumari A, Singh SK, Mathpal B, Verma KK, Garg VK, Bhattacharyya M, Bhatt R. (2023b). The biosynthesis, mechanism of action, and physiological functions of melatonin in horticultural plants: A review. *Horticulturae* 9, 913. https://doi.org/10.3390/horticulturae9080913

Li A-M, Liao F, Wang M, Chen Z-L, Qin C-X, Huang R-Q, Verma KK, Li Y-R, Que Y-X, Pan Y-Q, et al. (2023). Transcriptomic and proteomic landscape of sugarcane response to biotic and abiotic stressors. *Int. J. Mol. Sci.* 24, 8913. https://doi.org/10.3390/ijms24108913

Li Q, Song J, Peng S, Wang JP, Qu GZ, Sederoff RR, Chiang VL. (2014). Plant biotechnology for lignocellulosic biofuel production. *Plant Biotechnol. J.* 12(9), 1174–1192.

Loqué D, Scheller HV, Pauly M. (2015). Engineering of plant cell walls for enhanced biofuel production. *Curr. Opin. Plant Biol.* 25, 151–161.

Malviya MK, Li C-N, Lakshmanan P, Solanki MK, Wang Z, Solanki AC, Nong Q, Verma KK, Singh RK, Singh P, Sharma A, Guo D-J, Dessoky ES, Song X-P, Li Y-R. (2022). High-throughput sequencing-based analysis of rhizosphere and diazotrophic bacterial diversity among wild progenitor and closely related species of sugarcane (*Saccharum* spp. inter-specific hybrids). *Front. Plant Sci.* 13, 829337. https://doi.org/*10.3389/fpls.2022.829337.*

Malviya MK, Solanki MK, Li CN, Wang Z, Zeng Y, Verma KK, Singh RK, Singh P, Huang HR, Yang LT, Song XP, Li YR. (2021). Sugarcane-legume intercropping can enrich the soil microbiome and plant growth. *Front. Sustain. Food Syst.* 5, 606595. https://doi.org/10.3389/fsufs.2021.606595.

Mosier N, Wyman C, Dale B, Elander R, Lee YY, Holtzapple M, Ladisch M. (2005). Features of promising technologies for pretreatment of lignocellulosic biomass. *Bioresour. Technol.* 96, 673–686.

Oliveira MA, Verma KK, Bhatt R, Oliveira TBA. (2022). Impact of green and organic fertilizers on soil fertility and sugarcane productivity. In: Verma KK, et al. (Ed.),

Agro-industrial Perspectives on Sugarcane Production Under Environmental Stress. https://doi.org/10.1007/978-981-19-3955-6_11.

Park S-H, Ong RG, Sticklen M. (2016). Strategies for the production of cellwall-deconstructing enzymes in lignocellulosic biomass and their utilization for biofuel production. *Plant Biotechnol. J.* 14, 1329–1344.

Sharma A, Singh RK, Singh P, Vaishnav A, Guo DJ, Verma KK, Li DP, Song XP, Malviya MK, Khan N, Lakshmanan P, Li YR. (2021). Insights into the bacterial and nitric oxide-induced salt tolerance in sugarcane and their growth-promoting abilities. *Microorganisms* 9, 2203. https://doi.org/10.3390/microorganisms9112203.

Sharma A, Song X-P, Singh RK, Vaishnav A, Gupta S, Singh P, Guo D-J, Verma KK, Li Y-R. (2022). Impact of carbendazim on cellular growth, defence system and plant growth promoting traits of *Priestia megaterium* ANCB-12 isolated from sugarcane rhizosphere. *Front. Microbiol.* 13, 1005942. https://doi.org/10.3389/fmicb.2022.1005942

Singh P, Singh RK, Li HB, Guo DJ, Sharma A, Verma KK, Solanki MK, Upadhyay SK, Lakshmanan P, Yang LT, Li YR. (2023). Nitrogen fixation and phytohormone stimulation of sugarcane plant through plant growth promoting diazotrophic *Pseudomonas*. *Biotechnol. Genetic Eng. Rev.* 40(1), 15–35. https://doi.org/10.1080/02648725.2023.2177814

Singh P, Singh RK, Song QQ, Li HB, Guo DJ, Malviya MK, Verma KK, Song XP, Lakshmanan P, Yang LT, Li YR. (2021). Comparative analysis of protein and differential responses of defense-related gene and enzyme activity reveals the long-term molecular responses of sugarcane inoculated with *Sporisorium scitamineum*. *J. Plant Interact.* 16(1), 12–29.

Singh RK, Singh P, Li HB, Song QQ, Guo DJ, Solanki MK, Verma KK, Malviya MK, Song XP, Lakshmanan P, Yang LT, Li YR. (2020). Diversity of nitrogen-fixing rhizobacteria associated with sugarcane: A comprehensive study of plant-microbe interactions for growth enhancement in *Saccharum* spp. *BMC Plant Biol.* 20, 220. https://doi.org/10.1186/s12870-020-02400-9.

Solanki MK, Verma KK, Dastogeer KMG, Mora-Poblete F, Mundra S. (2023). Microbial resilience in plant nutrient management towards sustainable farming. *Front. Microbiol.* 14, 1280811. https://doi.org/10.3389/fmicb.2023.1280811

Song X-P, Mo F, Verma KK, Wei J, Zhang X, Yang L, Li YR. (2019). Effect of sugarcane smut (*Ustilago scitaminea* Syd.) on ultrastructure and biochemical indices of sugarcane. *Biomed. J. Sci. Tech. Res.* 002950. https://doi.org/10.26717/BJSTR.2019.17.002950.

Song XP, Mo F, Yan M, Zhang X, Zhang B, Huang X, Huang D, Pan Y, Verma KK, Li YR. (2022). Effect of smut infection on the photosynthetic physiological characteristics and related defense enzymes of sugarcane. *Life* 12, 1201. https://doi.org/10.3390/life12081201

Song XP, Verma KK, Tian DD, Zhang XQ, Liang YJ, Huang X, Li CN, Li YR. (2021). Exploration of silicon functions to integrate with biotic stress tolerance and crop improvement. *Biol. Res.* 54, 19. https://doi.org/10.1186/s40659-021-00344-4

Sticklen MB. (2008). Plant genetic engineering for biofuel production: Towards affordable cellulosic ethanol. *Nat. Rev. Genet.* 9, 433–443.

Tian D, Qin L, Verma KK, Wei L, Li J, Li B, Zhou W, He Z, Wei D, Huang S, Long S, Huang Q, Li C, Wei S. (2023). Transcriptomic and metabolomic differences between banana varieties which are resistant or susceptible to *Fusarium wilt. PeerJ* 11, e16549. https://doi.org/10.7717/peerj.16549

United Nations Environment Programme. (2007). *Global Environment Outlook: Environment for Development*, 4th edn., UNEP: New York, NY.

Verma KK, Anas M, Chen Z, Rajput VD, Malviya MK, Verma CL, Singh RK, Singh P, Song XP, Li YR. (2020f). Silicon supply improves leaf gas exchange, antioxidant defense system and growth in sugarcane responsive to water limitation. *Plants* 9, 1032, https://doi.org/10.3390/plants9081032.

Verma KK, Joshi A, Song X-P, Liang Q, Xu L, Huang H-r, Wu K-C, Seth CS, Arora J and Li Y-R. (2024a). Regulatory mechanisms of plant rhizobacteria on plants to the adaptation of adverse agroclimatic variables. *Front. Plant Sci.* 15, 1377793. https://doi.org/10.3389/fpls.2024.1377793

Verma KK, Joshi A, Song X-P, Singh S, Kumari A, Arora J, Singh SK, Solanki MK, Seth CS, Li Y-R. (2024b). Synergistic interactions of nanoparticles and plant growth promoting rhizobacteria enhancing soil-plant systems: A multigenerational perspective. *Front. Plant Sci.* 15, 1376214. https://doi.org/10.3389/fpls.2024.1376214

Verma KK, Li DM, Singh M, Rajput VD, Malviya MK, Minkina T, Singh RK, Singh P, Song XP, Li YR. (2020a). Interactive role of silicon and plant-rhizobacteria mitigating abiotic stresses: A new approach for sustainable agriculture and climate change. *Plants* 9, 1055. https://doi.org/10.3390/plants9091055.

Verma KK, Rajput VD, Singh M, Verma CL, Minkina T. (2022a). Influence of waterlogging on plant physiology and molecular biology. In: *Understanding Abiotic Stresses*. Nova Science Publishers, Inc.: New York.

Verma KK, Siddiqui N, Singh M. (2022b). Understanding the plant physiological and molecular adaptation strategies to soil salinity. In: *Understanding Abiotic Stresses*. Nova Science Publishers, Inc.: New York. https://doi.org/10.52305/UBUJ4024

Verma KK, Singh P, Song X-P, Malviya MK, Singh RK, Chen G-L, Solomon S, Li YR. (2020b). Mitigating climate change for sugarcane improvement: Role of silicon in alleviating abiotic stresses. *Sugar Tech.* 22(5), 741–749. https://doi.org/10.1007/s12355-020-00831-0.

Verma KK, Singh RK, Song QQ, Singh P, Zhang B-Q, Song X-P, Chen G-L, Li YR. (2019a). Silicon alleviates drought stress of sugarcane plants by improving antioxidant responses. *Biomed. J. Sci. Tech. Res.* 002957. https://doi.org/10.26717/BJSTR.2019.17.002957

Verma KK, Song X, Degu HD, Guo D, Joshi A, Huang H, Xu L, Singh M, Huang D, Rajput VD, Li YR. (2023a). Recent advances in nitrogen and nano-nitrogen fertilizers for sustainable crop production: A mini-review. *Chem. Biol. Technol. Agricult.* 10, 1–14. https://doi.org/10.1186/s40538-023-00488-3

Verma KK, Song X, Verma CL, Huang H, Singh M, Xu L, Wu J, Khan ZH, Zhang B, Li Y. (2023b). Mathematical modeling of climate and fluoride effects on sugarcane photosynthesis with silicon nanoparticles. *Plant Physiol. Biochem.* 204, 108089. https://doi.org/10.1016/j.plaphy.2023.108089

Verma KK, Song XP, Budeguer F, Nikpay A, Enrique R, Singh M, Zhang B, Wu, JM, Li YR. (2022c). Genetic engineering: An efficient approach to mitigating biotic and abiotic stresses in sugarcane cultivation. *Plant Signal. Behav.* 17(1), e2108253.

Verma KK, Song XP, Joshi A, Rajput VD, Singh M, Sharma A, Singh RK, Li DM, Arora J, Minkina T, Li YR. (2022d). Nanofertilizer possibilities for healthy soil, water and food in future: An overview. *Front. Plant Sci.* 13, 865048. https://doi.org/10.3389/fpls.2022.865048

Verma KK, Song XP, Joshi A, Tian DD, Rajput VD, Singh M, Arora J, Minkina TM, Li YR. (2022e). Recent trends in nano-fertilizer for advancing sustainable agriculture under the era of climate change to ensure food security for future generations. *Nanomaterials* 12, 173. https://doi.org/10.3390/nano12010173.

Verma KK, Song XP, Kumari A, Jagadesh M, Singh SK, Bhatt R, Singh M, Seth CS, Li YR. (2024c). Climate change adaptation: Challenges for agricultural sustainability. *Plant, Cell Environ.* 48(4), 2522–2533. https://doi.org/10.1111/pce.15078

Verma KK, Song XP, Li DM, Singh M, Wu JM, Singh RK, Sharma A, Zhang BQ, LI YR. (2022f). Silicon and soil microorganisms improves rhizospheric soil health with bacterial community, plant growth, performance and yield. *Plant Signal. Behav.* 17(1), e2104004.

Verma KK, Song XP, Lin B, Guo DJ, Singh M, Rajput VD, Singh RK, Singh P, Sharma A, Malviya MK, Chen GL, Li YR. (2022g). Silicon induced drought tolerance in crop plants: Physiological adaptation strategies. *Silicon* 14(6), 2473–2487. https://doi.org/10.1007/s12633-021-01071-x

Verma KK, Song XP, Rajput VD, Boldyreva V, Zhang BQ, Minkina T, Li YR. (2022h). Morpho-physiological, biochemical and ultrastructural modifications on sugarcane to prolonged water deficit. Verma KK, et al. (Eds.), *Agro-industrial Perspectives on Sugarcane Production under Environmental Stress*. https://doi.org/10.1007/978-981-19-3955-6_11.

Verma KK, Song XP, Rajput VD, Solomon S, Li YR, Rao GP. (2022i). *Agro-industrial Perspectives on Sugarcane Production Under Environmental Stress*. Springer Nature Singapore, https://doi.org/10.1007/978-981-19-3955-6

Verma KK, Song XP, Singh M, Tian DD, Rajput VD, Minkina T, Li YR. (2023c). Association of silicon and soil microorganisms induces stress mitigation, increasing plant productivity. In: de Mello Prado R. (Ed.), *Benefits of Silicon in the Nutrition of Plants*; pp. 299–328. https://doi.org/10.1007/978-3-031-26673-7_17

Verma KK, Song XP, Tian DD, Guo DJ, Chen ZL, Zhong CS, Nikpay A, Singh M, Rajput VD, Singh RK, Minkina T, Li YR. (2021a). Influence of silicon on biocontrol strategies to manage biotic stress for crop protection, performance and improvement. *Plants* 10, 2163. https://doi.org/10.3390/plants10102163

Verma KK, Song XP, Tian DD, Singh M, Verma CL, Rajput VD, Singh RK, Sharma A, Singh P, Malviya MK, Li YR. (2021b). Investigation of defensive role of silicon during drought stress induced by irrigation capacity in sugarcane: Physiological and biochemical characteristics. *ACS Omega* 6, 19811–19821.

Verma KK, Song XP, Verma CL, Chen ZL, Rajput VD, Wu KC, Liao F, Chen GL, Li YR. (2021c). Functional relationship between photosynthetic leaf gas exchange in response to silicon application and water stress mitigation in sugarcane. *Biol. Res.* 54, 15. https://doi.org/10.1186/s40659-021-00338-2

Verma KK, Song XP, Verma CL, Malviya MK, Guo DJ, Rajput VD, Sharma A, Wei KJ, Chen GL, Solomon S, Li YR. (2021d). Predication of photosynthetic leaf gas exchange of sugarcane (*Saccharum* spp.) leaves in response to leaf positions to foliar spray of potassium salt of active phosphorus under limited water irrigation. *ACS Omega* 6, 2396–2409.

Verma KK, Song X-P, Xu L, Huang H-R, Liang Q, Seth CS, Li Y-R. (2023d). Nano-microplastic and agro-ecosystems: A mini-review. *Front. Plant Sci.* 14, 1283852. https://doi.org/10.3389/fpls.2023.1283852

Verma KK, Song XP, Yadav G, Degu HD, Parvaiz A, Singh M, Huang HR, Mustafa G, Xu L, Li YR. (2022j). Impact of agroclimatic variables on proteo-genomics in sugarcane (*Saccharum* spp.) plant productivity. *ACS Omega* 7, 22997–23008.

Verma KK, Song XP, Zeng Y, Guo DJ, Sing M, Rajput VD, Malviya MK, Wei KJ, Sharma A, Li DP, Chen GL, Li YR. (2021e). Foliar application of silicon boosts growth, photosynthetic leaf gas exchange, antioxidative response and resistance to limited water irrigation in sugarcane (*Saccharum officinarum* L.). *Plant Physiol. Biochem.* 166, 582–592.

Verma KK, Song XP, Zeng Y, Li DM, Guo DJ, Rajput VD, Chen GL, Barakhov A, Minkina TM, Li YR. (2020c). Characteristics and correlation of leaf stomata and its relationship with photosynthesis on *Saccharum* spp. under different irrigation and silicon application. *ACS Omega* 5, 24145–24153.

Verma KK, Vatsal S, Singh M, Rajput VD, Minkina T. (2022k). Coping with water stress: Effects and adaptive mechanisms to plants. In: *Understanding Abiotic Stresses*. Nova Science Publishers, Inc.: New York.

Verma KK, Wu KC, Verma CL, Li DM, Malviya MK, Singh RK, Singh P, Chen GL, Song XP, Li YR. (2020d). Developing mathematical model for diurnal dynamics on photosynthetic responses in sugarcane responsive to different irrigation and silicon application. *PeerJ* 8, e10154. https://doi.org/10.7717/peerj.10154.

Verma KK, Liu X-H, Wu K-C, Singh RK, Song QQ, Malviya MK, Song X-P, Singh P, Verma CL, Li YR. (2020e). The impact of silicon on photosynthetic and biochemical responses of sugarcane under different soil moisture levels. *Silicon* 12, 1355–1367.

Verma KK, Wu K-C, Singh P, Malviya MK, Singh RK, Song X-P, Li YR. (2019b). The protective role of silicon in sugarcane under water stress: Photosynthesis and antioxidant enzymes. *Biomed. J. Sci. Tech. Res.* 002685. https://doi.org/0.26717/BJSTR.2019.15.002685

Wang J, Nayak S, Koch K, Ming R. (2013). Carbon partitioning in sugarcane (Saccharum species). *Front. Plant Sci.* 4, 201.

Wu K-C, Huang C-M, Verma KK, Deng Z-N, Huang H-R, Pang T, Cao H-Q, Luo H-B, Jiang S-L, Xu L. (2022). Transcriptomic responses of *Saccharum spontaneum* roots in response to polyethylene glycol–6000 stimulated drought stress. *Front. Plant Sci.* 13, 992755. https://doi.org/10.3389/fpls.2022.992755

Xu L, Deng Z-N, Wu K-C, Malviya MK, Solanki MK, Verma KK, Pang T, Li Y-J, Liu X-Y, Kashyap BK, Dessoky ES, Wang W-Z and Huang H-R. (2022). Transcriptome analysis reveals a gene expression pattern that contributes to sugarcane bud propagation induced by Indole-3-Butyric *Acid. Front. Plant Sci.* 13, 852886. https://doi.org/10.3389/fpls.2022.852886

Zheng F-J, Lin B, Yang Y-X, Fang X-C, Verma KK, Chen G-L. (2023). Efficacy and functionality of sugarcane original vinegar on mice. *Front. Microbiol.* 14, 1224666. https://doi.org/10.3389/fmicb.2023.1224666

CHAPTER 11

Economic Benefits of Biofuel Production for Sustainable Agroclimate

SWETA KATARIA, SIMRAN KAUR, and ARTI GOEL*

Amity Institute of Microbial Technology, Amity University, Noida, Uttar Pradesh, India

*Corresponding author

ABSTRACT

With the rise of population and industrial sectors, there has been a consistent rise in energy demand. By 2060, we will run out of our current conventional fuel reserves, which has created an alarming situation and pressure to find or create a sustainable renewable energy source that also supports our goal of environmental protection. The main energy source for several years was nonrenewable fuels, which accelerated global warming and released a number of air pollutants harmful to the environment and the public's health. An alternative to conventional fuel is the production and usage of biofuels, which is increasing rapidly and has a constant impact on the diverse sectors of a sustainable economy, which in turn leads to sustainable advances by optimizing social, economic, and environmental advantages. Bioethanol, which is one of the biofuels, has become a desirable alternative to conventional fuels. The increased production of such fuels can help farmers to be more lucrative, financially stable, and less reliant on conventional markets. Farmers and agrobusiness owners benefit the most from it as it increases the output of energy crops and opens up new opportunities.

Sugarcane Transformation: Biofuels, Biotechnology, and Beyond.
Krishan K. Verma, Xiu-Peng Song, Hai-Rong Huang, Aradhna Kumari, and Yang-Rui Li (Eds.)

DOI: 10.1201/9781779641571-11

11.1 INTRODUCTION

The current global economy is heavily reliant on fossil fuels, including coal, oil, natural gas, and others. These are employed in the manufacturing of things like gasoline and electricity (Ruan et al., 2019). Humanity has been using fossil fuels made of carbon for ages. The amount of oil in the globe is reportedly declining by four billion tons a year (Mathur et al., 2022). Due to excessive usage, greenhouse gas emissions (GHG), which include carbon dioxide (CO_2) and other harmful gases, including methane, carbon monoxide, and chlorofluorocarbons, have raised worries about both human health and global warming (Malode et al., 2021). Rising global energy demand and oil prices and increase in harmful gases in the environment in recent years have compelled energy-consuming countries to concentrate increasingly on alternative energy sources (Hasan et al., 2023). To discover new alternative resources is possible during this period, extending the deadline in order to maintain the current standard of living. It is important to look into ecologically friendly, renewable, and sustainable energy sources (Saleem, 2022; Chen et al., 2020, 2023). Biofuels are considered an alternative to conventional fuels, due to their nontoxic, sulfur-free, biodegradable nature and ability to come from renewable sources (Osman et al., 2021a; Zheng et al., 2023). As a new alternative resource, rapid growth in the production and use of biofuels has a constant impact on the various sectors of a sustainable economy, which, by maximizing benefits on the social, economic, and environmental levels, ultimately results in sustainable advancements (Correa et al., 2019; Verma et al., 2024a, 2024b, 2024c; Huang et al., 2024). It seems possible to put our world on the path toward energy-related sustainable development by using biofuels. This sustainable energy source, which can be produced by consuming different feedstocks including plant, algae, or animal dung, helps in making two of the most popular biofuels, biodiesel and bioethanol (Bhatt et al., 2023a, 2023b; Jagadesh et al., 2024). Since the transportation industry is dominated by liquid fossil fuels, substituting renewable energy for these fuels will play a major role in achieving the holistic energy and sustainability goals (Verma et al., 2019a, 2019b; Khan et al., 2021).

11.2 CONVENTIONAL AND ALTERNATIVE FUELS

Conventional fuels, fossil fuels, which are the main sources of nonrenewable energy cannot be prolonged over a long period of time (Singh et al.,

2020, 2021, 2023; Mahapatra et al., 2021). These conventional fuels significantly contribute to several forms of pollution, particularly air and water pollution. However, due to the increasing demand for energy worldwide and the release of CO_2 as a result of the use of fossil fuels, modern society is experiencing a severe problem (Stančin et al., 2020; Anas et al., 2021a, 2021b; Deng et al., 2023). These increased rates are a set off alarm to humankind and the environment.

The need for energy sources is rising along with the population. The quantity of energy required varies from nation to nation worldwide (Song et al., 2019, 2021, 2022; Li et al., 2023). Compared to developing countries, developed countries use more energy, and for social and economic progress, energy is essential (Mahapatra et al., 2021). People are currently more concerned about renewable energy sources being abundant on Earth, pollution-free, easily accessible, and less expensive. They are currently the source of great worry for many people as an alternative source (Sharma et al., 2021, 2022; Guo et al., 2021, 2022, 2023; Dey et al., 2022). Traditional fuel sources are natural gas and petroleum, while alternative fuel sources include any materials or substances that may be used as fuel but are not traditional fuels (Stančin et al., 2020).

Biofuels, alternative or renewable energy sources, have the possibility to be less polluting and emit less CO_2. Biofuel is developed and produced from organic material, such as plants and their residues, agricultural crops, and byproducts (Malviya et al., 2021, 2022; Oliveira et al., 2022). They encompass a wide range of vegetation and waste products, including agroforestry, farming residue (bran, stubble, animal blubber byproducts), waste and sewage from the paper industry, waste material from construction sites (wood), and decomposable waste products (human waste, sewage sludge and manure, edible oils). These fuels can serve as a sufficient replacement for gasoline derived from petroleum (Joshi et al., 2022, 2023). They work well in opposition to the emissions of greenhouse gases (GHGs) from transportation and vehicles, as well as the consequences of climate change (Mahapatra et al., 2021). They are called energy crops which are grown in order to produce energy and extract fuels, from which biofuels are derived (Chaudhary et al., 2023; Solanki et al., 2023). These plants, some of which include algae, microalgae, and seaweeds, are referred to as energy crops. The CO_2 which is released by photosynthesis is absorbed by the plants throughout their growth phase, and because of this, they are considered carbon-neutral energy sources (Huang et al., 2022; Wu et al., 2022; Xu et al., 2022) (Figure 11.1).

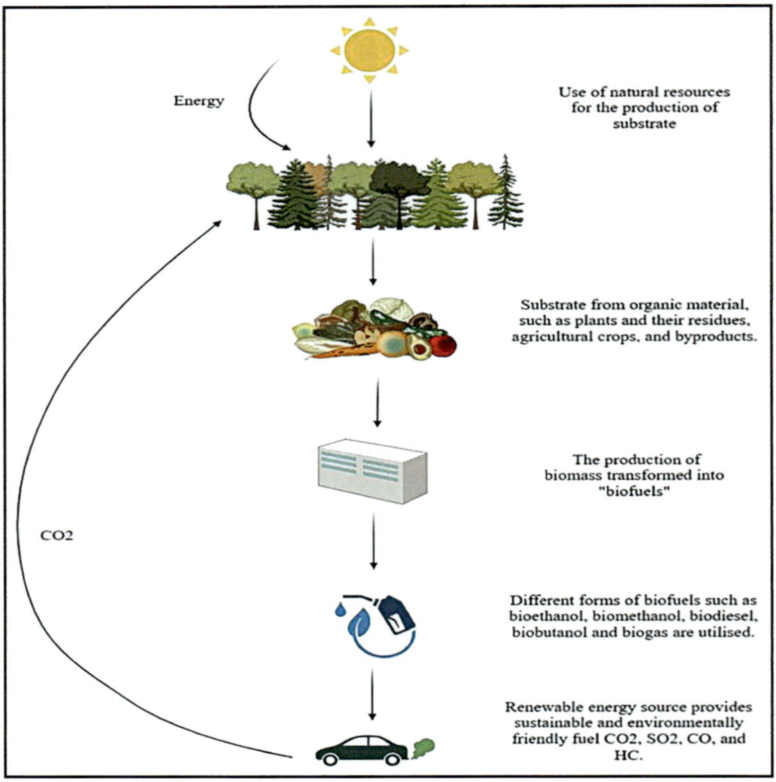

FIGURE 11.1 Life cycle of biofuel production and utilization.

Biofuels are divided into four categories based on the feedstock: Generations 1–4 of biofuels. First-generation biofuels are made from sugar, starch, and plants based on oil. Since their debut in 1996–1997, the creation of genetically modified yields has increased steadily (Liu et al., 2021; Krishna et al., 2022). Current methods of converting plant oils or carbohydrates from food products into ethanol and biodiesel—first-generation bioenergy technologies—are ineffective and have only a limited capacity to lower emissions, according to recent evaluations of energy crops (Abreu et al., 2022). Second-generation biofuels will supply energy in the form of cellulose and pure hydrocarbon-based nonoxygenated fuels, which include biomass to liquid fuels (Thanigaivel et al., 2022). It is expected that the energy crops and biofuel-producing technologies of the second generation would be more effective than the first generation (Abreu et al., 2022). Algal-used-derived third-generation (3G) biofuels are highly recognized, easily refined, and can be generated on a massive scale while absorbing CO_2.

11.2.1 GENERATION OF BIOFUELS

Biofuels technology is still an ongoing technology, which results in higher-quality production, less power usage, and less CO_2 emissions. The two methods that are often used are fermentation and transesterification. In which fermentation converts sugar or starch into ethanol and transesterification turns waste cooking oil or animal fat into biodiesel. While comparing the production of solid or gaseous biofuels, the production of fluid biofuels is said to be constant in innovating. When compared to gas-based biofuel manufacturing techniques, fluid biofuel manufacturing methods demonstrate a potential for larger transformations, reduced waste output, and utilization of low space and water (Anwar, 2021). Essential factors that determine the commercial viability of the expansion of biofuels include feedstock rates, industrial infrastructure, commodity quality, and market demand. Even though using organic waste feedstocks for biofuel production has potential, maintaining a sizable supply of natural deposits for continuous development requires effective management (Stephen and Periyasamy, 2018; Kumari et al., 2022; Tian et al., 2023). The substrate used in each generation of biofuel is mentioned in Figure 11.2 and Table 11.1.

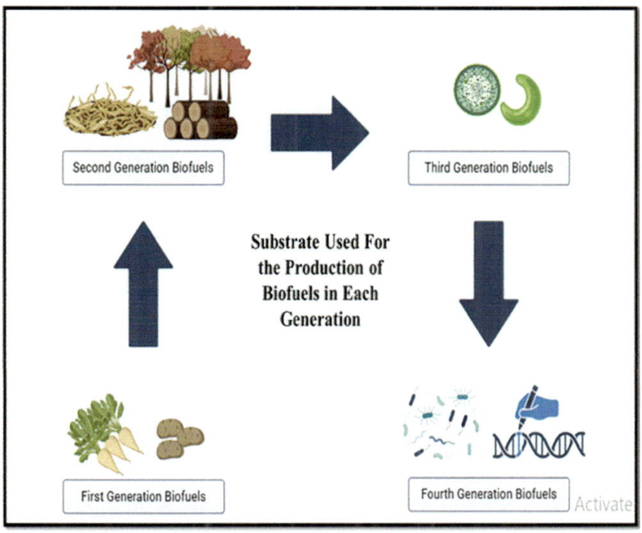

FIGURE 11.2 Substrate used for the production of biofuels in each generation.

TABLE 11.1 Generation of Biofuels and Their Substrate

Generation of Biofuels	Substrate	Source
First-generation biofuel	Edible biomass, such as sugar, oil crops, cereals, and starch	Why et al. (2022)
	Feedstock—sugarbeets, whey, barley, and potato wastes	Akbas and Stark (2016)
Second-generation Biofuels	Nonfood feedstocks which are lignocellulosic plants, agricultural and forest wastes, waste products, and specific energy crops	Jeswani et al. (2020)
Third-generation biofuels	Biomass from cyanobacteria and microalgae	Hoang et al. (2023)
Fourth-generation biofuels	Genetically modified microalgae, bacteria, yeast, and cyanobacteria	Khan et al. (2021)

11.2.2 FIRST-GENERATION BIOFUELS

First-generation biofuels are produced using edible biomass, such as sugar, oil crops, cereals, and starch, which raises production costs and results in inefficient use of the energy and resources used to cultivate crops (Why et al., 2022). Commercially produced biofuels from the first generation include biodiesel, bioethanol, and biogas. Oil transesterification of natural sources and residual fats and oils yields biodiesel, a fuel replacement. Simultaneously, fermentation of sugar or starch yields bioethanol, which is a fuel replacement. Other biofuels which are produced by the anaerobic processing of manure and other waste are biogas. Nonetheless, the amounts of biogas utilized in transportation nowadays are rather minimal (Mitkidis et al., 2018). Two primary assertions are being used to evaluate first-generation biofuels: for example, they actively try to outcompete crops in the feed market. Second, they would not be able to balance out the energy, economy, and environment. Many studies predict that by 2025, there will be roughly 1.2 billion people living in wealthy nations who are food insecure if the development of biofuel has the same impact on food prices (Khan et al., 2021).

These biofuels are also known as "conventional biofuels" since they are produced using several well-known technologies and procedures, including fermentation, distillation, and transesterification (Jeswani et al., 2020). First-generation bioethanol is really produced using only a few distinct feedstocks, primarily corn or sugarcane. Other less common feedstocks that are being utilized or being explored for the production of first-generation bioethanol are sugarbeets, whey, barley, and potato wastes, but one of the most popular feedstocks for the manufacture of biofuel is sugarcane, with Brazil being one of the top users. The sugarcane may be converted into ethanol through a rather straightforward method. After the sugarcane is crushed and the sucrose is extracted with water, it is filtered to provide either raw sugar or ethanol (Akbas and Stark, 2016). The only other biofuel generated on an industrial basis is biodiesel. This biofuel's creation might be classified as a chemical process, which makes it significantly distinct from ethanol. Naturally, biomass like oily plants and seeds are used in the process, but the main idea is to extract the oils and transform them into biodiesel by breaking the bonds that bind long-chain fatty acids to glycerol and substituting methanol for the glycerol through a procedure known as transesterification (Mat Aron et al., 2020).

The major drawback of this generation is that biofuel production is restricted due to worries about the effects on the environment and carbon balances. The primary drawback of first-generation biofuels is the dispute between food and fuel; because of this fuel's increased production, food prices have increased (Osman et al., 2021b).

11.2.3 SECOND-GENERATION BIOFUELS

Second-generation biofuels include all of the nonfood feedstocks which are lignocellulosic plants, agricultural and forest wastes, waste products, and specific energy crops (Jeswani et al., 2020). Relying on more cost-effective renewable substitutes, this type of biofuel is made from indigestible lignocellulosic biomass, including sawdust, switchgrass, inexpensive wood, crop residues, and municipal wastes (Alalwan et al., 2019). The second generation of biofuel production is an enhanced technique that prioritizes the creation of secondary raw materials as well as enhanced fuel recovery over the first generation. It is an economically viable strategy because, in contrast to the first-generation process, it concentrates on creating useful fuels while lowering the total energy cost and quantity of waste produced (Hoang et al., 2021).

Different approaches to producing second-generation biofuels are being explored globally. However, the two main pathways—the thermo or bio route that are produced by the biomass of cellulose and lignin, tree excess, and seasonal fodder crop—are the main focus of attention (Khan et al., 2021). The “thermo” technique refers to certain procedures that include heating biomass while using very little oxidizing chemical (Mat Aron et al., 2020). The “bio” method entails pretreating lignocellulosic material, enzymatic hydrolysis, and the fermentation of sugars by certain microbial strains (Khan et al., 2021). Even with their extremely advantageous potential to valorize waste streams, second-generation biofuels would not be enough on their own to meet the present global need for energy. Similar to food crops that provide first-generation biofuels, the biomass utilized in these procedures has a finite supply. In order to guarantee enough fuel supply, second-generation biofuels must be coupled with other technologies (Cavelius et al., 2023).

11.2.4 THIRD-GENERATION BIOFUELS

The 2G biomass from the biofuel generation, due to the high cost and need for sophisticated improvements in the manufacturing process, is not commercially viable. Following this, scientists focused on creating 3G biomass from microalgae, which is currently considered a viable alternative for producing biofuel sustainably by addressing the drawbacks of both 1G and 2G biofuel sources simultaneously (Chowdhury and Loganathan, 2019). The main source of 3G biofuels is biomass from cyanobacteria and microalgae, which are utilized to naturally produce lipids and alcohols that can be converted into biodiesel or any other high-energy fuel product (Hoang et al., 2023). Algal lipids can be converted to biodiesel using the transesterification method and to boost lipid productivity in microalgae; genetically modified organism (GMO) technology can also be utilized. A limited number of microbe species can be genetically engineered (GE) to enhance the lipid (oil) production of microalgal biomass, according to some research findings (Chowdhury and Loganathan, 2019).

Bioethanol and biodiesel are the two sustainable biofuels that are produced from microalgae. Given that microalgae may yield up to 20 times more biooil per acre than conventional crops, creating biofuels from them has a number of positive benefits (Chowdhury and Loganathan, 2019). It also have a shorter harvesting cycle and a rapid growth rate.

It can produce 10,000 L/hectare/year of biooil, a rate much higher than that of other biofuel sources, including soybean, sunflower, canola, Jatropha, and palm. Compared to other biomass, algal biomass is more similar to the final fuel product.

Fresh water and fertile land are not necessary for the growth of algae. A variety of cultures can be cultivated using wastewater, brackish water, or salt water. This method is economical and avoids competing with agricultural practices (Abdullah et al., 2019). Above all, a direct supply of CO_2 is necessary for effective algal culture. This can be obtained via industrial emitters or by atmospheric carbon capture. Approximately, 70% of the CO_2 provided is utilized for photosynthesis in typical cultivation methods, which results in the production of biomass (Ganesan at al., 2020). Large-scale algae processing has logistical and technological difficulties as well, but responders think algae-based biofuels have a big potential to improve the field (Abdullah et al., 2019).

11.2.5 FOURTH-GENERATION BIOFUELS

The fourth generation of biofuels has the greatest promise for improved biofuels. The genetically modified microalgae, bacteria, yeast, and cyanobacteria are engineered microorganisms that make up the feedstocks for fourth-generation biofuels (FGBs). All microbial species cannot be genetically manipulated for the purpose of selecting appropriate strains because of their complicated structures, high nutritional requirements, and environmental intolerance (Khan et al., 2021).

Common techniques for GMO include raising light penetration, decreasing photoinhibition, and enhancing photosynthetic efficiency (Abdullah et al., 2019). It has been successfully possible to produce ethanol and other fuel products, including butanol, isobutanol, and modified fatty acids, using FGB methods. There are still a lot of hurdles to overcome before moving to FGB from fossil fuels (Zhu et al., 2022).

To reduce the risk of large-scale genetically modified algae being released into the environment, two primary containment measures are being examined: first, physically preventing the algae from escaping into the atmosphere, and second, genetically preventing the algae from growing and competing in the environment (Zewdie and Ali, 2020; Kumari et al., 2023a, 2023b). Moreover, with context to cost-effectiveness and ecological compatibility, GE algae perform better than native strains, increasing the viability of algal biofuels. Proof of their superiority and lack of major negative aspects on a lab and prototype scale exists, but a commercial demonstration of this is required for the genetic stability of GE algae (Shokravi et al., 2021).

11.3 TYPES OF ALTERNATIVE BIOFUELS

Biomass which is plant-based organic matter and animal's waste required to generate biofuels can be renewed over time; they are regarded as renewable energy sources. Unlike other renewable energy sources, immediately after the production of biomass, they may be transformed into liquid fuels called "biofuels" for use in transportation. There are different forms in which biofuels come, such as bioethanol, biomethanol, biodiesel, biobutanol, and biogas in Figure 11.3 and Table 11.2. Biodiesel and ethanol, which come under first-generation biofuels, are the two most widely utilized types of it (Sharma et al., 2023).

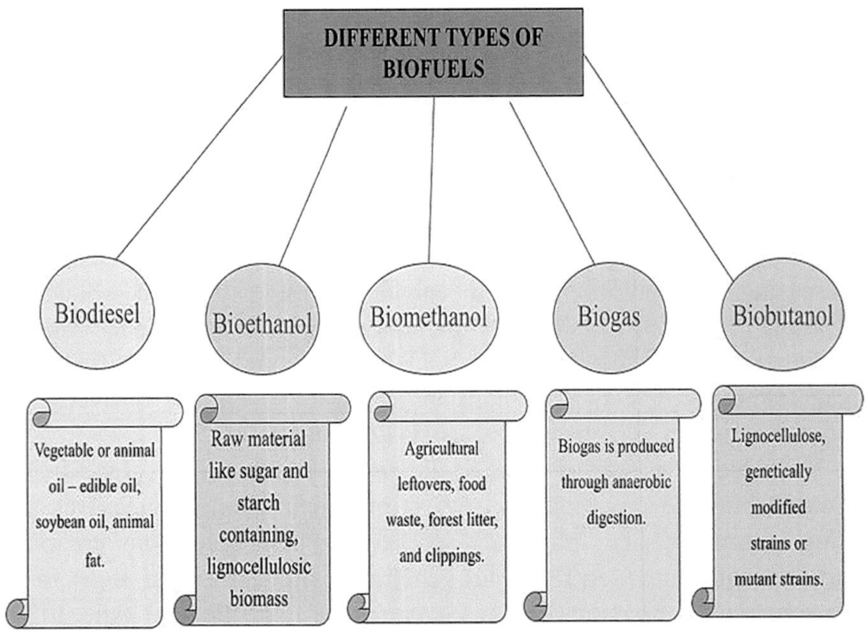

FIGURE 11.3 Different types of biofuels and their substrates.

TABLE 11.2 Alternative Biofuels with Benefits and Their Uses

Alternative Fuels	Biomass	Benefits	Usage
Biodiesel	Vegetable or animal oil—edible oil, soybean oil, animal fat (Kondrasheva and Eremeeva, 2023)	It can lessen CO_2, SO_2, CO, and HC, safeguarding the environment, less expensive, and be acquired from sustainable sources (Brahma et al., 2022)	Substitution for regular diesel oil (Singh et al., 2020)
Bioethanol	Raw resources that include sugar, such as sugarcane, sugarbeet, molasses, whey. Starch-containing feedstocks, such as cereals like corn and wheat and root vegetables like cassava. lignocellulosic biomass, which includes crop and wood wastes, straw, and agricultural waste (Gohain et al., 2021)	It includes oxygen, ethanol burns cleaner during combustion (Gohain et al., 2021)	May be utilized as pure fuels in adapted engines or as additives in petrol mixes (Erdiwansyah et al., 2019)
Biomethanol	Agricultural leftovers, food waste, forest litter, and clippings (Gautam et al., 2020)	Solves waste management issues but also lessens greenhouse gas emissions (Khan et al., 2023)	Provides sustainable and environmentally friendly fuel (Tabibian and Sharifzadeh, 2023)
Biogas	Biogas is produced through anaerobic digestion (Feiz and Ammenberg, 2017)	Emit 80% less CO_2 than cars powered by fossil fuels, 20% less global warming (Korberg et al., 2020)	Fuel for vehicles and injection into the natural gas system; other uses include the manufacture of steam and heat for business and homes, and the production of electricity (Korberg et al., 2020)
Biobutanol	Lignocellulose, genetically modified strains or mutant strains (Pugazhendhi et al., 2019)	It has a greater energy density, lowers the viscosity, and also improves resistance to water contamination (Mahalingam et al., 2022)	Replace kerosene, diesel and petrol (Mahalingam et al., 2022)

11.3.1 BIODIESEL

The "Biodiesel," also called as diesel fuel, is produced from vegetable or animal oil and contains long-chain alkyl (CH_3, C_2H_5, $CH_2CH_2CH_3$) esters. It is formed by the reaction of lipids such as edible oil, soybean oil, animal fat with an alcohol, thus resulting in fatty acid esters formation (Kondrasheva and Eremeeva, 2023). Fossil fuels in diesel engines have negative effects on the environment, and biodiesel has emerged as one viable substitute for petroleum and the fast depletion of petroleum resources. Recently, it has gained a lot of attention due to its advantages for the environment and also because they can be manufactured from less expensive, sustainable sources (Singh et al., 2020).

Fossil fuels can be easily substituted with biodiesel called a "clean energy source." It can lessen CO_2, SO_2, CO, and HC, therefore safeguarding the environment. Compared to the usage of fossil fuels, utilizing biodiesel may more successfully lower CO_2 emissions, save the environment, and preserve ecological balance (Brahma et al., 2022). The decreased sulfur content of biodiesel results in much lower SO_2 emissions during combustion when compared to regular diesel oil. Therefore, substituting biodiesel for regular diesel oil will significantly lessen acid rain, which poses a major risk to human infrastructure and the environment through soil acidification, damage to forests and other vegetation, increased corrosion of buildings and historical monuments made of calcium-containing stones, and surface and groundwater pollution. Additionally, biodiesel's ester components include oxygen, which promotes clean burning, less CO, HC, particulate matter will be released (Singh et al., 2020).

The production of biodiesel uses a wide variety of feedstock resources. These feedstocks, which are selected for the manufacturing of biodiesel, involve several critical considerations, including feedstock availability, quality, physicochemical characteristics, and cost of production (Abreu et al., 2022). To make biodiesel, many kinds of animal fats and edible or nonedible vegetable oils are used. Edible oils have a lot of promise as a diesel engine substitute fuel (Idowu et al., 2022). Nevertheless, several issues with this fuel have been observed include its high viscosity, low cetane number, low oxidative stability, and poor cold flow qualities, which cause a variety of byproducts to persist (Linganiso et al., 2022)

Edible sources—It is anticipated that 34% of vegetable oils would be utilized to produce biodiesel, and this percentage will continue to rise (Rizwanul Fattah et al., 2020). One important part in the manufacturing of biodiesel is the selection of feedstock and to have substantial molecular weighted substances in comparison to diesel, vegetable oils have a higher viscosity. Automotive engines may consume edible oils directly or in modified form, such as methyl/ethyl ester (Veza et al., 2021). Various types of vegetable oils are used in the production of biodiesel, from which the most studied are palm oil, rapeseed, sunflower, and soyabean. Unfortunately, these vegetable oils cannot be used as they are much more expensive than petroleum fuels.

Nonedible sources—The use of waste raw materials and nonedible oils to produce biodiesel has several advantages, including lowering the cost of raw materials and avoiding competition from the food market (Silviana et al., 2022). Over the years, the biofuel sector has made consistent progress in identifying raw material alternatives for biofuel production. The agronomic crops that contain sugar are sugarcane, maize, seeds, wheat, etc., which are favored as the feedstock for "first-generation biofuel" manufacturing. Nonedible oils including *Calophyllum inophyllum, Jatropha curcas, Croton megacarpus,* and *Pongamia pinnata* are used with prudence (Elgharbawy et al., 2021). Due to its low cost of manufacturing, more competent and cheaper resources for biodiesel production, the primary goal is to raise awareness of nonedible oil instead of edible oil.

11.3.2 BIOETHANOL

Fuels that come from sustainable biological sources and can be solid, liquid, or gaseous are referred to as biofuels, and the most common biofuel used is bioethanol (Fivga et al., 2019). Both bioethanol and biodiesel have the ability to be utilized in internal combustion engines because of their high vaporization heat and high-octane numbers. They may be utilized as pure fuels in adapted engines or as additives in petrol mixes (Erdiwansyah et al., 2019). Oil gets substituted with ethanol, which is regarded as a clean and sustainable fuel. Because it includes oxygen, ethanol burns cleaner during combustion, even if its energy equivalent is 68% less than that of petroleum fuel (Gohain et al., 2021). Reducing greenhouse gas emissions through the production of bioethanol from different biomass methods has provided significant advantages. It also spurred the growth of sustainably designed technologically sound sectors for the manufacture of bioethanol from sugar or biomass in the meantime (Rosen et al., 2021).

Various forms of biomass have the potential to be used as inputs to produce bioethanol. The three categories on the basis of chemical composition or the source of carbohydrates are the following: Raw materials like sugar, such as sugarcane, sugarbeet, molasses, whey, and sweet sorghum; feedstocks containing starch, such as cereals like corn and wheat and root vegetables like cassava; and lignocellulosic biomass includes crop and wood wastes, straw, and agricultural waste (Gohain et al., 2021). As a clean and sustainable fuel, bioethanol is seen to be a viable substitute for oil. Despite having a 68% lower energy equivalent than petroleum fuel, ethanol burns cleaner due to the presence of oxygen in its combustion (Comelli et al., 2020).

11.3.3 BIOMETHANOL

A renewable alternative to methanol made from biomass feedstocks, biomethanol provides a way forward for a more sustainable and environmentally friendly future. With several advantages for the environment and the economy, biomethanol has become a popular fuel choice (Tabibian and Sharifzadeh, 2023). Biomethanol is an appropriate biofuel because of its high volumetric energy density, it is liquid at room temperature, and it is easy to store and transport. It may be easily utilized as a raw material to create a wide range of beneficial organic compounds with significant industrial applications (Gautam et al., 2020). Using waste biomass as a feedstock is one potential method for producing biomethanol. Waste biomass like agricultural leftovers, food waste, forest litter, and clippings is a massive and inadequately used resource that may be used to produce sustainable fuels.

In the process of producing biomethanol, the carbon absorbed during the growth of the biomass feedstock easily balances out the CO_2 emitted during the burning of biomethanol. Its production from waste biomass not only solves waste management issues but also lessens greenhouse gas emissions and dependency on fossil fuels. The making of biomethanol from biomass can be attained by two methods—thermochemical and biochemical methods. Thermochemical procedures employ heat and catalysts to disassemble large biomass molecules into smaller parts that may subsequently be converted into methanol. Although it has been accomplished, the economic appeal of producing methanol on an industrial scale by thermochemical techniques is currently lacking. On the other hand, biochemical methods depend on the utilization of microbes and enzymes to speed up the conversion of biomass into methanol (Khan et al., 2023). A proper understanding of the bioconversion process and knowledge of various parameters are essential for scaling-up of the laboratory results to full-size production units (Gautam et al., 2020).

11.3.4 BIOGAS

Bioenergy is compatible with the present energy infrastructure since it offers temporal and geographical independence as compared to other renewable energy technologies. The primary uses of bioenergy in underdeveloped countries are for heating, cooking, and lighting. To improve the range of feedstocks available for the production of bioenergy, a few methods for converting biomass to bioenergy have been studied and refined throughout time. Biogas is a naturally renewable gas which is a bioenergy source since it is produced from biomass and has the ability to reduce greenhouse gas emissions by maintaining a closed carbon cycle (Gupta et al., 2023). It can replace natural gas and fossil fuels, which are created when organic substances decompose anaerobically. Nitrogen (N_2), CO_2, and methane (CH_4) makeup the major constituents. The growing market for renewable energy may benefit greatly from biogas, and its global utilization is anticipated to increase in the upcoming future. Fuel for vehicles and injection into the natural gas system are only a few uses for purified biogas. Other uses include the manufacture of steam and heat for business and homes, and the production of electricity. Despite being a firmly established technology, the commercial application of biogas as a fuel for vehicles is still currently limited due to the high level of purification required (Korberg et al., 2020).

CNG for cars can be replaced by pure bio-CNG. Utilizing biogas as a fuel for cars is advantageous since cars operated by bio-CNG emit 80% less CO_2 than cars powered by fossil fuels. Countries that utilize these are using either pure methane or a mixture of natural gas—as fuel for automobiles—and are also using bio-CNG, which results in around 20% less global warming overall than using CNG (Korberg et al., 2020). There is also a possibility of using biogas as a spark-ignition (SI) engine fuel. One of the major difficulties of using biogas as an SI engine, which includes the variation in the composition of the biogas, is the high CO_2 content's detrimental effects on the combustion process, and the need for huge storage tanks. According to some researchers, if the technical issues with using biogas as fuel are resolved, using SI engines powered by the fuel can be a sustainable technology (Gupta et al., 2023).

11.3.5 BIOBUTANOL

In taking into account the advantages and disadvantages associated with energy security, economic stability, environmental advantages, technological obstacles, substrate availability, and upkeep of the primary biofuels, biobutanol is becoming more and more recognized as an effective biofuel. It is thought to be the most advanced than any other biofuel and is used to replace kerosene, diesel, and petrol and is seen to be more effective than ethanol. It has a greater energy density, lowers the viscosity of vegetable oils, mixes at larger concentrations without requiring vehicles to be retrofitted, and can dissolve in vegetable oils at any proportion and also improves resistance to water contamination (Mahalingam et al., 2022).

There are two main approaches used to produce butanol: (1) the Oxo process—which is a two-step petrochemical procedure. Initially, propylene is hydroformylated, and the resulting is hydrogenated to make butanol. Second, microbial fermentation—a biological process—forms butyraldehyde. However, the price of petrochemical synthesis is very volatile and closely tied to the propylene market, which is determined by the price of crude oil. So, microbial fermentation as a butanol production technique could be a viable substitute to satisfy the present butanol demand (Liu et al., 2022). However, because some microbial strains are intolerant of butanol accumulation, the current frontiers in biobutanol production are genetically modified strains or mutant strains with the potential for maximum biobutanol yield, enhanced resistance to butanol accumulation, exploration of inexpensive substrates for anaerobic fermentation, and selection of a suitable fermentation system (Pugazhendhi et al., 2019). It also has a serious drawback, which is its limited yield and high manufacturing costs. Several initiatives were undertaken to address the issue, such as looking for an inexpensive substrate. Lignocellulose is thought to be the ideal precursor for the synthesis of biobutanol since it is less costly, more plentiful, and can be easily accessible (Liu et al., 2022).

11.4 BIOFUEL PRODUCTION AND SUSTAINABLE AGRICULTURE

The conventional fossil fuels are finite, resulting in the production of large amounts of CO_2 and greenhouse gases (Maciejczyk et al., 2021). Utilizing agricultural biomass can provide a solution for both, usage and reduction of agricultural waste and the production of sustainable fuel (Verma et al., 2020a, 2020b; 2021a). Biofuels present a safer and renewable source of energy (Verma et al., 2021b, 2021c). By highlighting the feasibility and advantages of incorporating biofuels and other renewable energy sources into national energy portfolios, countries such as Denmark, Sweden, and Poland serve as pioneering examples of successful renewable energy transitions. These countries have significantly accelerated their contribution in reducing greenhouse gas emissions, enhancing energy independence, and promoting a transition toward a low-carbon future by investing in renewable energy infrastructure, implementing supportive policies, and encouraging public–private partnerships (Verma et al., 2020c, 2020d; Saleem, 2022).

The conversion of chemical energy obtained from organic material into thermal energy, which can be utilized as an energy source, is known as bioenergy. It is also important to understand that crop products that are utilized as food sources cannot be used as a precursor of energy production; therefore, the usage of agricultural waste is ideal for the production of biofuels (Verma et al., 2020e, 2020f; 2021d, 2021e; Zhang et al., 2021). Lignocellulose is the major component of biofuel production. The type of crop determines the amount of energy that will be produced. Asia dominates the production of wheat and paddy crops, whereas in USA; sugarcane bagasse and corn straw are grown in larger quantities (Verma et al., 2022a, 2022b; 2023a, Saleem, 2022). In developed countries, biomass typically accounts for less than 10% of the total energy sources. However, in developing nations, this percentage can be as high as 20–30% (Zhang et al., 2021; Verma et al., 2022c, 2022d; 2023b). The USA and Brazil contribute up to 62% of the world's ethanol production. Cost is also a determining factor for bioethanol production on a large scale.

Characteristics of crops that can be used for biofuel production:

1. Crops should have a short period of growth.
2. The yield of usable biomass should be high.
3. It should not contradict the principles of food security.
4. The crop should survive harsh weather conditions.
5. Resistant to pests and diseases (Zhang et al., 2021)

11.5 INNOVATIVE CONVERSION OF AGRICULTURAL RESIDUES

11.5.1 RAW MATERIAL

Lignocellulose materials are the best suited for bioethanol production. Wheat straw, rice straw, corn straw, and sugarcane bagasse are the most preferred agricultural waste for biofuel production (Inyang et al., 2022). Out of the raw materials mentioned above, rice straw has the highest potential to produce bioethanol up to 205 billion liters per year (Dong, 2020). The raw material undergoes three major steps: pretreatment, hydrolysis, and then fermentation (Figure 11.4).

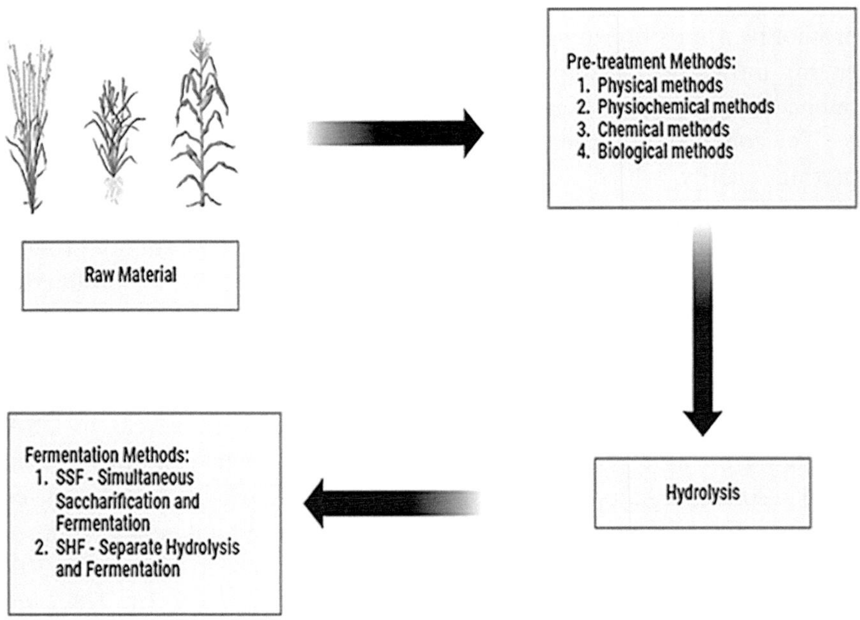

FIGURE 11.4 Summarization of the process of biofuel production from agricultural waste.

11.5.2 PRETREATMENT

Lignocellulose biomass comprises cellulose, lignin, and hemicellulose. Pretreatment methods involve the dissolution and separation of these biomass components, enhancing the accessibility of the remaining solid biomass for further chemical or biological process (Singh, 2021). The main objective of a successful pretreatment procedure is to minimize expenses, restrict the production of inhibitory products, generate sugars, and prevent the degeneration of them. For pretreatment, various methods such as physical, chemical, physicochemical, and biological treatments are employed.

Physical methods include mechanical size reduction, pyrolysis, electronic beam irradiation, and microwave oven pretreatment. The type of mechanical treatment depends on the type of wood we are using; for softwood, a cutter mill is used, and for hardwood, ball mill or hammer mill is advised (Sun et al., 2016). Other methods include wet milling, dry milling, and compression milling. Pyrolysis includes the usage of high temperature, around 300°C, to degrade cellulose into products like residual char, H_2, and CO. The leftover char undergoes additional treatment through either water leaching or mild acid leaching. Approximately, loss of the total biomass weight during water leaching typically is 55% (Kersten and Garcia-Perez, 2013). The electronic beam radiation is a nonthermal method, in which polar bonds vibrate which accelerates other physical, chemical, and biological processes. Microwave oven, on the other hand, generates heat using radiation, resulting in autohydrolysis.

Physicochemical treatment includes the autohydrolysis or stream explosion, ammonia fiber explosion, liquid hot water method, and CO_2 explosion. Among these methods, CO_2 explosion does not lead to the formation of inhibitors and is the most cost-effective method of chemical pretreatment (Kim et al., 2022).

In the autohydrolysis approach, the biomass is subjected to a high-pressure stream (ranging from 20 to 50 bar) and elevated temperature (between 160 and 290°C) for a brief duration. The reaction is subsequently halted by rapid decompressing the system to atmospheric pressure (Robak and Balcerek, 2020). This method has the advantage of high recovery of xylose, which makes this method a little economically affordable (Dhungana et al., 2022). Ammonia fiber explosion is a treatment that leads to the formation of inhibitory substances. In this method, lignocellulosic material is subjected to high temperature and pressure treatment, followed by a swift release of pressure. This method is less efficient and not economic (Zeng et al., 2021). The hot water is used to hydrolyze hemicellulose in the liquid hot water method.

The chemical treatment includes wet oxidation, acid pretreatment, alkaline pretreatment, and the use of organic solvents. Wet oxidation treats biomass with water and oxygen at high temperatures. It transfers hemicellulose to the liquid phase without hydrolyzing them, yielding sugar oligomers (Saha et al., 2019). Acid treatments utilize a wide variety of either diluted or concentrated acids to produce sugars from lignocellulose, but it also has a disadvantage as it also produces inhibitory substances. Acids that are used include nitric acid, sulfuric acid, hydrochloric acids, and phosphoric acid. Alkaline pretreatment results in the digestion of lignin, which makes cellulose and hemicellulose vulnerable to digestion by enzymes (Vilela et al., 2021). The organic solvents like performic acid, methanol, acetone, ethanol, acetic acid, and peracetic acid are used for delignification, and then lignin is separated out using distillation (Vaidya et al., 2022).

Biological treatment involves the use of microorganisms to degrade the lignin and cellulosic components. Mostly used organisms include soft rot fungi, brown rot, and white rot fungi. It generally needs a long period of time.

11.5.3 HYDROLYSIS USING ENZYMES

In this step, complex carbohydrates are converted to monomers and known as saccharification. It is a crucial step for the production ofbioethanol (Raina et al., 2022). It is advantageous as compared to alkaline or acid hydrolysis as it is cost-effective and low in toxicity (Hawrot-Paw and Stańczuk, 2023). The enzymes used are very substrate specific. The substrate specificity of these enzymes ensures precise and efficient hydrolysis of complex carbohydrates present in lignocellulosic biomass, maximizing the yield of fermentable sugars for subsequent ethanol fermentation. Cellulase enzymes exhibit optimum activity within a narrow pH range of 4–5 and a temperature range of 40–50°C, highlighting their sensitivity to environmental conditions (Khanahmadi et al., 2018). These enzymes target the beta-1,4-glycosidic bonds present in cellulose, efficiently cleaving them into glucose monomers. Similarly, xylanase enzymes function optimally within a pH range of 4–5 and a temperature range of 50°C. They target the backbone of xylan, a major component of hemicellulose, facilitating its depolymerization into xylose and other oligosaccharides.

11.5.4 FERMENTATION

The industrial application of lignocellulosic materials for bioethanol production faces challenges due to the absence of optimal microorganisms capable of effectively fermenting both hexose and pentose sugars (Hammond and McCann, 2023). Hence, GMOs are employed to ensure the comprehensive utilization of sugars in the hydrolysate, leading to enhanced production benefits. The common methods used in fermenting lignocellulosic hydrolysate are simultaneous saccharification and fermentation (SSF) and separate hydrolysis and fermentation (SHF). Traditionally, SHF has been the preferred process. However, SSF offers superior ethanol production capabilities as it enhances ethanol yields by alleviating end-product inhibition and negating the necessity for separate reactors (Devi et al., 2021). Moreover, it proves to be cost-effective. Nonetheless, the disparity in optimal temperature conditions for enzyme activity during hydrolysis and fermentation presents certain limitations.

11.6 BENEFITS TO THE ENVIRONMENT AND ECONOMY

Biofuel production holds significant economic benefits due to its potential to reduce dependency on fossil fuels, mitigate greenhouse gas emissions, create jobs, agricultural market expansion, technology innovation, and stimulate economic growth (Verma et al., 2022e, 2022f; 2023c, 2023d; Kocaturk et al., 2023). By harnessing renewable biological resources to produce transportation fuels, biofuels contribute to a more sustainable and resilient energy future while driving economic growth and prosperity (McCarl et al., 2017; Verma et al., 2022g, 2022h). Through collaborative efforts involving government, industry, academia, and civil society, we can overcome these challenges and accelerate the transition to a bio-based economy that benefits both the people and the planet. Reducing the dependence on fossil fuels creates energy security, stabilizes energy prices, and reduces geopolitical risks associated with oil-producing regions (Kour et al., 2019; Verma et al., 2022i). Not only it reduces the greenhouse gas emission like CO_2 and other pollutants, biofuel generation also increases job availability and economic growth (Yang et al., 2021). The production of biofuels requires a workforce for cultivating feedstock crops, operating refineries, conducting research, and distributing and marketing the final products (Holmatov et al., 2019). Biofuel production increases the development of rural areas and agricultural markets. The summarization of biofuel production is depicted in Figure 11.4.

Previously, the primary technique for removing crop residue from a field is by deliberately setting it on fire. This process not only clears the field of residue but also reintroduces vital nutrients such as potassium and phosphorus back into the soil (Verma et al., 2022j). One significant benefit of discontinuing burning off practices is the ability to utilize the thermal energy that would otherwise be lost as heat into the environment. One of the best alternatives of burning the residues is the conversion of them to biofuels (Holmatov et al., 2019; Verma et al., 2022k).

Production and usage of biofuels leads to a reduction in the disastrous effects of climate change and reduces the emission of pollutants. It has the capacity to change the agricultural and transport sectors by reducing the carbon emissions of societies. The easiest route to production of biofuels is the production of ethanol from sugarcane syrup (Andoga et al., 2021). It is predicted that biofuels can fulfil about one-third of the primary energy demand by 2050, and eventually, the world can fully depend on renewable energy sources with the development of more efficient technologies in future. It is being predicted that by 2040, biofuels can replace 40% of conventional fuels (Jegan et al., 2021), and by 2030, there will be 24 million people working in the biofuel sector. Not only it protects the environment, but it also creates more jobs and better business efficiency. Gathering the crop residues can contribute to the overall income of farmers (Holmatov et al., 2019). Compared to nonrenewable energy resource industries, biofuel industries generate roughly 100 times more jobs per unit of energy. Several policies have been introduced and implemented by both developed and developing countries to encourage the production and utilization of biofuels. Many of the advantages referred to above are particularly appealing to developing countries in Asia due to their significant agricultural base and their increasing status as energy importers fueling their rapid economic growth (Pryshliak and Tokarchuk, 2020). To address the development of sustainable fuels, India has a separate ministry known as the Ministry of New and Renewable Energy. Worldwide, a total of 15% of energy requirements is fulfilled by biomass, and in developing countries, it accounts for 38% of the energy requirements (Prasad and Ingle, 2019).

The study of biofuel in Mozambique highlights the potential for economic growth and poverty reduction. By analyzing the impacts of biofuel investments using a Computable General Equilibrium model, it emphasizes the benefits of the outgrower approach for smallholders and the importance of technology spillovers in enhancing gains. The findings suggest that biofuels can contribute to GDP growth, job creation, and poverty alleviation. Additionally, the study addresses concerns regarding the impact of biofuels on food prices, poverty levels, and the environment. By assessing the potential trade-offs between biofuel investments and development priorities, such as health and education, the study provides valuable insights for policymakers seeking to balance competing demands and optimize resource allocation for sustainable development (Iwaszczuk et al., 2019).

Liu et al. (2019) suggested a case study about using the waste nitrogen, which can be collected using constructed wetlands, for the production of biofuels. The use of autotrophic microalgae for the production of biofuel is also being explored as an opportunity. Currently, *Chlorella protothecoides* is considered a perfect alga for biofuel production due to its high lipid content (approx. 55%). These organisms can also be genetically modified to increase the production or extraction of desired products. Furthermore, the scalability and adaptability of microalgae cultivation make it suitable for both small-scale applications, such as community-level biofuel production, and large-scale industrial operations, addressing diverse energy needs across different sectors (Peng et al., 2020).

However, it must be noted that production of biofuels has raised concerns regarding the threat to food production, soil health, and biodiversity, and this led to a food versus fuel debate (Brown et al., 2021). Additionally, the removal of residues for biofuel production can disrupt crucial soil processes, such as organic matter decomposition and nutrient cycling, leading to long-term degradation of soil health and productivity. This underscores the importance of adopting sustainable agricultural practices that prioritize soil conservation and carbon sequestration to mitigate the adverse effects of biofuel production on soil quality. So, increasing the biofuel production can lead to a nonreturnable damage to soil, but some reports have stated that 20–30% of residues can be removed from the fields, although we also need to add fertilizer to fulfill the nutritional requirements. The second- and third-generation biofuels offer promising alternatives to conventional biofuels feedstocks, as they utilize nonfood biomass sources, such as agricultural residues, forestry waste, and algae. By harnessing these waste streams, second- and third-generation biofuels reduce reliance on dedicated cropland, alleviate pressure on food production systems, and mitigate environmental risks associated with land-use change and ecosystem degradation. Hereby, the second and third biofuel generations are considered as better replacements of nonrenewable fuels as they utilize waste products (Sikarwar et al., 2017).

11.7 CONCLUSION

The economic benefits of biofuel production for sustainable agriculture are vast and multifaceted, offering promising opportunities for enhancing agricultural productivity, rural development, and environmental sustainability. Throughout this disclosure, we have delved into various aspects of biofuel production, highlighting its potential to transform the agricultural landscape and contribute to a more resilient and prosperous future. It becomes evident that maximizing the economic benefits of biofuel production requires a comprehensive and integrated approach that addresses the diverse needs and challenges of agricultural stakeholders, while promoting environmental stewardship and social equity. One of the primary economic benefits of biofuel production lies in its potential to diversify income streams for farmer and rural communities. By cultivating energy crops and participating in biofuel supply chains, farmers can generate additional revenue sources, reduce dependence on volatile commodity markets, and improve financial resilience against economic shocks. This diversification of income can also contribute to poverty alleviation and rural development, empowering farmers to invest in sustainable agricultural practices, education, healthcare, and infrastructure, thereby fostering inclusive growth and prosperity in rural areas. Biofuel production offers opportunities for value addition and downstream processing, creating jobs and stimulating economic growth along the entire biofuel value chain. From crop cultivation and biomass harvesting to biofuel conversion and distribution, each stage of the biofuel production process generates employment opportunities in rural and urban areas, supporting livelihoods and stimulating local economies. Moreover, the development of biofuel infrastructure, such as biorefineries, storage facilities, and distribution networks, spurs investment and innovation in related industries, driving technological advancements and enhancing competitiveness in the global bioenergy market.

In addition to its economic benefits, biofuel production can also contribute to sustainable agricultural intensification and land-use optimization, thereby enhancing productivity, resilience, and food security. By integrating energy crops into crop rotation systems, agroforestry practices, and marginal lands, farmers can improve soil health, conserve water resources, and reduce erosion. Biofuel production can play a pivotal role in mitigating greenhouse gas emissions and combating climate change, thereby creating economic value through environmental stewardship and carbon market opportunities. As we continue to navigate the transition toward a more sustainable and equitable agricultural future, biofuel production stands poised to play a central role in shaping the economic, social, and environmental dynamics of tomorrow's agrarian society.

KEYWORDS

- **biomass**
- **biofuels**
- **fermentation**
- **sustainable**
- **microalgae**

REFERENCES

Abdullah B, ad Syed Muhammad SAF, Shokravi Z, Ismail S, Kassim KA, Mahmood AN, Aziz MMA. (2019). Fourth generation biofuel: A review on risks and mitigation strategies. *Renew. Sustain. Energy Rev.* 107. https://doi.org/10.1016/j.rser.2019.02.018

Abreu M, Silva L, Ribeiro B, Ferreira A, Alves L, Paixão SM, Gouveia L, Moura F, Carvalheiro F, Duarte LC, Fernando AL, Reis A, Gírio F. (2022). Low indirect land use change (ILUC) energy crops to bioenergy and biofuels—A review. *Energies* 15(12), 4348. https://doi.org/10.3390/en15124348

Akbas MY, Stark BC. (2016). Recent trends in bioethanol production from food processing byproducts. *J. Industrial Microbiol. Biotechnol.* 43(11), 1593–1609. https://doi.org/10.1007/s10295-016-1821-z

Alalwan HA, Alminshid AH, Aljaafari HAS. (2019). Promising evolution of biofuel generations. Subject review. *Renew. Energy Focus* 28, 127–139. https://doi.org/10.1016/j.ref.2018.12.006

Anas M, Verma KK, Haq IU, Naeem M, Li Q, Liao F, Xiang YX, Li YR. (2021a). Characterization of exotic, native and wild-type genotypes of sugarcane (*Saccharum* spp. Hybrids) for internal nitrogen use efficiency under different nitrogen levels. *Sugar Tech*. 23, 1258–1267. https://doi.org/10.1007/s12355-021-01014-1

Anas M, Verma KK, Riaz M, Qiang L, Liao F, Liu Y, Li YR. (2021b). Physio-morphological and biochemical mechanism of nitrogen use efficiency in sugarcane (*Saccharum* spp.) genotypes under different growth stages and nitrogen levels. *J. Plant Interact.* 16, 332–343.

Andoga R, Fozo L, Schrötter M, Szabo, S. (2021). The use of ethanol as an alternative fuel for small turbojet engines. *Sustainability*. *13*(5), 2541. https://doi.org/10.3390/su13052541

Anwar M. (2021). Biodiesel feedstocks selection strategies based on economic, technical, and sustainable aspects. *Fuel* 283, 119204. https://doi.org/10.1016/j.fuel.2020.119204

Bhatt R, Verma KK, Jamwal NS. (2023a). Can paddy-compost eliminate N fertilization for climate-smart sugarcane? *Sugar Tech.* 25, 1036–1046. https://doi.org/10.1007/s12355-023-01270-3

Bhatt R, Verma KK, Kumar R, Sanghera GS. (2023b). Foliar application of potassium salt of active phosphorus (PSAP) mitigates insect pests and improves yield along with sugarcane quality in response to agroclimatic conditions of Punjab. *Sugar Tech*. 25(3), 660–669.

Brahma S, Nath B, Basumatary B, Das B, Saikia P, Patir K, Basumatary, S. (2022). Biodiesel production from mixed oils: A sustainable approach towards industrial biofuel production. *Chem. Engineer. J. Adv.* 10, 100284. https://doi.org/10.1016/j.ceja.2022.100284

Brown B, Schoney R, Nolan, J. (2021). Assessing the food vs. fuel issue: An agent-based simulation. *Energy Policy*. 159, 112553, https://doi.org/10.1016/j.enpol.2021.112553

Cavelius P, Engelhart-Straub S, Mehlmer N, Lercher J, Awad D, Brück, T. (2023). The potential of biofuels from first to fourth generation. *PLoS Biol.* 21(3). https://doi.org/10.1371/JOURNAL.PBIO.3002063

Chaudhary P, Xu M, Ahamad L, Chaudhary A, Kumar G, Adeleke BS, Verma KK, Hu D-M, Siric I, Kumar P, Popescu SM, Fayssal SA. (2023). Application of synthetic consortia for improvement of soil fertility, pollution remediation, and agricultural productivity: A review. *Agronomy* 13, 643. https://doi.org/10.3390/agronomy13030643.

Chen GL, Zheng FJ, Lin B, Lao SB, He J, Huang Z, Zeng Y, Sun J, Verma KK. (2020). Phenolic and volatile compounds in the production of sugarcane vinegar. *ACS Omega* 5, 30587–30595.

Chen G-L, Zheng F-J, Lin B, Yang Y-X, Fang X-C, Verma KK, Yang L-F. (2023). Vinegar: A potential source of healthy and functional food with special reference to sugarcane vinegar. *Front. Nutr*. 10, 1145862. https://doi.org/10.3389/fnut.2023.1145862

Chowdhury H, Loganathan, B. (2019). Third-generation biofuels from microalgae: A review. *Curr. Opin. Green Sustain. Chem.* 20, 39–44. https://doi.org/10.1016/j.cogsc.2019.09.003

Comelli RN, Seluy LG, Benzzo MT, Isla MA. (2020). Combined utilization of agro-industrial wastewaters for non-lignocellulosic second-generation bioethanol production. *Waste Biomass Valor.* 11(1), 265–275. https://doi.org/10.1007/s12649-018-0391-x

Correa DF, Beyer HL, Fargione JE, Hill JD, Possingham HP, Thomas-Hall SR, Schenk PM. (2019). Towards the implementation of sustainable biofuel production systems. *Renew. Sustain. Energy Rev.* 107, 250–263. https://doi.org/10.1016/j.rser.2019.03.005

Deng ZN, Huang HR, Huang CM, Verma KK, Luo HB, Jiang SL, Li YJ, Cao HQ, Wu KC, Xu L. (2023). Short-term limited water irrigation influences the photosynthetic pigments, enzymatic and non-enzymatic activities in *Saccharum spontaneum* L. at vegetative stage. *Sugar Tech.* 25, 110–118. https://doi.org/10.1007/s12355-022-01175-7

Devi A, Singh A, Bajar S, Pant D, Din ZU. (2021). Ethanol from lignocellulosic biomass: An in-depth analysis of pre-treatment methods, fermentation approaches and detoxification processes. *J. Environ. Chem. Eng.* 9, 105798. https://doi.org/10.1016/j.jece.2021.105798

Dey S, Sreenivasulu A, Veerendra GTN, Rao KV, Babu PSSA. (2022). Renewable energy present status and future potentials in India: An overview. *Innov. Green Dev.* 1(1), 100006. https://doi.org/10.1016/j.igd.2022.100006

Dhungana P, Prajapati B, Maharjan S, Joshi, J. (2022). Current trends in lignocellulosic bioethanol production. *Int. J. Appl. Sci. Biotechnol.* 10(1), 1–11 https://doi.org/10.3126/ijasbt.v10i1.41139

Dong, S. (2020). Food waste as a raw material for biofuel production. *Centria University of Applied Sciences Environmental Chemistry and Technology*.

Elgharbawy AS, Sadik WA, Sadek OM, Kasaby MA. (2021). A review on biodiesel feedstocks and production technologies. *J. Chilean Chem. Soc.* 65(1). https://doi.org/10.4067/S0717-97072021000105098

Erdiwansyah, Mamat R, Sani MSM, Sudhakar K, Kadarohman A, Sardjono RE. (2019). An overview of Higher alcohol and biodiesel as alternative fuels in engines. *Energy Rep.* 5, 467–479. https://doi.org/10.1016/j.egyr.2019.04.009

Fivga A, Speranza LG, Branco CM, Ouadi M, Hornung, A. (2019). A review on the current state of the art for the production of advanced liquid biofuels. *AIMS Energy* 7(1), 1523–1535. https://doi.org/10.3934/ENERGY.2019.1.46

Ganesan R, Manigandan S, Samuel MS, Shanmuganathan R, Brindhadevi K, Lan Chi NT, Duc PA, Pugazhendhi, A. (2020). A review on prospective production of biofuel from microalgae. *Biotechnology Reports* 27, e00509. https://doi.org/10.1016/j.btre.2020.e00509

Gautam P, Neha, Upadhyay SN, Dubey SK. (2020). Bio-methanol as a renewable fuel from waste biomass: Current trends and future perspective. *Fuel* 273, 117783. https://doi.org/10.1016/j.fuel.2020.117783

Gohain M, Hasin M, Eldiehy KSH, Bardhan P, Laskar K, Phukon H, Mandal M, Kalita D, Deka, D. (2021). Bio-ethanol production: A route to sustainability of fuels using bio-based heterogeneous catalyst derived from waste. *Process Saf. Environ. Protect.* 146, 190-200. https://doi.org/10.1016/j.psep.2020.08.046

Guo D-J, Li D-P, Chen Z-D, Yang B, Verma KK, Singh RK, Singh P, Khan Q, Sharma A, Qin Y, Zhang B-Q, Song X-P, Li Y-R. (2023). Effect of endophytic diazotroph *Enterobacter roggenkampii* ED5 on nitrogen-metabolism-related microecology in the sugarcane rhizosphere at different nitrogen levels. *Front. Microbiol.* 14, 1132016. https://doi.org/10.3389/fmicb.2023.1132016

Guo D-J, Li D-P, Singh RK, Singh P, Sharma A, Verma KK, Qin Y, Khan Q, Lu Z, Malviya MK, Song X-P, Xing Y-X, Li Y-R. (2021). Differential protein expression analysis of two sugarcane varieties in response to diazotrophic plant growth-promoting endophyte

Enterobacter roggenkampii ED5. *Front. Plant Sci.* 12, 727741. https://doi.org/10.3389/fpls.2021.727741

Guo DJ, Li DP, Singh RK, Singh P, Verma KK, Sharma A, Qin Y, Khan Q, Song XP, Malviya MK, Xing YX, Li YR. (2022). Comparative transcriptome analysis of two sugarcane varieties in response to diazitrophic plant growth promoting endophyte *Enterobacter roggenkampii* ED5. *J. Plant Interact.* 17(1), 75–84.

Gupta P, Kurien C, Mittal, M. (2023). Biogas (a promising bioenergy source): A critical review on the potential of biogas as a sustainable energy source for gaseous fuelled spark ignition engines. In: *International Journal of Hydrogen Energy;* Vol. 48, Issue 21. https://doi.org/10.1016/j.ijhydene.2022.11.195

Hammond GP, McCann, N. (2023). Bioethanol processing from wheat straw: Investment appraisal of a full-scale UK biofuel refinery. *Biofuels*. https://doi.org/10.1080/17597269.2022.2132722

Hasan M, Abedin MZ, Amin, M. bin, Nekmahmud M, Oláh, J. (2023). Sustainable biofuel economy: A mapping through bibliometric research. In: *Journal of Environmental Management*. Academic Press; Vol. 336. https://doi.org/10.1016/j.jenvman.2023.117644

Hawrot-Paw M, Stańczuk, A. (2023). From waste biomass to cellulosic ethanol by Separate Hydrolysis and Fermentation (SHF) with trichoderma viride. *Sustainability 15*(1), 168. https://doi.org/10.3390/su15010168

Hoang AT, Ong HC, Fattah IMR, Chong CT, Cheng CK, Sakthivel R, Ok YS. (2021). Progress on the lignocellulosic biomass pyrolysis for biofuel production toward environmental sustainability. In: *Fuel Processing Technology*; Vol. 223. https://doi.org/10.1016/j.fuproc.2021.106997

Hoang AT, Sirohi R, Pandey A, Nižetić S, Lam SS, Chen WH, Luque R, Thomas S, Arıcı M, Pham VV. (2023). Biofuel production from microalgae: Challenges and chances. In: *Phytochemistry Reviews*; Vol. 22, Issue 4. https://doi.org/10.1007/s11101-022-09819-y

Holmatov B, Hoekstra AY, Krol MS. (2019). Land, water and carbon footprints of circular bioenergy production systems. *Renew. Sustain. Energy Rev.* https://doi.org/10.1016/j.rser.2019.04.085

Huang CM, Wu KC, Verma KK, Luo HB, Deng ZN, Shi GY, Cao HQ, Wu XJ, Ye LP, Wei YW, Xu L. (2024). Isolation and expression analysis of cellulose synthase 3 (*Ces3*) genes from sugarcane (*Saccharum officinarum* L.). *Chem. Biol. Technol. Agric*. 11, 71. https://doi.org/10.1186/s40538-024-00597-7

Huang HR, Gao YJ, Malviya MK, Verma KK, Solanki MK, Huang YX, Li X., Deng Y, Yan J, Tang SY, Wang LW, Xu L. (2022). Genetic diversity analysis of sugarcane (*Saccharum* spp. Hybrids) among high-sucrose clones of GT series and commonly used parents by using microsatellite markers in Guangxi. *China. Sugar Tech.* 24, 397–407. https://doi.org/10.1007/s12355-021-01033-y

Idowu IA, Hashim K, Shaw A, Nunes LJR. (2022). Enhancing the fuel properties of beverage wastes as non-edible feedstock for biofuel production. *Biofuels* 13(6), 763–770. https://doi.org/10.1080/17597269.2021.1923934

Inyang V, Laseinde OT, Kanakana GM. (2022). Techniques and applications of lignocellulose biomass sources as transport fuels and other bioproducts. *Int. J. Low-Carbon Technol* 17. https://doi.org/10.1093/ijlct/ctac068

Iwaszczuk N, Szyba M, Iwaszczuk A, Yakubiv, V. (2019). Production of agricultural biogas from waste—An element of socially responsible actions in the food sector. *Acta Innovations*. https://doi.org/10.32933/ActaInnovations.33.5

Jagadesh M, Dash M, Kumari A, Singh SK, Verma KK, Kumar P, Bhatt R, Sharma SK. (2024). Revealing the hidden world of soil microbes: Metagenomic insights into plant, bacteria, and fungi interactions for sustainable agriculture and ecosystem restoration. *Microbiol. Res*. 285, 127764. https://doi.org/10.1016/j.micres.2024.127764

Jegan TMC, Chitra R, Glivin, G. (2021). Testing of ethanol as an alternative fuel for IC engine. *Energy Res. J.* 12(1), 13–21. https://doi.org/10.3844/erjsp.2021.13.21

Jeswani HK, Chilvers A, Azapagic, A. (2020). Environmental sustainability of biofuels: A review: Environmental sustainability of biofuels. In: *Proceedings of the Royal Society A: Mathematical, Physical and Engineering Sciences*; Vol. 476, Issue 2243. https://doi.org/10.1098/rspa.2020.0351

Joshi A, Verma KK, Rajput VD, Arora J. (2023). Metabolic engineering of microorganisms in advancing biofuel production. In: *Advances in Lignocellulosic Biofuel Production Systems*; pp. 181–202. https://doi.org/10.1016/B978-0-323-91192-4.00016-X

Joshi A, Verma KK, Rajput VD, Minkina T, Arora J. (2022). Recent advances in metabolic engineering of microorganisms for advancing lignocellulose-derived biofuels. *Bioengineered* 13(4), 8135–8163. https://doi.org/10.1080/21655979.2022.2051856.

Kersten S, Garcia-Perez, M. (2013). Recent developments in fast pyrolysis of ligno-cellulosic materials. *Curr. Opin. Biotechnol*. 24(3), 414–420. https://doi.org/10.1016/j.copbio.2013.04.003

Khan N, Sudhakar K, Mamat, R. (2021). Role of biofuels in energy transition, green economy and carbon neutrality. In: *Sustainability (Switzerland)*; Vol. 13, Issue 22. https://doi.org/10.3390/su132212374

Khan S, Jain G, Srivastava A, Verma PC, Pande V, Dubey RS, Khan M, Haque S, Ahmad, S. (2023). Enzymatic biomethanol production: Future perspective. In: *Sustainable Materials and Technologies*; Vol. 38. https://doi.org/10.1016/j.susmat.2023.e00729

Khanahmadi M, Arezi I, Amiri, M. sadat, Miranzadeh, M. (2018). Bioprocessing of agro-industrial residues for optimization of xylanase production by solid-state fermentation in flask and tray bioreactor. *Biocatalysis Agric. Biotechnol*. 13, 272–282. https://doi.org/10.1016/j.bcab.2018.01.005

Kim Y, Il Kim W, Min B, Seo J, Lee, K. (2022). Experimental investigation of combustion characteristics of ethanol–gasoline blended fuel in a T-GDI engine. *Appl. Ther. Engineer*. https://doi.org/10.1016/j.applthermaleng.2022.118168

Kocaturk E, Salan T, Ozcelik O, Alma MH, Candan, Z. (2023). Recent advances in lignin-based biofuel production. In: *Energies*. https://doi.org/10.3390/en16083382

Kondrasheva NK, Eremeeva, A. (2023). Production of biodiesel fuel from vegetable raw materials. *J. Mining Inst.* 260. https://doi.org/10.31897/PMI.2022.15

Korberg AD, Skov IR, Mathiesen BV. (2020). The role of biogas and biogas-derived fuels in a 100% renewable energy system in Denmark. *Energy* 199, 117426. https://doi.org/10.1016/j.energy.2020.117426

Kour D, Rana KL, Yadav N, Yadav AN, Rastegari AA, Singh C, Negi P, Singh K, Saxena AK. (2019). *Technologies for Biofuel Production: Current Development, Challenges, and Future Prospects*. https://doi.org/10.1007/978-3-030-14463-0_1

Krishna D, Sachan HK, Verma KK. (2022). Abiotic stresses in plants: An overview. In: *Understanding Abiotic Stresses*. Nova Science Publishers, Inc.: New York.

Kumari A, Dash M, Singh SK, Jagadesh M, Mathpal B, Mishra PK, Pandey SK, Verma KK. (2023a). Soil microbes: A natural solution for mitigating the impact of climate change. *Environ. Monit. Assess*. 195, 1436. https://doi.org/10.1007/s10661-023-11988-y

Kumari A, Lakshmi GA, Krishna GK, Patni B, Prakash S, Bhattacharyya M, Singh SK, Verma KK. (2022). Climate change and its impact on crops: A comprehensive investigation for sustainable agriculture. *Agronomy* 12, 3008. https://doi.org/10.3390/agronomy12123008

Kumari A, Singh SK, Mathpal B, Verma KK, Garg VK, Bhattacharyya M, Bhatt R. (2023b). The biosynthesis, mechanism of action, and physiological functions of melatonin in horticultural plants: A review. *Horticulturae* 9, 913. https://doi.org/10.3390/horticulturae9080913

Li A-M, Liao F, Wang M, Chen Z-L, Qin C-X, Huang R-Q, Verma KK, Li Y-R, Que Y-X, Pan Y-Q, et al. (2023). Transcriptomic and proteomic landscape of sugarcane response to biotic and abiotic stressors. *Int. J. Mol. Sci*. 24, 8913. https://doi.org/10.3390/ijms24108913

Linganiso EC, Tlhaole B, Magagula LP, Dziike S, Linganiso LZ, Motaung TE, Moloto N, Tetana ZN. (2022). Biodiesel production from waste oils: A South African outlook. *Sustainability*, 14(4), 1983. https://doi.org/10.3390/su14041983

Liu D, Zou C, Xu, M. (2019). Environmental, ecological, and economic benefits of biofuel production using a constructed wetland: A case study in China. *Int. J. Environ. Res. Public Health 16*(5), 827. https://doi.org/10.3390/ijerph16050827

Liu Y, Cruz-Morales P, Zargar A, Belcher MS, Pang B, Englund E, Dan Q, Yin K, Keasling JD. (2021). Biofuels for a sustainable future. In: *Cell*; Vol. 184, Issue 6. https://doi.org/10.1016/j.cell.2021.01.052

Liu Y, Yuan Y, Ramya G, Mohan Singh S, Thuy Lan Chi N, Pugazhendhi A, Xia C, Mathimani, T. (2022). A review on the promising fuel of the future – Biobutanol; the hindrances and future perspectives. *Fuel* 327, 125166. https://doi.org/10.1016/j.fuel.2022.125166

Maciejczyk P, Chen LC, Thurston, G. (2021). The role of fossil fuel combustion metals in PM2.5 air pollution health associations. In: *Atmosphere*. https://doi.org/10.3390/atmos12091086

Mahalingam L, Abdulla R, Sani SA, Sabullah MK, Faik AAM, Misson, M. (2022). Lignocellulosic biomass – A sustainable feedstock for acetone-butanol-ethanol fermentation. *Periodica Polytechnica Chem. Engineer.* 66(2). https://doi.org/10.3311/PPch.18574

Mahapatra S, Kumar D, Singh B, Sachan PK. (2021). Biofuels and their sources of production: A review on cleaner sustainable alternative against conventional fuel, in the framework of the food and energy nexus. In: *Energy Nexus*; Vol. 4. https://doi.org/10.1016/j.nexus.2021.100036

Malode SJ, Prabhu KK, Mascarenhas RJ, Shetti NP, Aminabhavi TM. (2021). Recent advances and viability in biofuel production. *Energy Conv. Manag. X* 10, 100070. https://doi.org/10.1016/j.ecmx.2020.100070

Malviya MK, Li C-N, Lakshmanan P, Solanki MK, Wang Z, Solanki AC, Nong Q, Verma KK, Singh RK, Singh P, Sharma A, Guo D-J, Dessoky ES, Song X-P, Li Y-R. (2022).

High-throughput sequencing-based analysis of rhizosphere and diazotrophic bacterial diversity among wild progenitor and closely related species of sugarcane (*Saccharum* spp. inter-specific hybrids). *Front. Plant Sci.* 13, 829337. https://doi.org/10.3389/fpls.2022.829337.

Malviya MK, Solanki MK, Li CN, Wang Z, Zeng Y, Verma KK, Singh RK, Singh P, Huang HR, Yang LT, Song XP, Li YR. (2021). Sugarcane-legume intercropping can enrich the soil microbiome and plant growth. *Front. Sustain. Food Syst.* 5, 606595. https://doi.org/10.3389/fsufs.2021.606595.

Mat Aron NS, Khoo KS, Chew KW, Show PL, Chen WH, Nguyen THP. (2020). Sustainability of the four generations of biofuels – A review. *Int. J. Energy Res.* 44(12). https://doi.org/10.1002/er.5557

Mathur S, Waswani H, Singh D, Ranjan, R. (2022). Alternative fuels for agriculture sustainability: Carbon footprint and economic feasibility. In: *AgriEngineering*. MDPI; Vol. 4, Issue 4, pp. 993–1015. https://doi.org/10.3390/agriengineering4040063

McCarl BA, Jones JPH, Kapilakanchana M, Wang MZ, Zhao, G. (2017). Economical assessment of biofuel production. In: *Biofuels Production and Processing Technology*. https://doi.org/10.1201/9781315155067

Mitkidis G, Magoutas A, Kitsios, F. (2018). Market and economic feasibility analysis for the implementation of 2nd generation biofuels in Greece. *Energy Strategy Rev.* 19, 85–98. https://doi.org/10.1016/j.esr.2017.12.006

Oliveira MA, Verma KK, Bhatt R, Oliveira TBA. (2022). Impact of green and organic fertilizers on soil fertility and sugarcane productivity. In: Verma KK, et al. (Ed.), *Agro-industrial Perspectives on Sugarcane Production under Environmental Stress*. https://doi.org/10.1007/978-981-19-3955-6_11.

Osman AI, Mehta N, Elgarahy AM, Al-Hinai A, Al-Muhtaseb AH, Rooney DW. (2021a). Conversion of biomass to biofuels and life cycle assessment: A review. *Environ. Chem. Lett.*, 19(6). https://doi.org/10.1007/s10311-021-01273-0

Osman AI, Qasim U, Jamil F, Al-Muhtaseb AH, Jrai AA, Al-Riyami M, Al-Maawali S, Al-Haj L, Al-Hinai A, Al-Abri M, Inayat A, Waris A, Farrell C, Maksoud MIAA, Rooney DW. (2021b). Bioethanol and biodiesel: Bibliometric mapping, policies and future needs. *Renew. Sust. Energy Rev.*, 152. https://doi.org/10.1016/j.rser.2021.111677

Peng L, Fu D, Chu H, Wang Z, Qi, H. (2020). Biofuel production from microalgae: A review. *Environ. Chem. Lett.* https://doi.org/10.1007/s10311-019-00939-0

Prasad S, Ingle AP. (2019). Impacts of sustainable biofuels production from biomass. *Sust. Bioenergy Adv. Impacts*. https://doi.org/10.1016/B978-0-12-817654-2.00012-5

Pryshliak N, Tokarchuk, D. (2020). Socio-economic and environmental benefits of biofuel production development from agricultural waste in Ukraine. *Environ. Socio-economic Stud.* https://doi.org/10.2478/environ-2020-0003

Pugazhendhi A, Mathimani T, Varjani S, Rene ER, Kumar G, Kim SH, Ponnusamy VK, Yoon JJ. (2019). Biobutanol as a promising liquid fuel for the future—Recent updates and perspectives. *Fuel* 253, 637–646. https://doi.org/10.1016/j.fuel.2019.04.139

Raina N, Slathia PS, Sharma, P. (2022). Response surface methodology (RSM) for optimization of thermochemical pretreatment method and enzymatic hydrolysis of deodar sawdust (DS) for bioethanol production using separate hydrolysis and co-fermentation (SHCF). *Biomass Conv. Bioref.* 12, 5175–5195. https://doi.org/10.1007/s13399-020-00970-0

Rizwanul Fattah IM, Ong HC, Mahlia TMI, Mofijur M, Silitonga AS, Ashrafur Rahman SM, Ahmad, A. (2020). State of the art of catalysts for biodiesel production. In: *Frontiers in Energy Research*; Vol. 8. https://doi.org/10.3389/fenrg.2020.00101

Robak K, Balcerek, M. (2020). Current state-of-the-art in ethanol production from lignocellulosic feedstocks. In: *Microbiological Research*. https://doi.org/10.1016/j.micres.2020.126534

Rosen Y, Mamane H, Gerchman, Y. (2021). Immersed ozonation of agro-wastes as an effective pretreatment method in bioethanol production. *Renew. Energy* 174, 382–390. https://doi.org/10.1016/j.renene.2021.04.047

Ruan R, Zhang Y, Chen P, Liu S, Fan L, Zhou N, Ding K, Peng P, Addy M, Cheng Y, Anderson E, Wang Y, Liu Y, Lei H, Li, B. (2019). Chapter 1—Biofuels: Introduction. In: *Biomass, Biofuels, Biochemicals*.

Saha K, Maharana A, Sikder J, Chakraborty S, Curcio S, Drioli, E. (2019). Continuous production of bioethanol from sugarcane bagasse and downstream purification using membrane integrated bioreactor. *Catalysis Today* 331, 68–77. https://doi.org/10.1016/j.cattod.2017.11.031

Saleem, M. (2022). Possibility of utilizing agriculture biomass as a renewable and sustainable future energy source. *Heliyon*. 8(2). https://doi.org/10.1016/j.heliyon.2022.e08905

Sharma A, Singh RK, Singh P, Vaishnav A, Guo DJ, Verma KK, Li DP, Song XP, Malviya MK, Khan N, Lakshmanan P, Li YR. (2021). Insights into the bacterial and nitric oxide-induced salt tolerance in sugarcane and their growth-promoting abilities. *Microorganisms* 9, 2203. https://doi.org/10.3390/microorganisms9112203.

Sharma A, Song X-P, Singh RK, Vaishnav A, Gupta S, Singh P, Guo D-J, Verma KK, Li Y-R. (2022). Impact of carbendazim on cellular growth, defence system and plant growth promoting traits of *Priestia megaterium* ANCB-12 isolated from sugarcane rhizosphere. *Front. Microbiol.* 13, 1005942. https://doi.org/10.3389/fmicb.2022.1005942

Sharma A, Pavani Srikavya B, Urade AD, Joshi A, Shyam Narain R, Dwarakanath V, Alam T, Shankar Prasad, R. (2023). Economic and environmental impacts of biofuels in Indian context. *Mater. Today: Proc.* https://doi.org/10.1016/j.matpr.2023.05.118

Shokravi H, Shokravi Z, Heidarrezaei M, Ong HC, Rahimian Koloor SS, Petrů M, Lau WJ, Ismail AF. (2021). Fourth generation biofuel from genetically modified algal biomass: Challenges and future directions. *Chemosphere*, 285. https://doi.org/10.1016/j.chemosphere.2021.131535

Sikarwar VS, Zhao M, Fennell PS, Shah N, Anthony EJ. (2017). Progress in biofuel production from gasification. *Progr. Energy Combust. Sci.* https://doi.org/10.1016/j.pecs.2017.04.001

Silviana S, Anggoro DD, Hadiyanto H, Salsabila CA, Aprilio K, Utami AW, Sa'adah AN, Dalanta, F. (2022). A review on the recent breakthrough methods and influential parameters in the biodiesel synthesis and purification. *Int. J. Renew. Energy Dev.* 11(4). https://doi.org/10.14710/ijred.2022.43147

Singh P, Singh RK, Li HB, Guo DJ, Sharma A, Verma KK, Solanki MK, Upadhyay SK, Lakshmanan P, Yang LT, Li YR. (2023). Nitrogen fixation and phytohormone stimulation of sugarcane plant through plant growth promoting diazotrophic *Pseudomonas*. *Biotechnol. Genetic Eng. Reviews* 1–21. https://doi.org/10.1080/02648725.2023.2177814

Singh P, Singh RK, Song QQ, Li HB, Guo DJ, Malviya MK, Verma KK, Song XP, Lakshmanan P, Yang LT, Li YR. (2021). Comparative analysis of protein and differential responses of defense-related gene and enzyme activity reveals the long-term molecular responses of sugarcane inoculated with *Sporisorium scitamineum*. *J. Plant Interact.* 16(1), 12–29.

Singh RK, Singh P, Li HB, Song QQ, Guo DJ, Solanki MK, Verma KK, Malviya MK, Song XP, Lakshmanan P, Yang LT, Li YR. (2020a) Diversity of nitrogen-fixing rhizobacteria associated with sugarcane: A comprehensive study of plant-microbe interactions for growth enhancement in *Saccharum* spp. *BMC Plant Biol.* 20, 220. https://doi.org/10.1186/s12870-020-02400-9.

Singh, A. (2021). A review on bioethanol from cellulosic materials: A biomass-based renewable motor fuel. *Academicia.* 11. https://doi.org/10.5958/2249-7137.2021.02499.x

Singh D, Sharma D, Soni SL, Sharma S, Kumar Sharma P, Jhalani, A. (2020b). A review on feedstocks, production processes, and yield for different generations of biodiesel. In: *Fuel*; Vol. 262. https://doi.org/10.1016/j.fuel.2019.116553

Solanki MK, Verma KK, Dastogeer KMG, Mora-Poblete F, Mundra S. (2023). Microbial resilience in plant nutrient management towards sustainable farming. *Front. Microbiol.* 14, 1280811. https://doi.org/10.3389/fmicb.2023.1280811

Song X-P, Mo F, Verma KK, Wei J, Zhang X, Yang L, Li YR. (2019). Effect of sugarcane smut (*Ustilago scitaminea* Syd.) on ultrastructure and biochemical indices of sugarcane. *Biomed. J. Sci. Tech. Res.* 002950. https://doi.org/10.26717/BJSTR.2019.17.002950.

Song XP, Mo F, Yan M, Zhang X, Zhang B, Huang X, Huang D, Pan Y, Verma KK, Li YR. (2022). Effect of smut infection on the photosynthetic physiological characteristics and related defense enzymes of sugarcane. *Life* 12, 1201. https://doi.org/10.3390/life12081201

Song XP, Verma KK, Tian DD, Zhang XQ, Liang YJ, Huang X, Li CN, Li YR. (2021) Exploration of silicon functions to integrate with biotic stress tolerance and crop improvement. *Biol. Res.* 54, 19. https://doi.org/10.1186/s40659-021-00344-4

Stančin H, Mikulčić H, Wang X, Duić, N. (2020). A review on alternative fuels in future energy system. *Renew. Sustain. Energy Rev.* 128, 109927. https://doi.org/10.1016/j.rser.2020.109927

Stephen JL, Periyasamy, B. (2018). Innovative developments in biofuels production from organic waste materials: A review. In: *Fuel*; Vol. 214. https://doi.org/10.1016/j.fuel.2017.11.042

Sun S, Sun S, Cao X, Sun, R. (2016). The role of pretreatment in improving the enzymatic hydrolysis of lignocellulosic materials. *Biores. Technol.* https://doi.org/10.1016/j.biortech.2015.08.061

Tabibian SS, Sharifzadeh, M. (2023). Statistical and analytical investigation of methanol applications, production technologies, value-chain and economy with a special focus on renewable methanol. *Renew. Sustain. Energy Rev.* 179, 113281. https://doi.org/10.1016/j.rser.2023.113281

Thanigaivel S, Priya AK, Dutta K, Rajendran S, Vasseghian, Y. (2022). Engineering strategies and opportunities of next generation biofuel from microalgae: A perspective review on the potential bioenergy feedstock. *Fuel* 312, 122827. https://doi.org/10.1016/j.fuel.2021.122827

Tian D, Qin L, Verma KK, Wei L, Li J, Li B, Zhou W, He Z, Wei D, Huang S, Long S, Huang Q, Li C, Wei S. (2023). Transcriptomic and metabolomic differences between banana varieties which are resistant or susceptible to *Fusarium wilt. PeerJ* 11, e16549. https://doi.org/10.7717/peerj.16549

Vaidya AA, Murton KD, Smith DA, Dedual, G. (2022). A review on organosolv pretreatment of softwood with a focus on enzymatic hydrolysis of cellulose. *Biomass Conv. Bioref.* https://doi.org/10.1007/s13399-022-02373-9

Verma KK, Anas M, Chen Z, Rajput VD, Malviya MK, Verma CL, Singh RK, Singh P, Song XP, Li YR. (2020f). Silicon supply improves leaf gas exchange, antioxidant defense system and growth in sugarcane responsive to water limitation. *Plants* 9, 1032, https://doi.org/10.3390/plants9081032.

Verma KK, Joshi A, Song X-P, Liang Q, Xu L, Huang H-r, Wu K-C, Seth CS, Arora J, Li Y-R. (2024a). Regulatory mechanisms of plant rhizobacteria on plants to the adaptation of adverse agroclimatic variables. *Front. Plant Sci.* 15, 1377793. https://doi.org/10.3389/fpls.2024.1377793

Verma KK, Joshi A, Song X-P, Singh S, Kumari A, Arora J, Singh SK, Solanki MK, Seth CS, Li Y-R. (2024b). Synergistic interactions of nanoparticles and plant growth promoting rhizobacteria enhancing soil-plant systems: A multigenerational perspective. *Front. Plant Sci.* 15, 1376214. https://doi.org/10.3389/fpls.2024.1376214

Verma KK, Li DM, Singh M, Rajput VD, Malviya MK, Minkina T, Singh RK, Singh P, Song XP, Li YR. (2020a). Interactive role of silicon and plant-rhizobacteria mitigating abiotic stresses: A new approach for sustainable agriculture and climate change. *Plants* 9, 1055. https://doi.org/10.3390/plants9091055.

Verma KK, Rajput VD, Singh M, Verma CL, Minkina T. (2022a). Influence of waterlogging on plant physiology and molecular biology. In: *Understanding Abiotic Stresses*. Nova Science Publishers, Inc.: New York.

Verma KK, Siddiqui N, Singh M. (2022b). Understanding the plant physiological and molecular adaptation strategies to soil salinity. In: *Understanding Abiotic Stresses*. Nova Science Publishers, Inc.: New York. https://doi.org/10.52305/UBUJ4024

Verma KK, Singh P, Song X-P, Malviya MK, Singh RK, Chen G-L, Solomon S, Li YR. (2020b). Mitigating climate change for sugarcane improvement: Role of silicon in alleviating abiotic stresses. *Sugar Tech.* 22(5), 741–749. https://doi.org/10.1007/s12355-020-00831-0.

Verma KK, Singh RK, Song QQ, Singh P, Zhang B-Q, Song X-P, Chen G-L, Li YR. (2019a). Silicon alleviates drought stress of sugarcane plants by improving antioxidant responses. *Biomed. J. Sci. Tech. Res.* 002957. https://doi.org/10.26717/BJSTR.2019.17.002957

Verma KK, Song X, Degu HD, Guo D, Joshi A, Huang H, Xu L, Singh M, Huang D, Rajput VD, Li YR. (2023a). Recent advances in nitrogen and nano-nitrogen fertilizers for sustainable crop production: A mini-review. *Chem. Biol. Technol. Agric.* 10, 1–14. https://doi.org/10.1186/s40538-023-00488-3

Verma KK, Song X, Verma CL, Huang H, Singh M, Xu L, Wu J, Khan ZH, Zhang B, Li Y. (2023b). Mathematical modeling of climate and fluoride effects on sugarcane photosynthesis with silicon nanoparticles. *Plant Physiol. Biochem.* 204, 108089. https://doi.org/10.1016/j.plaphy.2023.108089

Verma KK, Song XP, Budeguer F, Nikpay A, Enrique R, Singh M, Zhang B, Wu, JM, Li YR. (2022c). Genetic engineering: An efficient approach to mitigating biotic and abiotic stresses in sugarcane cultivation. *Plant Signal. Behav.* 17(1), e2108253.

Verma KK, Song XP, Joshi A, Rajput VD, Singh M, Sharma A, Singh RK, Li DM, Arora J, Minkina T, Li YR. (2022d). Nanofertilizer possibilities for healthy soil, water and food in future: An overview. *Front. Plant Sci.* 13, 865048. https://doi.org/10.3389/fpls.2022.865048

Verma KK, Song XP, Joshi A, Tian DD, Rajput VD, Singh M, Arora J, Minkina TM, Li YR. (2022e). Recent trends in nano-fertilizer for advancing sustainable agriculture under the era of climate change to ensure food security for future generations. *Nanomaterials* 12, 173. https://doi.org/10.3390/nano12010173.

Verma KK, Song XP, Kumari A, Jagadesh M, Singh SK, Bhatt R, Singh M, Seth CS, Li YR. (2024c). Climate change adaptation: Challenges for agricultural sustainability. *Plant, Cell Environ.* 48, 2522–2533 https://doi.org/10.1111/pce.15078

Verma KK, Song XP, Li DM, Singh M, Wu JM, Singh RK, Sharma A, Zhang BQ, LI YR. (2022f). Silicon and soil microorganisms improves rhizospheric soil health with bacterial community, plant growth, performance and yield. *Plant Signal. Behav.* 17(1), e2104004.

Verma KK, Song XP, Lin B, Guo DJ, Singh M, Rajput VD, Singh RK, Singh P, Sharma A, Malviya MK, Chen GL, Li YR. (2022g). Silicon induced drought tolerance in crop plants: Physiological adaptation strategies. *Silicon* 14(6), 2473–2487. https://doi.org/10.1007/s12633-021-01071-x

Verma KK, Song XP, Rajput VD, Boldyreva V, Zhang BQ, Minkina T, Li YR (2022h). Morpho-physiological, biochemical and ultrastructural modifications on sugarcane to prolonged water deficit. In: Verma KK, et al. (Ed.), *Agro-industrial Perspectives on Sugarcane Production under Environmental Stress*. https://doi.org/10.1007/978-981-19-3955-6_11.

Verma KK, Song XP, Rajput VD, Solomon S, Li YR, Rao GP. (2022i). *Agro-industrial Perspectives on Sugarcane Production under Environmental Stress*. Springer Nature Singapore, https://doi.org/10.1007/978-981-19-3955-6

Verma KK, Song XP, Singh M, Tian DD, Rajput VD, Minkina T, Li YR. (2023c). Association of silicon and soil microorganisms induces stress mitigation, increasing plant productivity. In: de Mello Prado R. (Ed.), *Benefits of Silicon in the Nutrition of Plants*; pp. 299–328. https://doi.org/10.1007/978-3-031-26673-7_17

Verma KK, Song XP, Tian DD, Guo DJ, Chen ZL, Zhong CS, Nikpay A, Singh M, Rajput VD, Singh RK, Minkina T, Li YR. (2021a). Influence of silicon on biocontrol strategies to manage biotic stress for crop protection, performance and improvement. *Plants* 10, 2163. https://doi.org/10.3390/plants10102163

Verma KK, Song XP, Tian DD, Singh M, Verma CL, Rajput VD, Singh RK, Sharma A, Singh P, Malviya MK, Li YR. (2021b). Investigation of defensive role of silicon during drought stress induced by irrigation capacity in sugarcane: Physiological and biochemical characteristics. *ACS Omega* 6, 19811–19821.

Verma KK, Song XP, Verma CL, Chen ZL, Rajput VD, Wu KC, Liao F, Chen GL, Li YR. (2021c). Functional relationship between photosynthetic leaf gas exchange in response to silicon application and water stress mitigation in sugarcane. *Biol. Res.* 54, 15. https://doi.org/10.1186/s40659-021-00338-2

Verma KK, Song XP, Verma CL, Malviya MK, Guo DJ, Rajput VD, Sharma A, Wei KJ, Chen GL, Solomon S, Li YR. (2021d). Predication of photosynthetic leaf gas exchange of sugarcane (*Saccharum* spp.) leaves in response to leaf positions to foliar spray of potassium salt of active phosphorus under limited water irrigation. *ACS Omega* 6, 2396–2409.

Verma KK, Song X-P, Xu L, Huang H-R, Liang Q, Seth CS, Li Y-R. (2023d). Nano-microplastic and agro-ecosystems: A mini-review. *Front. Plant Sci.* 14, 1283852. https://doi.org/10.3389/fpls.2023.1283852

Verma KK, Song XP, Yadav G, Degu HD, Parvaiz A, Singh M, Huang HR, Mustafa G, Xu L, Li YR. (2022j). Impact of agroclimatic variables on proteo-genomics in sugarcane (*Saccharum* spp.) plant productivity. *ACS Omega* 7, 22997–23008.

Verma KK, Song XP, Zeng Y, Guo DJ, Sing M, Rajput VD, Malviya MK, Wei KJ, Sharma A, Li DP, Chen GL, Li YR. (2021e). Foliar application of silicon boosts growth, photosynthetic leaf gas exchange, antioxidative response and resistance to limited water irrigation in sugarcane (*Saccharum officinarum* L.). *Plant Physiol. Biochem.* 166, 582–592.

Verma KK, Song XP, Zeng Y, Li DM, Guo DJ, Rajput VD, Chen GL, Barakhov A, Minkina TM, Li YR. (2020c). Characteristics and correlation of leaf stomata and its relationship with photosynthesis on *Saccharum* spp. under different irrigation and silicon application. *ACS Omega* 5, 24145–24153.

Verma KK, Vatsal S, Singh M, Rajput VD, Minkina T. (2022k). Coping with water stress: Effects and adaptive mechanisms to plants. In: *Understanding Abiotic Stresses*. Nova Science Publishers, Inc.: New York.

Verma KK, Wu KC, Verma CL, Li DM, Malviya MK, Singh RK, Singh P, Chen GL, Song XP, Li YR. (2020d). Developing mathematical model for diurnal dynamics on photosynthetic responses in sugarcane responsive to different irrigation and silicon application. *PeerJ* 8, e10154. https://doi.org/10.7717/peerj.10154.

Verma KK, Liu X-H, Wu K-C, Singh RK, Song QQ, Malviya MK, Song X-P, Singh P, Verma CL, Li YR. (2020e). The impact of silicon on photosynthetic and biochemical responses of sugarcane under different soil moisture levels. *Silicon* 12, 1355–1367.

Verma KK., Wu K-C., Singh P., Malviya MK., Singh RK., Song X-P., Li YR. (2019b). The protective role of silicon in sugarcane under water stress: Photosynthesis and antioxidant enzymes. *Biomed. J. Sci. Tech. Res.* 002685. https://doi.org/0.26717/BJSTR.2019.15.002685

Veza I, Said MFM, Djamari DW, Abas MA, Latiff ZA, Perang MRM. (2021). Future Direction of Microalgae Biodiesel in Indonesia. *J. Adv. Res. Appl. Sci. Engineer. Technol.* 25(1). https://doi.org/10.37934/araset.25.1.16

Vilela RS, Fuess LT, Saia FT, Silveira CRM, Oliveira CA, Andrade PA, Langenhoff A, van der Zaan B, Cop F, Gregoracci GB, Damianovic MHRZ. (2021). Biofuel production from sugarcane molasses in thermophilic anaerobic structured-bed reactors. *Renew. Sustain. Energy Rev.* 144, 110974. https://doi.org/10.1016/j.rser.2021.110974

Why ESK, Ong HC, Lee HV, Chen WH, Asikin-Mijan N, Varman M, Loh WJ. (2022). Single-step catalytic deoxygenation of palm feedstocks for the production of sustainable bio-jet fuel. *Energy* 239. https://doi.org/10.1016/j.energy.2021.122017

Wu K-C, Huang C-M, Verma KK, Deng Z-N, Huang H-R, Pang T, Cao H-Q, Luo H-B, Jiang S-L, Xu L. (2022). Transcriptomic responses of *Saccharum spontaneum* roots in response to polyethylene glycol–6000 stimulated drought stress. *Front. Plant Sci.* 13, 992755. https://doi.org/10.3389/fpls.2022.992755

Xu L, Deng Z-N, Wu K-C, Malviya MK, Solanki MK, Verma KK, Pang T, Li Y-J, Liu X-Y, Kashyap BK, Dessoky ES, Wang W-Z and Huang H-R. (2022). Transcriptome analysis reveals a gene expression pattern that contributes to sugarcane bud propagation induced by Indole-3-Butyric *Acid. Front. Plant Sci.* 13, 852886. https://doi.org/10.3389/fpls.2022.852886

Yang Y, Tian Z, Lan Y, Wang S, Chen, H. (2021). An overview of biofuel power generation on policies and finance environment, applied biofuels, device and performance. In: *Journal of Traffic and Transportation Engineering (English Edition)*. https://doi.org/10.1016/j.jtte.2021.07.002

Zeng ZK, Jang JC, Shurson GC, Thakral S, Urriola PE. (2021). Ammonia fiber expansion increases in vitro digestibility and fermentability of corn distillers dried grains with solubles with or without carbohydrases. *Animal Feed Sci. Technol.* 273, 114824 https://doi.org/10.1016/j.anifeedsci.2021.114824

Zewdie DT, Ali AY. (2020). Cultivation of microalgae for biofuel production: Coupling with sugarcane-processing factories. *Energy Sustain. Soc.* 10(1), 27. https://doi.org/10.1186/s13705-020-00262-5

Zhang Y, Kusch-Brandt S, Salter AM, Heaven, S. (2021). Estimating the methane potential of energy crops: An overview on types of data sources and their limitations. *Processes 9*(9), 1565. https://doi.org/10.3390/pr9091565

Zheng F-J, Lin B, Yang Y-X, Fang X-C, Verma KK, Chen G-L. (2023). Efficacy and functionality of sugarcane original vinegar on mice. *Front. Microbiol.* 14, 1224666 https://doi.org/10.3389/fmicb.2023.1224666

Zhu Z, Sun J, Fa Y, Liu X, Lindblad, P. (2022). Enhancing microalgal lipid accumulation for biofuel production. *Front. Microbiol.* 13. https://doi.org/10.3389/fmicb.2022.1024441

CHAPTER 12

Significance of Sugarcane Germplasm: An Overview

S. ALARMELU,* R. M. SHANTHI, S. SHEELAMARY, and H. K. MAHADEVASWAMY

Division of Crop Improvement, ICAR-Sugarcane Breeding Institute, Coimbatore, Tamil Nadu, India

*Corresponding author

ABSTRACT

Sugarcane (*Saccharum* spp.) is cultivated across the globe for its ability to produce sugar and ethanol. Nowadays, farmed sugarcane is a complex combination of interspecific hybrids originating from *Saccharum spontaneum* and *Saccharum officinarum* with additional resources from *Saccharum robustum*, *Saccharumsinense*, *Saccharum barberi*, and allied grasses like *Miscanthus*, *Narenga*, and *Erianthus*. Wild species are the reservoirs which provides large amount of genetic diversity for the improvement of specific character and generate new allelic variability. Sugarcane breeding is widespread but depends upon working with the sugarcane germplasm pool. Base-broadening programs and intergeneric hybridization have played an important role in development of high-yielding cultivars in sugarcane through gene introgression. Germplasm diversity is essential for sustained cane improvement, with large number of accessions collected from separate explorations and conserved in the two global collections as genetic reservoirs including many members of the "*Saccharum* Complex." To sustain genetic progress, advance breeding pools with increased genetic diversity and new germplasm resources are explored by sugarcane breeders. This chapter elaborates natural history of wild *Saccharum* spp. and allied

Sugarcane Transformation: Biofuels, Biotechnology, and Beyond.
Krishan K. Verma, Xiu-Peng Song, Hai-Rong Huang, Aradhna Kumari, and Yang-Rui Li (Eds.)

DOI: 10.1201/9781779641571-12

genera, germplasm resources, their contribution in sugarcane genetic improvement and utilization in the breeding programs.

12.1 INTRODUCTION

Sugarcane (*Saccharum* spp.) is a high yielding C_4 crop cultivated world-wide approximately 26 million hectares of land (Henry and Kole, 2010) in about 95 countries, which accounts for about 80% of global production of sugar, with a total output of 1.86 billion MTs (FAO, 2023; Huang et al., 2024). It is an important food and bioenergy source which contributes to the economy of different nations in the arid and semiarid areas (Verma et al., 2023a, 2023b, 2023c, 2023d). It is best grown in latitude ranging from 35°N to 30°S and altitude from sea level to 1000 m. It also serves as the raw material for the production of white sugar, jaggery, khandsari and meet 80% of sugar demands globally and as well biofuel and other industrial products (Chen et al., 2020, 2023; Zheng et al., 2023). It is also one of the world's most efficient crops which produce lignocellulosic based ethanol to supplement fossil fuels for nearly 40% of the global bioethanol requirement (Lam et al., 2009; Bhatt et al., 2023a, 2023b; Deng et al., 2023).

The crop is also identified as potential candidate feedstock for the production of lignocellulosic biofuel (Kandel et al., 2018; Joshi et al., 2022, 2023; Guo et al., 2023; Li et al., 2023). In India, it is grown in five million hectares of land with a production of about 511 lakh tonnes of yield and plays an important role in Indian agricultural economy. The average yield of sugarcane in India is more than 75,000 kg/ha with the total production of 511 lakh tonnes in 2022–2023. The per capita consumption in Afro-Asian countries is expected to increase up to 11% in the next 40 years (Sugarcane Breeding Institute [SBI] vision 2050). In this respect, ultimately the cultivar will take a leading role to ensure sugar and energy security in the country. Traditionally, sugarcane breeding has contributed regularly enhanced productivity by producing improved genotypes for usage in India as well across sugarcane growing countries. The development of improved varieties with higher yield, sucrose content, disease and pest resistance, improved ratooning ability, and adaptability to environmental stresses through conventional and molecular approaches by utilizing the genetic resources is a challenge for sugarcane-breeding programs due to the rising demand for sugar and biofuel (Song et al., 2019, Wu et al., 2022; Xu et al., 2022).

The population of the world is expected to rise from nearly 8 billion in 2022 to over 10 billion by 2050 (United Nations, 2022). As a result, there will be a continued growth in the demand for energy and sugar worldwide, along with the threats such as decreasing arable land resources, increasing temperatures, and a highly unstable climate (Verma et al., 2021a, 2021b; 2022a, 2021b, 2021c; Oliveira et al., 2022; Jaiswal et al., 2023; Olsson et al., 2023). Economic losses at greater levels are incurred annual due to the attack of pathogenic diseases and new diseases (Song et al., 2021, 2022; Gopi et al., 2024). The theortical maximum yield of 380 t/ha (Waclawovsky et al., 2010) suggests enormous room for improvement. A new paradigm for energy production is also encouraging based on biomass production making thic crop more effective.

Genetic diversity is very important in development of varieties to meet the varietal needs of the country. Genetic enhancement of sugarcane through exploration and exploitation of the genetic diversity present in its genetic resources in the form of cultivars, wild relatives, and related genera (Guo et al., 2021, 2022; Huang et al., 2022; Singh et al., 2023). A huge collection of genetic resources is conserved in the ICAR-SBI Research Centre, Kannur and Sugarcane Research Institute, Miami, *United States Department of Agriculture* (USDA); the two world repositories of sugarcane germplasm of 3377 and 749 (Hale et al., 2022; Chandran et al., 2023). The farmed sugarcane known as the "*Saccharum* Complex" evolved from different species of *Saccharum* and associate generas. These genera and species comprise the fundamental genetic resources of sugarcane (Mukherjee, 1957; Daniels et al., 1975). The cultivated species of *Saccharum* are *S. officinarum* (noble canes), *S. barberi* and *S. sinense*, whereas the wild species are *S. robustum*, *S. spontaneum,* and *S. edule*. Indonesia–New Guinea region is the main diversity *center* for *S. officinarum* and *S. robustum*, with abundant diversity of *S. spontaneum*, *Erianthus*, *Narenga*, and *Sclerostachya* in the northeastern states, India for exploitation. Sugarcane breeding in India has contributed profoundly to sugarcane improvement globally. It has been possible to successfully introduce genes from wild germplasm, such as *S. pontaneum, Narenga porphyrocoma, Erianthus rockii,* and *Erianthus fulvus*, by intra- and interspecific crossing between *Saccharum* spp. and related species. The development of sugarcane cultivars with high levels of biotic and abiotic stress tolerance can greatly benefit from the conserved germplasm.

Breeding has been successful in all of the nations growing sugarcane countries by biparental mating and back cross to the original interspecific hybrids in sugarcane and their derived progenies and nobilization process helped in development of a large number of recombinants. Progress in breeding sugarcane cultivars for optimal yield under adverse agroclimatic conditions has been slow and unreliable. This is due to complex genotype–environment interactions, the large number of genotypes involved, and the lack of trustworthy markers linked to quantitative trait loci that influence stress tolerance. A wealth of genetic diversity is available in this crop, which holds promise for further genetic improvement of yield, quality, and fiber in sugarcane. The development of new cutting-edge technologies like genomic selection, which considers G × E interaction, will facilitate sugaracne development by combined stresses, such as biotic and abiotic, and increased yield potential and better climatic resilience. The genetic diversity observed in living germplasm provides chances for the development of sugarcane cultivars that significantly increase yield in a variety of conditions and can be used to produce a future wide variety of products. Advances in understanding of sugarcane genome and gene expresiion in this complex crop will further assist in genetic improvement. This viewpoint offers a summary of the notable achievements in sugarcane breeding over the past 100 years, as well as an outlook on upcoming challenges and opportunities that will be helpful for breeding initiatives.

12.2 ORIGIN AND DISTRIBUTION OF THE MEMBERS OF SACCHARUM COMPLEX

India is one of the principal centers of origin of the *Saccharum* germplasm complex (*Saccharum, Erianthus, Sclerostachya, Narenga,* and *Miscanthus*) (Mukherjee, 1957), and clones bred in India are ancestral to all commercial sugarcane varieties worldwide. Sugarcane was grown in India and later in China more than 2000 years ago. Indian texts from the 1400–1000 BC period contain the oldest references to sugarcane farming, which dates back to the Vedic era. India is the original home of *Saccharum species* with New Guinea is the origin center of *S. officinarum.*

Sugarcane domestication occurred across a large area of Southeast Asia, including the Malayan Archipelago, New Guinea, India, and some Melanesia island groups. Previous research suggests that the selection of *S. officinarum* from the wild species *S. robustum* is likely what resulted to the domestication of sugarcane in New Guinea (Brandes, 1956; Daniels and Roach, 1987; Grivet et al., 2006). Subsequently the origin of genera other than *Saccharum*, specifically *Erianthus*, *Sclerostachya*, *Narenga*, and *Miscanthus* (Barber, 1920; Jeswiet, 1930; Parthasarathy, 1948; Brandes, 1956; Mukherjee, 1957), were established based on cytological and breeding evidences. These hypotheses were supplemented by molecular data (Grivet et al., 2004, 2006). The original classification of cultivated sugarcane as *S. officinarum* L. by Linnaeus in 1753 and the extensive Kew GrassBase database (http://www.kew.org/data/grasses-db.html) lists 37 species of *Saccharum,* of which only four of which are currently accepted as species of sugarcane.

Saccharum include as many as 30–40 cultivars of plants, although various biological researchers acknowledge six or even less species of domesticated sugarcane and their wild relatives. The wild species of sugarcane are resilient (*S. robustum*), native only to New Guinea island and wild sugarcane (*S. spontaneum*), and native to semiarid regions of tropical Africa, Asia, and Australasia. Domesticated cane cultivars grown for sugar include the original domesticated sugarcane or *S. officinarum* and hybrid domesticated *S. barberi* in India, *S. sinense* in China, and modern sugarcane genotypes. *S. edule*, also native to New Guinea, is grown for edible green (referred as Figi asparagus, *duruka*, or other names). Domesticated sugarcane originated in New Guinea and quickly spread around the globe. Contributions from related grasses of the *Saccharum* complex (Brandes, 1956; Daniels and Roach, 1987; Roach and Daniels, 1987; Sreenivasan et al., 1987) were proposed but with limited molecular evidence available (Irvine 1999; Grivet et al., 2004, 2006). The native distribution, of the species *S. officinarum*, with high sucrose content, believed to have developed from the *S. robustum* in New Guinea (Brandes, 1956, 1958).

It has been postulated by Brandes 1956 and widely accepted by earlier workers (Daniels and Roach, 1987; Roach and Daniels, 1987) that *S. officinarum* spread during prehistoric times from New Guinea to Indonesia, the Malay Peninsula, China, India, Micronesia, and Polynesia, and by 500 AD, from Indonesia to southern Arabiaand east Africa. *S. barberi* and

S. sinense were originated from the *S. officinarum* and *S. spontaneum* between interspecific hybridization (Grivet et al., 2004, 2006) and possible introgression from other species (Brown et al., 2007). The distribution of *S. officinarum* from Polynesia to Hawaii probably took place with native migrations around 500–1000 AD. In the wild, *Saccharum* spontaneum can be found in eastern and northern Africa, the Middle East, India, China, Taiwan, and Malaysia, as well as the Pacific region, which includes New Guinea. The migrations of people to the Pacific islands helped the spread of the plant which spread to Indonesia, the Philippines and northern India. Records of sugarcane in history have been in existence since 510 BC where 'reeds which produce honey without bees' were first indicated by the soldiers of the Emperor Darius near the river Indus. With conquest of India in 327 BC by Alexander the Great, sugar spread westward (Purseglove, 1972). The Indians started pressing and boiling cane to crystallize sugar around 300 AD, and by 540 AD the Persians had learnt the art of sugar making. During 11th century, cane was brought to Europe, which spread rapidly in the 1400s and in 1420 reached the Canary slands, from where it was explained to the New World by Columbus in 1493 and from then sugar entered the world market as an important commodity.

Since prehistoric times, sugarcane native to North India and China, known as *S. barberi* and *S. sinense*, has been farmed. These canes were most likely the source of sugar extraction technology (Daniels and Daniels, 1975). According to Parthasarathy (1946), *S. barberi is* evolved from *S. spontaneum and S officinarum.* However, Grassl (1977) suggested that *Erianthus procerus* and *Sclerostachya* contributed to origin of *Barberi* clones. Grassl (1977) modified the concept of Brandes (1958) that *S. officinarum* introgressed with *S. spontaneum* to develop *S. sinense* and suggested that *Miscanthus sacchariflorus* initially introgressed with *S. officinarum* and that *S. spontaneum* introgression was at a later stage. These two species thought to be in cultivation since prehistory in northern India and China. Each species also comprise the chromosomes homologous to *S. officinarum* and *S. Spontaneum,* and those from the genera *Erianthus* and *Miscanthus*, giving evidence to the complex relationships between members of the *Saccharum* complex (Daniels and Roach, 1987; D'Hont et al., 1996). Although these canes are out of cultivation, there had been continuous interest in understanding its origin as the hypotheis indicates that these two species originated separately from *S. officinarum* which implies the possibility of two or more genetic pathways for sugaracne imrprovement (Daniels and Daniels 1975). Genomic *in situ* hybridization (GISH) studies (D'Hont et al., 1993) indicated the interspecific development of *S. barberi* and *S. sinense* in agreement with origin of these species. Chromosomal variations of *S. barberi* and *S. sinense* were variable with 61% of *S. officinarum* and 39% of *S. spontaneum.* Molecular studies suggest that both North Indian and Chinese sugarcanes represents a variety of horticultural crops than others. *S. barberi* and *S. sinense* canes were out of cultivation, after the introductionof the newly developed hybrid cultivars in 1920s, and conserved in field gene banks.

12.3 TAXONOMY AND CYTOGENETICS

Sugarcane belongs to *Saccharum* genus, tribe Andropogoneae and the subtribe Saccharinae. *Saccharum* spp. belongs to the family Poaceae and it is related to other genera, namely, *Miscanthus, Sclerostachya, Erianthus,* and *Narenga*, and thereby forming the *Saccharum* complex (Mukherjee, 1957). The term "*Saccharum*" complex, originally coined by Mukherjee (1957) was used by geneticists to refer a group of genera within Saccharinae that are directly associated to *Saccharum* and contributed to its genetic basis. The genus *Saccharum* L., which includes sugarcane, was first described by Linnaeus in Species plantarum in 1753. The other tropical grasses like Sorghum and Maize are belong to grass family. The genus has two species: *S. officinarum* and *S. spicalum*. The *Saccharum* complex, which includes the closely related genera *Erianthus* section *Ripidum, Miscanthus* section *Diandra, Narenga*, and *Sclerostachya*, is known to interbreed effectively (Daniels and Roach, 1987). Six species were included in Linnaeus' original classification, which has since been revised (Buzacott, 1965; Daniels and Roach, 1987; D'Hont and Layssac, 1998). *S. officinarum*, also known as the noble cane; *S. spontaneum, S. robustum,* and *S. edule* are considered wild species and *S. sinense* and *S. barberi* are considered ancient hybrids. The two primary categories of cultivated canes are the thick, juicy noble canes *S. officinarum* and the thin, resilient North Indian forms *S. barberi* and *S. sinense*. Of these, two are wild and are known as *S. robustum* Brandes and Jewiet ex Grassl and *S. spontaneum* L.

Traditional sugaracne taxonomy divides genus *Saccharum* into six species, namely, *Saccharum officinarum, S. robustum, S. spontaneum, S.*

sinense, S. barberi, and *S. edule* (Table 12.1). All the species are polyploids with high heterozygosity and morphological diversity. *S. spontaneum* and *S. robustum* are considered as wild type and others as cultivated. *S. officinarum* ($2n$=80) and *S. spontaneum* ($2n$=40–128), have played an important role in sugarcane breeding (Irvine, 1999). Noble canes with x = 10 basic chromosome number contribute alleles for high sugar content and all advance cultivars of sugarcane contains its lineage. *S. spontaneum* is a wild species that is characterized by its high drought, salinity, waterlogging resistance and is being the most utilized wild resource in the initial nobilization stage. It is highly polymorphic with a wide range of chromosome numbers of genus *Saccharum* (Panje and Babu, 1960). Sweet chewing cane (*S. officinarum*; x = 10, $2n$ = 80) is found in native farms in New Guinea and other South Pacific Islands; putative ancestor of *S. officinalrum,* and *S. robustum* (x = 10, $2n$ = 60, 80) are most commonly found in the same region of river banks. Tassels (*S. edule*; $2n$ = 60–80) is produced in the same region and is considered a delicacy; semisweet Indian cane (*S. barberi*; $2n$ = 111–120), semisweet Chinese cane (*S. sinense*; $2n$ = 81–124), and semisweet Chinese cane (*S. spontaneum*; x = 8, $2n$ = 40–128) is a wild cane found throughout Asia.

TABLE 12.1 Summary of Members of *Saccharum* spp. (Buzacott, 1965; Daniels et al., 1987)

Species	Feature	Sugar Level	Number of Chromosomes
S. barbari	Ancient hybrid	Low	$2n$=11–120
S. edule	Wild species	Used as a vegetable	$2n$=60–80 with aneuploid forms
S. officinarum	Noble canes	High	$2n$=80
S. robustum	Wild species	Nil	$2n$=60–200
S. sinense	Ancient hybrid	Low	$2n$=80–140
S. spontaneum	Wild species	Nil	$2n$=40–128

Erianthus, Miscanthus, Narenga, Saccharum, and *Sclerostachya* are among the genera that comprise the *Saccharum* complex (Amalraj and Balasundaram, 2006). An important genus related to *Saccharum* is *E. arundinaceus* which can be utilized in the nobilization process. The clones from this species are known to possess traits related to optimum fiber, biomass and stress resistance efficiency, that is, limited water supply and waterlogging. Hodkinson et al. (2002) reported on a significant molecular phylogenetic study of the group, assessed the intercorrelation of genera within the *Saccharum* complex using DNA sequences and indicated that *Saccharum* and *Miscanthus* are significantly related to each other than to other genera. Based on the examination of restriction fragment length polymorphisms (RFLP) markers, Lu et al. (1994) postulated a hybrid genesis for *S. barberi* and *S. sinense* from natural hybridization between *S. spontaneum* and *S. officinarum* (Irvine, 1999). Through the distribution of species-specific repeated sequences and RFLP data, D'Hont et al. (2002) also showed that no other genus has been closely linked to the origin of *S. barberi* and *S. sinense*, which are both derived through interspecific hybridization between *S. spontaneum* and *S. officinarum.*

Numerous studies employing traditional cytology have been conducted on sugarcane (Sreenivasan et al., 1987). These investigations have provided valuable insights into the genome's complexity and have been essential in interpreting the *Saccharum* genus's taxonomy. The sugarcane genome is a complex, polyploid and highly repetitive. The basic genome is $x = 10$ with high ploidy levels. The nuclear genome of modern sugarcane cultivars (allopolyploid hybrids) are composed of subgenomes and are derived mainly from the hybridization of two species, *S. officinarum* and *S. spontaneum* (Daniels and Roach, 1987) with $2n+n$ constitution (Bremer, 1961; Piperidis et al., 2010). The uneven inheritance of both parents during hybridization makes the genome more complex (D'Hont et al., 1996). *S. officinarum* genome is duplicated along with a copy of the *S. spontaneum* genome (Cesnik and Miocque, 2004). This phenomenon is known as nobilization, which made easier the work of breeders, because nobilization increases the ratio of *S. officinarum* to *S. spontaneum* genome. *S. officinarum* is an octoploid ($2n=80$) (D'Hont et al., 1995; Cesnik and Miocque, 2004; Nobrega and Dornelas, 2006) with thick canes of high sugar (Daniels and Roach, 1987; D'Hont and Glaszmann, 2001). *S. spontaneum* is thin stalked with low sugar and has $x = 8$ (Irvine, 1999) with variation in chromosome numbers (D'Hont et al., 1996), the five main cytotypes being $2n=62$, 80, 96, 112, or 128 (Daniels and Roach, 1987; Sreenivasan et al., 1987). Total size of the sugarcane genome is about 10 Gb (D'Hont and Glaszmann, 2001; Le Cunff et al., 2008); 10 times larger than *Sorghum* and nearly 20-folds more than the genome of rice (Aitken et al., 2016). Total genome size of *S. officinarum* is approximately about 7.88 Gb with eight homologous copies of each gene and hence the genome is autopolyploid (Zhang et al., 2012). The basic monoploid genome size of *S. spontaneum* is 750–843 Mb and total genome size range from 3.36 to

12.64 Gb (Panje and Babu, 1960; Daniels and Roach, 1987; Sreenivasan et al., 1987; da Silva et al., 1995; D'Hont et al., 1998; Zhang et al., 2012). The autopolyploidy levels of *S. spontaneum* in hybrids are reported in the range between 5*x* and 16*x* and the most frequently observed ploidy level is 8*x* (Irvine, 1999).

The random sorting of the chromosomes from parents during nobilization in sugarcane hybrids have resulted in difference for the total number of chromosomes in sugarcane genome (Heinz, 1987; Piperidis et al., 2010; Aitken et al., 2016). Between 100 and 130 chromosomes have been found in sugarcane scrop (Simmonds, 1976; Sreenivasan et al., 1987) with 8–14 homologous copies of a gene (Heinz, 1987; Aitken et al., 2007, 2016; Souza et al., 2011). Hence the hybrid genome is considered the mixture of aneuploid and uneven homologous chromosomes (Grivet et al., 1996). The authors also demonstrated that *S. officinarum* as female parent contributes about 80% of the chromosomes and *S. spontaneum* with 10–20% contribution to the genome of the hybrids. Piperidis et al. (2001), D'Hont and Glaszmann (2001), and D'Hont (2005) reported 5% to 17% of recombinant chromosomes in hybrids through *in situ* hybridization studies. Since sugarcane hybrids get chromosomes from autopolyploid progenitors, the hybrid genome has high repetitive content (Figueira et al., 2012), which makes the understanding of genome functions and construction of sugarcane reference genome (Souza et al., 201; Figueira et al., 2012). The interaction of different *Saccharum* species like *S. robustum, S. barberi*, and *S. Sinense* and their contribution toward hybridity also remains unclear, because of high and variable ploidy levels. Earlier workers, namely, Bremer (1961), Price (1965, 1968) conducted chromosomal studies in this complex autopolyploid crop had observed many cytogenetic complexities like high chromosome number, variability in chromosome number, en block elimination and female restitution in interspecific crosses. Sreenivasan et al. (1987) identified that the commercial cultivars chromosome number in the range between 100 and 130. Bremer (1961, 1962) reported that $2n+n$ transmission continues only up to the second back cross. Unreduced egg cells, endoduplication, postmeiotic fusion, endomitosis, differential survival of gametes, selective zygotic failure were identified as the most possible reasons for $2n+n$ transmission as suggested by Bhatt and Gill (1985) and Sreenivasan et al. (1987). Piperidis et al. (2010) also reported that $2n + n$ transmission of chromosomes in *S. officinarum* × cultivar crosses using GISH technique in second back cross. In the mid-1990s, it was shown that distinguishing the chromosome contributions made by *S. officinarum* and *S. spontaneum* was possible using GISH (D'Hont et al., 1995, 1996), about 10% of chromosomes were identified as having been contributed by *S. spontaneum* and about 80% from *S. officinarum*. The remaining 10% are recombinants of both species (D'Hont et al., 1995). While these figures indicate that recombination of chromosomes does occur in contrast to previous assumptions to the contrary (Price, 1963), the fact that 10% of the *S. spontaneum* chromosomes did not recombine indicates that recombination is not a frequent event (D'Hont et al., 1996).

12.4 THE GENOME STRUCTURE OF CUTTING-EDGE SUGARCANE CULTIVARS

It is widely accepted and has been established for a considerable amount of time that modern sugarcane cultivars developed interspecifically (Arceneaux, 1965; Price, 1965). Current cultivars of sugarcane are the result of hybridizations between the wild *S. spontaneum* and the domesticated *S. officinarum* species that occurred a century ago. Until sugarcane researchers utilized molecular cytogenetic techniques, it was unknown how much of the chromosomes from the wild species, *S. spontaneum*, remained after nobilization and contributed to present cultivars. Applying GISH to the French cultivar R570, D'Hont et al. (1996) contradicted the previous theory that no exchange between these two species occurs during meiosis by demonstrating the initial appearance of chromosomal exchange between the *S. officinarum* and *S. spontaneum* genomes (Price, 1963, 1965; Berding and Roach, 1987). They were also able to distinguish the chromosomes from the two species. It was demonstrated that R570 included around $2n = 115$ chromosomes, of which 80% were obtained from *S. officinarum*, 10% from *S. spontaneum*, and 10% from interspecific exchange between the two species. Since then, other scientists have examined the chromosomal content of contemporary cultivars or breeding clones using GISH that are significant for their breeding initiatives. In their analysis of seven contemporary cultivars, Piperidis and D'Hont (2001) found that the percentage of *S. spontaneum* chromosomes ranged from 10% to 20%, whereas the percentage of mixed chromosomes was between 5% and 17%. In their investigation, Cuadrado et al. (2004) found that less than 5% of the chromosomes investigated to be recombined, and

16% of the chromosomes were from *S. spontaneum* chromosomes. The success of prebreeding in sugarcane is due to differential chromosome transmission pattern with $2n+n$, $n+n$, $n+2n$, $2n+2n$ transmission with aneuploidy and it diversified the genetic base in the prebreeding stocks. The effective utilization of these prebreeding stock in backcrosses will produce new commercial varieties with enlarged genetic base and resistance to environmental stress condition.

12.5 GERMPLASM AND GENE POOL

Advance sugarcane cultivars are originated from the interspecific hybridization of five *Saccharum* species except for *S. edule*. *S. officinarum* ($2n = 80$) and *S. spontaneum* ($2n = 40$–128) have played an important role in sugarcane breeding. Noble cane (*S. officinarum*) is the primary source of alleles that regulate high sugar content; practically all current cultivars of sugarcane originate from this species. Though all members of *Saccharum* complex are crossable, so far, the major breakthrough in sugarcane improvement in sugarcane was achieved only through the utilization of *S. spontaneum*, which possess higher levels of disease resistance, ratoonability, adaptability to stress conditions Introgression of genome from *S. robustum* toward identification of modern cultivars is always challenging due to occurrence of low fertility in hybrids.

12.5.1 SACCHARUM OFFICINARUM (SUGAR CANE, NOBLE CANE, CULTIVATED SUGARCANE)

S. officinarum, being high in sucrose content, has been primarily used for sugar production (Verma et al., 2020a, 2020b; Anas et al., 2021a, 2021b). It is characterized by being moderately tall, of various colors, thick canes with low fiber. Over 400 clones of *S. officinarum* have been recorded. It is generally characterized by having a chromosome number of $2n=80$, with a basic chromosome number of $x=10$ (D'Hont et al., 1998). It is thought to be derived from $2n=80$ forms of the wild species *S. robustum* in New Guinea, the center of origin (Brandes, 1958). All the present-day cultivars are interspecific hybrids, of *S. officinarum* crossed with *S. spontaneum* and backcrossed twice with *S. officinarum* as the recurrentparent. *S. officinarum* transmitted its somatic chromosome number during the first hybridization andthe first backcross contrarily to the *S. spontaneum* clones.

The chromosome number after the second backcross became normal (Bremer, 1961). This process, referred to as "nobilization," has resulted in modern varieties with genomes that comprise multiple sets of homologous chromosomes derivedfrom a single species (autopolyploid); having two or more unlike groups of chromosomes (allopolyploid) (Sreenivasan et al., 1987); and having a high total chromosome number ($2n$=100–130) with 80% of the genome derived from *S. officinarum* and the rest from *S. spontaneum*.

12.5.2 SACCHARUM SPONTANEUM L. (WILD CANE)

It is a wild relative of the high-sugar producing *S. officinarum*, has played a vital role in the proper development of advance sugarcane cultivars. It is a complex polyploid with a chromosome number varying from $2n$=40–128. Its basic chromosome number was initially suggested as being either 8 or 10 (Sreenivasan et al., 1987) but has been validated by GISH as x=8 (D'Hont et al., 1998). Approximately 80% of the genome of modern sugarcane cultivars is believed to be contributed from *S. officinarum* (D'Hont et al., 1996; Piperidis et al., 2010), with the remainder from S. *spontaneum* which has led to the idea that cultivated sugarcane lacks diversity and has only captured a small proportion of the *S. spontaneum* genome. *S. spontaneum* has thin canes, highly polymorphic and high in fiber and produces little sugar. The appearance of *Saccharum spontaneum* plants vary from short to large and from bushy types with no stalk to stemmed types more than 5 m in height. It is highly adaptable to cold, drought, and other extreme climatic conditions; is highly vigorous and has resistance to several diseases. Apart from the considerable variation that is found in *S. spontaneum*, there has been evidence that at some locations, introgression with other genera has occurred (Aitken et al., 2018). Only a small number of original progenitor clones, particularly *S. spontaneum*, are believed to have contributed to the ancestry of sugarcane. The proportion of diversity within *S. spontaneum* captured within sugarcane-breeding programs is not clear.

12.5.3 SACCHARUM ROBUSTUM

This species was discovered in New Guinea and is thought to be the intermediate form in the evolutionary pathway from *S. spontaneum* to *S. officinarum*. Brandes and Jeswiet during 1928 on expedition to New

Guinea, collected 154 clones (Daniels and Daniels, 1975). Its stems are large in diameter, tall (up to 10 m) and vigorous but, like *S. officinarum,* is susceptible to similar diseases, such as sugarcane mosaic virus. Due to its growth form, *S. robustum* is often used as either wind breaks or fences in New Guinea. A variety of DNA sequence categories that correspond to distinct genomic compartments have been employed in the comparison of wild *Saccharum* accessions and noble cultivars. According to D'Hont et al. (1993), there are several lines of molecular evidence that indicate Noble clones originated directly from the wild species S. robustum.

12.5.4 S. BARBERI JESW. (2N–111–120)

These are called as North Indian canes, which are thin with short cylindrical internodes, grown in North India. Classified it into five groups (Mungo, Nargori, Saretha Sunnabile, and Pansahi) by Barber (1922).

S. sinense Roxb. ($2n$=81–124): Called as Chinese canes, are tall with bobbin-shaped internodes and broader leaves, in similarity to Pansahi group. Each species also present chromosomes homologous to *S. officinarum* and *S. spontaneum*, and those from the genera *Erianthus* and *Miscanthus*, giving evidence to the complex relationships between members of the *Saccharum* complex (Daniels and Roach, 1987; D'Hont et al., 1996).

Both *S. sinense* and *S. barberi* are characterized by thin to medium stalk, low sucrose, high fiber and tolerance to stress conditions. They are of limited fertility and their use in breeding is restricted. These two cultivated sugarcanes were probably the result of natural hybrids of *S. officinarum* and *S. spontaneum* with other genera about 1000 BCE. *S. barberi* subsequently spread from India to the Middle East, the Mediterranean, and to the New World beginning with the second voyage of Columbus (1493). *S. sinense* and *S. barberi* exist as collections (Heinz et al., 1994). According to genetic data, North Indian and Chinese cultivars are the outcome of interspecific hybridizations between members of the two *Saccharum* species: wild and domesticated clones at one end of the hybridization complex, and representatives of *S. spontaneum* at the other (Lu et al., 1994; D'Hont et al., 2002).

12.5.5 *SACCHARUM EDULE*

Cultivated in New Guinea and the Fiji Islands, *S. edule* ($2n = 60, 70, 80$) is an ornamental plant that did not contribute to the development modern cultivars. It is similar in appearance to *S. robustum* except that its inflorescence is compacted and it is cultivated as a vegetable on the islands of the Pacific and New Guinea. It is thought to be an intergeneric hybrid of *Miscanthus floridus* and *S. robustum* or *S. officinarum*.

12.6 GERMPLASM COLLECTION AND CONSERVATION

The SBI, located in Coimbatore, India (maintained at Cannanore), is the world's largest collection of sugarcane germplasm. USDA-Agricultural Research Service (ARS), Canal Point, USA (Collection at Coral Gables outside Miami, Florida), National Nursery for Sugarcane Germplasm Resources at Sugarcane Research Institute, Yunnan Academy of Agriculture Sciences, Yunnan, China and 11 secondary germplasm collections in various countries are being maintained as sugarcane genetic resources. High-altitude forms of *Erianthus fulvus* and *Miscanthus nepalensis* are being maintained at IARI Regional Research Station, Wellington, Nilgiris, India, Among the *Saccharum* spp., *S. officinarum* is the original sugarcane species which has its origin in the Indonesian Archipelago and had been conserved for a long time by the natives of these islands which then spread in other parts of the globe. The species, *S. barberi* and *S. sinense* were believed to be hybrid derivatives involving *S. officinarum* and wild species. *S. barberi* was grown in Northern parts of India and *S. sinense* in China. These two species were replaced with the introduction of new varieties.

Later in the 1900s, efforts were made to collect *Saccharum* germplasm from the regions of diversity. The main areas of diversity, the Indonesia–New Guinea region and the Indian subcontinent, were the focus of the collection activities. The area between Indonesia and New Guinea has been suggested to be the origin of farmed sugarcane, or *S. officinarum*. Beginning in 1892, there was a great deal of interest in gathering *Saccharum* germplasm from the New Guinea region in order to introduce cultivars. Later research, carried out after 1921, intended to obtain cultivated and wild-type *Saccharum* species. New Guinea was the location of collections in the years 1892–1895, 1921–1928, 1951–1957, and 1977. Between 1976 and, 1984, research on collecting *Saccharum* germplasm was conducted in Indonesia. Additionally, attempts were carried out to gather germplasm from the Philippines, Taiwan, and Thailand. Over the years, germplasm associated with *Saccharum* has also been collected in China and Japan.

12.7 SACCHARUM GERMPLASM COLLECTION IN INDIA

Saccharum and related genera *S. spontaneum* are considered to be extremely diverse in India, where they are found almost everywhere from the Southern Peninsula to the sub-Himalayan areas in a variety of eco-geographic environments. The species is highly polymorphic and show wide variation in morphology and chromosome number. More than 20 different cytotypes of *S. spontaneum* have been identified in India, including the $2n$=40 form with the lowest chromosome. The most extensive distribution of *Saccharum* and related genera is found in Northeast India. Attempts to collect *Saccharum* genetic resources began after the establishment of Breeding Station of Sugarcane, Coimbatore (1912), India. Dr. Barber collected 112 North Indian sugarcane clones from the subtropical regions of the country during 1912 for evaluation and selective introduction. The Spontaneum Expedition Scheme (SES) was launched in 1948 by the SBI to acquire *Saccharum* germplasm from India and other Asian countries. In a span of 10 years from 1948 to 1957, more than 500 clones were obtained from the different zones in this program. It renewed in, 1981–1990, 1999–2004 to collect *Saccharum* germplasm from the northeast India and a total of 535 clones were collected from the various states of the country. In order to develop sugarcane and assist breeders, a broad and diversified germplasm collection is necessary. These clones are being maintained for and kept at ICAR-SBI in Coimbatore. These materials serve as fresh sources for particular requirements, such as those related to quality, yield, and resilience to the era of climate change in order to increase the genetic foundation for the sugarcane.

A total of 38 expeditions were made by Indian scientists, and a wealth of 2400 wild accessions belonging to different species and genera of *Saccharum* complex have been collected and maintained at Coimbatore. There are two major World collections of *S. spontaneum,* the World Collection of Sugarcane and Related Grasses (World Collection) situated at the USDA-ARS Subtropical Horticulture Research Station, Miami, FL is one of two such germplasm collections in the world. The other is at the India. The part of the collections is maintained at ICAR-SBI, Research Centre at Kannur (Kerala, India) and another part mostly of wild collections is maintained at ICAR-SBI, Coimbatore. At ICAR-SBI, Coimbatore is maintaining world largest *ex situ* germplasm collection to conserve the individual accessions to maintain the genetic integrity. A total of 2256 wild germplasm clones comprising of *S. spontaneum, E. arundinaceus, Erianthus* spp., allied genera, and other *Saccharum* clones are being maintained at the field gene bank at SBI, Coimbatore and are clonally maintained by replanting every year. The high-altitude collections from Arunachal Pradesh; 47 *S. spontaneum*, six *E. procerus,* one *E. fulvus* and two *Miscanthus* spp. have been maintained at ICAR – IARI, Regional Station, Wellington, The Nilgiris. In addition to these collections, a number of breeding programs have their own smaller collections (Hemaprabha et al., 2022). These materials are the new sources for specific needs especially for yield, quality, and for biotic and abiotic stress resistance so as to broaden the genetic base of the sugarcane.

The world collection of germplasm maintained at ICAR—SBI comprises the following:

Site	Species/Cultivar	Indian Accessions	ISSCT Accessions
SBI, Coimbatore	*S. spontaneum*	878	–
	Erianthus spp.	313	–
	Allied genera	41	–
SBI, Kannur	*S. officinarum*	–	757
	S. robustum	–	145
	S. barberi	–	42
	S. sinense	–	30
	S. spontaneum	305	79
	Erianthus spp.	-	132
	Allied genera	88	20
SBI, Agali	*S. officinarum*	130	-
IARI, Wellington	*S. spontaneum*	47	-
	Erianthus rufipilus	5	-
	Miscanthus nepalensis	3	-

Source: ICAR-SBI, website.

12.8 CHARACTERIZATION OF AGRI-MORPHOLOGICAL ASPECTS

Multiple diversity evaluation research was carried out in the *Saccharum spontaneum* sets and a large range of variants for important quantitative traits. Xu et al. (2020) Phenotype *Saccharum spontaneum* L. in China, *Saccharum spontaneum* L. from the World Collection of Sugarcane were described by Tai and Miller (2001) as representing the great diversity that can improve preservation studies and take use of the potential value for breeding. In order to determine the degree of variability and to have a reference database for preserving their identity, 304 collections of *Saccharum* spontaneum from four northeastern Indian states that were kept in the field gene bank at SBI-RC, Kannur, were characterized using 21 qualitative and 12 quantitative descriptors. According to Manjunatha et al. (2018), the accessions shown significant variation in both quantitative and qualitative characteristics. Phenotypic measurements from *S. spontaneum* and *S. officinarum* groups showing that *S. officinarum* had a small number of large-diameter stalks with wide leaves, high Brix, and high water content, whereas *S. spontaneum* clones had a high stalk population with small diameter, narrow leaves, and low Brix. By employing the *Saccharum* spontaneum species, scientists are investigating cutting-edge technologies, such as marker-assisted breeding and genomic tools, to accelerate the generation of sugarcane variants with desirable traits. Extensive morphological variations in the *S. spontaneum* collection were reported by many sugarcane breeders.

12.9 CLASSICAL BREEDING ACHIEVEMENTS AND UTILIZATION OF SUGARCANE GERMPLASM

Sugarcane (*S. officinarum*) was grown in India, China, and Papua New Guinea for the production of sugar over centuries. Sugarcane breeding was situated in Java and Barbados in the late 1800s after Softwedel discovered that sugarcane could produce viable seeds. Sugarcane breeding started with identifying and intercrossing between *S. officinarum* cultivars, which resulted in progenies but were susceptible to both biotic and abiotic stresses. This was followed by the interspecific hybridization between *S. officinarum* and *S. spontaneum*, and regular backcrossing possibilities with *S. officinarum* cultivars. These interspecific hybrids and the notable hybrid Proefstation Oost Java (POJ) 2878 possessed high cane yield, disease resistance, and improved ratooning ability which revolutionized the sugar industry in Java.

Most extensive work in noble cane breeding was carried out in Barbados during 1887–1928 and a number of important varieties were developed from the noble cane breeding. After the work of Moquette and Walker in 1893, crosses between *S. officinarum* and *S. barberi* (clone-chunnee) also were attempted by Kobus in 1897. Subsequently these hybrids were back crossed and POJ 2878, one of the outstanding hybrids was produced by Jesweit in 1921. Co 205, the first interspecific hybrid was produced in 1912 in India through nobilization program. PoJ 2878 and Co 205 are the origins of several commercial cultivars used today all over the world. As a result, improved sugarcane cultivars have been essential to the growth of the global sugar industry.

12.10 CULTIVATED CANES (NOBLE AND NORTH INDIAN) PRIOR TO NOBILIZATION

During the late 18th century, navigators such as Bougainville or Cook contributed to spread of the widely cultivated Otaheite cane variety around the world for until the years 1840 to 1890 for sugar production. Before sugarcane-breeding programs were started in 1888, the most important noble cultivars were the Otaheite (Lousier in Mauritius, Bourbon in West Indes, Lahaina in Java and Hawaii and in India as Vellai), Cheribon (Louisiana Purple) of Java, and "Caledonia" of New Hebrides (Stevenson, 1965) which were broadly distributed in established cane agro-industries globally. After Otaheite, Cheribon became the major commercial cultivar around the globe in the late nineteenth and early twentieth centuries (James, 2004) until the identification of Java variety POJ, 2878. Tanna, known as Caledonian cane, was another important noble clone cultivated in Hawaii from 1881 to 1925 under unirrigated conditions and was also popular in Fiji.

The native sugarcanes, *S. barberi* and *S. sinense*, were significant in subtropical regions and grown in India and China, respectively. Significant Chinese clones (*S. sinense*) included Uba, Cayania, Oshima, and Zwinga; Indian clones (*S. barberi*) included Katha, Hemja, Pathri, Chin, Saretha, Laldi, Dhaul, Mungo, Matna, Nargori, Reora, Sarauti, and Baraukh (Hemaprabha et al., 2020). Canes of *S. barberi* were only found in India, while canes of *S. sinense* were widespread in China, Burma, and India. These two groups also achieved commercial importance in South Africa, Brazil, and the USA. Uba was the most commonly grown *S. sinense* in many parts of the World including Brazil, Louisiana, Mauritius, Puerto Rico, South Africa, and India due to its mosaic resistance and good ratoonability with up to 32 ratoon crops reported in South Africa (Zhou, 2013)

12.11 NOBILIZATION BREEDING

In order to effectively use sugarcane as a feedstock for the production of both sugar and fiber, varieties must be developed with genetic variety in consideration. Due to interspecific crossings between sugar-rich *S. officinarum* and *S. spontaneum*, which have vigor, disease resistance, and other agronomic qualities, the majority of present sugarcane cultivars are complex polyploids ($2n = 4x$ to $12x$, totaling 100–128 chromosomes) (Piperidis et al., 2010). One of the historical milestones in plant breeding era has been the successful utilization of *Saccharum spontaneum* in the development of cultivars. The origin of modern sugarcane cultivars was mainly from interspecific hybridizations of *S. officinarum* and *S. spontaneum* with repeated backcross to *S. officinarum*, as a recurring female parent to reset optimum sucrose level (Price, 1963).

The first sugarcane-breeding programs began in Java and Barbados in 1888. In 1893, Moquette and Walker crossed the *S. officinarum* clone Black Cheribon with Kassover, the latter being a natural hybrid between *S. officinarum* × *S. spontaneum*. Crosses between *S. officinarun and S. barberi* (Clone Chunnee) also were attempted by Kobus in 1897. Subsequently these hybrids were back crossed to reduce the undesirable attributes of the exotic species. Wilbrink and Jeswiet pursued the work and POJ 2878, one of the outstanding hybrids was produced by Jeswiet in 1921 which became the most successful variety in Java in 1929, occupying 90% of the cane area. Nobilization breeding in Java resulted in the production of the nobilized cultivar "POJ2878" in 1921 which became the most universal breeding cane ever developed, present in the genealogy of almost all of the dominant cane varieties grown around the world. Varieties produced by the POJ was the basic germplasm for varietal development in other countries. Hybridization of POJ 2878 with other nobilized POJ canes produced cultivars "POJ3016" and "POJ3067" which were popular varieties in Java, H32-8560 in Hawaai and Co 419 for the tropical area of India. These cultivars from the collection at Java dominated global commercial production which were subsequently replaced by new hybrids selected from sugarcane-breeding programs.

Early sugarcane breeding efforts for producing varieties with wide adaptation was at Coimbatore, India (1912), and several Co and NCo varieties were identified. In India in 1912, Dr C.A. Barber and Sir T.S. Venkataraman made a cross between the *S. officinarum* clone Vellai and a wild *S. spontaneum* clone (Coimbatore) and selected the clone Co 205 (Nair, 2008). Co 205 became the most popular variety in Punjab Province and revolutionized sugarcane across the globe. Co derivatives like Co 213, Co 281, Co 290, etc., were released and became popular commercial varieties in many countries.

The genotypes utilized in earlier breeding programs were limited number to few flowering *S. spontaneum* and *S. officinarum* clones, which was due to narrow genetic base of sugarcane cultivars (Lima et al., 2002). Promising noble cultivars were found susceptible to disease and insects and limited to particular tropical environments. Breeders soon realized that the genetic base of the noble canes need to be broadened and to use diverse germplasm for commercial sugarcane production and for wider adaptabilities and disease and insect resistance (Stevenson, 1965). Hence sugarcane breeders around the world started exploiting basic germplasm in breeding programs since 1992 till date (Miller and Tai, 1992).

The uneven inheritance of both parents during hybridization makes the genome more complex (D'Hont et al., 1996). The random sorting of the chromosomes from parents during nobilization in sugarcane hybrids have resulted in difference for the total number of chromosomes in sugarcane genome (Heinz, 1987; Piperidis et al., 2010; Aitken et al., 2016). The authors also demonstrated that *S. officinarum* as female parent contributes about 80% of the chromosomes and *S. spontaneum* with 10–20% contribution to the genome of the hybrids. Piperidis et al. (2001), D'Hont and Glaszmann (2001) and D'Hont (2005) reported 5% to 17% of recombinant chromosomes in hybrids through *in situ* hybridization studies. The interaction of different *Saccharum* species like *S. robustum, S. barberi* and *S. Sinense* and their contribution toward hybridity also remains unclear, because of high and variable ploidy levels.

The nobilization process involves chromosome nonreduction plus introgression of additional genes through a system of crossing the noble canes with wild clones of *S. spontaneum. S. officinarum*, in intraspecific crosses shows normal $n+n$ inheritance but when crossed with *S. spontaneum* it has the unusual pattern of transmission ($2n+n$) of chromosome number (Bremer, 1925; Price, 1961; Roach, 1969). Hemaprabha et al. (2017) indicated that *S. officinarum* clones Black Cheribon, Bandjermasin Hitam, Loethers, Lahiana and Fidji were exploited while Kaludai Boothan, Vellai, Ashy Mauritius, Green Sport, Zwart Cheribon, Striped Mauritius, Korpi, and White Transparent were less represented. Black Cheribon was common in the pedigree of all CP varieties and 1–6 clones of *S. officinarum*, and two clones each of *S. barberi* and *S. spontaneum* were present in varying frequencies. In contrast, the common parents of early generation Co canes, namely, Co 285, Co 312. Co 313, Co 419, and Co 449, had 1–7 clones of *S. officinarum*, 1–2 clones of *S. barberi* and two clones of *S. spontaneum* occurring 2–15 times which demonstrates the diversity of *S. officinarum* in the successful parents for quality enhancement. The study also suggested that trends in genealogy could provide guidance in parental choice for achieving genetic gain, particularly for sugar content.

12.12 INTRASPECIFIC IMPROVEMENT

Attempts to improve sugarcane varieties through breeding were started soon after the discovery of seed set in sugarcane by Soltwedel in 1888 in Java. Initial attempts were to improve the noble canes by intercrossing them. Most extensive work in noble cane breeding was carried out in Barbados between 1887 and 1928. Buke, B147, B156, and B208 in Barbados, POJ 100 in Java, Q813 and HQ 409 in Australia, Diamond 10, D109, D145, and D625 in Demerara, and H109 in Hawaii were among the significant varieties that resulted from the noble cane breeding. Improving the species before using them for interspecific hybridization was considered to be a possible strategy for producing better interspecific hybrids.

By sequential hybridization and selection within each species, the population upgradation of *S. officinarum* and *S. spontaneum* for the development of widespread parental lines was started in Barbados in the 1960s. Characters like high sugar content and cane thickness and low fiber content in *S. officinarum* and high adaptability in *S. spontaneum* formed the basis of selection of specific clones to be used as parents. Walker (1987) reported significant gains after three cycles of intercrossing and selection to develop a range of improved clones. Walker (1982, 1984) reported considerable progress from this program of generation wise improvement of *Saccharum* species through polycrossing and selection under specified environments.

SBI was the basis for the population improvement of *S. officinarum* and *S. spontaneum* in order to generate broad-based parental lines. Since these clones were obtained from broad latitudes and atmospheric situations, *S. spontaneum* was selected based on adaptability, whereas nobles were particularly selected for their high sugar content, low fiber content, and thick stalks. Intraspecific betterment of *S. officinarum. S. robustum*, and *S. spontaneum* at SBI since, 1988 Nair et al. (2002) had resulted in populations with improved sucrose content. With two cycles of intra-specific polycrossing and selection significant characteristics is essential features could be obtained. Improved S*officinarum* populations with more than 50% betterment in the production of cane than the base population with marginal improvement for sucrose were obtained. Intrapsecific populations of *S. robustum* with over 11% juice sucrose could be produced, which is about 4–5 units higher than the average sucrose levels in natural populations of *S. robustum*. Some of the selected clones recorded as high as 14% sucrose, which so far has not been recorded in the species. These improved intraspecific hybrids were used in interspecifie hybridization and the resultant hybrids were found to have better yield and quality potential.

12.13 INTERSPECIFIC IMPROVEMENT

Interspecific hybridization has contributed significantly to the varietal improvement progams of many countries across the globe. Among the members of *Saccharum* complex; *S. officinarum*, only few genotypes of *S. spontaneum* (wild species with the widest distribution) (Tai and Miller, 2002), one natural hybrid of *S. spontaneum* and *S. officinarum* and two *S. sinense* have been used as parents for the development of modern sugarcane cultivars (Berding and Roach, 1987). *S. spontaneum* has the most effective as a source of genetic variability for stress tolerance along with the scope to improve biomass production (Aitken and McNeil, 2010). Initially *S. robustum* was extensively utilized in the cane-breeding programs in Hawaii (Heinz, 1967). Interspecific hybridization gained momentum in Barbados during the early 1960s. Fresh crosses involving *S. officinaruum* and *S. spontaneum* were attempted during this period. There were also attempts toward intraspecific improvement of these two species prior to interspecific hybridization. It was assumed that the improved species clones produced through repeated cycles of intraspecific hybridization and selection for specific attributes will be a better genetic source than the original species clones. The interspecific hybrids were screened for drought and waterlogging apart from other agronomic characters like tonnage, sugar content, and low fiber. Selected bybrids are suitably integrated into the commercial breeding programs. Emphasis is also given for the exploitation of *Erianthus arundinaceus* and the hybrids are reported to possess drought tolerance (Berding and Roach, 1987). In Australia, an example of successful introgression breeding since the 1960s was the development of cultivars from the *S. spontaneum* clone "Mandalay," collected from Burma to develop a series of widely cultivars and parental material for utilization in Australian breeding programs. In Louisiana, USA, the use of the *S. spontaneum* clone US56-15-8, collected from northern Thailand, led to a range of important cultivars and parental germplasm, including the major cultivar LCP85-384 (Milligan et al., 1994; Arro et al., 2006).

The first interspecific hybrids of *Sacchurum* in India were produced by Barber in 1912 from a cross involving *S. officinarum* (clone Vellai) with the local *S. spontaneum*. From this cross, Co 205, the first commercial hybrid of India was released in 1918 which became a prominent variety in the Punjab Province. *S. officinarum* and *S. barberi* crosses were also attempted and trispecific bybrids involving *S. officinarum, S. spontaneum*, and *S. barberi* were developed. These hybrids and their derivatives were the early commercial varieties of tropical and subtropical India.

Remarkable success of the early interspecific hybrids and their derivatives as commercial cultivars, sustained continued interest in the utilization of wild germplasm in the sugarcane-breeding programs in India. A large collection of wild germplasm collected under the Spontaneum Expedition Scheme provided further thrust to the exploitation of wild germplasm. The PL 480 Schemes at SBI, during the 1960s were aimed at transferring the disease resistance and other desirable traits of Indian *S. spontaneum* to foreign commercial varieties from Canal Point, Clewiston, Hawaii, and Puerto Rico. The hybrids were screened for resistance to mosaic, rust, red rot, smut, RSD, cold, drought, and agronomic potential and nearly 140 hybrids were selected as IA (Indo American) Clones and successfully used in the breeding programs of the Institute to develop commercial varieties.

S. robustum species clones were used to produce F1, backcross hybrids of three generations and F1 intercrossed hybrids, and these were high yielding and sugar rich (Kandasamy and Thiraviam, 1987). *S. robustum* was also used to produce hybrids in Hawaii, USA, Barbados, and Australia (Roach, 1986). High fiber content, waterlogging resistance, and pest resistance are features of this species. The first successful interspecific hybridization using the red fleshed *Sanguineum* group of *S. robustum* has been accomplished using NG 77–76 and NG 77–84 with progeny showing increased growth, better brix content, reddish flesh color, and increased early shoot height (Chandran et al., 2017).

12.14 BASE-BROADENING PROGRAM

The production of sugarcane is seriously threatened by the current cultivars' limited genetic foundation, which can be traced back to original sugarcane clones created in Java and India. These clones included less than 20 *S. officinarum*, 2 *S. spontaneum*, and a few *S. barberi* and *S. sinense* clones. Attempts have been made and in recent years, the importance of genetic base broadening by exploring unutilized clones of wild *Saccharum* species and allied genera in the germplasm introgression programs has been strengthened across the globe.

New genetic diversity is to be incorporated in sugarcane-breeding programs from unutilized genetic resources to exploit desirable genes for important economic traits. Base-broadening programs were seriously taken up during the 1970s, to improve productivity of the cane varieties and also to impart better resistance to biotic and abiotic factors. Using the unused material, advanced cycles of interspecific and intergeneric crosses were emerging. Louisiana started base-broadening programs in the 1960s, with a view to improve the productivity and transferring resistance to mosaic virus and sugarcane borer and cold tolerance, from the wild species, particularly from *S. spontaneum* (Legendre, 1989).

Wild germplasm introgression takes a long period, involves a lot of work and resources, and makes progressive progress. In comparison, breeders found intercrossing of the already existing commercial hybrids a faster and more viable option to generate the variability required to sustain the varietal improvement programs. This resulted in the development of commercial varieties with limited genetic base world over, possibly contributing to the plateaning of yield levels over the years. Attempts to broaden the genetic base of sugarcane cultivars were seriously taken up during the 1960s at most of the research stations once the narrow genetic base of the commercial varieties became apparent (Areceneaux, 1967; Price, 1967). Apart from *S. spontaneum* and *S. robustum*, introgression of *Erianthus* spp. is also considered important in imparting high productivity and adaptability (Hogarth et al., 1997). New approaches being pursued include intraspecific improvement of the individual *Saccharum* species prior to interspecific hybridization (Nair et al., 1998) and the cytoplasmic diversification of the new hybrids by using *S. barberi, S. robustum*, and *Spontaneum* and *Erianthus* as female parents. It is expected that the new varieties that are emerging now will be complex hybrids of interspecific and intergeneric origin with a broad and diverse genetic base.

Introgression of useful genes through interspecific and intergeneric hybridization is a potential approach for improving and promoting food sustainability and energy development. In sugarcane (*Saccharum* spp.) introgression of wild germplasm is time bound program which requires substantial efforts and resources. *Saccharum* spp. hybrid cultivars are highly heterozygous, complex, and polyploid ($2n = 100–130$) products derived from the interspecific crosses between *S. officinarum* ($2n = 10x = 80$), the noble canes and the wild species *S. spontaneum* ($2n = 40–128$). This first interspecific hybridization played an evolutionary role in yield improvement, disease tolerance, and adapts to environmental stresses in sugarcane plants. A limited number of parental clones belonging to either the POJ or Coimbatore canes have contributed to this interspecific hybridization, thereby due to narrow genetic base and reducing the scope for further improvement in sugarcane production. Due to the narrow genetic base of modern varieties (Hogarth, 1987), sugarcane breeders attempted to use the genetic variation within the wild species and allied genera for traits such as sugar content, tolerance to drought and cold, pest and disease resistance, erectness, fiber content, tiller number, stalk size, and ratooning ability to broaden the gene base for genetic enhancement in sugarcane.

Prebreeding provides a unique opportunity through the introgression of desirable genes from wild germplasm into genetic background readily available for use by breeders with minimum linkage drag. Among the breeding methods, gene pyramiding involves convergent of genes identified from a number of parents through pedigree breeding, backcross breeding and recurrent selection. Backcross method is one of the classical breeding methods that had been adapted in sugarcane since 1912 during nobilization process for transferring only the desirable genes from donor species. Exploitation of new and diverse sources of variation will be useful for genetic enhancement to meet the future challenges and wild species offers scope for future growth with enhanced levels of resistance/tolerance to multiple stresses and forms the important source of genetic diversity for crop improvement. In sugarcane breeding programs, *S. spontaneum, S. robustum, S. barberi, S. sinense, Miscanthus sinensis, Erianthus arundinaceus,* and *Erianthus procerus* have been used in introgression to broaden the genetic diversity of sugarcane population and emphasis was made to create a new gene pool through introgression of interspecific and intraspecific clones and pyramiding of genes through backcross breeding at ICAR-SBI, Coimbatore, India. While the *S. officinarum* clones attribute for yield traits for quality. *S. barberi* clones can be exploited for developing clones with tolerance to water logging and drought, respectively.

As evidenced from earlier studies, *Saccharum* forms the primary source of red rot resistance. *Saccharum officinarum* can be utilized as a source of smut resistance, as *S. robustum, S. spontaneum,* and *S. barberi* are susceptible to smut. Among the *S. barberi, S. robustum, and S. spontaneum* exhibit high levels of resistance compared to *S. officinarum and S. sinense. S. barberi* could easily be exploited in the breeding programs for developing red rot resistant clones. It has been determined that the two most effective natural sources for sugarcane varietal development are *S. spontaneum* and *E. arundinaceus. Erianthus* species produce a lot of biomass, are highly resistant to pests, and tolerate salinity, drought, and flooding. They are also a significant source of fiber.

It is found that *S. barberi* and *S. robustum* accessions have good potential of providing resistance to the pink borer (SBI, 2001) and can be utilized as possible donor parents for breeding pink borer resistance. Among the species, IM 76-255 (*S. robustum*), Khadya and Pansahi were found be source of internode borer resistance. Three clones of *S. robustum* and *Erianthus* spp. were found to have resistance to multiple pests (SBI, 2001). Among the allied genera, *Erianthus* clones are resistant to nematodes and root parasites and other pests. *E. arundinaceus* showed exceptional tolerance to drought and imparted resistance to many pests. Narenga clones are resistant to almost all diseases, pests and root parasites and *Miscanthus* clones are resistant to diseases and tolerant to cold.

Base-broadening program with basic species is a continuous process and, along with improved *S. officinarum* and improved *S. robustum* clones, has yielded several hybrids with improved yield and diverse genetic background (Alarmelu et al., 2017) and identified as trait specific genetic stocks for further utilization. *S. robustum* had been properly applied in many breeding programs in Hawaii and in India. This species has been exploited for the development of clones with better sucrose percentage and resistance to red rot. Though utilization of *Saccharum barberi* is restricted, the derivatives show considerable increase for yield and quality and resistance to red rot for further exploitation. Success of genetic base broadening programs in breeding through introgression of *S. robustum* (Naidu and Sreenivasan, 1987; Alarmelu et al., 2014) have been achieved.

12.15 UTILIZATION OF SACCHARUM SPONTANEUM GENETIC RESOURCES IN PREBREEDING

The study of the evolutionary development of sugarcane varieties grown worldwide has clearly highlighted the limited genetic foundation of the contemporary variants. Through hybridization, primarily from parents who are not distantly related, the commercial sugarcane types developed. The primary reason of concern is still the limited genetic base of the already grown types and losses brought on by biotic and abiotic stressors. To expand the genetic base and verify their transfer into the real hybrids, several researchers have suggested introducing desirable genes from wild species that are present in the germplasm and, when needed, using biotechnological methods. In this context, the use of species-specific gentic markers and GISH has been shown to be successful in detecting real hybrids of *S. spontaneum* with commercial cane. In the case of F1 hybrids, where selection decisions are frequently subjective and based on the morphology of genotypes evaluated for a year or less in nonreplicated plots in the early stage of selection, the type of conventional selection is inefficient because many, if not all, economic traits are quantitative in nature. The confirmation of F1 hybrids as most of the earlier studies, the molecular, cytomorphological characterization of the hybrids is essential.

The major achievements of ICAR-SBI has been the successful utilization of *S. spontaneum*. In India in 1912, Dr C.A. Barber and Sir T.S. Venkataraman made a cross between the *S. officinarum* clone Vellai and wild *S. spontaneum* clone (Coimbatore) and derived the clone Co 205 (Nair, 2008). This Co 205 clone was hardy, high yielding and erect. It was a popular variety particularly suitable for subtropical regions and released for the regions of Punjab in 1918, which greatly increased yield in northern India. The hybrid developed by *E. arundinaceus* clone IK 76–62 and *S. spontaneum* clone Iritty-2 crossed with sugarcane clones by which hybrids with *Erianthus* cytoplasm could be produced. These hybrids could only be used as female parents in subsequent crosses because of the male sterility present in it (Premachandran et al., 2011). Given that *Saccharum spontaneum* is the wild species in the *Saccharum* complex with the highest degree of cytological diversity, Suganya et al. (2014) investigated the offspring of commercial cultivars × *S. spontaneum*, which offers higher genetic diversity. Therefore, the diversity of several progenies resulting from crossings between the Co 8371 and Co 86249 clones and various *S. spontaneum* cytotypes ($2n = 64$, 72, and 80) was evaluated.

In order to introduce the characters from *E. arundinaceus* to sugarcane cultivars, Premachandran et al. (2017) developed a rare hybrid between *Saccharum spontaneum* (Iritty-2, $2n = 64$) × *E. arundinaceus* (IK 76–78, $2n = 60$). After conducting *in vitro* colchicine treatment, the hybrid, CYM 04-420, with somatic chromosome number $2n = 62$, produced amphiploids with $2n = 124$. Through genomic *in situ* hybridization, they were able to determine for the first time the chromosome composition and chromosome pairing actions of an intergeneric amphiploid of *S. spontaneum* × *E. arundinaceus*. *Saccharum officinarum* × *E. arundinaceus* and *S. spontaneum* × *E. arundinaceus* backcross (BC2) hybrids, where *E. arundinaceus* was the pollen parent, produced progeny with plants that resembled E. arundinaceus morphologically (Premachandran et al., 2019).

At international level, the interspecific *Saccharum* hybrids in the cytoplasm of *S. spontaneum* were developed and confirmed with the aid of both conventional and molecular breeding approaches. The presence of both random amplified polymorphic DNA (RAPD) and microsatellite alleles from the paternal parents in these clones confirmed their hybrid nature. These F1 hybrids were backcrossed as maternal parents with elite sugarcane clones to produce BC1 and BC2 lines for evaluation and selection and they have made the biparental crosses between *S. spontaneum* and commercial-type sugarcane to widen sugarcane genetic base. They have classified the hybrid progenies based on the inheritance of microsatellite markers. The hybrid offspring were regarded as F1 hybrids since they gained microsatellite alleles from both the commercial type parents and *S. spontaneum*. (Pan et al., 2004, 2006). The use of molecular markers in the development and breeding of new sugarcane germplasm: *Saccharum* spontaneum cytoplasm. After surviving harsh frost in March 2003, one F1 offspring from the cross (SES 234A 9 LCP 85-384) was commercially marketed as the energy cane cultivar Ho 02-113 (Table 12.2).

TABLE 12.2 Achievements Through Introgression

Country	Research Work	References
Hawaii	*S. robustum* for utilization in hybridization	Heinz (1967
USA	Agronomic traits studied to identify apparently mosaic-resistant *S. spontaneum* genotypes	Dunckelman and Breaux (1972)
West Indies	Intra noble cane improvement	Walker (1974)
Australia	*S. spontaneum in introgression programs* *Erianthus arundinaceus* as a source of genes to rectify the root syndrome affecting the cane in Northern Queensland	Berding and Roach (1987)
China	F_1 hybrids of commercial hybrid and *Miscanthus* spp. showed improved hybrid vigor for stalk length and tolerance to drought and salinity	Chen et al. (1986)
China	Reported a steady increase for sugar content in successive nobilized generations of *Saccharum* × *Miscanthus hybrid*	Chen and Lo (1989)
Lousiana	*S. spontaneum* based program and derivatives of *S. spontaneum* accessions developed through introgression breeding attained commercial status	Milligan et al. (1994)
Florida	Characterization of sugar composition of *S. spontaneum* genotypes in World Collection of *Saccharum* (Miami, Florida) and identified the species as a potential source of positive alleles for sugar content.	Tai and Miller, (2002)
West Indies	Incorporation of wild germplasm and clones with diverse genetic background and promising as parents or commercial varieties combining agronomic traits and disease resistance were identified.	Kennedy (2001)
UK	Introgression breeding programs to combine tolerance genes for drought and cold and with high fiber of *Miscanthus* with the photosynthetic capacity of commercial sugarcane to produce a low input, high-biomass, drought- and cold-tolerant energy crop	Clifton-Brown et al. (2002, 2004)

TABLE 12.2 *(Continued)*

Country	Research Work	References
Australia	Success of genetic base-broadening programs in breeding through introgression with related genera such as *Erianthus* and *Miscanthus*	Jackson et al. (2004)
Australia	Eight species of *Erianthus* identified as a potential donor for stress-resistant genes.	Aitken et al. (2007)
	E. rockii was used in intergeneric hybridization with *S. officinarum and* $(n + n)$ transmission of gametes was reported in the introgressed hybrid	
Brazil	*S. officinarum* was crossed with *Erianthus* and *Miscanthus* and the introgressed hybrids were backcrossed to *S. spontaneum* to increase sugarcane biomass yield and productivity and also helped to identify disease resistant sources	Matsuoka et al. (2009)
USA	Screening of sugarcane wild germplasm and related species of USA for starch content, which indicated that screening for the trait before the selection of parents for introgression is beneficial.	Zhou et al. (2013)
China	Introgression study of *S. officinarum* L. cv. Ludashi × *Erianthus rockii* Yunnan 95-19, identified more than 30% of the introgression lines with resistance to sugarcane brown rust and the resistant lines identified were also further used as sources of brown rust resistance in breeding	Wang et al. (2013)
USA	Established a core collection of 130 *S. spontaneum* accessions from world collection of sugarcane and related grasses with phenotypic and genotypic markers for selection of diverse parents in sugarcane improvement programs.	Silva et al. (2018)

12.16 CORE COLLECTION WITH WILD SUGARCANE

Crop plant genetic resources have been assembled in gene banks all over the world for long-term preservation and convenient access by researchers, plant breeders, and other users. In fact, size and management are two main issues that many plant germplasm collections presently encounter suggested that an existing collection could be used to establish a restricted or "core collection." The core collection, which has limited similarities, is small in size and aims to represent the genetic variety of a crop, a wild species, a group of species, or a huge collection. It does not take the place of the present group or the material it originates from. Core collection development is recommended as one of the necessary efforts to improve the use of plant genetic resources in the Global Plan of Action for the Conservation and Sustainable Utilization of Plant Genetic Resources for Food and Agriculture.

12.16.1 SACCHARUM CORE COLLECTION

From the large collection in ICAR-SBI, Coimbatore about <1% is exploited by breeders. Hence, a diverse reference set will be a beneficial resource for the sugarcane breeders. Formation of a core collection of sugarcane genetic resource in India is very much important for assessing and capturing maximum diversity for identifying the source of variation to be utilized in breeding program. Approximately 60 genetically diverse *S. spontaneum* accessions in the form of core collections have been retained in ICAR-SBI genebanks (Amalraj and Balasundaram, 2006). There is the first comprehensive genetic diversity research on all accessions accessible in the USA-maintained *World Collections of Sugarcane and Related Grasses* (WCSRG). The WCSRG's highest genetic diversity was found in a core collection of 300 accessions, which can be utilized further for sugarcane and energy cane breeding. Three-hundred accessions representing most of the diversity within the WCSRG (Miami collection) was selected for developing diversity panel based on phenotypic traits. To demonstrate the diversity of desirable features and the breeding potential of accessions in the WCSRG for the creation of energy and sugarcane cultivars assembled 110 accessions that were not positively differ Brix content from the commercial standards for sugarcane. These included 10 *S. spontaneum* accessions as well as 17 and 6 accessions that were optimum compared to commercial traits in plant biomass, respectively.

12.16.2 DEVELOPMENT OF CORE COLLECTION FROM ICAR-SBI GERMPLASM HOLDINGS

Because there are so many accessions in the field gene bank, one of the main issues with plant breeding efforts is the inadequate use of the entire germplasm. There is an urgent need to assess the morphological and molecular diversity of the entire sugarcane germplasm at ICAR-SBI, indicating the importance of a complete evaluation of the accessions in the entire collection. A core collection is a small group of accessions that exemplify the highest level of genetic variation and the least amount of repetition among crop vultivars and their wild cousins. About 10% of the overall collection should be included in a core collection.

The improvement programs aim at development of new hybrids by the use of established Co canes and elite breeding lines. This approach has narrowed the genetic base of sugarcane in developing new breeding populations, which has to be broadened by better utilizing genetic resources. Due to the huge size of the germplasm collection, the available diversity has not been meticulously explored and appreciably utilized in sugarcane breeding. Morphological characteristics have been essential in the investigation of genetic diversity and in the selection of accessions that form a core collection over the years. Proper evaluation is feasible only for the traits which can be scored easily and do not show genotype by environment (G × E) interactions. In plant breeding programs, phenotypic characterization of the complete germplasm collection is becoming a more crucial phase because it allows breeders to more effectively assess variances and identify high-performing accessions. Characterization of entire germplasm of *S. spontaneum* available in field genebank for morphological, qualitative and molecular characterization is in progress for the development of core set.

12.17 MOLECULAR MARKERS IN SUGARCANE GERMPLASM

For the reason to enable breeders to analyze the variance found in germplasm collections, the evaluation of genetic diversity among cultivars has the potential to be a valuable tool in plant breeding. Breeders of sugarcane are able to choose a wider variety of germplasm to incorporate into their breeding programs because of this diversity. Molecular markers have the potential to provide a more accurate indicator of genetic variation and have been extensively employed for the characterization of germplasm within *Saccharum* complex species. Markers have been employed in several studies to analyze genetic diversity, interspecies interactions, and phylogeny among the *Saccharum* species, related taxa, and their hybrids.

When it comes to recognizing the domestication of crop plants, molecular markers are invaluable. Regarding sugarcane, they have assisted in selecting between earlier hypotheses that were based on historical and botanical knowledge rather than providing novel evolutionary strategies. Nowadays, it is almost certain that sugarcane developed independently of the lineages of *Erianthus* and *Miscanthus*, inside a particular *Saccharum* lineage, and that the scenario suggested by Brandesto to explain the genesis of classic cultivars is essentially realistic. Without a doubt, additional research focused on investigating the relationships between *S. robustum, S. spontaneum*, and noble cultivars in Kalimantan, Sulawesi, and New Guinea will contribute to a full comprehension of the dynamic process of sugarcane domestication.

12.18 MARKERS IN DIVERSITY STUDIES AND SPECIES IDENTIFICATION

DNA markers are commonly used for genotyping the individual germplasm accessions which is helpful in grouping them, discovering their kinship, population structure, discovery of identical types, duplicates, and their origin (Dufresne et al., 2014). Among the several marker system available, namely, RFLP, RAPD, AFLP, SSR, ISSR, TRAP, EST-SSR, SNP, etc., in the literature, microsatellites or SSRs and SNPs are the markers that are best suited for germplasm studies and provide the robust data necessary in scientific understandings and also useful in overcome the ambiguity in providing the intellectual property rights and patent.

Markers can be readily used to study genetic diversity relationships among the accessions and species available in the germplasm (Nair et al., 1998), studied *Saccharum* complex available at ICAR-SBI, Coimbatore using RAPD markers and found limited diversity of genetic among the *S. officinarum*. The markers were able to separate all the species present in the *Saccharum* complex, cluster analysis using genotypic data that grouped *S. officinarum, S. robustum, S. spontaneum*, and *Erianthus* spp., into discrete groups, while *S. barberi* and *S. sinense* formed a single cluster. Markers were also used to distinguish *S. robustum* from the *S. spontaneum*, and their relationship with *S. officinarum* (Nair et al., 1998). Use of RAPD, ISSR, and telomere primers were able to distinguish the *S. spontaneum* collection collected from various geographical region (Mary et al., 2006). Molecular markers have been applied to assess the diversity of genetic of *Saccharum* spp. accessions and beneficial for directly detecting differences in the genome (Dos Santos et al., 2012; Manechini et al., 2018; Barreto et al., 2019). Park et al. (2024) used single-dose SNP markers to characterize the *Saccharum* germplasm available at USDA and identified mislabeled clone, duplicates, near-identical genotypes, and homonymous groups. They were also able to clarify the species membership and parentage of clones that had ambiguous passport records.

While morphological and agronomical traits are crucial for classifying and organizing germplasm accessions, errors can still happen because vegetative traits are highly variable and influenced by environmental factors; therefore, they frequently do not accurately represent the genetic diversity of the *Saccharum*. The profile from the molecular markers are helpful and complement the morphological characterization to identify better combinations between accessions to make crosses according to breeding goals in a more reliable way (Lima et al., 2002; Singh et al., 2018; Alwala et al., 2006). Prebreeding is an essential phase in increasing the probability of producing more productive cultivars through choosing parents for crosses (Medeiros et al., 2020). The wild species and related genera of *Saccharum* were commonly used in sugarcane varietal improvement. The interspecific and intergeneric hybridizations were routinely used. The wild species of *Saccharum* and the related genera are considered to be potential sources for genes conferring high productivity and resistance to biotic and abiotic stresses (Nair et al., 2012). Molecular markers were highly helpful in identification of F_1s in wild crosses. The 5s rDNA primers were widely utilized to identify the true F_1 progenies derived from the *Erianthus* spp. and *Sccharum* spp. (Nair et al., 2017). They were also helpful in confirming the presence of donor segment in the backcross progenies of intergenic hybrids (Sobhakumari et al., 2020). Identification of core collection is very important in utilization of germplasm, the molecular markers are robust tool and compliment the phenotypic data and passport data in identification of a representative core of a germplasm for further utilization. A core of 300 accessions were formed from the 1002 accessions, using 36 microsatellite markers at world collections of sugarcane and related grasses in Miami, Florida (Nayak et al., 2014). According to Zhang et al. (2018), the genome of the *S. spontaneum* genotype (SES 208) has been sequenced. The reference genome will give researchers an array of sequencing information and be useful for characterising and using *Saccharum* germplasm.

12.19 ALLIED GENERA FROM SACCHARUM COMPLEX

The term "*Saccharum* complex" was designated by Mukherjee (1957) to represent the genera *Saccharum*, *Erianthus* (sect. Ripidium), *Sclerostachya* and *Narenga* constituting closely related cross-breeding groups associated in the sugarcane origin. All the species exist in the Indo-Burma and China border region and this area was the center of origin of *Saccharum* complex to which Daniels et al. (1975) added the genus *Miscanthus* sect. Diandra Keng. Presently, the *Saccharum* complex has five genera: *Saccharum*, *Erianthus* Michx. sect. *Ripidium* Henrard, *Sclerostachya* (Hack.) A. Camus, *Narenga* Bor. and *Miscanthus* Anderss. sect. Diandra Keng. Among these, *Erianthus* and *Miscanthus* are thought be the most ancient types (Daniels et al., 1975). Although, genera such as *Erianthus*, *Sclerostachya*, *Narenga*,

Miscanthus, and *Imperata* are freely crossable with *S. officinarum*, but their utilization for varietal improvement is very limited. In a recent study carried out in Indian commercial sugarcane cultivars generated over 100 years indicated that they were derived from 82 *Saccharum* species clones and three accessions *of Erianthus arundinaceus*, one each of *S. bengalense*, *Narenga porphyrocoma*, *Sorghum durrah*, and African sorghum. This Indian varietal pool is reported to be the most diverse collection across the globe as nine different genera were introgressed into the commercial cultivars.

In order to strengthen modern sugarcane cultivars' genetic basis, breeders have resorted to intergeneric hybridization utilizing related genera belonging to "*Saccharum* complex." Two high-altitude types of *Erianthus fulvus* and *Miscanthus nepalensis* are maintained in the world collections of sugarcane germplasm at SBI, Coimbatore, India (large collection at SBI Regional Centre, Kannur) and USDA-ARS, Canal Point, USA (bulk collection at Coral Gables outside Miami, Florida).

12.20 ERIANTHUS

Saccharum and *Erianthus* are the two genera in the *Saccharum* complex that are most closely related. Few species of *Erianthus*, looks very similar to sugarcane in their appearance. *Erianthus* has been recognized as a separate genus by few authors. It revealed that *Erianthus*, which has species from North and South America, Europe, Africa, and Asia, *Saccharum* s.s. contains taxa from India, Malaysia, and China (Mukherjee, 1957). Nonphylogenetic molecular studies have revealed the distinguishing features between *Saccharum* s.s. and *Erianthus*, as well as between the *Erianthus* and *Ripidium* (Trin.) Henrard of the genus *Erianthus*. Established the phylogenetic origin of *Tripidium*b ased on complete plastome sequences and acknowledged as a separate genus from *Saccharum* and *Erianthus*. Also, this finding is well described by morphologicaland biogeographical aspects. The two genera are morphologically differentiated by the awnless spikelets in *Saccharum* s.s and the awned spikelets in *Erianthus* s.s. While the spikelets in Tripidium (= *Erianthus* sect. Ripidium) can have one or many awned. Three distinct groups were identified by clustering (unweighted pair group method with arithmetic mean, UPGMA) and discriminant analyses: (1) *Saccharum* from the ancient era, (2) Tripidium along with some species of *Saccharum* that are frequently recognized as *Miscanthus, Miscanthidium,* or *Narenga* (all clones from the ancient era), and (3) *Erianthus* from the modern era. The findings confirmed that the natural species of the New World (*Erianthus* s.s.) differ morphologically from the commonly farmed species of *Saccharum* s.s., suggesting that *Erianthus* need to be classified as a separate genus from *Saccharum*. Additionally, the study established the differences in features between the two species of Tripidium based on DNA evidence as well as the physical attributes that differentiated them apart from *Saccharum* and *Erianthus*.

12.20.1 SPECIES DIVERSITY IN ERIANTHUS

The genus *Erianthus* comprises of several species and most of them are located in the Indian subcontinent, including *E. arundinaceus, E. procerus, E. longisetosus, E. bengalense, E. ravennae, E. fulvus, E. elephantinus,* and *E. hookeri. E. arundinaceus* represented by robust growth, thick canes capable of very high biomass production under low input conditions and a high level of adaptation ability in response to adverse agroclimatic conditions. These characteristics make *E. arundinaceus* a desirable candidate for exploitation in sugarcane breeding for vigor, high yielding ability, better ratoonability, tolerance to environmental stress and resistance to diseases and pests. Germplasm exploration trips by Nair and coworkers in 1993 in India surveyed Lohit and Tawang districts of Arunachal Pradesh that were surveyed during October 1990. Lohit District displayed abundant diversity in *Erianthus* while *Erianthus fulvus* and *Miscanthus nepalensis* were broadly distributed in Tawang district.

The reported chromosome numbers of *Erianthus* species are $2n$=20, 22, 24, 30, 40, and 60. Clones belonging to *E. elephantinus* and *E. ravennae* ($2n$=20) from India, Nepal, Pakistan, and USSR have terminal chromosome knobs on either one or both arms of most of the chromosomes. B chromosomes were also identified in a few clones with $2n$=20 chromosomes. The first *Saccharum* × *Erianthus* hybrid was produced by crossing *S. officinarum* clone EK 28 ($2n = 80$ with *E. sara* ($2n = 60$). The hybrids had $2n$=61 to 69 chromosomes made hybrids between *S. spontaneum* ($2n = 112$) with *E. ravennae* ($2n = 20$). Hybrids had slightly thicker stalks, longer leaves; long dense inflorescence and the florets were fertile. All F_1 hybrids had $2n = 66$ chromosomes and formed bivalents and trivalents and quadrivalents with some univalents. The bivalents arose from autosyndetically paired *S. spontaneum* chromosomes and univalents from *Erianthus* genomes. The frequent presence of seven univalents along with one quadrivalent and one trivalent indicated that atleast three chromosomes of *Erianthus* and *Saccharum* were homologous.

12.20.2 E. ARUNDINACEUS

Erianthus arundinaceus, a wild relative of sugarcane characterized by high yield, high fiber, requires low nutrients, resistant to nematodes and root feeders, tolerance to drought and salinity. *E. arundinaceus*, the only cane forming species has a widespread distribution in India, China, Myanmar, Thailand, Philippines, Indonesia and New Guinea. A majority of the *E. arundinaceus* germplasm collections in India are either from the Indian subcontinent or from the Indonesia–Papua New Guinea region. In Barbados, hybrids derived from *E. arundinaceus* showed exceptional tolerance to drought, cold, salinity and imparted resistance to many insect pests (Daniels and Roach, 1987). Some of the *Erianthus* clones, namely, IJ76-332, IJ76-365, IJ76-383, IJ76-384, IJ76-400, IK76-48, IK76-76, IK76-88, IK76-99, and IS76-199 are resistant to red rot (*Colletotrichum falcatum*). In Australia, crosses involving *Erianthus* clone, IK64-41 gave high yielding progenies that were tall and having high stalk weight. Attempts to develop intergeneric hybrids at SBI, Coimbatore generated a large number of hybrids. Early research initiatives to introgress the *E. arundinaceus* characters into the sugarcane varieties resulted in sterile hybrids in few occasions, problems in establishing the genuine hybrids through the morphological traits, chromosome elimination during successive back crossing and lack of recombination between the chromosomes of the two genera (D'Hont et al., 1995). Bridge crosses were successful in overcoming hybrid sterilty. Utilising *S. spontaneum* as a bridging species in crosses between *S. spontaneum* × *E. arundinaceus* hybrids when crossed with sugarcane produced fertile hybrids which could be further backcrossed to sugarcane. Hybrids derived from crosses between *S. spontaneum* × *Erianthus* and *E. arundinaceus* × *S. spontaneum* were fertile which enabled the introgression of *Erianthus* characters (Premachandran et al., 2011). The crosses between *S. officinarum* × *Erianthus* were found to be more advantageous for sucrose content than their reciprocal crosses (Nair et al., 2006; Nair and Mary Sheji, 2006). Besides vigor and improved yield, hybridizing *Erianthus* with sugarcane facilitated the introgression of cold tolerant genes combining genes for red rot resistance. One elite selection Co 87008 identified from ICAR-SBI, Coimbatore is a hybrid of Co 6304 and *Erianthus* clone (Hemaprabha et al., 2022).

12.20.3 EXTENT OF GENETIC DIVERSITY IN E. ARUNDINACEUS

Sugarcane breeders were successful in generating viable *Saccharum* × *Erianthus* intergeneric hybrids which are repeatedly crossed to sugarcane clones to produce wide intergeneric hybrids. *E. arundinaceus* species has been targeted for genetic improvement in sugarcane breeding and holds enormous potential toward the development of high biomass intergeneric hybrids. China has started a breeding program to develop bioenergy feed stocks involving *E. arundinaceus* at the Sichuan Academy of Grassland Science, China during 2011. In an earlier study, a high level of variation was reported based on morphological traits in *E. arundinaceus* accessions from China as compared to the clones from Indonesia that was relatively low. Karyotype analyses of Chinese *E. arundinaceus* indicated that majority of clones had $2n= 4x= 40$ and $6x= 60$ somatic chromosomes while $2n =2x =20$ types were not common. Using amplified fragment length polymorphisms (AFLP) markers, 207 accessions of *E. arundinaceus* from China, India, New Guinea, Indonesia, Vietnam, and Philippines were analyzed sampled for genetic diversity. Three AFLP primer combinations produced 286 markers with a high level of polymorphism (99.3%). Results on dendrogram analysis through UPGMA method by NTSYS software indicated a high level of diversity observed within the Chinese collection against the Indonesian collection. The clones collected from eastern and western China were grouped in two separate clusters and the clones collected from Yunnan province of western China was reported to be the diverse group as compared to collections from another region.

The extent of genetic variability in *E. arundinaceus* was studied in 164 progenies belonging to 18 populations collected from different Chinese provinces using 20 sequence-related amplified polymorphism (SRAP) primerpairs producing 277 polymorphic bands (Zhang et al., 2013). This study confirmed the presence of large genetic variation in the native Chinese *E. arundinaceus* germplasm depends on the SRAP analysis. This diverse collection would be highly beneficial for targeted genetic improvement programs in sugarcane.

Fifty germplasm collections of *E. arundinaceus* (Retz.) Jesw. from the Guangxi province in China were assessed for their genetic diversity using start codon targeted polymorphism (SCoT) marker primers. Fifteen primers generated 284 amplicons with 83.97% polymorphism in a germplasm panel of 336 clones explaining that the collections from Guangxi province of China exhibited high level of diversity. The Jaccard's similarity coefficients among these accessions varied from 0.61 to 0.88 with an average of 0.71. Clustering through UPGMA indicated that the 50 accessions could be grouped under four major clusters. The grouping of *E. arundinaceus* clones by dendrogram analysis corresponded to their geographic origin in China.

12.21 GENETIC DIVERSITY ANALYSIS BETWEEN SACCHARUM AND E. ARUNDINACEUS

Sugarcane breeders world over have shown continued interest in understanding the genetic diversity of *Saccharum* spp. and commercial hybrid varieties to broaden the genetic base by tapping the gene pool from wild relatives. Studies were conducted to estimate the genetic diversity of 79 accessions from *S. officinarum, S. spontaneum, S. robustum, S. barberi, S. sinense*, 6 accessions of *E. arundinaceus*, and 30 *Saccharum* spp. hybrids utilizing 21 pairs of fluorescence-labeled more polymorphic SSR primers 167 polymorphic SSR alleles were identified with an average value of polymorphic information content of 0.92. The estimates on genetic diversity characteristics among these 115 accessions showed that *Saccharum* spp. hybrids were most significant than *Saccharum* and *Erianthus* species. Phylogenetic investigation and principal component depends on the SSR datasets grouped the 115 accessions into seven phylogenetic groups, which responded to the genera of *Saccharum* and *Erianthus*. The authors suggested that seven SSR primer pairs (SMC31CUQ, SMC336BS, SMC597CS, SMC703BS, SMC24DUQ, mSSCIR3, and mSSCIR43) employed in this study is likely to have wide applications in genotype assessment of *Saccharum* spp. hybrids.

Using SRAP markers, genetic diversity of 208 world collection of *E. arundinaceus* germplasm clones from Indonesia, New Guinea, Fiji, Pakistan, Burma, and Philippines maintained at ICAR-SBI, Coimbatore, India was analyzed. Polymorpism survey of the entire germplasm with 30 combinations of SRAP marker produced distinct DNA profiles with fragments ranging from 120 to 1000 bp. Compilation of marker data of selected primer pairs gave 237 polymorphic bands with 75.9% mean percentage of polymorphic bands. The optimum distance of genetic (0.49) was observed between the Indonesian (IK 76) and Indian collections (SES), and the lowest distance (0.040) was monitored between the Fiji and Philippines population and within the Philippines population.

The phylogenetic clustering using dissimilarity index formed 10 major clusters and the clusters corresponded with geographical origin of the clones. The highest level of structure, or optimal K value, was found to be nine when structure output utilizing the SRAP data set was performed. There is significant genetic differentiation between the populations, as demonstrated by the structural analysis, which showed low gene flow (N_m) and high F_{st} values. The eight fragment sequences' similarity to functional genes revealed by blast analysis suggested that the SRAP markers amplified the coding areas. Among the 10 fragments, one showed homology to *Saccharum officinarum* ribosomal protein. L22-like mRNA and one fragment from combination E1M14 matched with *S. spontaneum* xyloglucan endotransglucosylase/hydrolase protein 6.

12.21.1 CHARACTERIZATION OF INTERGENERIC HYBRIDS OF *E. ARUNDINACEUS*

Earlier efforts to characterize intergeneric hybrids prior to the invention of sophisticated molecular techniques by D'Hont et al. (1995), suggested that simple molecular tools such as isozyme electrophoresis and a sequence-tagged PCR could be employed to identify the genuine intergeneric hybrids of *S. officinarum* and *E. arundinaceus.* Characterization of sugarcane intergeneric hybrids using RFLP enabled the assessment of different *Erianthus*-specific RFLP bands. Genomic DNA in *in situ* hybridization technique facilitated the molecular variations of the chromosomes assisted by *S. officinarum* and *E. arundinaceus*. Thus, the two molecular techniques (RFLP and GISH) were found to be promising in monitoring the *Erianthus* genome in the intergeneric hybrids during introgression.

Recovery of fertile *Saccharum* × *Erianthus* hybrids is the major bottleneck challenging the utilization of this genus *Erianthus* in sugarcane-breeding programs. Studies to understand the pattern of chromosome transmission in the fertile *Saccharum officinarum* × *E. arundinaceus* hybrid (F1) was attempted through GISH (Piperidis et al., 2010). The chromosome number in the F1, BC1 (F1 × sugarcane cultivar), and BC2 (BC1 × sugarcane cultivar) hybrids were documented for the first time.

The results on GISH indicated *n*+*n* chromosome transmission in F1 and BC2, while 2*n*+*n* transmission was observed in BC1 hybrids. The total chromosome number in BC1 hybrids was reported to range from 21 to 30 of which the number of *E. arundinaceus* chromosomes numbered in BC2 clones ranged from 14 to 15 indicating the loss of chromosomes during backcrossing. However, there was no occurrence of recombination between the chromosomes of *Saccharum* and *Erianthus* in both BC1 and BC2 clones. Based on the results obtained, the authors suggested identifying the agronomic merit of the chromosomes from *E. arundinaceus* to conduct targeted breeding from the knowledge gained through this study.

Reported on the successful development of fertile intergeneric hybrids of *Saccharum* spp. and *Erianthus arundinaceus*. Applying 5S rDNA, five intergeneric hybrids were identified with a production efficiency of approximately 3%. The intergeneric hybridization between *Saccharum* spp. hybrid and *E. arundinaceus* changed the root characters, polyphenol content, shoot/root ratio, and root dry weight, etc.

Fertile *Saccharum* × *Erianthus* hybrids was also reported by Chinese breeders (Deng et al., 2002; Cai et al., 2005; Qi et al., 2021) and the same was backcrossed with commercial sugarcane up to BC4 progeny. The pattern of transmission of *E. arundinaceus* chromosomes in hybrids of sugarcane was *n*+*n* in F1 (*S. officinarum* × *E. arundinaceus*), BC2 (BC1 × sugarcane variety) and BC3 (BC2 × sugarcane variety) generations and 2*n*+*n* in BC1 (F1 × sugarcane variety). Intergeneric crosses between *E. arundinaceus* × *S. Spontaneum* were advanced up to BC2 generations by backcrossing with commercial varieties (Zhou et al., 2019; Qi et al., 2021). Potential clones emerging from this program is likely to be used for commercial variety development. Substantial efforts have also been made in China toward the introgression of drought tolerance genes from *Narenga* germplasm to commercial varieties.

GISH analysis revealed the somatic chromosomes number of the hybrids as $2n = 53$ to 82 with the *Erianthus* chromosomes ranging from 1 to 29. Field evaluation of the intergeneric hybrids indicated a significant and positive correlation between the number of *Erianthus* chromosomes and dry mass, millable stalk weight, diameter and weight of stalk.

Modern commercial varieties have received the maternal cytoplasm from relatively few number of *S. officinarum* clones, while *S. spontaneum, S. barberi, S. sinense* or other hybrids used as male parent. Limited cytoplasmic diversity could be a possible challenge in the days to come as we are still unaware whether the cytoplasm from allied genera will be able to provide more trait plasticity besides improved cane yield and disease resistance. At ICAR-SBI, Coimbatore, Premachandran et al. (2011) in their efforts to diversify the maternal cytoplasm developed new cytoplasm substitution lines in sugarcane with the cytoplasm from *S. spontaneum* and *E. arundinaceus*. In this process, repeated backcrossing of F1 intergeneric hybrids of *E. arundinaceus* × *S. spontaneum* and *S. spontaneum* × *E. bengalensis* with sugarcane (as male parent) was carried out to develop near commercial clones. RFLP analysis using mitochondrial DNA segment nad4 for *S. spontaneum* and chloroplast DNA segment psbC-trnL intron for *Erianthus*, was done to verify the *S. spontaneum* and *Erianthus* cytoplasm of such hybrids. Despite the problems such as sterility and pistillody in the hybrids, this backcrossing scheme was successful in introgressing both nuclear and cytoplasmic genes of *Erianthus* into sugarcane. Both the *Erianthus* and *S. spontaneum* cytoplasm were found to be compatible with sugarcane nuclear genome with normal meiosis and chromosome segregation. The pattern of transmission of gametes in the first, third, and fourth generation followed the n+n pattern, whereas it was 2n+n in the second generation as reported in a previous study by Piperidis et al. (2010) with *S officinarum* cytoplasm. GISH technique was applied to study the transmission of chromosomes in the F1 and four successive generations of backcross hybrids using labelled *E. arundinaceus* probe. In the hybrids, *Erianthus* chromosomes got eliminated in each generation of backcrossing. The *E. arundinaceus* × *S. spontaneum* hybrid had somatic chromosome number (2n = 62) with 30 *Erianthus* chromosomes, the BC1 (2n = 118) had 24 *Erianthus* chromosomes, and the BC2 (2n = 108) had only 12 *Erianthus* chromosomes. Co 15015, a BC3 hybrid identified from this program with 2n = 106 had two *E. arundinaceus* chromosomes. CYM11-511, a BC5 hybrid between *S. spontaneum* × *E. bengalense*, has recorded substantially higher cane yield than the best performing checks. CYM 08-922 is yet another hybrid clone that registered the highest average cane yield under drought and identified as a potential donor parent in breeding programs for drought tolerance (Mohanraj et al., 2017).

Using highly diverse prebred clones and derivatives of different basic species of *Saccharum* and *Erianthus* (*S. officinarum, S. robustum, S. spontaneum, S. barberi/S. sinense*, and *E. arundinaceus, E. bengalense, E. procerus*) maintained at ICAR-SBI, Coimbatore, novel research program was initiated to develop Multiparent Advanced Generation Intercross population in cane plants. These hybrids developed from diverse sources are expected to have a mosaic of different genomes that would serve as a valuable and unique genetic resource with immense scope for genetic improvement of drought tolerance besides serving as a mapping population for other major traits (Mohanraj et al., 2017).

A novel research initiative was undertaken at ICAR-SBI toward the development of Multiparent Advanced Generation Inter-Cross population in sugarcane by using the most potential prebred clones which are highly diverse and derivatives of different basic species of *Saccharum* and *Erianthus* (*S. officinarum*, *S. robustum*, *S. spontaneum*, *S. barberi*/*S. sinense*, and *E. arundinaceus*, *E. bengalense*, *E. procerus*). These hybrids with a mosaic of different genomes would be valuable and diverse gene pool for genetic improvement of drought tolerance besides serving as a reservoir for many novel genes (Mohanraj et al., 2017).

Few genes associated with drought resistance have been cloned from *E. arundinaceus*. The gene HSP70 from *E. arundinaceus* has a significant function in sugarcane (*Saccharum* spp. hybrid) in response to drought stress. Over expression of *EaDREB2* gene in sugarcane enhanced drought resistance to more than untransformed plants. Moreover, *E. arundinaceus* was shown to develop arbuscular mycorrhizal fungi association in their natural habitats, which could further enhance drought tolerance.

12.21.2 SURVEY OF E. ARUNDINACEUS GENOME

Understanding the genomic information of crop species has become possible using the next generation sequence technology carried out a genome survey sequence in the *E. arundinaceus clone* HN92-77 with 60 chromosomes. Fertile genuine hybrids had been produced by crossing *S. officinarum* and HN92-77, which were backcrossed successfully with cultivated sugarcane varieties 167.36 Gb raw data were collected into 3161 Mb genome sequences the estimated genome size is 3.23 Gp. Mining the *E. arundinaceus* genome yielded 36,616 SSR motifs that comprises about 61% dinucleotide, 34% trinucleotide, 3% tetranucleotide, 0.98% pentanucleotide, and 0.71% hexanucleotide repeats, respectively. The most common repeat motif was AG/CT while GC/CG was less frequent accounting for 54.16% and 2.69% of dinucleotide repeat motifs. About 20 SSR primer pairs amplified 365 SSR loci were observed in 20 clones of *E. arundinaceus* and its related cultivars, and 188 SSR loci were specific to *E.*

arundinaceus. Clustering analysis based on genetic similarity coefficient by UPMGA method grouped 20 clones of *E. arundinaceus* and its associated clones into two major clusters. Five clones of *E. arundinaceus* clustered individually while the remaining clones were grouped in another cluster. The high level of polymorphism and specificity of the SSR markers developed in this study revealed that these SSRs could be applied in genetic linkage mapping, marker-assisted selection, and identification of realistic hybrids between *E. arundinaceus* and sugarcane.

12.21.3 ERIANTHUS PROCERUS

E. procerus bears a close resemblance to *E. arundinaceus* and E. kanashiro*i* and has large droopy silky panicles. *E. procerus* is distributed mostly in India, Burma, China, Indonesia, and New Guinea (Amalraj and Balasundaram, 2006). To broaden the genetic base, our initial efforts to introgressthe desirable traits of *E. procerus* in sugarcane resulted in anovel intergeneric hybrid GU04 (28)EO-2 between *E. procerus* (IND 90-776) × *S. officinarum* (PIO96-435). The hybridity of the clone GU04 (28) EO-2 has been established using genus specific 5s rDNA markers. This novel hybrid has been registered with ICAR-National Bureau of Plant Genetic Resources, New Delhi, India as a unique germplasm. The hybrid was characterized with comparatively lesser stalk weight and thickness and the BC1 hybrids exhibited remarkable improvement over F1 for the production and quality of sugarcane juice (Nair et al., 2017). Substantial progress has been made toward the development of genus and species-specific molecular markers. Using TRAP markers of *S. officinarum*, *S. spontaneum*, *E. arundinaceus*, *E. bengalense* and *E. procerus*, nine *Erianthus* specific markers, four *Saccharum* specific markers and two *S. spontaneum* specific markers were developed.

E. procerus has been identified as a potential and novel source for modifying the genetic base of the sugarcane cultivars. Evaluated 24 backcross hybrids of *E. procerus* at ICAR-SBI Coimbatore and ICAR-SBI regional station (RS), Karnal for red rot resistance adopting controlled condition testing facility at Coimbatore, and plug method under field conditions for two seasons. F1 hybrids were resistant (R) and 70% of backcross hybrids showed resistance against CF06 except for one BC hybrid GU 12-36 exhibited highly susceptible reaction. For subtropical isolates of CF08 and Cf09, the clones GU 12-21, GU 12-22, GU 12-26, GU 12-27, GU

12-31, and GU 12-33 were highly resistant and these clones also showed resistant reaction against the tropical isolate CF 06. These diverse hybrids with *Erianthus* base constitute a valuable and novel source for developing sugarcane varieties with stable resistance to red rot.

Recent investigations by ICAR-SBI, to assess the *Erianthus procerus* genome-introgressed *Saccharum* hybrids in two environments (tropical and subtropical) reported on seven hybrids as significant donors for freeze resistance and optimum ratooning efficiency combining red rot resistance, under subtropical climates. The most promising *E. procerus* hybrid that recorded the highest winter sprouting potential was GU 12-19 with sprouting of winter index of 10.6, followed by GU 12-22 with a WSI of 8.5. In sugarcane, anthocyanin is supposed to act as an antioxidant and is capable of withstanding damages due to cold stress. The *E. procerus* derived hybrid GU12-19 has more than 22% fiber thereby making it a suitable candidate for industrial applications and bioenergy production. The above study demonstrated the potential of *E. procerus* genome-introgressed *Saccharum* hybrids which holds promise as a novel source in future sugarcane breeding programs to develop sugarcane genotypes with high winter ratooning potential alongwith red rot disease tolerance in response to semiarid regions.

12.21.4 *E. ROCKII*

Using 400 AFLP markers, intergeneric hybrids of *Erianthus rockii* and *Saccharum* were characterized and diversity was estimated (Aitkin et al., 2007). RAPD markers could be accurately identified the hybrids of *S. officinarum* × *Sorghum bicolor* and *S. officinarum* × *Zea mays* (Nair et al., 2006; Nair and Mary Sheji, 2006).

12.22 MISCANTHUS ANDERSS.: GENETIC RESOURCES

Miscanthus is classified as an East Asian perennial rhizomatous C_4 grass. Andersson established the genus *Miscanthus* in 1855. The genus can be easily identified from *Saccharum* and *Erianthus* by its tenacious nature of the rachis and by paired pedicellate spikelets. The inflorescences of the genus *Miscanthus* are easily identified by their prominent long racemes in relation to thefloral axis. It was dispersed from Tahiti in the Pacific througheastern Indonesia, Indochina to northern China, Siberia, and Japan (Daniels and Roach, 1987). Besides, a separate section, Diandra, is indigenous to north India, Nepal, and southwest China, the genus *Miscanthus* was classified into four sections: Triarrhena (*M. sacchariflorus*, *M.* × *giganteus*), Eumiscanthus (*M. sinensis, M. floridulus*, *M. condensatus),* Kariyasua (*M. oligostachyus*), and Diandra (*M. nepalensis*). Amalraj and Balasundaram (2006) gave a different perspective by classifying the genus *Miscanthus* into three sections based on the number of stamens. Members of the genus *Miscanthus* occur in low to high altitudes diverse environments such as river banks, deserts, and mountains which resulted in ecotypes within the same species. Cytomorphological studies indicate that *Miscanthus, Erianthus*, *and Sclerostachya* were involved in the origin of *Saccharum officinarum*. GrassI (1974) proposed that Chinese sugarcane researchers selected *S. sinense* from a natural cross of *S. officinarum* and *M. sacchariflorus* ($2n$=76). *Miscanthus* is freely crossable with *Saccharum*, *Erianthus, Narenga,* and *Sclerostachya* and produces fertile intergeneric hybrids. Investigations on chloroplast restriction enzyme site mutations in 32 genotypes from 8 genera and 19 species of Andropogoneae tribe resulted in a salient finding that the genera *Narenga, Miscanthus, Erianthus*, and *Sclerostachya* were involved in the origin and evolution of *S. officinarum.*

The members of the genus *Miscanthus* holds lot of promise as potential feedstocks for biofuel because of their high biomass yields per unit of planted area and can be exploited in marginal and abandoned lands. *Miscanthus* is having the highest energy-use efficiency because these grasses efficiently produce biomass and play avital role in sustainable agriculture of biofuel production in the near future. *Miscanthus* has several salient features such as high photosynthesis efficiency, high nutrient and WUE, and broad tolerability to adverse climatic conditions and soil. Chromosome number in *Miscanthus* species range from $2n$=28 to 114. The Indian forms of *M. nepalensis* has $2n$=40 chromosomes. Three *Miscanthus* species (*M. sinensis*, *M. sacchariflorus,* and *M. giganteus*) were identified as having the highest potential for biomass production. Cytogenetic analysis of *M.* × *giganteus* revealed that it is a triploid hybrid between *M. sinensis* and *M. sacchariflorus*. Phylogenetic analysis conducted in *Erianthus, Miscanthus, Saccharum, Sclerostachya*, and *Narenga* by analyzing DNA sequences of the organelles did not exhibit any distinct polymorphism between analyses. However, differences in mitochondrial loci were observed between the *M. sinensis* and a *Miscanthus* species from New Guinea with a high chromosome number of 192 while similar mitochondrial loci were reported in members of the *Saccharum* complex. Phylogenetic studies indicated that the *Miscanthus* species originating from New Guinea could have evolved by hybridization between *Miscanthus* and *Saccharum* genera.

12.22.1 SCLEROSTACHYA

The genus *Sclerostachya* (Hack.) A. Camus consists of three species, *S. fusca* (Roxb.) A. Camus, *S. milroyi* Bor. and *S. ridleyi* (Hack.) A. Camus, distributed throughout northern India to an altitude up to 1300 m in Himalayas. *Sclerostachya* is characterized by glabrous sheath and leaves, hollow stem, no root eyes, poorly developed buds, copper-brown panicles, and resistant to waterlogging. Among the allied members in *Saccharum* complex, *Sclerostachya* species are less explored and utilized until now.

12.23 WILD SPECIES AS AN ALTERNATIVE SOURCE FOR BIOMASS AND RENEWABLE BIOENERGY

Bioenergy crops are considered as predominant feedstock for bioenergy production, especially for liquid biofuel, as they are characterized with high biomass yields. Sugarcane is one of the most dependable biomass crops due to its high productivity and ethanol plays a significant role as a bioenergy. The worldwide ethanol supply is primary generate from the sugar and starch feedstocks and the requirement of biofuel has growing major issue about the security and safety of feedstock for the production of food and biofuel. Hybrid commercial cane and its wild cultivars with optimum fiber level and decreased fertilizer and required water, and suitable for cultivation in harsh environmental conditions without competing with food production (Diniz et al., 2019; Hale et al., 2013). In view of the urgent need to combat climate change and replace fossil fuels, sugarcane has become a valuable alternative energy source. The ultimate objective of the energy cane concept, which Alexander originally presented in Puerto Rico in, 1985, is to produce sugarcane cultivars with higher fiber content and biomass production (Bischoff et al., 2008; White et al., 2011). To meet the requirements for bioenergy, sugarcane breeding programs have been conducted globally to create better energy cane cultivars for the use of bioenergy production. Nearly 100 t/ha of biomass can be produced by sugarcane, which is double as much as any other crop. Related grasses such as *Erianthus* and *S. spontaneum* have an equally great capacity to produce large amounts of biomass; *Erianthus* yields about 100 t/ha of biomass and tolerant of suboptimal growing factors, including drought and waterlogging.

E. arundinaceus not only plays an important role in sugarcane breeding, but also has a great potential in environmental remediation and as a biofuel feed stock. Wang et al. (2019) while reviewing *E. arundinaceus* as a bioenergy crop with environmental pollution remediation efficiency, has presented a comprehensive information on its taxonomy, genetic diversity, habitat distribution, potential as feed stock for bioethanol, biogas production, and phytoremediation. November is the most favorable time for harvesting *E. arundinaceus* as a bioenergy feedstock as the calorific value per unit area reaches its maximum. The high biomass yield of *E. arundinaceus* is determined by its growth conditions, planting time, and harvest time. Compared to sugarcane, which yields 32 t/ha of wet bagasse, *Erianthus* produces 60 t/ha of wet bagasse and contains 25–30% fiber. With a larger biomass production and greater adaptability than other energy crops, *Erianthus*—basically a tropical grass—has a far greater potential. Multi ratoonability, tolerance to adverse stresses and maximum fiber level make *Erianthus* as a potential bioenergy crop. *E. arundinaceus* can also be used as feedstock for biogas production. The highest methane yield (based on total solids) from *E. arundinaceus* was reported as 83.5 L/kg. Studies in India have shown that mechanically compressed fibers of *E. arundinaceus* are used for paper production. The chemical oxygen demand of the resulting liquid was approximately 93,974 mg/L, indicating that 1 L of liquid can produce 40 m^3 biogas.

In addition, switch grass is widely used in the USA for energy generation, while miscanthus is already cultivated as an energy crop in Europe. Therefore, breeding programs utilize *S. spontaneum* and *S. officinarum* or other commercial cultivars to generate *Erianthus* arundinaceus clones that produce high fiber and high biomass in order to produce energy canes (Wang et al., 2008). L 79-1002, Ho 00-961, and Ho 02-113 belong to the energy cane varieties that have been made available for the use of feedstock for biofuels. Brazil has developed energy cane varieties under the name Vertix®, with 11 varieties registered in Brazil (Cursi et al., 2022). Improvements in energy cane varietals have also been reported by Argentina, Thailand, India, and Japan. Utilizing *S. spontaneum* and *Erianthus* arundinaceus bases into cultivated sugarcane lines, the energy breeding effort at the ICAR-SBI has resulted in the production of high biomass-producing energy cane (Mohanraj et al., 2020; Govindaraj 2020). These constitute two types of energy canes: type-I (*Saccharum spontaneum*) and type-II energy canes (*Erianthus arundinaceus),* and 18 energy canes were developed. It has been developed to produce type I energy canes with >20% fiber and >15% sucrose, as well as type II energy canes with >25% fiber and <15% sucrose. The success of the Brazilian sugarcane industry in ethanol production has increased interest in producing sugarcane for ethanol throughout the world. Energy cane is more resilient to grow well in a variety of climatic conditions and has a larger biomass potential compare to other plants, such as sorghum, elephant grass, and eucalyptus, according to several breeding programs. Energy cane clones that have been developed can be considered efficient, environmentally friendly substitutes for sugar and lignocelluloses in the manufacturing of ethanol that is both profitable and sustainable.

Though collection and exploitation of *E. arundinaceus* germplasm has been extensively done, more research initiatives are required in the systematic screening and breeding of traits related to biomass utilization. Further, there is a need to develop genotypes with low lignin and high cellulose content for the purpose of cellulosic ethanol production; that requires screening the collected germplasm resources and then improving it through conventional breeding or transgenic approaches (Wang et al., 2019). The potential for interspecific and intergeneric hybrids to provide biomass, energy derived from biomass, and feedstock to meet the increasing demand is considerable.

12.24 FUTURE OUTLOOK

Sugarcane is one of the important industrial crops because of its economic importance and by-products. A wealth of genetic diversity is available in this crop in terms of cane yield, sugar content, and adaptability to environmental stresses and pathogenic diseases. This hidden and unexplored genetic variation is promising for genetic improvement. Increasing the genetic variety of sugarcane germplasm is vital to improve the durability of sugarcane cultivars. Given the significance of identifying wild relatives as sources of novel genetic features, efforts to increase genetic diversity need to focus on crossable wild relatives like *Miscanthus* and *Tripidium* as well as the introgression of genes from wild species inside the *Saccharum* species complex. In current scenario, the diverse environmental stresses due to climate change, challenges the identification of novel varieties with diverse genetic basis and with resistance to biotic or abiotic stresses. In recent years, the exploitation of potential diversity stored in the unutilized genetic resources had become an important objective for crop improvement in sugarcane and the prospects to utilize the gentic variability are large. It is now important that conventional breeding can be integrated with genomic/biotechnological tools to harness the maximum advantage to manage the sugarcane stresses. Advancements in plant biotechnology, especially genomics have paved the way for a detailed understanding of the mechanisms underlying various biotic and abiotic stress responses. In this era of smart breeding, genetic enhancement in this highly polyploid crop is possible through appropriate utilization of germplasm resources integrated with molecular farming, information technology and artificial intelligence. By continuous exploration of the new and unutilized germplasm resources, appropriate breeding technologies and allele mining in the "*Saccharum* complex," it is possible to broaden the genetic base and identify new varieties combining high yield, high sucrose, disease and abiotic stress resistance, and widely suitable for mechanization and as well to meet climate change challenges for the sustainable development of cane-agro-industries. Emphasis should also be concentrated in developing energy canes with high biomass and varieties with high ethanol for sustainable and economic ethanol production.

KEYWORDS

- ***Saccharum* spp.**
- **hybrid**
- **wild type**
- **germplasm collection**
- **taxonomy**
- **cytogenetics**
- **genome structure**
- **gene pool**
- **intra-interspecific program**

REFERENCES

Aitken KS, McNeil MD, Hermann S, Bundock, PC, Kilian A, Heller Uszynska KA (2014). Comprehensive genetic map of sugarcane that provides enhanced map coverage and integrates high-throughput diversity array technology (DArT) markers. *BMC Genom.* 15, 152.

Aitken K., Berkman P., Rae A. (2016). The first sugarcane genome assembly: How can we use it? *Proc. Aust. Soc. Sugar Cane Technol.* 38, 193–199.

Aitken K, Li J, Piperidis G, Qing C, Yuanhong F, Jackson P. (2018). Worldwide genetic diversity of the wild species *Saccharum spontaneum* and level of diversity captured within sugarcane breeding programs. *Crop Sci.* 58(1), 218–229.

Aitken K, McNeil M. (2010). Diversity analysis. In: Henry R., Kole C. (Eds.), *Genetics, Genomics and Breeding of Sugarcane* (pp. 19–42). Science Publisher: Enfield, NH, USA.

Aitken KS, Jackson PA, McIntyre CL. (2007). Construction of a genetic linkage map for *Saccharum officinarum* incorporating both simplex and duplex markers to increase genome coverage. *Genome.* 50, 742–756.

Alarmelu S, Adhini SP, Jayabose C, Manjunatha T. (2017). Pre-breeding utilizing the improved intraspecific clones of *Saccharum* spp, In: Hemaprabha et al. (Eds.), *Proceedings of International Symposium on Sugarcane Research Since Co 205: 100 Years and Beyond (Sucrosym 2017)*. ICAR-Sugarcane Breeding Institute.

Alarmelu S, Nagarajan, R, Shanthi RM, Hemaprabha G, Nair NV. (2014). Development and evaluation of backcross progenies of improved *Saccharum* spp. for yield and quality traits. *J. Sugarcane Res.* 1, 19–32.

Alexander AG. (1985). *The Energy Cane Alternative*. Elsevier: Amsterdam.

Alwala S, Suman A, Arro JA, Veremis JC, Kimbeng CA. (2006). Target region amplification polymorphism (TRAP) for assessing genetic diversity in sugarcane germplasm collections. *Crop Sci.* 46(1), 448–455.

Amalraj VA, Balasundaram N. (2006). On the taxonomy of the members of '*Saccharum* complex'. *Genet. Resour. Crop Evol.* 53, 35–41.

Anas M, Verma KK, Haq IU, Naeem M, Li Q, Liao F, Xiang YX, Li YR. (2021a). Characterization of exotic, native and wild-type genotypes of sugarcane (*Saccharum* spp. Hybrids) for internal nitrogen use efficiency under different nitrogen levels. *Sugar Tech.* 23, 1258–1267. https://doi.org/10.1007/s12355-021-01014-1.

Anas M, Verma KK, Riaz M, Qiang L, Liao F, Liu Y, Li YR. (2021b). Physio-morphological and biochemical mechanism of nitrogen use efficiency in sugarcane (*Saccharum* spp.) genotypes under different growth stages and nitrogen levels. *J. Plant Interact.* 16, 332–343.

SBI. (2001). Annual Report 2000–2001. Sugarcane Breeding Institute, Coimbatore.

Arceneaux G. (1965). Cultivated Sugarcanes of the World and their Botanical Derivation.

Arceneaux G. (1967). Cultivated sugarcanes of the world and their botanical derivation. *Proc. Int. Soc. Sugare Cane Technol.* 12, 844–854.

Arro JA, Veremis JC, Kimbeng CA, Botanga C. (2006). Genetic diversity and relationships revealed by AFLP markers among *Saccharum spontaneum* and related species and genera. *J. Am. Soc. Sugar Cane Technol.*, 26,101–115.

Barber CA. (1922). The classification of indian canes. *Int. Sugar J.* 24, 18–20.

Barber CA. (1920). The origin of the sugarcane. *Int. Sugar J.* 22,249–251.

Barreto FZ, Rosa JRBF, Balsalobre TWA, Pastina MM, Silva RR, Hoffmann HP, de Souza AP, Garcia AAF, Carneiro MS. (2019). A genome-wide association study identified loci for yield component traits in sugarcane (*Saccharum* spp.). *PLoS One.* 14(7), e0219843.

Berding N, Roach BT. (1987). Germplasm collection, maintenance and use. In: Heinz DJ (Ed.), *Sugarcane Improvement Through Breeding* (pp. 143–210). Elsevier: Amsterdam.

Bhatt R, Verma KK, Jamwal NS. (2023a). Can paddy-compost eliminate N fertilization for climate-smart sugarcane? *Sugar Tech.* 25, 1036–1046. https://doi.org/10.1007/s12355-023-01270-3.

Bhatt R, Verma KK, Kumar R, Sanghera GS. (2023b). Foliar application of potassium salt of active phosphorus (PSAP) mitigates insect pests and improves yield along with sugarcane quality in response to agroclimatic conditions of Punjab. *Sugar Tech.* 25(3), 660–669.

Bischoff KP, Gravois KA, Reagan TE, Hoy JW, Kimbeng CA, LaBorde CM, Hawkins GL (2008). Registration of 'L 79-1002'sugarcane. *J. Plant Regist.* 2, 211–217.

Brandes E. (1958). Origin, classification and characteristics. In: Artschwager E, Brandes EW (Eds.), *Sugarcane (Saccharum officinarum* L.) (pp. 1–35). U.S. Department of Agriculture Handbook.

Brandes E. (1956). Origin, dispersal and use in breeding of the Melanesian garden sugarcane and their derivatives, *Saccharum officinarum* L. *Proc. Int. Soc. Sugar Cane Technol.*, 9, 709–750.

Bremer G. (1962). Problems in breeding and cytology of sugar cane. Chromosome increase in *Saccharum* hybrids in relation to interspecific and intergeneric hybrids in other genera. *Euphytica.* 11, 65–80.

Bremer G. (1925). The cytology of the sugarcane. The chromosomes of primitive forms of the genus *Saccharum*. *Genetica*. 7, 293–322.

Bremer G. (1961). Problems in breeding and cytology of sugar cane. III. The cytological crossing research of sugar cane. *Euphytica*. 10, 229–243.

Brown JS, Schnell RJ, Power EJ, Douglas SL, Kuhn DN. (2007). Analysis of Clonal Germplasm from Five *Saccharum species: S. barberi S. robustum, S. officinarum, S. sinense and S. spontaneum*. A study of inter- and intra-species relationships using microsatellite markers. *Genet. Res. Crop Evol.* 54, 627–648.

Buzacott JH. (1965). Cane varieties and breeding. In: King NJ, Mungomery RW, Hughes CG (Eds.), *Manual of Cane Growing* (pp. 220–253). Angus and Robertson: Sydney, Australia.

Cesnik R, Miocque J. (2004). *Improvement of Sugar Cane* (pp. 1–307), Embrapa: Brasilia.

Chandran K, Nisha M, Gireesan PP. (2017). Characterization of inter-specific progenies from polycrosses of two *Saccharum robustum* clones belonging to the forma '*sanguineum*'. In: *Proceedings of International Symposium on Sugarcane Research Since CO 205: 100 Years and Beyond (SucroSym 2017)*; Hemaprabha et al. (Eds.), ICAR-Sugarcane Breeding Institute, 2017.

Chandran K, Nisha M, Gopi R, Mahendran B, Chandran D, Mahesh P, Arun Kumar R, Krishnapriya V, Gomathi R, Malathi P, Viswanathan R, Hemaprabha G. (2023). Sugarcane genetic resources for challenged agriculture. *Sugar Tech.* 25, 1285–1302.

Chen GL, Zheng FJ, Lin B, Lao SB, He J, Huang Z, Zeng Y, Sun J, Verma KK. (2020). Phenolic and volatile compounds in the production of sugarcane vinegar. *ACS Omega.* 5, 30587–30595.

Chen G-L, Zheng F-J, Lin B, Yang Y-X, Fang X-C, Verma KK, Yang L-F. (2023). Vinegar: A potential source of healthy and functional food with special reference to sugarcane vinegar. *Front. Nutr*. 10, 1145862. https://doi.org/10.3389/fnut.2023.1145862.

Cuadrado A, Acevedo R, Moreno Diaz De La Espina S, Jouve N, De La Torre C. (2004). Genome remodeling in three modern *S. officinarum* × *S. spontaneum* sugarcane cultivars. *J. Exp. Bot.* 55, 847–854.

Cursi DE, Hoffmann HP, Barbosa GVS, Bressiani JA, Gazaffi R, Chapola RG, Fernandes Junior AR, Balsalobre TWA, Diniz CA, Santos JM, et al. (2022). History and current status of sugarcane breeding, germplasm development and molecular genetics in Brazil. *Sugar Tech*. 24, 112–133.

D'Hont A. (2005). Unraveling the genome structure of polyploids using FISH and GISH; Examples of sugarcane and banana. *Cytogenet. Genome Res.* 109, 27–33.

D'Hont A, Glaszmann JC. (2001). Sugarcane genome analysis with molecular markers: A first decade of research. *Proc. Int. Soc. Sug. Cane Technol.* 2, 556–559.

D'Hont A, Grivet L, Feldmann P, Rao P, Berding N, Glaszmann JC. (1996). Characterisation of the double genome structure of modern sugarcane cultivars (*Saccharum* spp.) by molecular cytogenetics. *Mol. Gen. Genet.*, 250, 405–413.

D'Hont A, Ison D, Alix K, Roux C, Glaszmann JC. (1998). Determination of basic chromosome numbers in the genus *Saccharum* by physical mapping of ribosomal RNA genes. *Genome*, 41, 221–225.

D'Hont A, Layssac M. (1998). Analysis of cultivars genome structure by molecular cytogenetics and the study of introgression mechanisms. In: Annual Crops Department. (Eds.), *Centre de Coopération International en Recherche Agronomique Pour le Développement* (pp. 9–10). CIRAD-CA.

D'Hont A, Lu YH, Feldmann P, Glaszmann JC. (1993). Cytoplasmic diversity in sugarcane revealed by heterologous probes. *Sugar Cane*. 1, 12–15.

D'Hont A, Rao PS, Feldmann P, Grivet L, Islam-Faridi N, Taylor P, Glaszmann JC. (1995). Identification and characterization of sugarcane intergeneric hybrids *Saccharum officinarum* × *Erianthus arundinaceus*, with molecular markers and DNA in situ hybridization. *Theor. Appl. Genet*. 91, 320–326.

da Silva J, Honeycutt RJ, Burnquist W, Al-Janabi SM, Sorrells ME, Tanksley SD. (1995). *Saccharum spontaneum* L 'SES 208' genetic linkage map combining RFLP- and PCR-based markers. *Mol. Breed.* 1, 165–179.

Daniels J, Roach BT. (1987). Taxonomy and evolution. In: Heinz DJ (Eds.), *Sugarcane Improvement Through Breeding* (pp. 7–84). Elsevier: Amsterdam.

Daniels J, Daniels CA. (1975). Geographical, historical and cultural aspects of the origin of the indian chinese sugarcanes *S. barberi* and *S. sinense*. *Int. Soc. Sugar Cane Technol. Sugargarcane Breed. Newslett.* 1028-1193, 36 4–23.

Daniels J, Smith P, Paton N, Williams CA. (1975). The origin of the genus *Saccharum. Sugarcane Breed. Newslett.* 36, 24–39.

Deng ZN, Huang HR, Huang CM, Verma KK, Luo HB, Jiang SL, Li YJ, Cao HQ, Wu KC, Xu L. (2023). Short-term limited water irrigation influences the photosynthetic pigments, enzymatic and non-enzymatic activities in *Saccharum spontaneum* L. at vegetative stage. *Sugar Tech.* 25, 110–118. https://doi.org/10.1007/s12355-022-01175-7.

D'Hont A, Lu, YH, Paulet F, Glaszmann, JC. (2002). Oligo clonal interspecific origin of 'North Indian' and 'Chinese' Sugarcanes. *Chromosome Res.* 10, 253–262.

Diniz AL, Ferreira SS, ten-Caten F, Margarido GRA, dos Santos JM, Barbosa GVd.S, Carneiro MS, Souza GM. (2019). Genomic resources for energy cane breeding in the post genomics era. *Comput. Struct. Biotechnol. J.* 2019, 17, 1404–1414.

Dos Santos JM, Duarte Filho LSC, Soriano ML, da Silva PP, Nascimento VX, de Souza Barbosa GV, Todaro AR, Neto CER, Almeida C. (2012). Genetic diversity of the main progenitors of sugarcane from the RIDESA germplasm bank using SSR markers. *Ind. Crop Prod.* 40, 145–150.

Dufresne F, Stift M, Vergilino R, Mable BK. (2014). Recent progress and challenges in population genetics of polyploid organisms: An overview of current state-of-the-art molecular and statistical tools. *Mol. Ecol.* 23(1), 40–69.

Duncelman PH, Breaux RD. (1972). Breeding sugarcane varieties for louisiana with new germplasm. *Proc. Int. Soc. Sugare Cane Technol.* 14, 233–239.

Duncleman PH, Breaux RD. (1970). New sugarcane breeding clones from indian crosses evaluated at Houma LA, 1966–1969. *Int. Sugar J.*, 72, 43–44.

FAO. (2021). FAOSTAT, FAO statistical databases. Retrived January 5, 2024. from http:// www.fao.org/-faostat/zh/#data/.

FAO (Food and Agriculture Organization). *Agricultural Production Statistics*; 2023. (fao. org).

Figueira T, Okura V, Rodrigues Da Silva F, Jose Da Silva M, Kudrna D, Ammiraju JS. (2012). A BAC library of the SP80-3280 sugarcane variety (*Saccharum* sp.) and its inferred microsynteny with the *Sorghum* genome. *BMC Res.* 5, 185.

Gopi R, Chandran K, Sundar AR, Viswanathan R, Jayakumar V, Mahendran B, Jayaraman S, Gireesan P. (2024). Thirty-two years of smut disease status in the world's largest sugarcane germplasm. *Indian Phytopathol.* 77, 51–59.

Govindaraj P. (2020). SBIEC 14006—A high biomass energycane for power, alcohol and paper industries. *J. Sugarcane Res.* 10, 100–106.

Grassl CO. (1977). The origin of the sugar producing cultivars of *Saccharum. ISSCT Sugarcane Breeders NewsLetters*, 39, 8–33.

Grivet L, Daniels C, Glaszmann JC, D'Hont A. (2004). A review of recent molecular genetics evidence for sugarcane evolution and domestication. *Ethnobot. Res. Appl.* 2, 9–17.

Grivet L, Glaszmann J-C, D'Hont A. (2006). Molecular evidence of sugarcane evolution and domestication. In: (Eds.), *Darwin's Harvest: New Approaches to the Origins, Evolution, and Conservation of Crops* (pp. 49–66). Columbia University Press.

Guo D-J, Li D-P, Chen Z-D, Yang B, Verma KK, Singh RK, Singh P, Khan Q, Sharma A, Qin Y, Zhang B-Q, Song X-P, Li Y-R. (2023). Effect of endophytic diazotroph *Enterobacter roggenkampii* ED5 on nitrogen-metabolism-related microecology in the sugarcane rhizosphere at different nitrogen levels. *Front. Microbiol.* 14, 1132016. https://doi.org/10.3389/fmicb.2023.1132016.

Guo D-J, Li D-P, Singh RK, Singh P, Sharma A, Verma KK, Qin Y, Khan Q, Lu Z, Malviya MK, Song X-P, Xing Y-X, Li Y-R. (2021). Differential protein expression analysis of two sugarcane varieties in response to diazotrophic plant growth-promoting endophyte enterobacter roggenkampii ED5. *Front. Plant Sci.* 12, 727741. https://doi.org/10.3389/fpls.2021.727741.

Guo DJ, Li DP, Singh RK, Singh P, Verma KK, Sharma A, Qin Y, Khan Q, Song XP, Malviya MK, Xing YX, Li YR. (2022). Comparative transcriptome analysis of two sugarcane varieties in response to diazitrophic plant growth promoting endophyte *Enterobacter roggenkampii* ED5. *J. Plant Interact.* 17(1), 75–84.

Hale AL, Todd JR, Gravois KA, Mollov D, Malapi-Wight M, Momotaz A, Laborde C, Goenaga R, Kimbeng C, Solis A, Waguespack H. (2022). Sugarcane breeding programs in the USA. *Sugar Tech.* 24, 97–111.

Hale AL, Dufrene EO, Tew TL, Pan Y-B, Viator RP, White PM, Veremis JC, White, WH, Cobill R, Richard EP, Jr., et al. (2013). Registration of 'Ho 02-113' sugarcane. *J. Plant Regist.* 7, 51–57.

Heinz DJ. (1987). *Sugarcane Improvement Through Breeding*, Elsevier: Amsterdam.

Heinz DJ, Osgood RV, Moore. PM. (1994). Sugarcane. *Encycl. Ag. Sci.* 1994, 4, 225–238.

Heinz DJ. (1967). Wild *Saccharum* species for breeding in Hawaii. *Proc. Int. Soc. Sug. Cane Technol.* 12, 1037–1043.

Hemaprabha G, Mohanraj K, Alarmelu S, Ram B. (2020). Assessment of breeding methods and parental value of Co canes developed during 1918 to 2017. *J. Sugarcane Res.* 10 (1), 24–31.

Hemaprabha G, Mohanraj K, Alarmelu S, Ram B. (2017). Relative performance of coimbatore canes (Co canes) for major component traits of yield and quality and an analysis of their genealogies to measure genetic gain over a century of sugarcane breeding at ICAR Sugarcane Breeding Institute. In: Hemaprabha et al. (Eds.), *International Symposium on Sugarcane Research Since Co 205: 100 Years and Beyond ScroSym* (pp. 98–102). Sugarcane Breeding Institute, Coimbatore, India.

Hemaprabha G, Ram B. (1993). Genetic variability in nobilization stages of *Saccharum robustum* Brandes et Jeswiet ex Grassl. *Sugarcane*. 6, 6–9.

Hemaprabha G, Mohanraj K, Jackson PA, Lakshmanan P, Ali GS, Li AM, Huang DL, Ram B. (2022). Sugarcane genetic diversity and major germplasm collections. *Sugar Technol.* 24, 279–297.

Henry RJ, Kole C (2010). In: (Eds.), *Genetics, Genomics and Breeding of Sugarcane*. Science Publishers: Enfield.

Hodkinson TR, Chase MW, Lledó MD, Salamin N, Renvoize SA. (2002). Phylogenetics of *Miscanthus*, *Saccharum* and related genera (Saccharinae, Andropogoneae, Poaceae) based on DNA sequences from ITS nuclear ribosomal DNA and plastid trnLintron and trnL-F intergenic spacers. *J. Plant. Res.* 115, 381–392.

Hogarth DM, Cox MC, Bull JK. (1997). Sugarcane improvement: Past achievements and future prospects. In: Kang, MS (Eds.), *Crop Improvement for the 21st Century* (pp. 29–56). Research Signpost: Trivandrum, India.

Hogarth M. (1987). Genetics of Sugarcane. In: Heinz DJ (Eds.), *Developments in Crop Science. II. Sugarcane Improvement through Breeding* (pp. 255–271). Elsevier: New York.

http://www.kew.org/data/grasses-db.html.

Huang CM, Wu KC, Verma KK, Luo HB, Deng ZN, Shi GY, Cao HQ, Wu XJ, Ye LP, Wei YW, Xu L. (2024). Isolation and expression analysis of cellulose synthase 3 (*Ces3*) genes from sugarcane (*Saccharum officinarum* L.). *Chem. Biol. Technol. Agric.* 11, 71. https://doi.org/10.1186/s40538-024-00597-7.

Huang HR, Gao YJ, Malviya MK, Verma KK, Solanki MK, Huang YX, Li X, Deng Y, Yan J, Tang SY, Wang LW, Xu L. (2022). Genetic diversity analysis of sugarcane (*Saccharum* spp. Hybrids) among high-sucrose clones of GT series and commonly used parents by using microsatellite markers in Guangxi, China. *Sugar Tech.* 24, 397–407. https://doi.org/10.1007/s12355-021-01033-y.

ICAR-SBI Vision 2050. ICAR-Sugarcane Breeding Institute, Coimbatore.

Irvine JE. (1999). *Saccharum* species as horticultural classes. *Theor. Appl. Genet.* 98, 186–194.

Jaiswal R, Mall RK, Patel S, Singh N, Mendiratta N, Gupta A. (2023). Indian sugarcane under warming climate: A simulation study. *Eur. J. Agron.* 2023, 144, 126760.

James GL. (2004). An introduction to sugarcane. In: James G. (Ed.), *Sugarcane*, 2nd edn (pp. 1–18). Blackwell Science Limited: United States.

Jeswiet J. (1930). The development of selection and breeding of the sugarcane in java. *Proc. Int. Soc. Sugar Cane Technol.* 3, 44–57.

Joshi A, Verma KK, Rajput VD, Arora J. (2023). Metabolic engineering of microorganisms in advancing biofuel production. In: *Advances in Lignocellulosic Biofuel Production Systems* (pp. 181–202). https://doi.org/10.1016/B978-0-323-91192-4.00016-X.

Joshi A, Verma KK, Rajput VD, Minkina T, Arora J. (2022). Recent advances in metabolic engineering of microorganisms for advancing lignocellulose-derived biofuels. *Bioengineered.* 13(4), 8135–8163. https://doi.org/10.1080/21655979.2022.2051856.

Kandasami PA, Thiraviam N. (1987). The Association Study in Sugarcane, M.Sc. (Agriculture) Thesis RAU, Pusa, Samastipur, Bihar.

Kandel R, Yang X, Song J, Wang J. (2018). Potentials, challenges, and genetic and genomic resources for sugarcane biomass improvement. *Front. Plant Sci.* 9, 151.

Lam E, Shine J, Silva JD, Lawton M, Bonos S, Calvino M, Carrer H, Silva-Filho MC, Glynn N, Helsel Z, Ma J, Richard E, Souza GM, Ming, R. (2009). Improving sugarcane for biofuel: Engineering for an even better feedstock. *GCB Bioenergy.* 1, 251–255.

Le Cunff L, Garsmeur O, Raboin LM, Pauquet J, Telismart H, Selvi A. (2008). Diploid/polyploid syntenic shuttle mapping and haplotype-specific chromosome walking toward a rust resistance gene (Bru1) in highly polyploid sugarcane (2n approximately 12x approximately 115). *Genetics.* 180, 649–660.

Legendre, BI. (1989). Use of ferral germplasm for sugarcane improvement in Lousiana. *Proc. Int. Soc. Sug. Technol.,* 20, 883–891.

Li A-M, Liao F, Wang M, Chen Z-L, Qin C-X, Huang R-Q, Verma KK, Li Y-R, Que Y-X, Pan Y-Q, et al. (2023). Transcriptomic and proteomic landscape of sugarcane response to biotic and abiotic stressors. *Int. J. Mol. Sci.* 24, 8913. https://doi.org/10.3390/ijms24108913.

Lima M, Garcia A, Oliveira K, Matsuoka S, Arizono H. (2002). Analysis of genetic similarity detected by AFLP and coefficient of parentage among genotypes of sugarcane (*Saccharum* spp.). *Theor. Appl. Genet.* 104, 30–38.

Lu YH, D'Hont, A, Paulet, F, Grivet, L, Arnaud M, Glaszmann JC. (1994). Molecular diversity and genome structure in modern sugarcane varieties. *Euphytica.* 78, 217–226.

Manechini JRV, da Costa JB, Pereira BT, Carlini-Garcia LA, Xavier MA, Landell MGDA, Pinto LR. (2018). Unraveling the genetic structure of Brazilian commercial sugarcane cultivars through microsatellite markers. *PLoS One*. 13(4), e0195623.

Mary S, Nair NV, Chaturvedi PK, Selvi A. (2006). Analysis of genetic diversity among *Saccharum spontaneum* L. from four geographical regions of India, using molecular markers. *Genet. Resour. Crop Evol.* 53, 1221–1231.

Medeiros C, Balsalobre TWA, Carneiro MS. (2020). Molecular diversity and genetic structure of *Saccharum* complex accessions. *PLoS One*. 15(5), e0233211.

Miller JD, Tai P. (1992). Use of plant introductions in sugarcane cultivar development. In: Shands HL, Weisner LE. (Eds.), *Use of Plant Introductions in Cultivar Development, Part 2* (pp. 137–149). Crop Science Society of America: Madison, WI, USA.

Milligan SB, Martin FA, Bischoff KP, Quebedeaux JP, Dufrene EO, Quebedeaux KL, Miller JD. (1994). Registration of 'LCP 85-384' Sugarcane. *Crop Sci.* 34, 819–820.

Mohanraj KSA. (2018). Broadening the genetic base of sugarcane: Introgression of red rot resistance from the wild relative *Erianthus procerus*. In: (Eds.), *Proceedings of XII International Society of Sugar Cane Technologists Pathology Workshop.* ICAR-SBI, Coimbatore.

Mukherjee SK. (1957). Origin and distribution of *Saccharum*. *Bot. Gaz.* 119, 55–61.

Naidu KM, Sreenivasan TV. (1987). Genetic resources of sugarcane and importance of wild relatives. In: Paroda et al. (Eds.), *Plant Genetic Resources-Indian Perspective*. NBPGR: New Delhi.

Nair NV. (2012). Sugarcane genetic resources—Status, potential and role in sugarcane improvement. *J. Sugarcane Res.* 2(2), 1–8.

Nair NV. (2008). Sugarcane breeding institute, coimbatore: A perspective. *Sugar Tech*. 10, 285–292.

Nair NV, Mohanraj K, Sunadaravelpandian K, Suganya A, Selvi A, Appunu C. (2017). Characterization of an intergeneric hybrid of *Erianthus* procerus × *Saccharum* officinarum and its backcross progenies. *Euphytica*. 213, 1–11

Nair NV, Nagarajan R, Amalraj VA (2006). *Saccharum* germplasm collection from the cauvery river basin and Coastal Tamil Nadu, India. *Plant Genet. Res. Newslett.*, 146, 56–59.

Nair NV, Nair S, Sreenivasan, TV, Mohan M. (1999). Analysis of genetic diversity and phylogeny in *Saccharum* and related genera using RAPD markers. *Genet. Resour. Crop Evol.* 46, 73–79.

Nair NV, Selvi A, Sreenivasan TV, Pushpalatha KN. (2002). Molecular diversity in Indian sugarcane cultivars as revealed by randomly amplified DNA polymorphisms. *Euphytica*. 127(2), 219.

Nair NV, Sreenivasan TV, William Jebadhas A. (1998). Intraspecific improvement in *Saccharum.* In: (Eds.), *Proceedings of the First National Plant Breeding Congress, Indian Society of Plant Breeders* (pp. 140–145). Tamil Nadu Agricultural University: Coimbatore.

Nobrega JCM, de Dornelas MC. (2006). Biotecnologia e Melhoramento da Cana-de-Açúcar. In: Segato SV, Pinto AS, Jendiroba E, Nóbrega JCM. (Eds.), *Atualização em Produção de Cana-de-açúcar* (pp. 39–56).

Oliveira MA, Verma KK, Bhatt R, Oliveira TBA. (2022). Impact of green and organic fertilizers on soil fertility and sugarcane productivity. In: Verma KK, et al. (Eds.),

Agro-Industrial Perspectives on Sugarcane Production under Environmental Stress. https://doi.org/10.1007/978-981-19-3955-6_11.

Olsson L, Cotrufo F, Crews T, Franklin J, King A, Mirzabaev A, Scown M, Tengberg A, Villarino S, Wang Y. (2023). The state of the world's Arable land. *Annu. Rev. Env. Resour.* 2023, 48, 451–475.

Panje RR, Babu CN. (1960). Studies in *Saccharum spontaneum* distribution and geographical association of chromosome numbers. *Cytologia*. 25, 152–72.

Park S, Zhang D, Ali GS. (2024). Assessing the genetic integrity of sugarcane germplasm in the USDA-ARS National Plant Germplasm System collection using single-dose SNP markers. *Front. Plant Sci.* 14, 1337736. https://doi.org/10.3389/fpls.2023.1337736.

Parthasarathy N. (1948). Origin of noble sugarcanes (*Saccharum officinarum* L.). *Nature*. 161, 608.

Parthasarathy N. (1946). The probable origin of North Indian Sugarcanes. MOP. Iyengar Commemoration Volume. *J. Indian Bot. Soc.* 1946, 133–150.

Piperidis G, D'Hont A, Hogarth D. (2001). Chromosome Composition Analysis of Various *Saccharum* Interspecific Hybrids by Genomic in Situ Hybridization (GISH). In: Hogarth DM. (Eds.), *International Society of Sugar Cane Technologists*, *Proceedings of the XXIV Congress, Vol.* 2 (pp. 565–566). Australian Society of Sugar Cane Technologists: Brisbane.

Piperidis G, Piperidis N, D'Hont A. (2010). Molecular cytogenetic investigation of chromosome composition and transmission in sugarcane. *Mol. Genet. Genom*. 284, 65–73.

Premachandran MN, Lekshmi M, Viola VR, Remadevi RK, Pazhany AS. (2019). Androgenic diploids of *Erianthus arundinaceus* with paternal chloroplast and mitochondrial genomes in backcross progeny of *Saccharum* × *Erianthus* intergeneric hybrids. *Trop. Plant Biol.* 12(1), 1–11. https://doi.org/10.1007/s12042-018-9214-2.

Premachandran MN, Prathima, PT, Lekshmi M. (2011). Sugarcane and polyploidy—A review. *J. Sugarcane Res.* 1(2), 1–15.

Premachandran MN, Sobhakumari VP, Lekshmi M, Viola R. (2017). Genome characterization of *in vitro* induced amphiploids of an intergeneric *Hybrid Erianthus arundinaceus* × *Saccharum spontaneum. Sugar Tech*. 19(4), 386–393. https://doi.org/10.1007/s12355-016-0482-6.

Price S. (1963). Cytogenetics of modern sugarcanes. *Econ. Bot.* 17, 97–105.

Price S. (1961). Cytological studies in *Saccharum* and allied genera. VII. Maternal chromosome transmission by *S. officinarum* in intra- and interspecific crosses. *Bot. Gaz.* 1961, 122, 298–305.

Price S. (1968). Cytology of Chinese and North Indian sugar canes. *Econ. Bot.* 22(2), 155–164.

Price S. (1965). Cytology of *Saccharum robustum* and related sympatric species and natural hybrids. USDep. *Agric. Res. Serv. 1965 Tech. Bull.* 1337.

Price S. (1967). Interspecific hybridization in sugarcane breeding. *Proc. Int. Soc. Sug. Cane Technol.* 12, 1021–1026.

Purseglove JW. (1972). Tropical crops: In: *Monocotyledons*. Longman, New York.

Qi Y, Gao X, Zeng Q, Zheng Z, Wu C, Yang R, Feng X, Wu Z, Fan L, Huang Z. (2021). Sugarcane breeding, germplasm development and related molecular research in China. *Sugar Tech.* 1, 21.

Qi YW, Deng HH, Li QW. (2012). Advance in utilization of sugarcane germplasm in China mainland. *Crop Res.*, 26 (5), 443–446.

Ram B, Kumar S. (2007). Screening of sugarcane clones for abiotic stresses under subtropical conditions. In: Ram B., Nair NV (Eds.), *Platinum Jubilee Souvenir* (pp. 20–24). Sugarcane Breeding Institute Regional Centre: Karnal.

Roach BT. (1969). Cytological studies in *Saccharum*. In: Liu K-C. (Eds.), *Chromosome Transmission in Interspecific and Intergeneric Crosses*. *Proceeding*s of the *International Society of Sugar Cane Technologists XIII* (pp. 901–920).

Roach BT. (1986). Evaluation and use of sugarcane germplasm. *Proc. Int. Soc. Sug. Cane Technol.* 16, 492–503.

Roach BT, Daniels J. (1987). A review of the origin and improvement of sugarcane. In: *Copersucar International Sugarcane Breeding Workshop* (pp. 1–31).

Simmonds NW. (1976). *"Sugarcane," in Evolution of Crop Plants* (pp. 104–108). Longmans: London.

Singh P, Singh RK, Li HB, Guo DJ, Sharma A, Verma KK, Solanki MK, Upadhyay SK, Lakshmanan P, Yang LT, Li YR. (2024). Nitrogen fixation and phytohormone stimulation of sugarcane plant through plant growth promoting diazotrophic *Pseudomonas*. *Biotechnol. Genetic Eng. Rev.* 40, 15–35. https://doi.org/10.1080/02648725.2023.2177814.

Singh RB, Singh B, Singh RK. (2018). Evaluation of genetic diversity in *Saccharum* species clones and commercial varieties employing molecular (SSR) and physiological markers. *Indian J. Plant Genet. Resour.*, 31(1), 17–26.

Sobhakumari VP, Mohanraj K, Nair NV, Mahadevaswamy HK, Ram B. (2020). Cytogenetic and molecular approaches to detect alien chromosome introgression and its impact in three successive generations of *Erianthus procerus* × *Saccharum*. *Cytologia*. 85(4), 341–346.

Song X-P, Mo F, Verma KK, Wei J, Zhang X, Yang L, Li YR. (2019). Effect of sugarcane smut (*Ustilago scitaminea* Syd.) on ultrastructure and biochemical indices of sugarcane. *Biomed J. Sci. Tech. Res.* 12546–12551. https://doi.org/10.26717/BJSTR.2019.17.002950.

Song XP, Mo F, Yan M, Zhang X, Zhang B, Huang X, Huang D, Pan Y, Verma KK, Li YR. (2022). Effect of smut infection on the photosynthetic physiological characteristics and related defense enzymes of sugarcane. *Life.* 12, 1201. https://doi.org/10.3390/life12081201.

Song XP, Verma KK, Tian DD, Zhang XQ, Liang YJ, Huang X, Li CN, Li YR. (2021). Exploration of silicon functions to integrate with biotic stress tolerance and crop improvement. *Biol. Res*. 54, 19. https://doi.org/10.1186/s40659-021-00344-4.

Souza GM, Berges H, Bocs S, Casu R, D'Hont A, Ferreira JE. (2011). The Sugarcane genome challenge: Strategies for sequencing a highly complex genome. *Trop. Plant Biol*. 4, 145–156.

Sreenivasan TV, Ahloowalia BS, Heinz DJ. (1987). Cytogenetics. In: Heinz DJ (Eds.), *Sugarcane Improvement Through Breeding* (pp. 211–253). Elsevier: Amsterdam.

Stevenson GC. (1965). *Genetics and Breeding of Sugar Cane* (pp. 44–45). Longmans, Green and Co.: London.

Suganya A, Nithyanandham R, Andriya Sathish. (2014). Variability studies in the interspecific hybrids of sugarcane derived with different cytotypes of *Saccharum spontaneum*.L. *J. Sugar. Res.* 4(2), 76–79.

Tai P, Miller J. (2002). Germplasm diversity among four sugarcane species for sugar composition. *Crop Sci.* 42, 958–964.

Verma KK, Li DM, Singh M, Rajput VD, Malviya MK, Minkina T, Singh RK, Singh P, Song XP, Li YR. (2020a). Interactive role of silicon and plant-rhizobacteria mitigating abiotic stresses: A new approach for sustainable agriculture and climate change. *Plants.* 9, 1055. https://doi.org/10.3390/plants9091055.

Verma KK, Rajput VD, Singh M, Verma CL, Minkina T. (2022a). Influence of waterlogging on plant physiology and molecular biology. In: *Understanding Abiotic Stresses.* Nova Science Publishers, Inc.: New York.

Verma KK, Siddiqui N, Singh M. (2022b). Understanding the plant physiological and molecular adaptation strategies to soil salinity. In: *Understanding Abiotic Stresses*. Nova Science Publishers, Inc.: New York. https://doi.org/10.52305/UBUJ4024.

Verma KK, Singh P, Song X-P, Malviya MK, Singh RK, Chen G-L, Solomon S, Li YR. (2020b). Mitigating climate change for sugarcane improvement: Role of silicon in alleviating abiotic stresses. *Sugar Tech.* 22(5), 741–749. https://doi.org/10.1007/s12355-020-00831-0.

Verma KK, Song X, Degu HD, Guo D, Joshi A, Huang H, Xu L, Singh M, Huang D, Rajput VD, Li YR. (2023a). Recent advances in nitrogen and nano-nitrogen fertilizers for sustainable crop production: A mini-review. *Chem. Biol. Technol. Agric.* 10, 1–14. https://doi.org/10.1186/s40538-023-00488-3.

Verma KK, Song X, Verma CL, Huang H, Singh M, Xu L, Wu J, Khan ZH, Zhang B, Li Y. (2023b). Mathematical modeling of climate and fluoride effects on sugarcane photosynthesis with silicon nanoparticles. *Plant Physiol. Biochem.* 204, 108089. https://doi.org/10.1016/j.plaphy.2023.108089.

Verma KK, Song XP, Budeguer F, Nikpay A, Enrique R, Singh M, Zhang B, Wu JM, Li YR. (2022c). Genetic engineering: An efficient approach to mitigating biotic and abiotic stresses in sugarcane cultivation. *Plant Signal. Behav.* 17(1), e2108253.

Verma KK, Song XP, Singh M, Tian DD, Rajput VD, Minkina T, Li YR. (2023c). Association of silicon and soil microorganisms induces stress mitigation, increasing plant productivity. In: R de Mello Prado. (Eds.), *Benefits of Silicon in the Nutrition of Plants*. (pp. 299–328). https://doi.org/10.1007/978-3-031-26673-7_17.

Verma KK, Song XP, Tian DD, Guo DJ, Chen ZL, Zhong CS, Nikpay A, Singh M, Rajput VD, Singh RK, Minkina T, Li YR. (2021a). Influence of silicon on biocontrol strategies to manage biotic stress for crop protection, performance and improvement. *Plants.* 10, 2163. https://doi.org/10.3390/plants10102163.

Verma KK, Song XP, Tian DD, Singh M, Verma CL, Rajput VD, Singh RK, Sharma A, Singh P, Malviya MK, Li YR. (2021b). Investigation of defensive role of silicon during drought stress induced by irrigation capacity in sugarcane: Physiological and biochemical characteristics. *ACS Omega,* 6, 19811–19821.

Verma KK, Song X-P, Xu L, Huang H-R, Liang Q, Seth CS, Li Y-R. (2023d). Nano-microplastic and agro-ecosystems: A mini-review. *Front. Plant Sci.* 14, 1283852. https://doi.org/10.3389/fpls.2023.1283852.

Waclawovsky AJ, Sato PM, Lembke CG, Moore PH, Souza GM. (2010). Sugarcane for bioenergy production: An assessment of yield and regulation of sucrose content. *Plant Biotechnol. J.* 8, 263–276.

Walker DIT. (1987). Manipulating the genetic base of sugarcane. In: (Eds.), *Copersucar International Sugarcane Breeding Workshop* (pp. 321–334). Copersucar: Brazil.

Walker DIT. (1982). The West Indies central sugarcane breeding station–A golden jubilee review. *West Indies Central Sugarcane Breeding Station Bull.* 45, 1–24.

Walker DIT. (1974). Utilization of noble and *S. spontaneum* germplasm in the West Indies. *Proc. Int. Soc. Sug. Cane Technol.* 1974, 14: 224–232.

Wang L, Jackson PA, Lu X, Fan YH, Foreman JW, Chen XK, Deng HH, Fu C, Ma L, Aitken KS. (2008). Evaluation of sugarcane × *Saccharum spontaneum* progeny for biomass composition and yield components. *Crop Sci.* 48, 951–961.

Wang W, Li R, Wang H, Qui B, Jiang X, Zhu Q, Cai D, Tang X, Zhao Q. (2019). Sweetcane (*Erianthus arundinaceus*) as a native bioenergy crop with environmental remediation potential in Southern China: A review. *GCB Bioenergy*. 11, 1012–1025. https://doi.org/10.1111/gcbb.12600.

White WH, Tew TL, Cobill RM, Burner DM, Grisham MP, Dufrene EO, Pan Y-B, Richard EP, Jr., Legendre BL. (2011). Registration of 'Ho 00-961' sugarcane. *J. Plant Regist.* 5, 332–338.

Wu K-C, Huang C-M, Verma KK, Deng Z-N, Huang H-R, Pang T, Cao H-Q, Luo H-B, Jiang S-L, Xu L. (2022). Transcriptomic responses of *Saccharum spontaneum* roots in response to polyethylene glycol-6000 stimulated drought stress. *Front. Plant Sci.* 13, 992755. https://doi.org/10.3389/fpls.2022.992755.

Xu L, Deng Z-N, Wu K-C, Malviya MK, Solanki MK, Verma KK, Pang T, Li Y-J, Liu X-Y, Kashyap BK, Dessoky ES, Wang W-Z, Huang H-R. (2022). Transcriptome analysis reveals a gene expression pattern that contributes to sugarcane bud propagation induced by Indole-3-Butyric Acid. *Front. Plant Sci.* 13, 852886. https://doi.org/10.3389/fpls.2022.852886.

Zhang JS, Nagai C, Yu QY, Pan YB, Ayala-Silva T, Schnell RJ. (2012). Genome size variation in three *Saccharum* species. *Euphytica*. 185, 511–519.

Zhang J, Zhang X, Tang H, Zhang Q, Hua X, Ma X, Zhu F, Jones T, Zhu X, Bowers J, Wai CM. (2018). Allele-defined genome of the autopolyploid sugarcane *Saccharum spontaneum* L. *Nat. Genet.* 50(11), 1565–1573.

Zheng F-J, Lin B, Yang Y-X, Fang X-C, Verma KK, Chen G-L. (2023). Efficacy and functionality of sugarcane original vinegar on mice. *Front. Microbiol.* 14, 1224666. https://doi.org/10.3389/fmicb.2023.1224666.

Zhou CA, Kimbeng SJ, Edme., Hale AL. (2013). Characterization of *Saccharum* species germplasm for starch content. *J. Plant. Stud.* 2(1), 54–71.

CHAPTER 13

Environmental Sustainability: Attitude Changes in Response to Scientific Knowledge

MUNMUN DASH,[1] ARADHNA KUMARI,[2*] SANTOSH KUMAR SINGH,[3] APARNA JAISWAL,[2] and KRISHAN K. VERMA[4]

[1]*Odisha University of Agriculture and Technology, Bhubaneswar, Odisha, India*

[2]*College of Agriculture, Jawaharlal Nehru Krishi Vishwa Vidyalaya, Ganj Basoda, Vidisha, Madhya Pradesh, India*

[3]*Department of Soil Science, Dr. Rajendra Prasad Central Agricultural University, Pusa, Samastipur, Bihar, India*

[4]*Sugarcane Research Institute, Guangxi Academy of Agricultural Sciences/Key Laboratory of Sugarcane Biotechnology and Genetic Improvement (Guangxi), Ministry of Agriculture and Rural Affairs/ Guangxi Key Laboratory of Sugarcane Genetic Improvement, Nanning, Guangxi, China*

*Corresponding author

ABSTRACT

Environmental sustainability is a critical global concern, requiring comprehensive understanding and actionable insights to mitigate pressing environmental issues. This chapter delves into the transformative power of scientific knowledge in shaping attitudes toward sustainability. It highlights the pivotal role of scientific research in identifying environmental problems, assessing risks, and informing effective

Sugarcane Transformation: Biofuels, Biotechnology, and Beyond.
Krishan K. Verma, Xiu-Peng Song, Hai-Rong Huang, Aradhna Kumari, and Yang-Rui Li (Eds.)

DOI: 10.1201/9781779641571-13

policies and regulations. Through detailed examples, it showcases how groundbreaking discoveries, such as the greenhouse effect and ozone layer depletion, have heightened awareness and driven international policy responses. This chapter also emphasizes the importance of innovative technologies, sustainable practices, and effective communication in promoting environmental sustainability. By addressing the challenges in translating scientific knowledge into action, it provides strategies for overcoming resistance and fostering collaborative efforts. Emerging trends in sustainability research and practices are explored, illustrating the dynamic landscape of environmental innovation and the ongoing quest for a sustainable future.

13.1 INTRODUCTION TO ENVIRONMENTAL SUSTAINABILITY

As outlined by Hacking and Guthrie (2008), sustainability assessment is essentially a process aimed at guiding decision-making toward sustainable outcomes. Similarly, Devuyst et al. (2001) defined it as a tool aiding decision-makers and policymakers in determining actions conducive to enhancing societal sustainability. Sustainability assessment is a decision-making aid (Pope et al., 2017). However, it is important to note that individual perceptions and definitions of sustainability, along with the temporal boundaries established for monitoring progress, play significant roles in shaping the assessment process (Ness et al., 2007), thereby influencing expectations regarding the outcomes (Bond and Morrison-Sanders, 2011).

13.2 IMPORTANCE OF SCIENTIFIC RESEARCH IN UNDERSTANDING ENVIRONMENTAL ISSUES

Scientific research helps in detecting and understanding the root causes of environmental issues. For instance, climate change, pollution, and biodiversity loss are complex phenomena that require detailed study to comprehend their origins, impacts, and interactions (Kumari et al., 2022; Patni et al., 2022; Verma et al., 2024a, 2024b, 2024c). Research provides data on the levels of pollutants, patterns of climate change (Kumari et al., 2014, 2018; Kumari and Singh, 2022), and the health of ecosystems, which is crucial for identifying problems accurately. Through research, scientists can assess the impacts of environmental issues on ecosystems, human health, and the economy (Chen et al., 2020, 2023; Xu et al., 2022; Verma et al., 2022a, 2022b). Studies on the effects of air and water pollution, for example, reveal how these pollutants affect wildlife, natural habitats, and public health. This knowledge is essential for understanding the severity and breadth of environmental issues (Joshi et al., 2022, 2023; Huang et al., 2024). Research is the backbone of innovation. It enables the development of new technologies and strategies to mitigate environmental problems (Song et al., 2019, 2021, 2022; Li et al., 2023; Tian et al., 2023; Solanki et al., 2023). For instance, scientific advancements have led to the creation of renewable energy sources, waste reduction techniques, and pollution control methods (Oliveira et al., 2022; Bhatt et al., 2023a, 2023b).

Research also supports the development of sustainable practices in agriculture, industry, and urban planning. Sound scientific research provides the evidence base needed for effective environmental policies and regulations (Verma et al., 2020a, 2020b). Policymakers rely on scientific data and analyses to draft laws that protect the environment, such as emission standards, conservation laws, and regulations on hazardous substances (Singh et al., 2020, 2021, 2023). Without solid research, policies might be ineffective or counterproductive. Ongoing research and monitoring allow scientists to track changes in the environment and predict future trends. This is crucial for early warning systems and for preparing for events such as natural disasters, outbreaks of diseases linked to environmental factors, and long-term changes like sea-level rise (Verma et al., 2019a, 2019b; Sharma et al., 2021, 2022; Malviya et al., 2022). Accurate forecasting helps societies adapt and mitigate potential damage. Increased awareness can lead to behavioral changes and greater public support for environmental policies (Figure 13.1). For example, research on the effects of plastic pollution has significantly raised awareness and led to global movements to reduce plastic waste (Malviya et al., 2021; Guo et al., 2021, 2022, 2023). Collaborative research efforts, such as those led by the Intergovernmental Panel on Climate Change (IPCC) or the United Nations Environment Programme, bring together scientists from around the world to address global challenges (Wu et al., 2022; Huang et al., 2022; Krishna et al., 2022). This fosters a unified approach to tackling environmental problems.

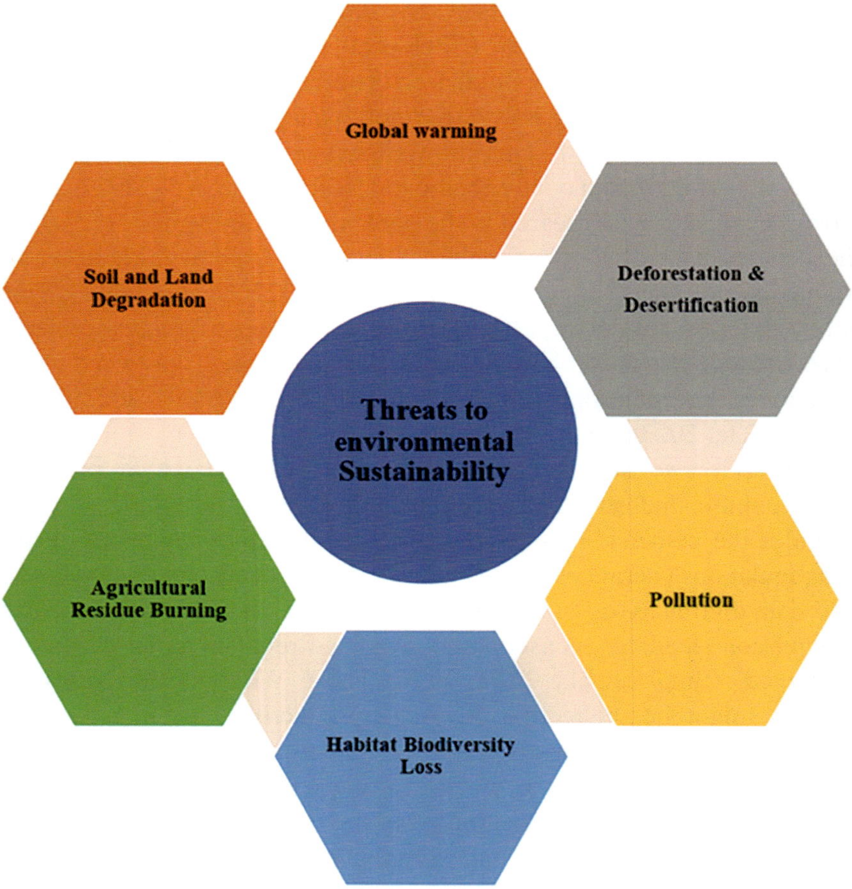

FIGURE 13.1 Threats to environmental sustainability.

13.3 EXAMPLES OF SCIENTIFIC DISCOVERIES SHAPING ATTITUDES TOWARD SUSTAINABILITY

The discovery of the greenhouse effect by Svante Arrhenius in the late 19th century, and subsequent research by scientists like Charles Keeling, who measured the rise in atmospheric carbon dioxide, has profoundly impacted our understanding of climate change. The Keeling Curve, which shows the increasing concentration of CO_2 in the atmosphere, has become a powerful symbol of human impact on the climate, leading to increased awareness and calls for sustainable practices to reduce carbon footprints (Anas et al.,

2021a, 2021b; Guo et al., 2021; Zheng et al., 2023). The discovery of the ozone layer's depletion by scientists such as Mario Molina and Sherwood Rowland in the 1970s highlighted the harmful effects of chlorofluoro-carbons. Their research led to the landmark Montreal Protocol in 1987, an international treaty to phase out the production of ozone-depleting substances. This discovery and the subsequent policy response have raised global awareness about the importance of protecting the ozone layer and the atmosphere (Deng et al., 2023). Research by ecologists like E.O. Wilson has emphasized the importance of biodiversity and the concept of ecosystem services—benefits that humans derive from ecosystems, such as pollination, water purification, and climate regulation.

This understanding has helped shift public attitudes toward conservation and sustainable use of natural resources, recognizing that biodiversity loss can have profound impacts on human well-being (Verma et al., 2020c, 2021a, 2021b). Studies revealing the extent of plastic pollution in oceans, such as those by marine biologist Richard Thompson and others, have highlighted the environmental and health hazards posed by plastic waste. This research has fuelled public campaigns to reduce plastic use, led to bans on single-use plastics in various countries, and spurred innovations in biodegradable materials. Advancements in renewable energy technologies, including solar and wind power, have been driven by scientific research. Innovations in photovoltaic cells and wind turbine efficiency have made these energy sources more viable and cost-effective. This has shifted public and political attitudes toward embracing renewable energy as a key component of sustainable development and reducing reliance on fossil fuels. Discoveries in sustainable agricultural practices, such as permaculture and agroecology, have demonstrated that farming can be productive while maintaining ecological balance (Verma et al., 2021c, 2021d, 2021e; Chaudhary et al., 2023). Research into crop rotation, organic farming, and soil health by scientists such as Wes Jackson and the Land Institute has promoted methods that enhance sustainability and reduce environmental impact, influencing both farmers and consumers.

Research on the effects of deforestation, such as that conducted by the Amazon Rainforest researchers, has shown the critical role forests play in carbon sequestration, climate regulation, and biodiversity. Findings on the consequences of deforestation have led to increased efforts to protect forests and reforestation initiatives, supported by both public sentiment and international policy frameworks such as Reducing Emissions from

Deforestation and Forest Degradation. The development of sustainable urban planning and green building technologies has been influenced by research into energy efficiency, water conservation, and sustainable materials. The Leadership in Energy and Environmental Design certification system, developed by the US Green Building Council, is based on scientific research and has promoted the construction of environmentally responsible buildings, shaping industry standards and public expectations.

13.4 HOW SCIENTIFIC KNOWLEDGE RAISES AWARENESS ABOUT ENVIRONMENTAL ISSUES

13.4.1 EDUCATIONAL OUTREACH

Scientific research often forms the basis of educational programs in schools, universities, and public forums. Curricula that include environmental science topics help students understand the importance of sustainability from a young age. Public lectures, documentaries, and books by scientists such as David Attenborough and Jane Goodall also play a significant role in spreading knowledge about environmental issues to a broader audience.

13.4.2 MEDIA COVERAGE

Scientific findings are frequently reported in the media, bringing attention to environmental issues. Newspapers, television, and online platforms cover research studies, making complex scientific concepts accessible to the general public. For instance, the publication of the IPCC reports on climate change receives extensive media coverage, highlighting the urgent need for climate action.

13.4.3 PUBLIC CAMPAIGNS AND ADVOCACY

Many environmental campaigns and advocacy efforts are grounded in scientific research. Organizations such as Greenpeace and the World Wildlife Fund use scientific data to advocate for policy changes and raise public awareness. These campaigns often translate scientific information into compelling messages and visuals that resonate with the public and policymakers.

13.4.4 CITIZEN SCIENCE PROJECTS

Citizen science projects involve the public in collecting data and contributing to research efforts. Examples include monitoring bird populations, measuring local air and water quality, and tracking the spread of invasive species. These projects educate participants about environmental issues and foster a sense of connection to scientific research and environmental stewardship.

13.4.5 INTERACTIVE AND VISUAL TOOLS

Advances in technology have enabled the development of interactive tools and visualizations that help to communicate scientific findings effectively. Maps, infographics, and simulations can illustrate the impact of phenomena such as deforestation, pollution, and climate change. Websites like Global Forest Watch provide real-time data on deforestation, making it easier for the public to understand and engage with these issues.

13.4.6 POLICY INFLUENCE

Scientific research informs policy decisions at local, national, and international levels. Policymakers often rely on scientific assessments to draft legislation and regulations aimed at addressing environmental problems. The scientific community's consensus on issues like climate change has been instrumental in shaping policies such as the Paris Agreement, which aims to limit global warming.

13.4.7 SOCIAL MEDIA AND DIGITAL PLATFORMS

Social media platforms allow scientists to share their research directly with the public. Researchers and institutions can use platforms such as Twitter, Instagram, and YouTube to disseminate information, engage in discussions, and raise awareness about their work. Hashtags such as #ClimateChange and #BiodiversityLoss help aggregate information and foster online communities focused on environmental issues.

13.4.8 INFLUENCE ON CONSUMER BEHAVIOR

Scientific research influences consumer behavior by highlighting the environmental impact of products and lifestyle choices. Studies on the ecological footprint of various goods, such as plastic products, meat, and fast fashion, have led to increased consumer demand for sustainable alternatives. Labels such as "organic," "fair trade," and "sustainably sourced" are backed by scientific criteria and appeal to environmentally conscious consumers.

13.4.9 EMERGENCY AND DISASTER RESPONSE

Scientific research is crucial in predicting and responding to environmental disasters. Early warning systems for hurricanes, floods, and wildfires rely on scientific data to forecast events and mitigate damage. Awareness campaigns about disaster preparedness and response are often based on scientific research, helping communities to better understand and react to natural hazards.

13.4.10 DOCUMENTARY AND FILM PRODUCTION

Documentaries and films based on scientific research, such as "An Inconvenient Truth," "Our Planet," and "The Blue Planet," have played a major role in raising awareness about environmental issues. These visual narratives can reach a global audience and inspire action by presenting scientific facts in an engaging and accessible format.

13.5 HOW SCIENTIFIC EVIDENCE INFORMS ENVIRONMENTAL POLICIES AND REGULATIONS

13.5.1 PROBLEM IDENTIFICATION AND UNDERSTANDING

Gathering data on environmental conditions, such as air and water quality, biodiversity, and climate parameters. Conducting studies to understand the sources, mechanisms, and impacts of environmental issues such as pollution, habitat destruction, and climate change. Research on the harmful effects of lead exposure led to policies that phase out leaded gasoline and regulate lead in paints and plumbing.

13.5.2 RISK ASSESSMENT

Scientific risk assessments evaluate the potential health and environmental risks associated with various pollutants and activities. Determining which substances or activities are harmful. Understanding the relationship between exposure levels and the severity of adverse effects (Verma et al., 2020d, 2020e; 2022c, 2022d, 2022e). Estimating how much, how often, and how long humans and the environment are exposed to hazards. Combining the above information to describe the nature and magnitude of the risk. The risk assessment of bisphenol A in plastics led to regulations limiting its use in food containers and baby bottles.

13.5.3 SETTING STANDARDS AND LIMITS

Based on scientific evidence, policymakers establish standards and permissible limits for pollutants and activities. Reviewing and synthesizing research findings to reach a consensus on safe levels of exposure (Verma et al., 2020f, 2023a, 2023b). Using scientific data to develop criteria for protecting human health and the environment. The Clean Air Act in the United States sets National Ambient Air Quality Standards for pollutants such as ozone and particulate matter based on scientific research on their health impacts.

13.5.4 DEVELOPING REGULATORY FRAMEWORKS

Scientific evidence informs the creation of comprehensive regulatory frameworks that outline the rules and procedures for managing environmental issues. Using scientific data to design regulations that effectively mitigate identified risks. Establishing monitoring and reporting requirements to ensure compliance with regulations. The European Union's REACH (Registration, Evaluation, Authorization, and Restriction of Chemicals) regulation is based on scientific assessments of chemical risks and requires companies to provide safety data for chemicals.

13.5.5 POLICY IMPLEMENTATION AND ENFORCEMENT

Setting up monitoring systems to track environmental conditions and pollution levels. Using scientific criteria to guide inspections and enforce compliance with regulations. The monitoring of water quality standards under the Clean Water Act involves regular sampling and analysis to ensure compliance with established limits.

13.5.6 EVALUATING POLICY EFFECTIVENESS

Studying the outcomes of policies to determine their effectiveness in reducing pollution or improving environmental health. Using scientific findings to adjust and refine policies and regulations as needed. Evaluations of the Montreal Protocol, which regulates substances that deplete the ozone layer, have shown significant recovery of the ozone layer, demonstrating the policy's effectiveness.

13.5.7 GUIDING TECHNOLOGICAL INNOVATIONS

Research leading to new pollution control technologies, renewable energy sources, and sustainable practices. Establishing regulatory guidelines that promote the adoption of best available technologies. Regulations promoting the use of catalytic converters in vehicles are based on scientific research showing their effectiveness in reducing harmful emissions.

13.5.8 SUPPORTING INTERNATIONAL AGREEMENTS

Conducting comprehensive assessments that provide a unified understanding of global environmental issues. Using scientific evidence to negotiate, draft, and implement international agreements. The Paris Agreement on climate change is informed by the scientific findings of the IPCC, which provides the basis for global commitments to reduce greenhouse gas emissions.

13.6 THE ROLE OF SCIENTIFIC INNOVATION IN DEVELOPING SUSTAINABLE TECHNOLOGIES

Advances in photovoltaic cell technology have increased the efficiency and reduced the cost of solar panels, making solar power a more accessible and affordable energy source. Improvements in turbine design and materials have enhanced the performance and reliability of wind energy systems. Innovations in battery technology, such as lithium-ion and emerging solid-state batteries, have improved energy storage solutions, crucial for the stability of renewable energy systems. Advances in building materials, insulation, and smart home technologies have made buildings more energy efficient. The development of fuel-efficient engines, electric vehicles, and hybrid systems has reduced energy consumption and emissions from the transportation sector. Use of GPS, drones, and Internet of Things (IoT) devices to monitor and optimize agricultural inputs such as water, fertilizers, and pesticides, enhancing productivity while reducing environmental impact. Development of genetically modified crops that are resistant to pests, diseases, and extreme weather conditions, leading to higher yields and reduced reliance on chemical inputs. Innovations in soil management techniques, such as biochar and cover cropping, improve soil fertility and carbon sequestration. Innovations in desalination technology, such as reverse osmosis and solar desalination, have made it possible to convert seawater into potable water more efficiently. Developing advanced treatment methods, including biological treatments and membrane filtration, has enhanced the ability to recycle and reuse wastewater. Integration of sensors and automation in irrigation systems has optimized water use in agriculture, reducing waste and conserving water resources. The development of biodegradable and compostable materials has provided alternatives to conventional plastics. Advances in sorting, processing, and re-manufacturing technologies have improved the efficiency and viability of recycling programs.

Innovations promoting the circular economy model, where products are designed for reuse, refurbishment, and recycling, reducing waste generation. Development of polymers that break down more easily in the environment. Innovation in catalysts that reduce the environmental impact of industrial chemical reactions. Use of water or other benign solvents in place of harmful organic solvents in chemical processes. Use of satellites, drones, and other remote sensing technologies to monitor environmental conditions such as deforestation, water quality, and air pollution. Application of big data analytics and artificial intelligence to analyze environmental data, predict trends, and optimize resource management. Development of more efficient methods to capture CO_2 from power plants and industrial sources. Research into safe and effective ways to store captured CO_2, such as geological sequestration or converting it into useful products. Implementation of smart city technologies that enhance urban living through improved energy management, transportation systems, and waste management. Incorporation of green roofs, urban forests, and other green infrastructure elements to improve air quality, manage stormwater, and increase urban biodiversity.

13.7 CHALLENGES AND OBSTACLES IN TRANSLATING SCIENTIFIC KNOWLEDGE INTO ACTION

Scientific information can be complex and difficult to understand for non-experts, including policymakers and the general public. This complexity can lead to misunderstandings or misinterpretations of the data. Scientists may struggle to communicate their findings clearly and compellingly, which resonates with policymakers and the public. Effective communication requires translating technical jargon into accessible language and emphasizing the relevance and urgency of the findings. Policymakers often operate on short-term election cycles, while environmental issues require long-term planning and commitment. This mismatch can result in a lack of political will to implement necessary but potentially unpopular measures. Industries that might be negatively impacted by environmental regulations often have significant lobbying power and can influence political decisions, leading to resistance against scientific recommendations (Verma et al., 2022f, 2023c, 2023d). Implementing sustainable technologies and practices often requires substantial upfront investment, which can be a barrier for businesses and governments with limited budgets. There is often a perception that environmental regulations and sustainability measures will harm economic growth, jobs, and competitiveness, creating resistance to change (Figure 13.2).

Sustainable practices often require significant changes in behavior and lifestyle, which can be difficult to achieve (Verma et al., 2022g, 2022h). People may be resistant to change due to habits, convenience, or lack of awareness. Low levels of public awareness and engagement with scientific issues can hinder the acceptance and adoption of policies based on scientific evidence. Scientific understanding evolves, and new findings can change the understanding of issues. This can create uncertainty and hesitation among policymakers and the public. Legitimate scientific debate

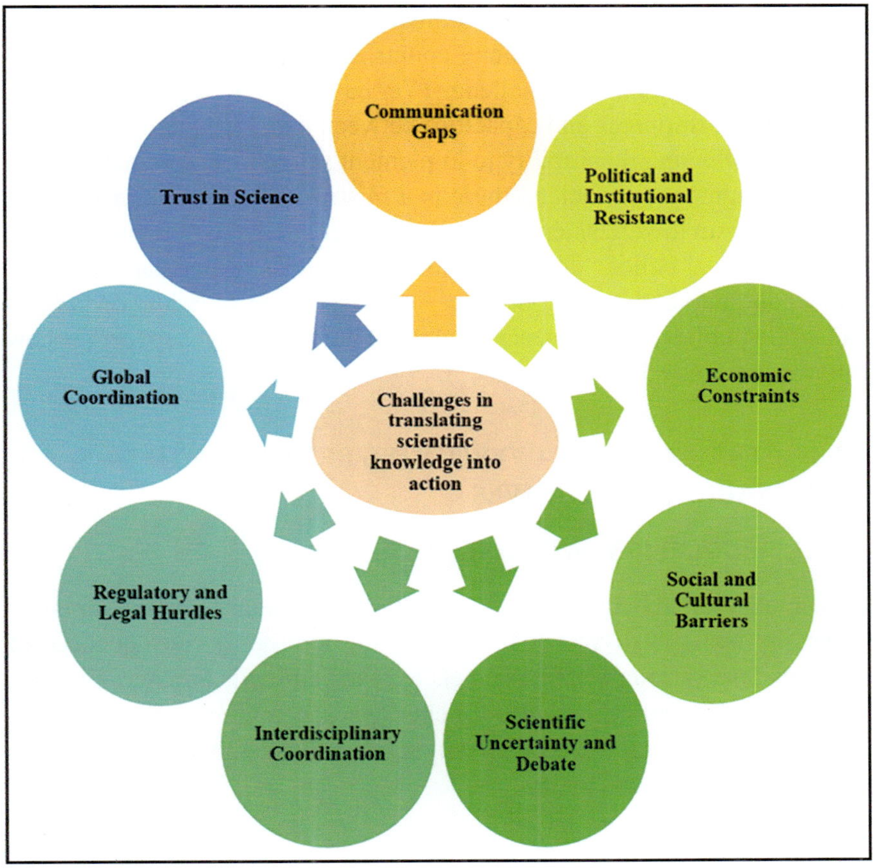

FIGURE 13.2 Challenges in translating scientific knowledge into action.

and differing interpretations of data can be exploited by interest groups to cast doubt on well-established findings, delaying action. Environmental issues are often interdisciplinary, requiring coordination across various scientific fields, as well as between scientists, policymakers, and other stakeholders. Silos within and between organizations can hinder effective collaboration and integration of knowledge (Verma et al., 2022i, 2022j). Existing regulations and institutional frameworks can be slow to adapt to new scientific insights, creating a lag between scientific discovery and policy implementation.

New regulations based on scientific knowledge can face legal challenges from affected industries or other stakeholders, delaying their implementation. Many environmental issues, such as climate change and ocean pollution, are global and require coordinated international action. Achieving consensus and cooperation among countries with differing priorities and capabilities is a significant challenge (Verma et al., 2022k). There is often a disparity in resources and capacities between developed and developing countries, affecting their ability to implement scientific recommendations. In some segments of society, there is a mistrust of scientific institutions and skepticism toward scientific findings, often fuelled by misinformation and ideological beliefs. This can undermine efforts to translate scientific knowledge into action. Past instances of scientific misconduct or failures can erode public trust in science, making it harder to gain support for new initiatives (Figure 13.2).

13.8 STRATEGIES FOR OVERCOMING RESISTANCE TO SUSTAINABILITY INITIATIVES

13.8.1 EDUCATION AND AWARENESS

Implement comprehensive public education campaigns to raise awareness about the benefits of sustainability and the long-term consequences of inaction. Use various media channels to reach a broad audience. Conduct workshops and training sessions for employees, community members, and industry stakeholders to explain the importance of sustainability initiatives and how they can contribute.

13.8.2 ECONOMIC INCENTIVES

Offer tax breaks, subsidies, and grants to businesses and individuals who adopt sustainable practices. Financial incentives can make the transition to sustainability more appealing and affordable. Communicate the long-term cost savings and economic benefits of sustainability initiatives, such as reduced energy costs, improved efficiency, and enhanced brand value.

13.8.3 LEGISLATIVE AND REGULATORY MEASURES

Develop and enforce regulations that mandate certain sustainability practices. Ensure these regulations are fair and transparent and provide adequate time for compliance. Promote standards and certifications that recognize and reward sustainable practices, such as LEED certification for green buildings or Energy Star ratings for appliances.

13.8.4 STAKEHOLDER ENGAGEMENT

Involve stakeholders, including employees, customers, suppliers, and community members, in the planning and decision-making process. This fosters a sense of ownership and commitment to the initiatives. Create platforms for open dialogue and collaboration among stakeholders to address concerns, share ideas, and develop mutually beneficial solutions. Highlight and publicize successful sustainability initiatives and their positive impacts. Success stories can serve as powerful examples and encourage wider adoption.

13.8.5 TECHNOLOGICAL INNOVATION

Support research and development of new sustainable technologies that can make practices more efficient and cost-effective. Innovations can reduce the perceived burden of sustainability initiatives. Identify and adopt best practices and technologies from other organizations or regions that have successfully implemented sustainability initiatives.

13.8.6 BEHAVIORAL CHANGE

Leverage social norms and peer pressure to encourage sustainable behavior. Campaigns that emphasize widespread adoption and social approval can be effective. Use behavioral economics principles to design incentives and nudges that encourage sustainable choices, such as defaulting to double-sided printing or offering discounts for using reusable bags.

13.8.7 TRANSPARENT COMMUNICATION

Provide clear, consistent, and transparent communication about the goals, benefits, and progress of sustainability initiatives. Address any misconceptions and provide evidence-based information. Establish mechanisms for receiving and addressing feedback from stakeholders. This helps to identify and mitigate concerns early on.

13.8.8 PARTNERSHIPS AND ALLIANCES

Form partnerships and alliances with other organizations, governments, and NGOs to share resources, knowledge, and best practices. Engage in public–private partnerships to leverage the strengths of both sectors in driving sustainability initiatives.

13.8.9 PILOT PROJECTS AND PHASED IMPLEMENTATION

Start with pilot projects to test and demonstrate the feasibility and benefits of sustainability initiatives on a smaller scale before scaling up. Implement initiatives in phases, allowing for gradual adjustment and learning. This reduces the perceived risk and resistance associated with large-scale changes.

13.9 EMERGING TRENDS IN ENVIRONMENTAL SUSTAINABILITY RESEARCH AND PRACTICE

13.9.1 CIRCULAR ECONOMY

The circular economy emphasizes designing products for longevity, reuse, remanufacturing, and recycling, thus minimizing waste and resource consumption. Companies are adopting circular business models such as product-as-a-service, where products are leased rather than sold, encouraging manufacturers to design for durability and recyclability (Figure 13.3).

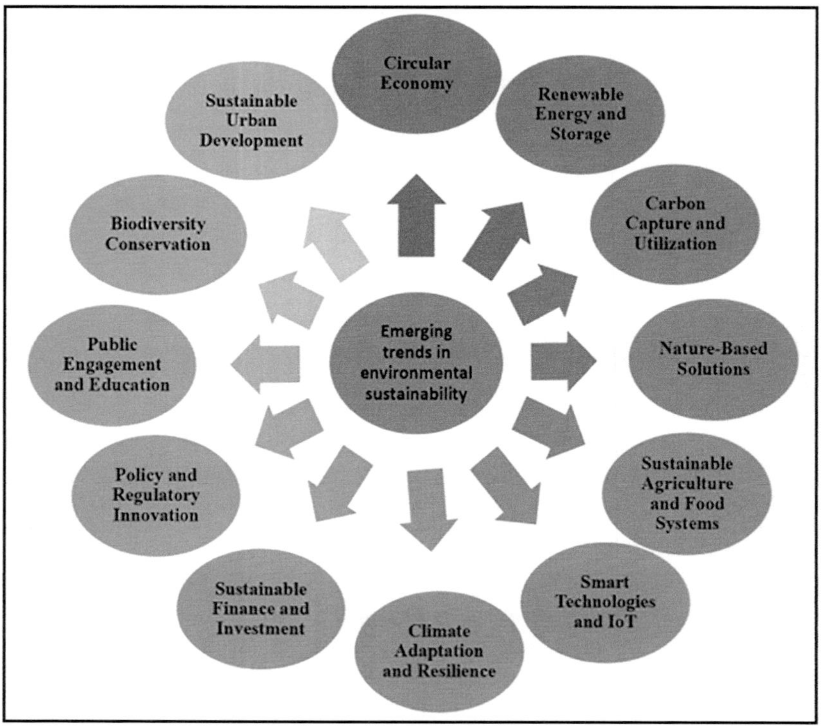

FIGURE 13.3 Emerging trends in environmental sustainability research and practice.

13.9.2 RENEWABLE ENERGY AND STORAGE

Continued innovation in solar, wind, and other renewable energy technologies is making them more efficient and cost-effective. Offshore wind and floating solar panels are notable areas of development. Improvements in battery technology, including lithium-ion and emerging solid-state batteries, are enhancing the storage capacity and efficiency of renewable energy systems, addressing the intermittency issue of renewable energy (Figure 13.3).

13.9.3 CARBON CAPTURE AND UTILIZATION

Research is advancing in technologies that capture carbon dioxide from industrial emissions and the atmosphere. Methods include direct air capture and bioenergy with carbon capture and storage (Kumari et al., 2022, 2023). Captured carbon dioxide is being used in various applications, such as producing synthetic fuels, enhancing concrete, and creating carbon-based products, turning a liability into a resource (Kumari et al., 2022, 2023).

13.9.4 NATURE-BASED SOLUTIONS

Initiatives focused on restoring natural ecosystems, such as reforestation, wetland restoration, and coral reef rehabilitation, are gaining traction for their ability to sequester carbon, enhance biodiversity, and provide ecosystem services. Cities are integrating green infrastructure, such as green roofs, walls, and urban forests, to improve air quality, manage stormwater, and enhance urban resilience (Figure 13.3).

13.9.5 SUSTAINABLE AGRICULTURE AND FOOD SYSTEMS

Practices that restore soil health, increase biodiversity, and sequester carbon, such as cover cropping, no-till farming (Kumari and Singh, 2016), and agroforestry, are becoming more widespread. The development and adoption of plant-based and lab-grown meat alternatives aim to reduce the environmental impact of traditional livestock farming (Kumari et al., 2022, 2023).

13.9.6 SMART TECHNOLOGIES AND IOT

The implementation of smart grids enables more efficient distribution and use of electricity, integrating renewable energy sources and enhancing grid stability. IoT devices are being used to optimize water use in agriculture, monitor air and water quality, and manage waste systems more efficiently (Figure 13.3).

13.9.7 CLIMATE ADAPTATION AND RESILIENCE

Designing and building infrastructure that can withstand climate impacts, such as extreme weather events, sea-level rise, and temperature fluctuations, is a growing focus. Empowering local communities to develop and implement adaptation strategies that address their specific vulnerabilities and needs (Figure 13.3).

13.9.8 SUSTAINABLE FINANCE AND INVESTMENT

The issuance of green bonds is increasing, providing capital for projects that have positive environmental benefits, such as renewable energy installations and green infrastructure projects. Environmental, Social, and Governance criteria are being integrated into investment decision-making processes, driving capital toward more sustainable and responsible enterprises (Figure 13.3).

13.9.9 POLICY AND REGULATORY INNOVATION

Governments and corporations are setting ambitious net-zero carbon emissions targets, often backed by robust policy frameworks and international agreements. Increasing legislation aimed at reducing plastic waste, including bans on single-use plastics and requirements for extended producer responsibility (Figure 13.3).

13.9.10 PUBLIC ENGAGEMENT AND EDUCATION

Public participation in scientific research, known as citizen science, is expanding, particularly in areas such as biodiversity monitoring and pollution tracking. Educational institutions are incorporating sustainability into curricula at all levels, preparing future generations to tackle environmental challenges (Figure 13.3).

13.9.11 BIODIVERSITY CONSERVATION

Expansion and improved management of protected areas and wildlife corridors to preserve biodiversity and support ecosystem services. Advances in genetic technologies for the conservation of endangered species and the restoration of genetic diversity (Figure 13.3).

13.9.12 SUSTAINABLE URBAN DEVELOPMENT

Development of smart cities that leverage technology to improve energy efficiency, reduce emissions, and enhance the quality of life for residents. Planning and development strategies that prioritize public transportation and cycling to reduce reliance on personal vehicles (Figure 13.3).

13.10 CONCLUSION

Scientific knowledge is indispensable in understanding and addressing environmental challenges. This chapter underscores the profound impact of scientific discoveries on shaping public attitudes and driving policy changes toward sustainability. It highlights the crucial role of education, media, and advocacy in raising awareness and fostering behavioral changes. Despite the challenges in translating scientific knowledge into action, strategic approaches such as economic incentives, stakeholder engagement, and technological innovation offer viable solutions. The emerging trends in sustainability research and practice demonstrate the ongoing advancements and the collective efforts required to achieve environmental sustainability. As we move forward, we must continue leveraging scientific insights, fostering collaboration, and promoting sustainable practices to safeguard our planet for future generations.

KEYWORDS

- **environmental sustainability**
- **adverse atmospheric variables**
- **enviro-awareness**
- **policies and regulations**
- **advanced sustainable approaches**
- **sustainable research and applications**
- **smart technologies**

REFERENCES

Anas M, Verma KK, Haq IU, Naeem M, Li Q, Liao F, Xiang YX, Li YR. (2021a). Characterization of exotic, native and wild-type genotypes of sugarcane (*Saccharum* spp. hybrids) for internal nitrogen use efficiency under different nitrogen levels. *Sugar Tech*. 23, 1258–1267.https://doi.org/10.1007/s12355-021-01014-1

Anas M, Verma KK, Riaz M, Qiang L, Liao F, Liu Y, Li YR. (2021b). Physio-morphological and biochemical mechanism of nitrogen use efficiency in sugarcane (*Saccharum* spp.) genotypes under different growth stages and nitrogen levels. *J. Plant Interact.* 16, 332–343.

Angon P B, Anjum N, Akter MM, KC S, Suma, RP, Jannat, S. (2023). An overview of the impact of tillage and cropping systems on soil health in agricultural practices. *Adv. Agric.*, 2023(1), 8861216.

Balkenende R, Bocken N, Bakker C. (2017). Design for the circular economy. In: *Routledge Handbook of Sustainable Design*, Routledge; pp. 498–513.

Bang DY, Kyung M, Kim MJ, Jung BY, Cho MC, Choi SM, Lee BM. (2012). Human risk assessment of endocrine-disrupting chemicals derived from plastic food containers. *Compreh. Rev. Food. Sci. Food Saf.* 11(5), 453–470.

Bhandari MP. (2018). The role of international organization in addressing the climate change issues and creation of Intergovernmental Panel on Climate Change (IPCC). *Adv. Agr. Environ. Sci.* 1(1), 19.

Bhatt R, Verma KK, Jamwal NS. (2023a). Can paddy-compost eliminate N fertilization for climate-smart sugarcane? *Sugar Tech.* 25, 1036–1046. https://doi.org/10.1007/s12355-023-01270-3

Bhatt R, Verma KK, Kumar R, Sanghera GS. (2023b). Foliar application of potassium salt of active phosphorus (PSAP) mitigates insect pests and improves yield along with sugarcane quality in response to agroclimatic conditions of Punjab. *Sugar Tech.* 25(3), 660–669.

Bond A J, Morrison-Saunders A. (2011). Re-evaluating sustainability assessment: Aligning the vision and the practice. *Environ. Impact. Assess. Rev.* *31*(1), 1–7.

Brown L A. (2024). Implementing Sustainable Blue Carbon Initiatives as a Climate Change and Biodiversity Loss Solution: Restoration Considerations and Upending Conventional Top-Down Governance (Doctoral dissertation).

Bwambale E Abagale, F K, Anornu, G K. (2022). Smart irrigation monitoring and control strategies for improving water use efficiency in precision agriculture: A review. *Agric. Water Manage.* 260, 107324.

Chaudhary P, Xu M, Ahamad L, Chaudhary A, Kumar G, Adeleke BS, Verma KK, Hu D-M, Siric I, Kumar P, Popescu SM, Fayssal SA. (2023). Application of synthetic consortia for improvement of soil fertility, pollution remediation, and agricultural productivity: A review. *Agronomy.* 13, 643. https://doi.org/10.3390/agronomy13030643.

Chen GL, Zheng FJ, Lin B, Lao SB, He J, Huang Z, Zeng Y, Sun J, Verma KK. (2020). Phenolic and volatile compounds in the production of sugarcane vinegar. *CS Omega.* 5, 30587–30595.

Chen G-L, Zheng F-J, Lin B, Yang Y-X, Fang X-C, Verma KK, Yang L-F. (2023). Vinegar: A potential source of healthy and functional food with special reference to sugarcane vinegar. *Front. Nutr.* 10, 1145862. https://doi.org/10.3389/fnut.2023.1145862

Dao C, Kazemtabrizi B, Crabtree C. (2019). Wind turbine reliability data review and impacts on levelised cost of energy. *Wind Energy*. 22(12), 1848–1871.

Delbeke J, Runge-Metzger A, Slingenberg Y, Werksman J. (2019). The Paris agreement. In *Towards a Climate-Neutral Europe,* Routledge; pp. 24–45.

Deng ZN, Huang HR, Huang CM, Verma KK, Luo HB, Jiang SL, Li YJ, Cao HQ, Wu KC, Xu L. (2023). Short-term limited water irrigation influences the photosynthetic pigments, enzymatic and non-enzymatic activities in *Saccharum spontaneum* L. at vegetative stage. *Sugar Tech.* 25, 110–118. https://doi.org/10.1007/s12355-022-01175-7

Devuyst D. (2001). Introduction to sustainability assessment at the local level. In *How Green is the City? Sustainability Assessment and the Management of Urban Environments*, Columbia University Press, pp. 1–36.

Dickinson J. LShirk, J Bonter, DBonney, R Crain, R L, Martin, JPurcell K. (2012). The current state of citizen science as a tool for ecological research and public engagement. *Front. Ecol. Environ.*, *10*(6), 291–297.

Dover J W. (2015). *Green Infrastructure: Incorporating Plants and Enhancing Biodiversity in Buildings and Urban Environments*. Routledge.

Esworthy, R McCarthy, J E. (2013). The National Ambient Air Quality Standards (NAAQS) for Particulate Matter (PM): EPA's 2006 Revisions and Associated Issues. Library of Congress, Congressional Research Service.

Ferguson RS, Lovell ST. (2014). Permaculture for agroecology: Design, movement, practice, and worldview. A review. *Agron. Sustain. Dev.* 34, 251–274.

Foth, H., Hayes, A W. (2008). Concept of REACH and impact on evaluation of chemicals. *Human Exp. Toxicol.*, *27*(1), 5–21.

Glicksman RL, Buzbee WW, Mandelker DR, Hammond E, Camacho A. (2023). environmental protection: Law and policy [Connected EBook with Study Center]. Aspen Publishing.

Gonzalez M, Taddonio KN, Sherman, NJ. (2015). The Montreal protocol: How today's successes offer a pathway to the future. *J. Environ. Stud. Sci.* 5, 122–129.

Guo D-J, Li D-P, Chen Z-D, Yang B, Verma KK, Singh RK, Singh P, Khan Q, Sharma A, Qin Y, Zhang B-Q, Song X-P, Li Y-R. (2023). Effect of endophytic diazotroph *Enterobacter roggenkampii* ED5 on nitrogen-metabolism-related microecology in the sugarcane rhizosphere at different nitrogen levels. *Front. Microbiol.* 14, 1132016. https://doi.org/10.3389/fmicb.2023.1132016

Guo D-J, Li D-P, Singh RK, Singh P, Sharma A, Verma KK, Qin Y, Khan Q, Lu Z, Malviya MK, Song X-P, Xing Y-X and Li Y-R (2021). Differential protein expression analysis of two sugarcane varieties in response to diazotrophic plant growth-promoting endophyte *Enterobacter roggenkampii* ED5. *Front. Plant Sci.* 12, 727741. https://doi.org/10.3389/fpls.2021.727741

Guo DJ, Li DP, Singh RK, Singh P, Verma KK, Sharma A, Qin Y, Khan Q, Song XP, Malviya MK, Xing YX, Li YR. (2022). Comparative transcriptome analysis of two sugarcane varieties in response to diazitrophic plant growth promoting endophyte *Enterobacter roggenkampii* ED5. *J. Plant Interact.* 17(1), 75–84.

Hacking T, Guthrie P. (2008). A framework for clarifying the meaning of triple bottom-line, integrated, and sustainability assessment. *Environ. Impact Assess. Rev.* 28(2–3), 73–89.

Hoban S, Bruford MW, Funk WC., Galbusera P, Griffith MP, Grueber CE, Ernesi C. (2021). Global commitments to conserving and monitoring genetic diversity are now necessary and feasible. *Bioscience.* 71(9), 964–976.

Howe J P. (2015). This is nature; this is un-nature: Reading the Keeling curve. Environmental History.

Huang CM, Wu KC, Verma KK, Luo HB, Deng ZN, Shi GY, Cao HQ, Wu XJ, Ye LP, Wei YW, Xu L. (2024). Isolation and expression analysis of cellulose synthase 3 (*Ces3*) genes from sugarcane (*Saccharum officinarum* L). *Chem. Biol. Technol. Agric.* 11, 71. https://doi.org/10.1186/s40538-024-00597-7

Huang HR, Gao YJ, Malviya MK, Verma KK, Solanki MK, Huang YX, Li X., Deng Y, Yan J, Tang SY, Wang LW, Xu L. (2022). Genetic diversity analysis of sugarcane (*Saccharum* spp. Hybrids) among high-sucrose clones of GT series and commonly used parents by using microsatellite markers in Guangxi, China. *Sugar Tech.* 24, 397–407. https://doi.org/10.1007/s12355-021-

Hui C X, Dan G, Alamri S, Toghraie D. (2023). Greening smart cities: An investigation of the integration of urban natural resources and smart city technologies for promoting environmental sustainability. *Sustain. Cities Soc.* 99, 104985.

Joshi A, Verma KK, Rajput VD, Arora J. (2023). Metabolic engineering of microorganisms in advancing biofuel production. In: *Advances in Lignocellulosic Biofuel Production Systems.* pp. 181–202. https://doi.org/10.1016/B978-0-323-91192-4.00016-X

Joshi A, Verma KK, Rajput VD, Minkina T, Arora J. (2022). Recent advances in metabolic engineering of microorganisms for advancing lignocellulose-derived biofuels, *Bioengineered.* 13(4), 8135–8163. https://doi.org/10.1080/21655979.2022.2051856.

Koytsoumpa E I, Bergins C, Kakaras E. (2018). The CO2 economy: Review of co2 capture and reuse technologies. *J. Supercrit. Fluids.* 132, 3–16.

Krishna D, Sachan HK, Verma KK. (2022). Abiotic stresses in plants: An overview. In: *Understanding Abiotic Stresses*, Nova Science Publishers, Inc.: New York.

Kritsanaviparkporn E, Baena-Moreno FM, Reina TR. (2021). Catalytic converters for vehicle exhaust: Fundamental aspects and technology overview for newcomers to the field. *Chemistry*. 3(2), 630–646.

Kubba S. (2009). *LEED Practices, Certification, and Accreditation Handbook.* Butterworth-Heinemann.

Kumari A, Dash M, Singh SK, Jagadesh M, Mathpal B, Mishra PK, Pandey SK, Verma KK. (2023). Soil microbes: A natural solution for mitigating the impact of climate change. *Environ. Monit. Assess.* 195:1436. https://doi.org/10.1007/s10661-023-11988-y

Kumari A, Lakshmi GA, Krishna GK, Patni B, Prakash S, Bhattacharyya M, Singh SK, Verma KK. (2022). Climate change and its impact on crops: A comprehensive investigation for sustainable agriculture. *Agronomy.* 12, 3008.

Kumari A, Sairam RK, Singh SK. (2014). Effect on physiological, growth and yield parameters of wheat genotypes under drought stress. *Int. J. Agric. Stat. Sci.* 10(1), 167–174.

Kumari A, Singh AK, Singh S, Khan IM. (2018). Morphological, physiological and biochemical responses of Poplar plants under drought stress. *J. AgriSearch.* 5(3), 184–189.

Kumari A, Singh SK (2022). Role of plant hormones in combating biotic stress in plants. In: Aftab T., Roychoudhury A. (Eds.), *Plant Perspectives to Global Climate changes: Developing Climate-Resilient Plants*. Edited by: Published by: Academic Press, Elsevier, pp. 373–391

Kumari A, Singh SK. (2016). Impact of different tillage practices on soil organic carbon and nitrogen pool in rice-wheat cropping system. *J. AgriSearch. 3*(2), 82–86.

Kusch, S., Fiebelkorn F. (2019). Environmental impact judgments of meat, vegetarian, and insect burgers: Unifying the negative footprint illusion and quantity insensitivity. *Food Qual. Prefer. 78*, 103731.

Li A-M, Liao F, Wang M, Chen Z-L, Qin C-X, Huang R-Q, Verma KK, Li Y-R, Que Y-X, Pan Y-Q, et al. (2023). Transcriptomic and proteomic landscape of sugarcane response to biotic and abiotic stressors. *Int. J. Mol. Sci.* 24, 8913. https://doi.org/10.3390/ijms24108913

Malviya MK, Li C-N, Lakshmanan P, Solanki MK, Wang Z, Solanki AC, Nong Q, Verma KK, Singh RK, Singh P, Sharma A, Guo D-J, Dessoky ES, Song X-P, Li Y-R. (2022). High-throughput sequencing-based analysis of rhizosphere and diazotrophic bacterial diversity among wild progenitor and closely related species of sugarcane (*Saccharum* spp. inter-specific hybrids). *Front. Plant Sci. 13*, 829337. https://doi.org/10.3389/fpls.2022.829337.

Malviya MK, Solanki MK, Li CN, Wang Z, Zeng Y, Verma KK, Singh RK, Singh P, Huang HR, Yang LT, Song XP, Li YR. (2021). Sugarcane-legume intercropping can enrich the soil microbiome and plant growth. *Front. Sustain. Food. Syst. 5*, 606595. https://doi.org/10.3389/fsufs.2021.606595.

Manisalidis I., Stavropoulou E., Stavropoulos A., Bezirtzoglou E. (2020). Environmental and health impacts of air pollution: A review. *Front. Public. Health. 8,* 14.

Martin JL., Maris V., Simberloff DS. (2016). The need to respect nature and its limits challenges society and conservation science. *Proc. Nat. Acad. Sci.* 113(22), 6105–6112.

Maurrasse D. (2013). Strategic public private partnerships: Innovation and development. Edward Elgar Publishing.

Ness B, Anderberg S, Olsson L. (2010). Structuring problems in sustainability science: The multi-level dpsir framework. *Geoforum.* 41(3), 479–488.

Okonkwo E C, AlNouss, A, Shahbaz, M Al-Ansari, T. (2023). Developing integrated direct air capture and bioenergy with carbon capture and storage systems: Progress towards 2°C and 1.5°C climate goals. *Energy Convers. Manag.* 296, 117687.

Oliveira MA, Verma KK, Bhatt R, Oliveira TBA. (2022). Impact of green and organic fertilizers on soil fertility and sugarcane productivity. In: Verma KK., et al. (Ed.), *Agro-Industrial Perspectives on Sugarcane Production Under Environmental Stress*. https://doi.org/10.1007/978-981-19-3955-6_11.

Oliver TH, Morecroft MD. (2014). Interactions between climate change and land use change on biodiversity: attribution problems, risks, and opportunities. WIREs Clim. Change, 5(3), 317–335.

Panwar NL, Kaushik SC, Kothari S. (2011). Role of renewable energy sources in environmental protection: A review. *Renew. Sustain. Energy Rev.* 15(3), 1513–1524.

Patni B, Bhattacharyya M, Kumari A, Purohit VK. (2022). Alarming influence of climate change and compromising quality of medicinal plants. *Plant Physiol Rep.* 27, 1–10. https://doi.org/10.1007/s40502-021-00616-x.

Pope J, Bond A, Hugé J, Morrison-Saunders A. (2017). Reconceptualising sustainability assessment. *Environ. Impact. Assess. Rev.* 62, 205–215.

Rules, P. (2015). Clean Water Act Methods Update Rule for the Analysis of Effluent. *Federal. Register*. 80(33).

Salam A. (2024). Internet of things for water sustainability. In *Internet of Things for Sustainable Community Development: Wireless Communications, Sensing, and Systems,* Springer International Publishing: Cham; pp. 113–145.

Shabir I, Dash K K, Dar A H, Pandey V K, Fayaz U, Srivastava S, Nisha R. (2023). Carbon footprints evaluation for sustainable food processing system development: A comprehensive review. *Future. Foods*. 7, 100215.

Sharma A, Singh RK, Singh P, Vaishnav A, Guo DJ, Verma KK, Li DP, Song XP, Malviya MK, Khan N, Lakshmanan P, Li YR. (2021). Insights into the bacterial and nitric oxide-induced salt tolerance in sugarcane and their growth-promoting abilities. *Microorganisms.* 9, 2203. https://doi.org/10.3390/microorganisms9112203.

Sharma A, Song X-P, Singh RK, Vaishnav A, Gupta S, Singh P, Guo D-J, Verma KK, Li Y-R. (2022). Impact of carbendazim on cellular growth, defence system and plant growth promoting traits of Priestia megaterium ANCB-12 isolated from sugarcane rhizosphere. *Front. Microbiol.* 13, 1005942. https://doi:10.3389/fmicb.2022.1005942

Singh P, Singh RK, Li HB, Guo DJ, Sharma A, Verma KK, Solanki MK, Upadhyay SK, Lakshmanan P, Yang LT, Li YR. (2023). Nitrogen fixation and phytohormone stimulation of sugarcane plant through plant growth promoting diazotrophic Pseudomonas. *Biotechnol. Genetic Eng. Rev.* 1–21. https://doi.org/10.1080/02648725.2023.2177814

Singh P, Singh RK, Song QQ, Li HB, Guo DJ, Malviya MK, Verma KK, Song XP, Lakshmanan P, Yang LT, Li YR. (2021). Comparative analysis of protein and differential responses of defense-related gene and enzyme activity reveals the long-term molecular responses of sugarcane inoculated with *Sporisorium scitamineum*. *J. Plant Interact.* 16(1), 12–29.

Singh RK, Singh P, Li HB, Song QQ, Guo DJ, Solanki MK, Verma KK, Malviya MK, Song XP, Lakshmanan P, Yang LT, Li YR. (2020). Diversity of nitrogen-fixing rhizobacteria associated with sugarcane: A comprehensive study of plant-microbe interactions for growth enhancement in *Saccharum* spp. *BMC Plant Biol.* 20, 220. https://doi.org/10.1186/s12870-020-02400-9.

Solanki MK, Verma KK, Dastogeer KMG, Mora-Poblete F, Mundra S. (2023). Microbial resilience in plant nutrient management towards sustainable farming. *Front. Microbiol.* 14, 1280811. https://doi.org/10.3389/fmicb.2023.1280811

Song X-P, Mo F, Verma KK, Wei J, Zhang X, Yang L, Li YR. (2019). Effect of sugarcane smut (*Ustilago scitaminea* Syd.) on ultrastructure and biochemical indices of sugarcane. *Biomed J. Sci. Tech. Res.* 002950. https://doi.org/10.26717/BJSTR.2019.17.002950.

Song XP, Mo F, Yan M, Zhang X, Zhang B, Huang X, Huang D, Pan Y, Verma KK, Li YR. (2022). Effect of smut infection on the photosynthetic physiological characteristics and related defense enzymes of sugarcane. *Life*. 12, 1201. https://doi.org/10.3390/life12081201

Song XP, Verma KK, Tian DD, Zhang XQ, Liang YJ, Huang X, Li CN, Li YR. (2021). Exploration of silicon functions to integrate with biotic stress tolerance and crop improvement. *Biol. Res.* 54, 19. https://doi.org/10.1186/s40659-021-00344-4

Steer M, Thompson RC. (2020). Plastics and microplastics: Impacts in the marine environment. Mare Plasticum-the Plastic Sea: Combatting plastic pollution through science and art, 49–72.

Tian D, Qin L, Verma KK, Wei L, Li J, Li B, Zhou W, He Z, Wei D, Huang S, Long S, Huang Q, Li C, Wei S. (2023). Transcriptomic and metabolomic differences between banana varieties which are resistant or susceptible to Fusarium wilt. *PeerJ.* 11, e16549. https://doi.org/10.7717/peerj.16549

Verma KK, Anas M, Chen Z, Rajput VD, Malviya MK, Verma CL, Singh RK, Singh P, Song XP, Li YR. (2020f). Silicon supply improves leaf gas exchange, antioxidant defense system and growth in sugarcane responsive to water limitation. *Plants.* 9, 1032, https://doi.org/10.3390/plants9081032.

Verma KK, Joshi A, Song X-P, Liang Q, Xu L, Huang H-r, Wu K-C, Seth CS, Arora J and Li Y-R (2024a). Regulatory mechanisms of plant rhizobacteria on plants to the adaptation of adverse agroclimatic variables. *Front. Plant Sci.* 15, 1377793. https://doi.org/10.3389/fpls.2024.1377793

Verma KK, Joshi A, Song X-P, Singh S, Kumari A, Arora J, Singh SK, Solanki MK, Seth CS, Li Y-R. (2024b). Synergistic interactions of nanoparticles and plant growth promoting rhizobacteria enhancing soil-plant systems: A multigenerational perspective. *Front. Plant Sci.* 15, 1376214. https://doi.org/10.3389/fpls.2024.1376214

Verma KK, Li DM, Singh M, Rajput VD, Malviya MK, Minkina T, Singh RK, Singh P, Song XP, Li YR. (2020a). Interactive role of silicon and plant-rhizobacteria mitigating abiotic stresses: A new approach for sustainable agriculture and climate change. *Plants.* 9, 1055. https://doi.org/10.3390/plants9091055.

Verma KK, Rajput VD, Singh M, Verma CL, Minkina T. (2022a). Influence of waterlogging on plant physiology and molecular biology. In: *Understanding Abiotic Stresses*, Nova Science Publishers, Inc.: New York.

Verma KK, Siddiqui N, Singh M. (2022b). Understanding the plant physiological and molecular adaptation strategies to soil salinity. In: *Understanding Abiotic Stresses*, Nova Science Publishers, Inc.: New York. https://doi.org/10.52305/UBUJ4024

Verma KK, Singh P, Song X-P, Malviya MK, Singh RK, Chen G-L, Solomon S, Li YR. (2020b). Mitigating climate change for sugarcane improvement: Role of silicon in alleviating abiotic stresses. *Sugar. Tech.* 22(5), 741–749. https://doi.org/10.1007/s12355-020-00831-0.

Verma KK, Singh RK, Song QQ, Singh P, Zhang B-Q, Song X-P, Chen G-L, Li YR. (2019a). Silicon alleviates drought stress of sugarcane plants by improving antioxidant responses. *Biomed. J. Sci. Tech. Res.* 002957. https://doi.org/10.26717/BJSTR.2019.17.002957

Verma KK, Song X, Degu HD, Guo D, Joshi A, Huang H, Xu L, Singh M, Huang D, Rajput VD, Li YR. (2023a). Recent advances in nitrogen and nano-nitrogen fertilizers for sustainable crop production: A mini-review. *Chem. Biol. Technol. Agric.* 10, 1–14. https://doi.org/10.1186/s40538-023-00488-3

Verma KK, Song X, Verma CL, Huang H, Singh M, Xu L, Wu J, Khan ZH, Zhang B, Li Y. (2023b). Mathematical modeling of climate and fluoride effects on sugarcane photosynthesis with silicon nanoparticles. *Plant Physiol. Biochem.* 204, 108089. https://doi.org/10.1016/j.plaphy.2023.108089

Verma KK, Song XP, Budeguer F, Nikpay A, Enrique R, Singh M, Zhang B, Wu, JM, Li YR. (2022c). Genetic engineering: An efficient approach to mitigating biotic and abiotic stresses in sugarcane cultivation. *Plant. Signal. Behav.* 17(1), e2108253.

Verma KK, Song XP, Joshi A, Rajput VD, Singh M, Sharma A, Singh RK, Li DM, Arora J, Minkina T, Li YR. (2022d). Nanofertilizer possibilities for healthy soil, water and food in future: An overview. *Front. Plant Sci.* 13, 865048. https://doi.org/10.3389/fpls.2022.865048

Verma KK, Song XP, Joshi A, Tian DD, Rajput VD, Singh M, Arora J, Minkina TM, Li YR. (2022e). Recent trends in nano-fertilizer for advancing sustainable agriculture under the era of climate change to ensure food security for future generations. *Nanomaterials.* 12, 173. https://doi.org/10.3390/nano12010173.

Verma KK, Song XP, Kumari A, Jagadesh M, Singh SK, Bhatt R, Singh M, Seth CS, Li YR. (2024c). Climate change adaptation: Challenges for agricultural sustainability. *Plant. Cell. Environ.* https://doi.org/10.1111/pce.15078

Verma KK, Song XP, Li DM, Singh M, Wu JM, Singh RK, Sharma A, Zhang BQ, LI YR. (2022f). Silicon and soil microorganisms improves rhizospheric soil health with bacterial community, plant growth, performance and yield. *Plant. Signal. Behav.* 17(1), e2104004.

Verma KK, Song XP, Lin B, Guo DJ, Singh M, Rajput VD, Singh RK, Singh P, Sharma A, Malviya MK, Chen GL, Li YR. (2022g). Silicon induced drought tolerance in crop plants: Physiological adaptation strategies. *Silicon.* 14(6), 2473–2487. https://doi.org/10.1007/s12633-021-01071-x

Verma KK, Song XP, Rajput VD, Boldyreva V, Zhang BQ, Minkina T, Li YR. (2022h). Morpho-physiological, biochemical and ultrastructural modifications on sugarcane to prolonged water deficit. In: Verma KK., et al. (Eds.), *Agro-Industrial Perspectives on Sugarcane Production Under Environmental Stress*. https://doi.org/.

Verma KK, Song XP, Rajput VD, Solomon S, Li YR, Rao GP. (2022i). Agro-industrial perspectives on sugarcane production under environmental stress. Springer Nature Singapore, https://doi.org/10.1007/978-981-19-3955-6

Verma KK, Song XP, Singh M, Tian DD, Rajput VD, Minkina T, Li YR. (2023c). Association of silicon and soil microorganisms induces stress mitigation, increasing plant productivity. In: de Mello Prado R. (Ed.), *Benefits of Silicon in the Nutrition of Plants*. pp. 299–328. https://doi.org/10.1007/978-3-031-26673-7_17

Verma KK, Song XP, Tian DD, Guo DJ, Chen ZL, Zhong CS, Nikpay A, Singh M, Rajput VD, Singh RK, Minkina T, Li YR. (2021a). Influence of silicon on biocontrol strategies to manage biotic stress for crop protection, performance and improvement. *Plants.* 10, 2163. https://doi.org/10.3390/plants10102163

Verma KK, Song XP, Tian DD, Singh M, Verma CL, Rajput VD, Singh RK, Sharma A, Singh P, Malviya MK, Li YR. (2021b). Investigation of defensive role of silicon during drought stress induced by irrigation capacity in sugarcane: physiological and biochemical characteristics. *ACS Omega.* 6, 19811–19821.

Verma KK, Song XP, Verma CL, Chen ZL, Rajput VD, Wu KC, Liao F, Chen GL, Li YR. (2021c). Functional relationship between photosynthetic leaf gas exchange in response to silicon application and water stress mitigation in sugarcane. *Biol. Res.* 54, 15. https://doi.org/10.1186/s40659-021-00338-2

Verma KK, Song XP, Verma CL, Malviya MK, Guo DJ, Rajput VD, Sharma A, Wei KJ, Chen GL, Solomon S, Li YR. (2021d). Predication of photosynthetic leaf gas exchange of sugarcane (*Saccharum* spp.) leaves in response to leaf positions to foliar spray of potassium salt of active phosphorus under limited water irrigation. *ACS Omega.* 6, 2396–2409.

Verma KK, Song X-P, Xu L, Huang H-R, Liang Q, Seth CS, Li Y-R. (2023d). Nano-microplastic and agro-ecosystems: A mini-review. *Front. Plant Sci.* 14, 1283852. https://doi.org/10.3389/fpls.2023.1283852

Verma KK, Song XP, Yadav G, Degu HD, Parvaiz A, Singh M, Huang HR, Mustafa G, Xu L, Li YR. (2022j). Impact of agroclimatic variables on proteo-genomics in sugarcane (*Saccharum* spp.) plant productivity. *ACS Omega.* 7, 22997–23008.

Verma KK, Song XP, Zeng Y, Guo DJ, Sing M, Rajput VD, Malviya MK, Wei KJ, Sharma A, Li DP, Chen GL, Li YR. (2021e). Foliar application of silicon boosts growth, photosynthetic leaf gas exchange, antioxidative response and resistance to limited water irrigation in sugarcane (*Saccharum officinarum* L.). *Plant. Physiol. Biochem.* 166, 582–592.

Verma KK, Song XP,Zeng Y, Li DM, Guo DJ, Rajput VD, Chen GL, Barakhov A, Minkina TM, Li YR. (2020c). Characteristics and correlation of leaf stomata and its relationship with photosynthesis on *Saccharum* spp. under different irrigation and silicon application. *ACS Omega.* 5, 24145–24153.

Verma KK, Vatsal S, Singh M, Rajput VD, Minkina T. (2022k). Coping with water stress: Effects and adaptive mechanisms to plants. In: *Understanding Abiotic Stresses*, Nova Science Publishers, Inc.: New York.

Verma KK, Wu KC, Verma CL, Li DM, Malviya MK, Singh RK, Singh P, Chen GL, Song XP, Li YR. (2020d). Developing mathematical model for diurnal dynamics on photosynthetic responses in sugarcane responsive to different irrigation and silicon application. *PeerJ.* 8, e10154. https://doi.org/10.7717/peerj.10154.

Verma KK, Liu X-H, Wu K-C, Singh RK, Song QQ, Malviya MK, Song X-P, Singh P, Verma CL, Li YR. (2020e). The impact of silicon on photosynthetic and biochemical responses of sugarcane under different soil moisture levels. *Silicon.* 12, 1355–1367.

Verma KK, Wu K-C, Singh P, Malviya MK, Singh RK, Song X-P, Li YR. (2019b). The protective role of silicon in sugarcane under water stress: Photosynthesis and antioxidant enzymes. *Biomed J. Sci. Tech. Res.* 002685. https://doi.org/0.26717/BJSTR.2019.15.002685

Wallerstein N, Duran, B. (2010). Community-based participatory research contributions to intervention research: The intersection of science and practice to improve health equity. *Am. J. Public. Health.* 100(S1), S40-S46.

Wu K-C, Huang C-M, Verma KK, Deng Z-N, Huang H-R, Pang T, Cao H-Q, Luo H-B, Jiang S-L, Xu L. (2022). Transcriptomic responses of *Saccharum spontaneum* roots in response to polyethylene glycol-6000 stimulated drought stress. *Front. Plant Sci.* 13, 992755. https://doi.org/10.3389/fpls.2022.992755

Xu L, Deng Z-N, Wu K-C, Malviya MK, Solanki MK, Verma KK, Pang T, Li Y-J, Liu X-Y, Kashyap BK, Dessoky ES, Wang W-Z and Huang H-R. (2022). Transcriptome analysis reveals a gene expression pattern that contributes to sugarcane bud propagation induced by indole-3-butyric acid. *Front. Plant Sci.* 13, 852886. https://doi.org/10.3389/fpls.2022.852886

Yadav N, Sidana N. (2023). Precision agriculture technologies: Analysing the use of advanced technologies, such as drones, sensors, and GPS, in precision agriculture for optimizing resource management, crop monitoring, and yield prediction. *J. Adv. Zool. 44*.

Yang Z, Liu J, Baskaran S, Imhoff C H, Holladay J D. (2010). Enabling renewable energy—and the future grid—with advanced electricity storage. *JOM. 62*, 14–23.

Zailani, S., Amran, A., Jumadi, H. (2011). Green innovation adoption among logistics service providers in Malaysia: An exploratory study on the managers' perceptions. *Int. Bus. Manag.* 5(3), 104–113.

Zheng F-J, Lin B, Yang Y-X, Fang X-C, Verma KK, Chen G-L. (2023). Efficacy and functionality of sugarcane original vinegar on mice. *Front. Microbiol.* 14, 1224666. https://doi.org/10.3389/fmicb.2023.1224666

Ziolo M, Filipiak B Z, Bąk I, Cheba K. (2019). How to design more sustainable financial systems: The roles of environmental, social, and governance factors in the decision-making process. *Sustainability.* 11(20), 5604.

CHAPTER 14

Environmental Impacts of Sugarcane Biofuels

KRISHAN K. VERMA, XIU-PENG SONG, QIANG LIANG, HAI-RONG HUANG, LIN XU, and YANG-RUI LI*

Sugarcane Research Institute, Guangxi Academy of Agricultural Sciences/Key Laboratory of Sugarcane Biotechnology and Genetic Improvement (Guangxi), Ministry of Agriculture and Rural Affairs/ Guangxi Key Laboratory of Sugarcane Genetic Improvement, Nanning, Guangxi, China

*Corresponding author

ABSTRACT

The potential of sugarcane as a crop for renewable energy has attracted a lot of attention. Although manufacturing ethanol from sugar juice has been a long-standing process, there is still a growing interest in applying bagasse, a fibrous coproduct, to produce bio-based chemicals and renewable fuels. The development of new energy cane types and the success of these initiatives could significantly boost the use of sugarcane and sugarcane biomass as fuels while improving the sustainability and competitiveness of the industry. A variety of existing and emerging biofuels as well as other renewable products produced from sugarcane and sugarcane-based coproducts, like bagasse, are studied in sugarcane-based biofuels and bioproducts. Essential information about how environmental factors affect innovative fuels and sugarcane-derived bioproducts is compiled in this chapter.

Sugarcane Transformation: Biofuels, Biotechnology, and Beyond.
Krishan K. Verma, Xiu-Peng Song, Hai-Rong Huang, Aradhna Kumari, and Yang-Rui Li (Eds.)

DOI: 10.1201/9781779641571-14

14.1 INTRODUCTION

Sugarcane (*Saccharum* spp. hybrid) is a significant cash crop globally. This species is highly desirable in a biomass-dependent economy, not only because it is the world's primary source of sugar but also because it can generate large amounts of biomass in a comparatively short length of time (Moore and Botha, 2013; Chen et al., 2020; Singh et al., 2020, 2021, 2023; Huang et al., 2024). While ethanol fuel can be produced through many different methods, sugar fermentation is the most popular method for producing it on a large scale (Verma et al., 2019a, 2019b; Bhatt et al., 2023a, 2023b). In many tropical and subtropical nations, sugarcane has become the second-largest source of biofuel during the past three decades and has significant social, economic, and environmental significance (Carpio and Simone de Souza, 2017; Scheiterle et al., 2018; Song et al., 2019; Malviya et al., 2021, 2022; Li et al., 2023). It is currently the second most important C_4 species, after maize, and the sixth most commercially significant crop (Sage et al., 2014; Anas et al., 2021a, 2021b). According to FAOSTAT, 2018, more than 70% of the world's sugarcane crop is grown in Brazil, India, China, Thailand, and Pakistan. Brazil now produces almost half of the world's sugarcane, making it the top producer in the world (Guo et al., 2021, 2022, 2023; Huang et al., 2024).

According to the Renewable Fuels Association (RFA, 2017), Brazil and the United States produce 85% of the ethanol worldwide. Still, corn starch accounts for a large portion of US ethanol production, but sugarcane sucrose is the main source of ethanol production in Brazil. By using so-called second-generation methods, sugarcane juice, molasses, and, more recently, cellulose are fermented to generate sugarcane ethanol (Goldemberg et al., 2014; Sharma et al., 2021, 2022; Deng et al., 2023). The primary producer of cane ethanol is Brazil, where the National Bioethanol Program, also known as ProAlcool, solidified the sugarcane ethanol sector in the late 1970s. The industry has been developing since the early 20th century. The success of the program at that time was largely attributed to Brazil's implementation of mandatory laws requiring a 25% bioethanol blend with petrol. This development also encouraged companies to design engines that run exclusively on hydrated ethanol (Walter et al., 2014).

According to the International Energy Agency (IEA), the European Union is the world's third-largest producer of ethanol (5%), with a highly diverse supply of feedstock that includes 39% maize, 30% wheat, 20%

sugar beetroot, 7% other starchy cereals, and 3% lignocellulosic and other new substances (IEA, 2016; Joshi et al., 2022, 2023; Wu et al., 2022). China, a newly developed nation, started using ethanol derived from maize in 2002 and produced 1.92 million tons of the fuel in 2012, ranking it as the world's third-largest producer and consumer of ethanol worldwide. Presently, China produces ethanol in parts of the Huanghuai River Delta, mostly in saline-alkaline regions, as well as in Northeast, North, and Northwest China from sweet sorghum and cassava (Ge et al., 2014). One of the biggest challenges for developing nations is the expense of implementing the entire sugarcane-based ethanol production system, despite the fact that the yield of ethanol from this crop is nearly two times higher than that of maize and other energy crops (Goettemoeller and Goettemoeller, 2007; Xu et al., 2022; Oliveira et al., 2022). Furthermore, the development of alternative energy sources for common transportation vehicles, including electric automobiles, could reduce the need for ethanol in the years to come (Crago et al., 2010; Krishna et al., 2022; Huang et al., 2022; Zheng et al., 2023).

Economic, environmental, and social sustainability of ethanol production can be investigated, and this study can help ethanol-producing nations acquire useful information. Based on previously published studies, the authors provide an in-depth review of the primary sustainability indicators in this chapter, which include social, environmental, and economic indicators.

14.2 ENVIROSUSTAINABILITY OF SUGARCANE ETHANOL

When bioenergy is produced sustainably, among other things, carbon emissions from the burning of biofuels are absorbed by the photosynthesis process of plant growth in the bioenergy generation or utilization process (Goldemberg et al., 2008a, 2008b, 2017; Verma et al., 2020a, 2020b, 2023a, 2023b). Furthermore, since the development of biofuels involves agricultural and industrial stages, all carbon emissions must be considered, given that they are a worldwide process and the source of climate change (Verma et al., 2024a, 2024b, 2024c). The industrial phase's eventual use of fossil fuels contributes to improve the energy balance and prevent greenhouse gas (GHG) emissions (Chen et al., 2023; Jagadesh et al., 2024).

However, ethanol from other crops, sugarcane bagasse, can generate all of the energy required for the industrial process during the ethanol production process when it is burned in boilers. Accordingly, the only sources of carbon emissions are the use of fertilizer and the diesel oil employed for agricultural operations and transportation (Verma et al., 2020c, 2020d; Song et al., 2021, 2022). As a result, sugarcane ethanol has a higher energy balance than alternative solutions, and replacing petrol with it results in a higher reduction in carbon emissions. With ethanol replacing gasoline, carbon emissions have been reduced by 62–64% in Argentina (UNSAM, 2015); 89% in Brazil; 54% in Colombia; and 65% in Uruguay (Hernández et al., 2017). When ethanol from other crops is employed, less pollutants are eliminated than with this approach (Table 14.1). According to the most recent statistics, Vietnam's switch from petrol to cassava ethanol reduced carbon emissions by 59% (Quang Ha, 2017).

TABLE 14.1 Reduce Greenhouse Gas Emissions by Gasoline Alternative by Ethanol Form a Variety of Plants

Crops/Plants for Ethanol	Emissions Reduction (%)
Sugarcane (Saccharum spp. hybrid)	62–89
Cassava (manioc)	59–63
Maize	30–38
Wheat	19–47
Sugar beet	35–56
Lignocellulosic residues[a]	66.5–73

Note: [a]Theoretical estimates.

Source: Adapted with permission from Dai et al. (2006), EBAMM (2005), IEA (2004), Macedo et al. (2008), Nguyen et al. (2007), UNSAM (2015), and Quang Ha (2017).

Only Malawi GHG emissions have been assessed among African nations. Dunkelberg et al. (2014) showed that the ethanol produced in Malawi results in GHG emissions expressed as CO_2eq of 116 g/MJ of ethanol under current production situations. The ideal crop for producing ethanol might be selected depending on interesting findings, but regional factors also need to be taken into account. Conditions related to soil, climate, and economy are significant.

Water contamination and quality are mainly affected by how effectively byproducts can be disposed of. Vinasse disposal is one of the most significant problems associated with water contamination in the sugarcane economy (Chaudhary et al., 2023; Solanki et al., 2023; Verma et al., 2023c, 2023d). Vinasse, a dark liquid with an acidic pH of 4–5, is a byproduct of fermentation and distillation that is high in organic matter. Due to its composition and high temperature, it poses a possible hazard to water (Verma et al., 2020e, 2020f; Kumari et al., 2022, 2023a, 2023b). It was employed without control for this ferti-irrigation for a while. However, the effects of the improper disposal in the soil began to become visible in certain situations; local studies discovered that, depending on the type of soil, inappropriate disposal in the crops might lead to severe subterranean water contamination (Tian et al., 2023).

Large amounts of water are required for sugarcane growth; if it is rain-fed, it needs a substantial amount of rainfall annually, ideally distributed evenly throughout the growing season (about 1500–2500 mm). Rainfall is inadequate for sugarcane in many countries; thus, irrigation is required (Verma et al., 2021a, 2021b, 2021c, 2022a, 2022b, 2022c, 2022d). For example, irrigation is necessary for sugarcane crops in African nations. In the past, Brazil's industrial ethanol manufacturing required a lot of water. Recently, the amount of water used in the milling process has declined.

Several concerns pertaining to the preservation of natural forests and other significant biomes constantly come up when talking about the sustainability of biofuels; some studies even go so far as to claim that any crop utilized to produce bioenergy originates from deforestation (Fargione et al., 2008a; Searchinger et al., 2008; World Bank, 2008; Verma et al., 2021d, 2021e). However, since biofuel production does not to be seem generally moving into tropical forests, these studies only take into account the worst-case scenario, which cannot occur at the present time. Results from large-scale studies that have already been conducted and evaluated CO_2 emissions from nondeforestation agriculture techniques are less concerning (Cerri et al., 2007; Verma et al., 2022e, 2022f, 2022g, 2022h, 2022i). Other research, however, indicates that bioenergy crops are being grown in pasturelands without causing any deforestation (Goldemberg et al., 2008a, 2008b, 2017).

According to Souza et al. (2015), existing pastureland could support 3.8 times more animals globally if animal stocking densities were increased to currently achievable levels that are climate appropriate, as in the case of Brazil. Overdoubling the number of grazing animals worldwide would be possible if the worst-performing pastures received up to 50% of their maximum achievable density. The typical method of first-generation ethanol has been considered inferior to second-generation ethanol due to concerns about land availability and other environmental effects. More land that could be used for food production might become accessible by using cellulosic wastes to manufacture ethanol. Second-generation biofuel development failed to occur as expected, mainly in developing nations.

Soil quality is an important issue globally. Soil quality has a major impact on agricultural productivity in semiarid environments (Verma et al., 2022j, 2022k), such as the countries in sub-Saharan Africa. Since the sugar mills developed specific processes ensuring the proper application of fertilizers and soil protection against erosion, soil compaction, and moisture loss, sugarcane culture became more sustainable over time. In some parts of the nation, sugarcane has been grown on certain soils for nearly two centuries without suffering an overall decrease in yield. The production of agricultural products has really expanded dramatically. With the development of sustainable sugarcane harvesting, there is currently an intense debate about how much residue should be maintained in the field to maintain the soil.

Numerous inorganic substances are required for sugarcane crops, such as herbicides, that eliminate weeds, insects, mites, and fungi as well as defoliants and other chemicals that speed up the cane maturation. When it concerns agrochemicals, sugarcane production requires more of them than other crop cultivation. Compared to citrus, corn, coffee, and soybeans, sugarcane uses fewer pesticides per hectare. Even yet, sugarcane requires fewer herbicides per hectare than citrus, corn, and soybeans, but more than coffee requires. Moreover, sugarcane requires around the same amount of fertilizer as soybeans and less than cotton, coffee, oranges, and rice (Macedo, 2005). Using industrial waste as fertilizer, particularly the vinasse, is beneficial for sugarcane production. As a result, the soil's potassium content and production have significantly increased (Ripoli et al., 2005).

The emissions into the atmosphere during the manufacture of sugarcane ethanol must be taken into consideration as well as the environmental benefits of using sugarcane ethanol as opposed to petrol in engines, which reduces most pollutant emissions (mostly CO_2, SO_x, and particulates/PM). There are two issues to consider in this area: environmental emissions from burning bagasse in boilers for cogeneration and emissions from burning sugarcane before harvest. Studies that have been done on the subject of ethanol use in automobiles demonstrate the advantages associated with doing away with petrol in several nations.

14.3 ENVIRONMENTAL CHALLENGES

Producing sugarcane ethanol has drawn attention from around the world in the last 10 years since it is thought to be one of the most promising biofuels for short-term climate change mitigation (Fargione et al., 2008b; Goldemberg, 2007; Verma et al., 2024a, 2024c). For many tropical nations, the increased production and use of BE has had significant social, economic, and environmental effects (Goldemberg, 2007; Guo et al., 2015; Marin et al., 2016; Scarpare et al., 2016; Verma et al., 2024a, 2024b, 2024c; Huang et al., 2024). Supported by regulations like the Energy Independence and Security Act in the United States (Sissine, 2007) and RenovaBio in Brazil (Caldarelli and Gilio, 2018), the socioeconomic effects include local prosperity, food and energy security, and economic viability. In addition to lowering GHG emissions, sugarcane's environmental interactions concern biodiversity, land usage, soil conservation, and water resource utilization (Filoso et al., 2015; Guo et al., 2015; Wu et al., 2022; Deng et al., 2023).

The nation's energy safety and energy autonomy are generally improved by the production and use of biofuels, which also promote research and development, create possibilities for employment, and increase incomes for farmers. However, the production of feedstock establishes additional strain on water resources and necessitates land, fertilizers, and other inputs. Under some conditions, the production of biofuels might produce higher GHG emissions as well as utilize more fossil fuels in terms of energy equivalents (IPCC, 2014; Haberl et al., 2013). Compared to gasoline or diesel fuels, biofuels typically have direct fuel-cycle GHG emissions that are 30–90% lower per kilometer operated. Compared to oil consumption, the sugarcane ethanol system has the potential to compensate for 86% of CO_2 emissions from the transportation sector, like corn ethanol. However, emissions originating from the conversion of land to sugarcane require less time to recover, normally 2–4 years (Jaiswal et al., 2017). A significant and urgent scientific task is still determining sustainable bioenergy levels and figuring out how to combine bioenergy with food production and ecological preservation (Haberl et al., 2013). However, with careful planning, technology advancements, government incentives, suitable land use policies, and sustainable farming inputs, these negative effects can be reduced (IPCC, 2014). The production of cane ethanol poses sustainability and environmental challenges, such as the need to cease burning sugarcane in the fields, reduce water consumption in mills, control the application of vinasse in areas where groundwater contains contaminants with K, and preserve and restore riparian buffers and forest fragments within sugarcane farms.

14.4 SOCIOECONOMIC CHALLENGES

According to Moore and Botha (2013), sugarcane is a perennial grass that survives in tropical and subtropical environments. The production of it is limited to a large portion of developing nations due to this attribute. Additionally, sugarcane plants are highly susceptible to water scarcity; ideal locations for sugarcane growth should have an average annual rainfall (Scarpare et al., 2016). Even while irrigation appears like a good way to address the depletion of water, it can be difficult to develop the infrastructure needed for a mill and/or transport biomass in remote regions of developing nations. While evaluating possible future sugarcane results scenarios, climate change is an additional variable to take into consideration. This is due to the possibility that significant environmental problems that could impact the globe in the 21st century may arise from the variability and change of the global climate driven by anthropogenic and natural processes.

Biofuels have become more significant during the last 10 years as a result of social, economic, and environmental factors (Goldemberg, 2007). When ethanol was first produced, it had to overcome the difficulties associated with acquiring and embracing new technologies. Following the oil crisis, ethanol production as a biofuel gained acceptance and was viewed as a substitute liquid fuel to reduce dependency on fossil fuels (Hira and de Oliveira, 2009). Therefore, among other new technologies, ethanol was not revolutionary discovery, and it required some time for consumers to accept it. Ethanol has around 30% less efficiency than petrol, which decreases the autonomy of cars; also, problems with cold-weather start ignition have prevented the use of ethanol in temperate climates (Nakata et al., 2006). These issues were resolved by adopting required (5–27%) or optional (10–90%) ethanol combinations with petrol, regulating ethanol costs in relation to petrol efficiency, and modifying ignition timing for lower temperatures (Goldemberg, 2007).

The adverse opinion of hazardous conditions consistently associated with slavery on sugarcane plantations presented another difficulty for ethanol produced from sugarcane (Rocha et al., 2010). Furthermore, the massive plantations' burning of sugarcane for harvesting deteriorated air pollution and respiratory problems in nearby metropolitan areas (Barbosa et al., 2012; Paraiso and Gouveia, 2015). In many producing regions, burning sugarcane became illegal, and harvesting was mechanized in an effort to break down social barriers. However, the mechanization was primarily utilized by big and/or technologically advanced producers; as a result, Australia, the United States, China, Brazil, and South Africa were the countries that adopted it most completely. Harvest mechanization is not well established in India, the country that produces the second most sugarcane globally. This is because Indian agriculture, which includes sugarcane farming, is typified by tiny, scattered holdings (Singh et al., 2011).

In reality, the harvested biomass can be used by sugarcane mills to generate sugar, ethanol, energy, and feedstock for the paper and plastic sectors. Agronomic techniques and transportation quality have a significant impact on the production and overall expense of ethanol. In contrast to other crops, sugarcane yields between 60 and 100 t/ha/year, which presents a challenging economic challenge due to postharvest losses. This is because sugars in the stalks break rapidly after harvest, and the crop cannot be maintained like cereal crops. One of the main requirements of the industry is real-time transportation management since biomass is delivered by trucks with a load capacity of about 60 t and because crushing and harvesting are continual processes.

14.5 CONCLUSION

The world's sugarcane area is expanding rapidly because of the rising need for ethanol and sugar. For tropical and subtropical nations, sugarcane is a viable product since it generates significant foreign cash through exports and produces a completely local subsidiary that supports transportation, industry, and farming. Given the risks associated with climate change, sugarcane ethanol presents itself as a viable choice for lowering global GHG emissions while also promoting the growth of developing nations in tropical regions by providing a more affordable and sustainable source of energy. The difficulties arise from the fact that sugarcane grows best in tropical climates, and most arable land under these conditions cannot support significantly more sugarcane due to a lack of availability and future fuel and food shortages.

ACKNOWLEDGMENT

The authors would like to thank the Guangxi Academy of Agricultural Sciences, Nanning, Guangxi, China, for providing the necessary facilities for this study.

FUNDING

This research was financially supported by the Guangxi Natural Science Foundation (2021GXNSFAA220022; 2023GXNSFAA026459; 2023GXNSFAA026460), Guangxi Innovation Teams of Modern Agriculture Technology (nycytxgxcxtd-2021-03), Guangxi Characteristic Crop Experimental Station (GTS2022022), National Key Research and Development Project (2022YFD2301102-07), The National Natural Science Foundation of China (31760415; 32060468), Fund of Guangxi Academy of Agricultural Sciences (2021YT011), Science and Technology Major Project of Guangxi (Guike AA22117002-1), Fundamental Research Fund of Guangxi Academy of Agriculture Sciences (2023YM55), and Key Research and Development Program of Nanning (20232060).

KEYWORDS

- **socioeconomic impacts**
- **atmospheric variables**
- **biofuels**
- ***Saccharum* spp**
- **cane ethanol**
- **bioenergy**

REFERENCES

Anas M, Verma KK, Haq IU, Naeem M, Li Q, Liao F, Xiang YX, Li YR. (2021a). Characterization of exotic, native and wild-type genotypes of sugarcane (*Saccharum* spp. hybrids) for internal nitrogen use efficiency under different nitrogen levels. *Sugar Tech.* 23, 1258–1267. https://doi.org/10.1007/s12355-021-01014-1

Anas M, Verma KK, Riaz M, Qiang L, Liao F, Liu Y, Li YR. (2021b). Physio-morphological and biochemical mechanism of nitrogen use efficiency in sugarcane (*Saccharum* spp.) genotypes under different growth stages and nitrogen levels. *J. Plant Interact.* 16, 332–343.

Barbosa CMG, Terra-Filho M, de Albuquerque ALP et al. (2012). Burnt sugarcane harvesting—cardiovascular effects on a group of healthy workers, Brazil. *PLoS One* 7(1), e46142

Bhatt R, Verma KK, Jamwal NS. (2023a). Can paddy-compost eliminate N fertilization for climate-smart sugarcane? *Sugar Tech.* 25, 1036–1046. https://doi.org/10.1007/s12355-023-01270-3

Bhatt R, Verma KK, Kumar R, Sanghera GS. (2023b). Foliar application of potassium salt of active phosphorus (PSAP) mitigates insect pests and improves yield along with sugarcane quality in response to agroclimatic conditions of Punjab. *Sugar Tech.* 25(3), 660–669.

Caldarelli CE, Gilio L. (2018). Expansion of the sugarcane industry and its effects on land use in São Paulo: Analysis from 2000 through 2015. *Land Use Policy* 76, 264–274.

Carpio LGT, Simone de Souza F. (2017). Optimal allocation of sugarcane bagasse for producing bioelectricity and second generation ethanol in Brazil: Scenarios of cost reductions. *Renew. Energy* 111, 771–780.

Cerri CEP , Easter M, Paustian K, Killian K, Coleman K, Bernoux M, Falloon P , Powlson D, Batjes N, Milne E, Cerri CC. (2007). Predicted soil organic carbon stocks and changes in the Brazilian Amazon between 2000 and 2030. *Agric. Ecosyst. Environ.* 122, 58–72, https://doi.org/10.1016/j.agee.2007.01.008

Chaudhary P, Xu M, Ahamad L, Chaudhary A, Kumar G, Adeleke BS, Verma KK, Hu D-M, Siric I, Kumar P, Popescu SM, Fayssal SA. (2023). Application of synthetic consortia for improvement of soil fertility, pollution remediation, and agricultural productivity: A review. *Agronomy* 13, 643. https://doi.org/10.3390/agronomy13030643.

Chen GL, Zheng FJ, Lin B, Lao SB, He J, Huang Z, Zeng Y, Sun J, Verma KK. (2020). Phenolic and volatile compounds in the production of sugarcane vinegar. *ACS Omega* 5, 30587–30595.

Chen G-L, Zheng F-J, Lin B, Yang Y-X, Fang X-C, Verma KK, Yang L-F. (2023). Vinegar: A potential source of healthy and functional food with special reference to sugarcane vinegar. *Front. Nutr*. 10, 1145862. https://doi.org/10.3389/fnut.2023.1145862

Crago CL, Khanna M, Barton J et al. (2010). Competitiveness of Brazilian sugarcane ethanol compared to US corn ethanol. *Energy Policy* 38(11), 7404–7415.

Dai D, Zhiyuan H, Gengqiang P , Li H, Wang C. (2006). Energy efficiency and potentials of cassava fuel ethanol in Guangxi region of China. *Energy Convers. Manag.* 47(13–14), 1686–1699

Deng ZN, Huang HR, Huang CM, Verma KK, Luo HB, Jiang SL, Li YJ, Cao HQ, Wu KC, Xu L. (2023). Short-term limited water irrigation influences the photosynthetic pigments, enzymatic and non-enzymatic activities in *Saccharum spontaneum* L. at vegetative stage. *Sugar Tech.* 25, 110–118. https://doi.org/10.1007/s12355-022-01175-7

Dunkelberg E, Finkbeiner M, Hirschl B. (2014). Sugarcane ethanol production in Malawi: Measures to optimize the carbon footprint and to avoid indirect emissions. *Biomass Bioenergy* 71, 37–45. https://doi.org/10.1016/j.biombioe.2013.10.006

EBAMM. (2005). *ERG Biofuel Analysis Meta-model.* Energy and Resources Group and Richard & Rhoda Goldman School of Public Policy at UC Berkeley. http://rael.berkeley.edu/old_drupal/sites/default/files/EBAMM/. [Accessed 10 July 2024].

Fargione J, Hill J, Tilman D et al. (2008a). Land clearing and the biofuel carbon debt. *Science* 319(5867), 1235–1238

Fargione J, Hill J, Tilman D, Polasky S, Hawthorne P. (2008b). Land clearing and the biofuel carbon debt. *Science* 319, 1235–1238

Filoso S, do Carmo JB, Mardegan SF et al. (2015). Reassessing the environmental impacts of sugarcane ethanol production in Brazil to help meet sustainability goals. *Renew. Sustain. Energy Rev.* 52, 1847–1856

Food and Agriculture Organization Statistics (2018). http://www.fao.org/faostat/en/#data. [Accessed 15 June 2024].

Ge J, Lei Y, Tokunaga S. (2014). Non-grain fuel ethanol expansion and its effects on food security: A computable general equilibrium analysis for China. *Energy* 65, 346–356.

Goettemoeller J, Goettemoeller A. (2007). *Sustainable Ethanol: Biofuels, Biorefineries, Cellulosic Biomass, Flex-fuel Vehicles, and Sustainable Farming for Energy Independence*. Prairie Oak Publishing: Michigan; p. 195.

Goldemberg J. (2007). Ethanol for a sustainable energy future. *Science* 315(5813), 808–810.

Goldemberg J, Coelho ST, Guardabassi P. (2008b). The Sustainability of ethanol production from sugar cane. *Energy Policy* 36. https://doi.org/10.1016/j.enpol.2008.02.028

Goldemberg J, Coelho ST, Guardabassi P , Nastari P. (2017). Bioethanol from sugar: The Brazilian experience. In: Meyers RA (Ed.), *Encyclopedia of Sustainability Science and Technology*. Springer: New York. https://doi.org/10.1007/978-1-4939-2493-6_312-4

Goldemberg J, Mello FFC, Cerri CEP , Davies CA, Cerri CC. (2014). Meeting the global demand for biofuels in 2021 through sustainable land use change policy. *Energy Policy* 69, 14–18

Goldemberg J, Nigro FEB, Coelho ST. (2008a). *Bioenergia no Estado de São Paulo: Situação Atual. Perspectivas, Barreiras e Propostas*. Imprensa Oficial do Estado de São Paulo: São Paulo.

Guo D-J, Li D-P, Chen Z-D, Yang B, Verma KK, Singh RK, Singh P, Khan Q, Sharma A, Qin Y, Zhang B-Q, Song X-P, Li Y-R. (2023). Effect of endophytic diazotroph *Enterobacter roggenkampii* ED5 on nitrogen-metabolism-related microecology in the sugarcane rhizosphere at different nitrogen levels. *Front. Microbiol.* 14, 1132016. https://doi.org/10.3389/fmicb.2023.1132016

Guo D-J, Li D-P, Singh RK, Singh P, Sharma A, Verma KK, Qin Y, Khan Q, Lu Z, Malviya MK, Song X-P, Xing Y-X and Li Y-R. (2021). Differential protein expression analysis of two sugarcane varieties in response to diazotrophic plant growth-promoting endophyte

Enterobacter roggenkampii ED5. *Front. Plant Sci.* 12, 727741. https://doi.org/10.3389/fpls.2021.727741

Guo DJ, Li DP, Singh RK, Singh P, Verma KK, Sharma A, Qin Y, Khan Q, Song XP, Malviya MK, Xing YX, Li YR. (2022). Comparative transcriptome analysis of two sugarcane varieties in response to diazitrophic plant growth promoting endophyte *Enterobacter roggenkampii* ED5. *J. Plant Interact.* 17(1), 75–84.

Guo M, Song W, Buhain J. (2015). Bioenergy and biofuels: History, status, and perspective. *Renew. Sustain. Energy Rev.* 42, 712–725.

Haberl H, Erb K-H, Krausmann F, Running S, Searchinger TD, Smith WK. (2013). Bioenergy: How much can we expect for 2050? *Environ. Res. Lett.* 8, 031004

Hernández V, Castromán N, Rava C (2017). Uruguay: country presentation. presentation on sustainable bioenergy production and use in the Americas: lessons learnt and future opportunities. In: *Ministry of Industry Energy and Mining, Ministry of Livestock, Agriculture, and Fisheries*. FAO/GBEP Asuncion August 02, 2024.

Hira A, de Oliveira LG. (2009). No substitute for oil? How Brazil developed its ethanol industry.

Huang CM, Wu KC, Verma KK, Luo HB, Deng ZN, Shi GY, Cao HQ, Wu XJ, Ye LP, Wei YW, Xu L. (2024). Isolation and expression analysis of cellulose synthase 3 (*Ces3*) genes from sugarcane (*Saccharum officinarum* L.). *Chem. Biol. Technol. Agric.* 11, 71. https://doi.org/10.1186/s40538-024-00597-7

Huang HR, Gao YJ, Malviya MK, Verma KK, Solanki MK, Huang YX, Li X, Deng Y, Yan J, Tang SY, Wang LW, Xu L. (2022). Genetic diversity analysis of sugarcane (*Saccharum* spp. Hybrids) among high-sucrose clones of GT series and commonly used parents by using microsatellite markers in Guangxi, China. *Sugar Tech.* 24, 397–407. https://doi.org/10.1007/s12355-021-01033-y

IEA. (2004). *Biofuels for Transport: An International Perspective*. International Energy Agency: Paris, ISBN:92-64-01.

Intergovernmental Panel on Climate Change. (2014). *Climate Change 2013: The Physical Science Basis: Working Group I Contribution to the Fifth Assessment Report of the Intergovernmental Panel on Climate Change*. Cambridge.

International Energy Agency (2016). *Medium-term Renewable Energy Market Report 2016*. New York.

Jagadesh M, Dash M, Kumari A, Singh SK, Verma KK, Kumar P, Bhatt R, Sharma SK. (2024). Revealing the hidden world of soil microbes: Metagenomic insights into plant, bacteria, and fungi interactions for sustainable agriculture and ecosystem restoration. *Microbiol. Res.* 127764. https://doi.org/10.1016/j.micres.2024.127764

Jaiswal D, De Souza AP , Larsen S, LeBauer DS, Miguez FE, Sparovek G, Bollero G, Buckeridge MS, Long SP. (2017). Brazilian sugarcane ethanol as an expandable green alternative to crude oil use. *Nat. Clim. Chang.* 7(11), 788

Joshi A, Verma KK, Rajput VD, Arora J. (2023). Metabolic engineering of microorganisms in advancing biofuel production. In: *Advances in Lignocellulosic Biofuel Production Systems*; pp. 181–202. https://doi.org/10.1016/B978-0-323-91192-4.00016-X

Joshi A, Verma KK, Rajput VD, Minkina T, Arora J. (2022). Recent advances in metabolic engineering of microorganisms for advancing lignocellulose-derived biofuels. *Bioengineered* 13(4), 8135–8163. https://doi.org/10.1080/21655979.2022.2051856.

Krishna D, Sachan HK, Verma KK. (2022). Abiotic stresses in plants: An overview. In: *Understanding Abiotic Stresses.* Nova Science Publishers, Inc.: New York.

Kumari A, Dash M, Singh SK, Jagadesh M, Mathpal B, Mishra PK, Pandey SK, Verma KK. (2023a). Soil microbes: A natural solution for mitigating the impact of climate change. *Environ. Monit. Assess*. 195, 1436. https://doi.org/10.1007/s10661-023-11988-y

Kumari A, Lakshmi GA, Krishna GK, Patni B, Prakash S, Bhattacharyya M, Singh SK, Verma KK. (2022). Climate change and its impact on crops: A comprehensive investigation for sustainable agriculture. *Agronomy* 12, 3008. https://doi.org/10.3390/agronomy12123008

Kumari A, Singh SK, Mathpal B, Verma KK, Garg VK, Bhattacharyya M, Bhatt R. (2023b). The biosynthesis, mechanism of action, and physiological functions of melatonin in horticultural plants: A review. *Horticulturae* 9, 913. https://doi.org/10.3390/horticulturae9080913

Li A-M, Liao F, Wang M, Chen Z-L, Qin C-X, Huang R-Q, Verma KK, Li Y-R, Que Y-X, Pan Y-Q, et al. (2023). Transcriptomic and proteomic landscape of sugarcane response to biotic and abiotic stressors. *Int. J. Mol. Sci.* 24, 8913. https://doi.org/10.3390/ijms24108913

Macedo IC. (2005). *Sugarcane's Energy: Twelve Studies on Brazilian Sugar Cane Agribusiness and its Sustainability*. ÚNICA: São Paulo.

Macedo IC, Seabra J, Silva J (2008). Geen house gases emissions in the production and use of ethanol from sugarcane in Brazil: The 2005/2006 averages and a prediction for 2020. *Biomass Bioenergy* 32(7), 582–595.

Malviya MK, Li C-N, Lakshmanan P, Solanki MK, Wang Z, Solanki AC, Nong Q, Verma KK, Singh RK, Singh P, Sharma A, Guo D-J, Dessoky ES, Song X-P, Li Y-R. (2022). High-throughput sequencing-based analysis of rhizosphere and diazotrophic bacterial diversity among wild progenitor and closely related species of sugarcane (*Saccharum* spp. inter-specific hybrids). *Front. Plant Sci.* 13, 829337. *https://doi.org/10.3389/fpls.2022.829337.*

Malviya MK, Solanki MK, Li CN, Wang Z, Zeng Y, Verma KK, Singh RK, Singh P, Huang HR, Yang LT, Song XP, Li YR. (2021). Sugarcane-legume intercropping can enrich the soil microbiome and plant growth. *Front. Sustain. Food Syst.* 5, 606595. https://doi.org/10.3389/fsufs.2021.606595.

Marin FR, Martha GB Jr, Cassman KG, Grassini P. (2016). Prospects for increasing sugarcane and bioethanol production on existing crop area in Brazil. *Bioscience* 66(4), 307–316

Moore PH, Botha FC. (2013). *Sugarcane: Physiology, Biochemistry and Functional Biology*. Wiley: Ames; p. 750.

Nakata K, Utsumi S, Ota A, Kawatake K, Kawai T, Tsunooka T. (2006). *The Effect of Ethanol Fuel on a Spark Ignition Engine*. SAE technical paper. Toyota Motor Corporation: Warrendale.

Nguyen TL, Gheewala SH, Garivait S. (2007). Energy balance and GHG-abatement cost of cassava utilization for fuel ethanol in Thailand. *Energy Policy* 35(9), 4585–4596. https://doi.org/10.1016/j.enpol.2007.03.012

Oliveira MA, Verma KK, Bhatt R, Oliveira TBA. (2022). Impact of green and organic fertilizers on soil fertility and sugarcane productivity. In: Verma KK. et al. (Eds.),

Agro-industrial Perspectives on Sugarcane Production under Environmental Stress. https://doi.org/10.1007/978-981-19-3955-6_11.

Paraiso ML de S, Gouveia N. (2015). Health risks due to pre-harvesting sugarcane burning in São Paulo State, Brazil. *Rev. Bras. Epidemiol.* 18, 691–701.

Quang Ha P (2017). Building capacity for enhancing bioenergy sustainability through the use of GBEP indicators (GCP/GLO/554/GER). In *Vietnam Experiences for 24 Indicators Institute for Agricultural Environment/V AAS*. Rome. http://www.globalbioenergy.org/fileadmin/user_upload/gbep/docs/2017_events/9_GBEP_WGCB_30_November_2017/Vietnam_PQH_30NOV_2017sent.pdf. [Accessed 10 July 2024].

Renewable Fuels Association. (2017). *World Fuel Ethanol Production of 2016*. RFA: Washington DC.

Ripoli TCC, Ripoli MLC (2005). Biomassa da cana-de-açúcar: colheita, energia e ambiente. ESALQ/USP, Piracicaba, 27-38-51

Rocha FLR, Marziale MHP , Hong O-S. (2010). Work and health conditions of sugar cane workers in Brazil. *Rev. Esc. Enferm. USP* 44(4), 978–983.

Sage RF, Peixoto MM, Sage TL. (2014). Photosynthesis in sugarcane. In: *Sugarcane Physiology, Biochemistry and Functional Biology*, 1st Edn., Wiley: New York; pp 121–149.

Scarpare FV, Hernandes TAD, Ruiz-Corrêa ST, Picoli MCA, Scanlon BR, Chagas MF, Duft DG, Cardoso TF. (2016). Sugarcane land use and water resources assessment in the expansion area in Brazil. *J. Clean. Prod.* 133, 1318–1327.

Scheiterle L, Ulmer A, Birner R, Pyka A. (2018). From commodity-based value chains to biomass-based value webs: The case of sugarcane in Brazil's bioeconomy. *J. Clean. Prod.* 172, 3851–3863.

Searchinger T, Heimlich R, Houghton RA, Fengxia D, Elobeid A, Fabiosa J, Tokgoz S, Hayes D, Y u TH. (2008). Use of US croplands for biofuels increases greenhouse gases through emissions from land-use change. *Science* 319(5867), 1238–1240.

Sharma A, Singh RK, Singh P, Vaishnav A, Guo DJ, Verma KK, Li DP, Song XP, Malviya MK, Khan N, Lakshmanan P, Li YR. (2021). Insights into the bacterial and nitric oxide-induced salt tolerance in sugarcane and their growth-promoting abilities. *Microorganisms* 9, 2203. https://doi.org/10.3390/microorganisms9112203.

Sharma A, Song X-P , Singh RK, Vaishnav A, Gupta S, Singh P, Guo D-J, Verma KK, Li Y-R. (2022). Impact of carbendazim on cellular growth, defence system and plant growth promoting traits of *Priestia megaterium* ANCB-12 isolated from sugarcane rhizosphere. *Front. Microbiol.* 13, 1005942. https://doi.org/10.3389/fmicb.2022.1005942

Singh J, Singh AK, Sharma MP , Singh PR, Srivastava AC. (2011). Mechanization of sugarcane cultivation in India. *Sugar Tech.* 13(4), 310–314

Singh P, Singh RK, Li HB, Guo DJ, Sharma A, Verma KK, Solanki MK, Upadhyay SK, Lakshmanan P, Yang LT, Li YR. (2023). Nitrogen fixation and phytohormone stimulation of sugarcane plant through plant growth promoting diazotrophic *Pseudomonas*. *Biotechnol. Genetic Eng. Rev.* 40(1), 15–35. https://doi.org/10.1080/02648725.2023.2177814

Singh P, Singh RK, Song QQ, Li HB, Guo DJ, Malviya MK, Verma KK, Song XP, Lakshmanan P, Yang LT, Li YR. (2021). Comparative analysis of protein and differential responses of defense-related gene and enzyme activity reveals the long-term molecular responses of sugarcane inoculated with *Sporisorium scitamineum*. *J. Plant Interact.* 16(1), 12–29.

Singh RK, Singh P, Li HB, Song QQ, Guo DJ, Solanki MK, Verma KK, Malviya MK, Song XP, Lakshmanan P, Yang LT, Li YR. (2020). Diversity of nitrogen-fixing rhizobacteria associated with sugarcane: A comprehensive study of plant-microbe interactions for growth enhancement in *Saccharum* spp. *BMC Plant Biol.* 20, 220. https://doi.org/10.1186/s12870-020-02400-9.

Sissine F. (2007). *Energy Independence and Security act of 2007: A Summary of Major Provisions*. Library of Congress, Washington, DC, Congressional Research Service

Solanki MK, Verma KK, Dastogeer KMG, Mora-Poblete F, Mundra S. (2023). Microbial resilience in plant nutrient management towards sustainable farming. *Front. Microbiol.* 14, 1280811. https://doi.org/10.3389/fmicb.2023.1280811

Song X-P, Mo F, Verma KK, Wei J, Zhang X, Yang L, Li YR. (2019). Effect of sugarcane smut (*Ustilago scitaminea* Syd.) on ultrastructure and biochemical indices of sugarcane. *Biomed. J. Sci. Tech. Res.* 17, 002950. https://doi.org/10.26717/BJSTR.2019.17.002950.

Song XP, Mo F, Yan M, Zhang X, Zhang B, Huang X, Huang D, Pan Y, Verma KK, Li YR. (2022). Effect of smut infection on the photosynthetic physiological characteristics and related defense enzymes of sugarcane. *Life* 12, 1201. https://doi.org/10.3390/life12081201

Song XP, Verma KK, Tian DD, Zhang XQ, Liang YJ, Huang X, Li CN, Li YR. (2021). Exploration of silicon functions to integrate with biotic stress tolerance and crop improvement. *Biol. Res.* 54, 19. https://doi.org/10.1186/s40659-021-00344-4

Souza GM, Victoria R, Joly C, Verdade L (Eds.) (2015). *Bioenergy & Sustainability: Bridging the Gaps*. SCOPE: Paris; p. 779, vol 72. ISBN 978-2-9545557-0-6

Tian D, Qin L, Verma KK, Wei L, Li J, Li B, Zhou W, He Z, Wei D, Huang S, Long S, Huang Q, Li C, Wei S. (2023). Transcriptomic and metabolomic differences between banana varieties which are resistant or susceptible to *Fusarium wilt. PeerJ* 11, e16549. https://doi.org/10.7717/peerj.16549

Universidad Nacional de San Martín (2015). *Estudio Piloto de Indicadores GBEP de Sustentabilidad de la Bioenergia en Argentina*. GBEP/FAO. Centro Ideas de la Universidad Nacional de San Martín (UNSAM). https://www.agroindustria.gob.ar/sitio/areas/bioener-gia/sustentabilidad/_archivos//000001_Indicadores%20de%20Sustentabilidad%20de%20Bioenerg%C3%ADa%20Agentina/000001_Indicadores%20de%20Sustentabilidad%20de%20Bioenerg%C3%ADa%20Agentina.pdf. [Accessed 12 Jun 2024].

Verma KK, Anas M, Chen Z, Rajput VD, Malviya MK, Verma CL, Singh RK, Singh P, Song XP, Li YR. (2020f). Silicon supply improves leaf gas exchange, antioxidant defense system and growth in sugarcane responsive to water limitation. *Plants* 9, 1032, https://doi.org/10.3390/plants9081032.

Verma KK, Joshi A, Song X-P, Liang Q, Xu L, Huang H-r, Wu K-C, Seth CS, Arora J and Li Y-R. (2024a). Regulatory mechanisms of plant rhizobacteria on plants to the adaptation of adverse agroclimatic variables. *Front. Plant Sci.* 15, 1377793. https://doi.org/10.3389/fpls.2024.1377793

Verma KK, Joshi A, Song X-P, Singh S, Kumari A, Arora J, Singh SK, Solanki MK, Seth CS, Li Y-R. (2024b). Synergistic interactions of nanoparticles and plant growth promoting rhizobacteria enhancing soil-plant systems: A multigenerational perspective. *Front. Plant Sci.* 15, 1376214. https://doi.org/10.3389/fpls.2024.1376214

Verma KK, Li DM, Singh M, Rajput VD, Malviya MK, Minkina T, Singh RK, Singh P, Song XP, Li YR. (2020a). Interactive role of silicon and plant-rhizobacteria mitigating abiotic stresses: A new approach for sustainable agriculture and climate change. *Plants* 9, 1055. https://doi.org/10.3390/plants9091055.

Verma KK, Rajput VD, Singh M, Verma CL, Minkina T. (2022a). Influence of waterlogging on plant physiology and molecular biology. In: *Understanding Abiotic Stresses*. Nova Science Publishers, Inc.: New York.

Verma KK, Siddiqui N, Singh M. (2022b). Understanding the plant physiological and molecular adaptation strategies to soil salinity. In: *Understanding Abiotic Stresses*. Nova Science Publishers, Inc.: New York. https://doi.org/10.52305/UBUJ4024

Verma KK, Singh P, Song X-P, Malviya MK, Singh RK, Chen G-L, Solomon S, Li YR. (2020b). Mitigating climate change for sugarcane improvement: Role of silicon in alleviating abiotic stresses. *Sugar Tech.* 22(5), 741–749. https://doi.org/10.1007/s12355-020-00831-0.

Verma KK, Singh RK, Song QQ, Singh P, Zhang B-Q, Song X-P, Chen G-L, Li YR. (2019a). Silicon alleviates drought stress of sugarcane plants by improving antioxidant responses. *Biomed. J. Sci. Tech. Res.* 002957. https://doi.org/10.26717/BJSTR.2019.17.002957

Verma KK, Song X, Degu HD, Guo D, Joshi A, Huang H, Xu L, Singh M, Huang D, Rajput VD, Li YR. (2023a). Recent advances in nitrogen and nano-nitrogen fertilizers for sustainable crop production: A mini-review. *Chem. Biol. Technolog. Agricult.* 10, 1–14. https://doi.org/10.1186/s40538-023-00488-3

Verma KK, Song X, Verma CL, Huang H, Singh M, Xu L, Wu J, Khan ZH, Zhang B, Li Y. (2023b). Mathematical modeling of climate and fluoride effects on sugarcane photosynthesis with silicon nanoparticles. *Plant Physiol. Biochem.* 204, 108089. https://doi.org/10.1016/j.plaphy.2023.108089

Verma KK, Song XP, Budeguer F, Nikpay A, Enrique R, Singh M, Zhang B, Wu, JM, Li YR. (2022c). Genetic engineering: An efficient approach to mitigating biotic and abiotic stresses in sugarcane cultivation. *Plant Signal. Behav.* 17(1), e2108253.

Verma KK, Song XP, Joshi A, Rajput VD, Singh M, Sharma A, Singh RK, Li DM, Arora J, Minkina T, Li YR. (2022d). Nanofertilizer possibilities for healthy soil, water and food in future: An overview. *Front. Plant Sci.* 13, 865048. https://doi.org/10.3389/fpls.2022.865048

Verma KK, Song XP, Joshi A, Tian DD, Rajput VD, Singh M, Arora J, Minkina TM, Li YR. (2022e). Recent trends in nano-fertilizer for advancing sustainable agriculture under the era of climate change to ensure food security for future generations. *Nanomaterials* 12, 173. *https://doi.org/10.3390/nano12010173.*

Verma KK, Song XP, Kumari A, Jagadesh M, Singh SK, Bhatt R, Singh M, Seth CS, Li YR. (2024c). Climate change adaptation: Challenges for agricultural sustainability. *Plant, Cell Environ.* 48, 2522–2533, https://doi.org/10.1111/pce.15078

Verma KK, Song XP, Li DM, Singh M, Wu JM, Singh RK, Sharma A, Zhang BQ, LI YR. (2022f). Silicon and soil microorganisms improves rhizospheric soil health with bacterial community, plant growth, performance and yield. *Plant Signal. Behav.* 17(1), e2104004.

Verma KK, Song XP, Lin B, Guo DJ, Singh M, Rajput VD, Singh RK, Singh P, Sharma A, Malviya MK, Chen GL, Li YR. (2022g). Silicon induced drought tolerance in crop plants: Physiological adaptation strategies. *Silicon* 14(6), 2473–2487. https://doi.org/10.1007/s12633-021-01071-x

Verma KK, Song XP, Rajput VD, Boldyreva V, Zhang BQ, Minkina T, Li YR. (2022h). Morpho-physiological, biochemical and ultrastructural modifications on sugarcane to prolonged water deficit. In: Verma KK et al. (Ed.), *Agro-industrial Perspectives on Sugarcane Production under Environmental Stress*. https://doi.org/10.1007/978-981-19-3955-6_11.

Verma KK, Song XP, Rajput VD, Solomon S, Li YR, Rao GP. (2022i). *Agro-industrial Perspectives on Sugarcane Production under Environmental Stress*. Springer: Nature Singapore, https://doi.org/10.1007/978-981-19-3955-6

Verma KK, Song XP, Singh M, Tian DD, Rajput VD, Minkina T, Li YR. (2023c). Association of silicon and soil microorganisms induces stress mitigation, increasing plant productivity. In: de Mello Prado R. (Ed.), *Benefits of Silicon in the Nutrition of Plants*; pp. 299–328. https://doi.org/10.1007/978-3-031-26673-7_17

Verma KK, Song XP, Tian DD, Guo DJ, Chen ZL, Zhong CS, Nikpay A, Singh M, Rajput VD, Singh RK, Minkina T, Li YR. (2021a). Influence of silicon on biocontrol strategies to manage biotic stress for crop protection, performance and improvement. *Plants* 10, 2163. https://doi.org/10.3390/plants10102163

Verma KK, Song XP, Tian DD, Singh M, Verma CL, Rajput VD, Singh RK, Sharma A, Singh P, Malviya MK, Li YR. (2021b). Investigation of defensive role of silicon during drought stress induced by irrigation capacity in sugarcane: Physiological and biochemical characteristics. *ACS Omega* 6, 19811–19821.

Verma KK, Song XP, Verma CL, Chen ZL, Rajput VD, Wu KC, Liao F, Chen GL, Li YR. (2021c). Functional relationship between photosynthetic leaf gas exchange in response to silicon application and water stress mitigation in sugarcane. *Biol. Res*. 54, 15. https://doi.org/10.1186/s40659-021-00338-2

Verma KK, Song XP, Verma CL, Malviya MK, Guo DJ, Rajput VD, Sharma A, Wei KJ, Chen GL, Solomon S, Li YR. (2021d). Predication of photosynthetic leaf gas exchange of sugarcane (*Saccharum* spp.) leaves in response to leaf positions to foliar spray of potassium salt of active phosphorus under limited water irrigation. *ACS Omega* 6, 2396–2409.

Verma KK, Song X-P, Xu L, Huang H-R, Liang Q, Seth CS, Li Y-R. (2023d). Nano-microplastic and agro-ecosystems: A mini-review. *Front. Plant Sci*. 14, 1283852. https://doi.org/10.3389/fpls.2023.1283852

Verma KK, Song XP, Yadav G, Degu HD, Parvaiz A, Singh M, Huang HR, Mustafa G, Xu L, Li YR. (2022j). Impact of agroclimatic variables on proteo-genomics in sugarcane (*Saccharum* spp.) plant productivity. *ACS Omega* 7, 22997–23008.

Verma KK, Song XP, Zeng Y, Guo DJ, Sing M, Rajput VD, Malviya MK, Wei KJ, Sharma A, Li DP, Chen GL, Li YR. (2021e). Foliar application of silicon boosts growth, photosynthetic leaf gas exchange, antioxidative response and resistance to limited water irrigation in sugarcane (*Saccharum officinarum* L.). *Plant Physiol. Biochem*. 166, 582–592.

Verma KK, Song XP,Zeng Y, Li DM, Guo DJ, Rajput VD, Chen GL, Barakhov A, Minkina TM, Li YR. (2020c). Characteristics and correlation of leaf stomata and its relationship with photosynthesis on *Saccharum* spp. under different irrigation and silicon application. *ACS Omega* 5, 24145–24153.

Verma KK, Vatsal S, Singh M, Rajput VD, Minkina T. (2022k). Coping with water stress: Effects and adaptive mechanisms to plants. In: *Understanding Abiotic Stresses*. Nova Science Publishers, Inc.: New York.

Verma KK, Wu KC, Verma CL, Li DM, Malviya MK, Singh RK, Singh P, Chen GL, Song XP, Li YR. (2020d). Developing mathematical model for diurnal dynamics on photosynthetic responses in sugarcane responsive to different irrigation and silicon application. *PeerJ* 8, e10154. https://doi.org/10.7717/peerj.10154.

Verma KK, Liu X-H, Wu K-C, Singh RK, Song QQ, Malviya MK, Song X-P, Singh P, Verma CL, Li YR. (2020e). The impact of silicon on photosynthetic and biochemical responses of sugarcane under different soil moisture levels. *Silicon* 12, 1355–1367.

Verma KK, Wu K-C, Singh P, Malviya MK, Singh RK, Song X-P, Li YR. (2019b). The protective role of silicon in sugarcane under water stress: Photosynthesis and antioxidant enzymes. *Biomed J. Sci. Tech. Res.* 002685. https://doi.org/0.26717/BJSTR.2019.15.002685

Walter A, Galdos MV , Scarpare FV , Leal MRLV , Seabra JEA, Cunha MP , Picoli MCA, Oliveira COF. (2014). Brazilian sugarcane ethanol: Developments so far and challenges for the future. *WIREs Energy Environ*. 3(1), 70–92.

World Bank. (2008). Double jeopardy: Responding to high food and fuel prices. http:// web.world-bank.org/WBSITE/EXTERNAL/NEWS/0,,contentMDK:21827681~pageP K:64257043~piPK:437376~theSitePK:4607,00.html. [Accessed 21 June 2024].

Wu K-C, Huang C-M, Verma KK, Deng Z-N, Huang H-R, Pang T, Cao H-Q, Luo H-B, Jiang S-L, Xu L. (2022). Transcriptomic responses of *Saccharum spontaneum* roots in response to polyethylene glycol–6000 stimulated drought stress. *Front. Plant Sci*. 13, 992755. https://doi.org/10.3389/fpls.2022.992755

Xu L, Deng Z-N, Wu K-C, Malviya MK, Solanki MK, Verma KK, Pang T, Li Y-J, Liu X-Y, Kashyap BK, Dessoky ES, Wang W-Z and Huang H-R. (2022). Transcriptome analysis reveals a gene expression pattern that contributes to sugarcane bud propagation induced by Indole-3-Butyric acid. *Front. Plant Sci*. 13, 852886. https://doi.org/10.3389/fpls.2022.852886

Zheng F-J, Lin B, Yang Y-X, Fang X-C, Verma KK, Chen G-L. (2023). Efficacy and functionality of sugarcane original vinegar on mice. *Front. Microbiol*. 14, 1224666. https://doi.org/10.3389/fmicb.2023.1224666

Index

Note: Page numbers followed by 'f' and 't' indicate figures and tables, respectively.

C

D

E

F

G

S

T